Random Walk, Sequential Analysis and Related Topics

A Festschrift in Honor of
Yuan-Shih Chow

Random Walk, Sequential Analysis and Related Topics

A Festschrift in Honor of Yuan-Shih Chow

Fudan University, Shanghai, China 18 – 19 July 2004

Editors

Agnes Chao Hsiung
National Health Research Institutes, Taiwan

Zhiliang Ying
Columbia University, USA

Cun-Hui Zhang
Rutgers University, USA

World Scientific

NEW JERSEY • LONDON • SINGAPORE • BEIJING • SHANGHAI • HONG KONG • TAIPEI • CHENNAI

Published by

World Scientific Publishing Co. Pte. Ltd.
5 Toh Tuck Link, Singapore 596224
USA office: 27 Warren Street, Suite 401-402, Hackensack, NJ 07601
UK office: 57 Shelton Street, Covent Garden, London WC2H 9HE

British Library Cataloguing-in-Publication Data
A catalogue record for this book is available from the British Library.

RANDOM WALK, SEQUENTIAL ANALYSIS AND RELATED TOPICS
A Festschrift in Honor of Yuan-Shih Chow

ISBN 981-270-355-1

Printed in Singapore by World Scientific Printers (S) Pte Ltd

Statistical Science
2006, Vol. 21, No. 1, 99–112
DOI 10.1214/088342304000000224

A Conversation with Yuan Shih Chow

Zhiliang Ying and Cun-Hui Zhang

Abstract. Yuan Shih Chow was born in Hubei province in China, on September 1, 1924. The eldest child of a local militia and political leader, he grew up in war and turmoil. His hometown was on the front line during most of the Japanese invasion and occupation of China. When he was 16, Y. S. Chow journeyed, mostly on foot, to Chongqing (Chung-King), the wartime Chinese capital, to finish his high school education. When the Communist party gained power in China, Y. S. Chow had already followed his university job to Taiwan. In Taiwan, he taught mathematics as an assistant at National Taiwan University until he came to the United States in 1954. At the University of Illinois, he studied under J. L. Doob and received his Ph.D. in 1958. He served as a staff mathematician and adjunct faculty at the IBM Watson Research Laboratory and Columbia University from 1959 to 1962. He was a member of the Statistics Department at Purdue University from 1962 to 1968. From 1968 until his retirement in 1993, Y. S. Chow served as Professor of Mathematical Statistics at Columbia University. At different times, he was a visiting professor at the University of California at Berkeley, University of Heidelberg (Germany) and the National Central University, Taiwan. He served as Director of the Institute of Mathematics of Academia Sinica, Taiwan, and Director of the Center of Applied Statistics at Nankai University, Tianjin, China. He was instrumental in establishing the Institute of Statistics of Academia Sinica in Taiwan. He is currently Professor Emeritus at Columbia University.

Y. S. Chow is a fellow of the Institute of Mathematical Statistics, a member of the International Statistical Institute and a member of Taiwan's Academia Sinica. He has numerous publications, including *Great Expectations: The Theory of Optimal Stopping* (1971), in collaboration with Herbert Robbins and David Siegmund, and *Probability Theory* (1978), in collaboration with Henry Teicher. Y. S. Chow has a strong interest in mathematics education. He taught high school mathematics for one year in 1947 and wrote a book on high school algebra in collaboration with J. H. Teng and M. L. Chu. In 1992, Y. S. Chow, together with I. S. Chang and W. C. Ho, established the Chinese Institute of Probability and Statistics in Taiwan.

This conversation took place in the fall of 2003 in Dobbs Ferry, New York.

FAMILY BACKGROUND AND EARLY YEARS

Query: We have always been fascinated by the interesting stories you have told us about yourself, but there are certainly a lot more. It would be greatly appreciated if you could share your experience and thoughts with colleagues, students, and friends in and outside the statistical community. We wonder if you could start with your childhood.

Chow: I was born in 1924 into a family with a large estate in a rural town in central China. It was a time of war and turmoil. My family had been dependent on land for our livelihood since the Qing Dynasty. My father, a local militia leader, had more than 100 militia-

Zhiliang Ying is Professor and Co-Chair, Department of Statistics, Columbia University, New York, New York 10027, USA. Cun-Hui Zhang is Professor of Statistics, Rutgers University, New Brunswick, New Jersey 08854-8019, USA.

men to protect our hometown. When I was four, my father lost a battle against the communists and our home was burned down. We first left for Wuhan, but my father soon found the provincial capital too expensive for an extended family of 40 people without income from land. So we eventually settled in Xiangfan, a city near my hometown.

Query: How did you become interested in mathematics?

Chow: During the period of our "refugee" life, the family hired an old-fashioned tutor, who taught traditional Chinese language and literature. I started formal schooling in a highly scholastic primary school when I was eight, after we settled down in Xiangfan. I started as a second grader, but I was not able to recognize Arabic numerals. I got a big shock when I received a "0" mark on my first math exam, but it motivated me to work hard. My efforts paid off, resulting in excellent grades in subsequent exams. That is how I got interested in mathematics.

FIG. 1. *Y. S. Chow in 1940.*

MIDDLE SCHOOLS AND WORLD WAR II

Query: Your teenage years overlapped with the Sino–Japanese War. What happened to your education?

Chow: The war broke out in 1937, right after my graduation from the primary school. Soon schools were all shut down when the Japanese began bombing our area. Although I passed the entrance examination for a provincial middle school, I had to wait until the end of 1938 to attend a new combined junior high school.

Query: Were you still able to get a decent education?

Chow: The combined junior high was a good school, but the combined senior high lacked qualified teachers since many had left for mountainous areas to avoid the war. Upon completion of the junior high, three classmates and I decided to find a better high school outside our area. That turned out to be quite an adventure.

We first went to the wartime provincial capital. We walked for two whole weeks through mountains, including the Three Gorges, in the midst of war and cold winter weather. It was quite a dangerous trip. However, to our disappointment we were "drafted" into an agricultural school there, which was treated as a military school during the war.

Query: What happened then? Did you stay there?

Chow: Because it was a "military school," running away was punishable by imprisonment. Despite this, in July 1941, we decided to run away for Chongqing, the wartime capital of China. We left the school at night and walked for four days to arrive at a port city on the Yangtze River. By then we were penniless. We spent a night in a park under the open sky, and traded a gold ring for our boat passage to Chongqing the next morning. In Chongqing, we stayed in a youth center run by the government, which also served as a shelter for refugee students.

Query: What did you do in Chongqing?

Chow: We decided to take an entrance exam for the No. 2 National Middle School, located about 90 kilometers, or one day and a half by walking, outside of Chongqing. From the youth center, we got a small amount of money for the cost of the free meals we would miss for three days during our journey. However, upon our arrival, we learned that we had to arrive one day earlier to apply. We pleaded with the principal and argued that because we could afford food for only three days, we really were not able to come earlier. The principal finally allowed us to take the exam. He even let us sleep on the tables in the dining hall. I passed the exam, which was quite an honor and resulted in a cash award from the school district of my hometown. Better yet, I did not have to walk back to Chongqing!

Query: Were you able to get a good education there, especially in mathematics?

Chow: The No. 2 National Middle School, a combined school of some of the best schools around Shanghai, was one of the most envied among all the key schools in China during the war. The students there were all refugees from faraway, but they enjoyed very high status and were very proud of themselves. We had excellent teachers. Among the best was Mr. Mao, our mathematics instructor. Mr. Mao had a great influence on the way we studied mathematics. He emphasized careful reading of advanced textbooks, not working on numerous exercises. Students in his class enjoyed total freedom: free to decide whether to do homework and whether to attend classes! However, they all worked hard and excelled in their studies. The percentage of graduates admitted into top universities was very high. Mr. Mao was also the student advisor of our class. On holidays, he would invite us to his house for dinner, since we had no place else to go.

ZHEJIANG UNIVERSITY

Query: It is impossible to imagine nowadays what you went through seeking a good education at such young ages. What happened during your college years?

Chow: During the war, many universities moved with the central government to southwest China. Since our school was the best near Chongqing, they often sent professors there to recruit students. A common practice was to admit top students without entrance examinations. Zhejiang University was the first to come during my senior year. When I received an offer from it, I decided to seize the opportunity.

Query: Why did you choose mathematics as your major?

Chow: I had a childhood dream that someday I would found a private middle school in my own hometown. My family had approximately 1000 *mus* (about 200 acres) of land. I could draw out a certain portion of income from the land to keep the school going. Students could work part time on the land so they would not have to pay tuition. The main problem would be to hire qualified teachers, especially mathematics teachers. This motivated me to choose mathematics as my major.

Query: What was your college life like during the war?

Chow: In the summer of 1944, I left Chongqing by bus for the wartime campus of Zhejiang University. The trip was dangerous, as accidents were com mon on the bus route through the mountain paths. The bus, which burned charcoal to generate steam, was in terrible condition. Shortly after our arrival, in December 1944, the Japanese reached areas near the campus. School had to be suspended. Since we were worried that the Japanese could interrupt our financial support from the central government, our first responsibility was to find food. The second was to protect female students. Each student got a gun. Meanwhile, we organized Delegates to Battle Fields to visit the front to prop up the morale of soldiers. A few months later, the Yunnan–Burma road opened. We saw soldiers in large U.S. made ten-wheel trucks heading toward central China. We all cheered, "Americans have come to help! There is hope at last!" Although we did not see any American soldiers, that was a turning point in the Sino–Japanese War. I still vividly remember the end of the war. It was a hot summer day. We were swimming and heard someone crying, "Japanese have surrendered!" We were overjoyed. That was the beginning of my second year at Zhejiang University. A year later, the university moved back to its permanent campus in Hangzhou, the capital of Zhejiang province.

Query: Who were your teachers at Zhejiang University? How did they influence your life?

Chow: The most senior faculty members were Buqing Su and Jiangong Chen. When we returned to Hangzhou, Chen took a position in Academia Sinica in Shanghai. Ruiyun Xu, a student of Carathéodory, directed my thesis. She wrote an excellent book on real variables and later became Chairwoman of the Department of Mathematics at Hangzhou University.

The Chinese Civil War broke out shortly after World War II. When I was a junior, my hometown changed hands repeatedly between the Communist troops and the Nationalist government, so my family could no longer support me financially. I sought help from Su, who used his connection to secure a teaching job for me at a middle school. Su was very kind to students, but he was also a very strict teacher. Su's lectures were very serious. He would not speak a word before all the students arrived, so nobody dared to be tardy. Chen and Su were the most respected professors by colleagues and students alike at Zhejiang University. Every year, Chen and Su gave a formal party in a fine restaurant for students. Everyone had to have a drink to the health of each teacher. Since we had about a dozen teachers, we all learned to drink.

TAIWAN

Query: What happened after you graduated from college?

Chow: In 1947 and 1948, Su and Chen went to Taiwan for the formation of the National Taiwan University, which was Taipei Imperial University under Japanese rule. Chen became the first Provost of National Taiwan University and Su became the first Dean of the College of Science. Su also founded the mathematics department there. Because I took a one-semester leave to teach in the middle school, I graduated in January 1949 instead of July 1948. I planned to do my postgraduate work with Chen. The department waived my entrance examination, but the graduate program had only fall enrollment. What was I supposed to do to fill up the gap? Chen suggested that I spend six months in Taiwan to help the mathematics department there. However, I did not expect that I would not be able to return to the Mainland. Soon after I left for Taiwan, the Nationalist government was defeated and retreated to Taiwan.

I still dreamed about pursuing postgraduate study. At that time, students in Taiwan were not allowed to go overseas, especially male students since they had to be ready for military service if called upon. Later, the government began to lift the restrictions. H. C. Chow, a mathematician, suggested to the government that Taiwan should adopt an open policy so that students would have hope about their future; otherwise they could become a source of political instability. Specifically, he suggested that the government should allow a limited number of students to go abroad after passing certain examinations. The government accepted the suggestion. As a result, students became more motivated in their study. I believe this indeed helped the government to prevent student movements. The first exam took place in 1953, but I had no connection to learn about the exam in advance, so I came to America in 1954.

The Mathematics Department at Zhejiang University gave very low grades. Even though I did well, my transcript was not presentable. Fortunately, Chung-Tao Yang, who was two years ahead of me at Zhejiang, was then a postdoc at Illinois. He wrote a letter for me saying that he and I got exactly the same training at Zhejiang and that I would have no problem completing the Ph.D. program at Illinois. With his reference, I got a fellowship award from Illinois. I learned from this experience so that I was always generous in marking the papers of my students.

URBANA–CHAMPAIGN

Query: So, you came to Illinois. Why did you decide to study probability with Doob?

Chow: I arrived at Champaign more than two weeks late. It took me 33 days by ship to reach Oregon, and then a three-day ride on a Greyhound bus. The steamer ticket, costing 100 U.S. dollars after a 50% discount for students, was not so easy to purchase. Red tape and mandatory political training delayed my passport application.

I first went to see H. Ray Brahana, Chairman of the Mathematics Department. After examining my transcript, he found that I had taken differential geometry, real variables and other advanced courses as an undergraduate. He concluded that I had an extremely solid foundation and put me in all the hard courses least selected by graduate students: topology, Riemann geometry, measure theory and abstract algebra (taught by himself). At first, I had a hard time coping with the course materials, due in part to my problem with English. I could not remember how I survived, but my grades for mid-term exams turned out fine. I felt better and began to blend in with my classmates. I made many good friends, including Jim Abbott.

At the end of my first year, there was a summer camp in Indiana for graduate students from China. The camp invited Shiing-Shen Chern, then at the University of Chicago, to give a lecture. After his talk, I approached him for advice about choosing a specialty and a Ph.D. advisor. Chern told me to go with Doob if he was willing to accept me. When I returned to Illinois, the first thing I did was to find Doob, whom I had never met and knew very little about. Doob said O.K. and asked me what I intended to work on. I replied Fourier series, since I had studied the topic as an undergraduate. "Then, you should be my advisor," he joked. When I asked him what I should learn from him, he answered, "Probability."

Query: What was it like with your thesis research?

Chow: I took Doob's probability course in 1955, after I chose him as my advisor. There were four students in the class, including Jim Abbott and me. It was actually one of the largest probability classes. Among the auditors were Don Burkholder, who had just arrived at Illinois as an Assistant Professor, and Klaus Krickeberg from Germany.

From 1956 to 1957, Doob took a sabbatical leave in Europe. I wrote a paper on martingales while he was away. I felt very satisfied. After I mailed him the paper, I let myself relax and really had a good time. However, one day I realized that I still had to prove a set-theoretical lemma upon which my paper was based. I could not prove it although I had thought that it was obvious when I first wrote it down. I had to let Doob know the situation. He asked me if there was a counterexample. I did not have one, but I just could not prove the lemma. I decided to start afresh with a different approach. Krickeberg was also working on the topic. His advisor C. Y. Pauc was visiting Purdue University. Since Doob was in Europe, I went to see Pauc often. In this sense, Pauc was my quasi-advisor. I finished my thesis work when Doob came back. Two months later, in January 1958, I got my Ph.D. degree.

Query: Besides probability, were you involved in any other research project?

Chow: Illinois required a minor, so I selected physics. It was an easy decision. I had taken Riemann geometry in my first year, which required relativity theory. To fulfill the physics minor, I just needed to take relativity and hydrodynamics, which were very close. My advisor for my minor was A. H. Taub, who worked on the first computer at Princeton with von Neumann.

After sending my paper to Doob in Europe, I sought help from Taub to find a job for me. He gave me one right away. It was a Research Associate position with his NASA project. Sputnik had just gone into the space. Taub said it was one of the most important and urgent projects in the country. He offered to help me apply for a green card, but he also demanded that I stop working on probability if I took the job. It was very hard then for a Chinese person to get a green card, since the quota for China was only 105 per year, compared with 20,000 for each European nation. The job offer was hard to resist. I could get a green card while working on an important problem and I liked Illinois. When Doob came back, I handed him the thesis and told him that I would have to say goodbye to probability, not even to publish my thesis. I jumped into Taub's computer lab.

The specific problem I worked on came from hydrodynamics. The problem has five differential equations and six unknown quantities. By 1959, after over a year's effort, I came up with a principle, which could lead to a sixth equation. Taub was delighted. He said he would present the result in a talk in Europe and asked me to compute the solution. The capacity of our computer was 1K, so that we could compute the solution at no more than 50 points. It turned out that the domain of our solution was very narrow, requiring at least 250 points. We had to wait for a more powerful computer!

IBM AND COLUMBIA

Query: I remember you told me that you would probably also do very well if you stayed with computers. What else did you do in that field?

Chow: One day in 1959, IBM sent a recruiter to Illinois. He told me that I was exactly the man they were looking for. IBM had just begun to form the Watson Research Center, and it had appointed Herman Goldstine as Director of Scientific Development in its Data Processing Division. Goldstine was a deputy director of the computer lab under von Neumann in the Institute for Advanced Studies. After von Neumann's death, the Institute dissolved its computer group. As the head of a new division, Goldstine certainly wanted to hire. I was very knowledgeable with software and computer operations, and I was a student of Taub's, so I was a perfect match for IBM. I was pleased to get IBM's offer.

Since I was leaving Illinois anyway, I decided to explore the job market. Jacob Wolfowitz made offers to both Roger Farrell and me. Farrell, my contemporary at University of Illinois, accepted the offer, but I decided to choose IBM. I liked New York City. Cornell was an excellent university, but Ithaca seemed rather isolated. It did not hurt that IBM's salary was double what Cornell offered.

Since IBM had not yet built the Yorktown campus for the Watson Center, it put Goldstine's division at the Lamb estate near New York City. At that time, Purdue was the only university with a department of mathematics and statistics, so Goldstine lured Purdue's chair, Carl Kossack, to be in charge of the statistics group. When I arrived at IBM, Kossack asked me if I wanted to join his group, but I said statistics was not my field. Instead, I joined the pattern recognition group, which included two other people: one from Princeton specializing in topology and the other from Harvard specializing in physics. William Feller was our consultant. Our main assignment was to study the feasibility of developing computer-based methodology to read the Soviet newspaper *Pravda* and translate its content into English. That is, we wished to be able to place a copy of *Pravda* in the computer and automatically produce its English version. We spent a lot of time and eventually decided that the project would take at least 10 years. Because of our report, IBM did not participate in the project, while other big corporations, including GE and RCA, did and lost millions of dollars. This might be my greatest contribution to IBM.

Query: How did you return to probability?

Chow: After we submitted our report, I became idle again and began to think about my future. I had promised Taub not to work on probability. I had worked in hydrodynamics with him and then pattern recognition with IBM. However, I was no longer working for Taub and thus free to return to probability. I decided to attend seminars at Columbia University.

Ron Pyke was the seminar chair. After the first seminar, I introduced myself and asked Ron if I could attend the seminars. He said I was very welcome. Herb Robbins heard our conversation and asked me if I had to go back to IBM right after the seminar. When I replied no, he took me to his office and said that it was too bad I missed his more interesting seminar a week earlier. He went on to tell me that it was a secretary problem and provided details on the blackboard.

I worked on Robbins' problem afterward, using martingales. One week later, I went to see Robbins to show him my result. He went over it and asked me what he was supposed to do with it. When I realized he was talking about a joint paper, I told him that it was his stuff. However, Robbins said he did not prove his theorem and my proof was more general. This was the beginning of our collaboration. It was around the end of 1959.

Query: So how often did you go to Columbia?

Chow: I went to Columbia every week after Robbins and I started to collaborate. IBM had a unit called Watson Scientific Computing Laboratory at Columbia University, located in a small house right at 116th Street between Broadway and Claremont Avenue, with a staff of about a dozen people. Wallace Eckert, founder and Director of the Watson Lab, was a very good friend of Robbins'. Robbins felt it was too much trouble for me to come back and forth so often, so he called Eckert to suggest that IBM transfer me over to Manhattan. At that time, Columbia did not have its own computer. IBM provided computing services to Columbia through the Watson Lab. IBM provided the staff, while Columbia took care of the maintenance. The IBM staff members usually had adjunct titles at Columbia. In fact, they were in charge of the applied math program at Columbia. Peter Welch got his Ph.D. degree in statistics while being a student staff member at the Watson Lab.

At the Watson Lab, my duty was to offer a course in statistics at Columbia, and that was all. IBM explained to me that the Watson Lab was a window for IBM, through which IBM could keep informed of what was going on in the outside world. IBM later merged the Watson Lab into Watson Research Center, which was much larger in scale. The Watson Lab had many high-caliber people. It covered many fields, including chemistry, mathematics, physics and astrophysics.

Query: Who else was at Columbia at that time? What did you teach?

Chow: In the spring semester of 1960, I taught as an Adjunct Assistant Professor at Columbia. Ron Pyke, Don Ylvisaker, Jerry Sacks and Lajos Takacs were Assistant Professors. Joe Gani was probably a Visiting Associate Professor. These were the young people there. I suggested that the department should offer probability, but we needed to offer measure theory first. Ron agreed, so it was the first course I taught at Columbia. We called it "mathematics for statistics," since measure theory was in the territory of the mathematics department. The second course I offered was martingales, which attracted more people. Robbins sat in, as did S. Moriguti and G. Mariyama; both were visiting from Tokyo University.

Query: You started teaching at Columbia and working with Robbins before you went to Purdue. Why did you decide to go there?

Chow: In 1961, I was promoted to Adjunct Associate Professor, but IBM paid my salary anyway. IBM staff had to go through a formal evaluation process periodically. When Eckert asked Robbins about his impression of me, Robbins commented that I was more suitable for Columbia than for IBM. Eckert summoned me to his office and told me about his conversation with Robbins, but he said he could not figure out what Robbins actually intended to tell him.

In the same year, I went to Washington, DC, to attend an IMS meeting. I invited Joe Doob and Frank Spitzer for dinner. During our conversation, I mentioned to them that I was working for IBM, but my duty was just to teach a class at Columbia. Spitzer said that it was "too good to last," and his remark kept on bothering me afterward. When I asked Robbins what he meant about his conversation with Eckert, he said he was ready to offer me a job. The idea seemed fine. I would become an Associate Professor automatically. However, the Statistics Department at Columbia did not have a tenure-track line, so I was appointed as a Visiting Associate Professor. Robbins promised that my position would last indefinitely as long as Columbia had money. That was 1962.

After I took the position at Columbia, Shiing-Shen Chern happened to pass by New York. I had kept in touch with him since he advised me to work with Doob. He asked me how I had been doing. I said that I had left IBM to teach at Columbia. He remarked that the job at IBM was a very good one, better than a visiting position with year-to-year appointment. When I told him Robbins' promise, Chern commented, "What difference does that make?" It made me uneasy again.

Henry Teicher, who had been at Purdue since my student days at Illinois, was visiting New York University at that time. I knew him very well because Illinois and Purdue had a get-together every year. When I was a

FIG. 2. *Y. S. Chow in 1962.*

student at Illinois, he was an Associate Professor at Purdue. Since I was Doob's student, it was very easy for me to get to know people well. Henry had graduated from Columbia, started with Stanford and then went on to Purdue. When he visited NYU, he came to Columbia often to hang around with us and I played Go with him. I told him that I had resigned from IBM and was currently looking for a job, since my position at Columbia was not a permanent one. He called Felix Haas at Purdue, who immediately made me an offer: Associate Professor with tenure! I was very happy to get an offer without even having to apply, so I accepted. It was almost too good to be true. Since I was to visit Berkeley the following year with Robbins during Jerzy Neyman and Michel Loève's sabbatical leave, my appointment at Purdue would start in the fall of 1963.

Query: So you went to Berkeley with Robbins. What did you do there?

Chow: Robbins and I stayed in Neyman's house and shared the same office. Although we should have been able to see each other all the time, it was just the opposite. When I got up early in the morning, he would be fast asleep after coming home way beyond midnight. Of course, when I came back from the office, he was never in the house.

We were working together on the same project. The problem was optimal stopping for the relative frequency with coin tossing. One day, David Blackwell asked me what I was doing. I told him the problem and said we had no idea how to proceed. He said that it was very interesting and he would work on nothing else but this problem. I told Robbins about Blackwell's competing effort and suggested that we work harder to solve the problem first. However, during our entire visit to Berkeley, Robbins spent a lot of time dating girls. Blackwell was a serious researcher. He would come to our office, claiming the existence of the optimal rule at one moment and trying to give a counterexample the next. We were confused. By April or May, I found a proof for the existence of the optimal stopping rule. When Blackwell heard this, he focused his attention on finding the stopping boundary explicitly. It turned out that the square-root boundary is optimal.

I like to have conversations about mathematics and to collaborate with others. In a large sense, the purpose of research is to learn and to understand. Solving problems is far more important than who solves them. This attitude makes it easy for me to work with others.

PURDUE AND TWO BOOKS

Query: So, you went to Purdue after visiting Berkeley. Could you share with us your experience there?

Chow: Felix Haas was Chairman of the Department of Mathematics and Statistics at Purdue. But soon after I accepted his offer, Purdue decided to make the department even larger by establishing a new Division of Mathematical Sciences, including the Departments of Mathematics, Statistics and Computer Science. Shanti Gupta was hired from Bell Labs as head of the Statistics Department. When I arrived at Purdue, Gupta was also new there. Teicher had already become a full professor.

I had a joint appointment, half mathematics and half statistics. Traditionally, measure theory was a prerequisite for probability theory. Since measure theory was very hard, most students would be able to start to work on a thesis in functional analysis when they took it. That was why Doob did not have many students at Illinois. I proposed to offer probability theory without requiring measure theory. My colleagues in the Mathematics Department were not sure about my approach. I said I would write my own lecture notes. When they asked why, I replied that I would never get any students otherwise. This is how I started to teach probability without requiring measure theory. My first Ph.D. student was William Stout, whose undergraduate major was electrical engineering. His successful career demonstrates that my approach worked.

I offered my new course and started to write notes. A well-known publisher heard about this and proposed to publish a book based on my notes. However, writing a book is a different matter. My officemate at Purdue was Teicher, so I asked him if he would help by

FIG. 3. *Y. S. Chow and Yi Chow in 1966.*

being a co-author. He agreed. The publisher gave us a very attractive offer with a $7000 advance. We took the money without having to write down a single word, and I spent it on a new car. We were supposed to finish the book in three years. However, I left Purdue to visit Columbia in 1965. One year later, Teicher also left, first for Columbia and then Rutgers. The three-year contract was extended to five years, but then I started to visit Taiwan frequently. Since Henry and I were in different places, it seemed that we could never finish the book. The publisher finally gave up on us. They did not even want us to pay back the advance. They just wanted to forget about it. When Springer learned this, it got interested in the book. We signed up with Springer, but we did not see a penny until the publication of the book in 1978. This was how we wrote our book on probability theory.

In 1962, when I was at IBM, I went to the International Mathematics Congress in Stockholm. Doob was there too. When I told him about my work with Robbins on optimal stopping, he said it was an interesting topic and suggested that we write a book for Springer. I was reluctant since we did not have enough materials—the topic was very new. However, Doob replied that being short of materials made it even more sense to write a book. Robbins agreed with Doob when I told him about the conversation, so we decided to offer a course on the subject to force ourselves to write down some notes. I offered the course, and Robbins participated too. David Siegmund was in the class. We would have a draft if we put the class notes together, but then I went to Purdue and we only met during breaks after that. It was very difficult for us to complete the book. In 1966, at the IMS Annual Meeting, I suggested to Robbins, who was then the IMS President, that since we had been lazy for two years, we should invite David Siegmund to co-author. Robbins liked the idea, Siegmund agreed and the book on optimal stopping was finally published in 1972.

Query: Herb Robbins was also at Purdue for a while. How did that happen?

Chow: The Statistics Department at Columbia was full of uncertainty when I visited there in the spring of 1965 from Purdue. I told Robbins that Haas had become a dean at Purdue and he was very interested in building up statistics. I asked if he was willing to go there, but he replied that he had to think about it. I then asked Aryeh Dvoretzky and David Siegmund, who was writing his thesis with Robbins; they both said yes. I then called Shanti Gupta. Shanti was very supportive and offered everyone a position. All of sudden, Robbins, Dvoretzky, Siegmund and I all went to Purdue. It was like a package deal.

Robbins visited Purdue for one year. Although Purdue was very interested in keeping him, Robbins did not like a school mainly composed of science and engineering. He left for the Mathematics Department at University of Michigan, which did not have a statistics department when Robbins went. Michigan wanted him to stay and formed a new department, but Robbins left anyway after helping Michigan to establish its statistics

FIG. 4. *Herbert Robbins, Y. S. Chow, Yi Chow and Carol Robbins.*

department. He visited the Electrical Engineering Department at Berkeley and then returned to Columbia in 1968. I believe that he still found New York City to be the most attractive place after his grand tour. Of course, I also came back to New York in 1968.

ACADEMIA SINICA, TAIWAN

Query: You mentioned that you traveled to Taiwan often around that time. What did you do over there?

Chow: In 1968, Shiing-Shen Chern was recruiting scientists from overseas for Taiwan. He sent out letters asking if anyone would be willing to visit for a year. Columbia had given me credit for my service as an adjunct professor, so I qualified for a sabbatical leave in 1970, two years after my return to Columbia. I replied to Chern that I would be able to spend my sabbatical leave in Taiwan, so Chern wrote to Shichie Wang, then President of Academia Sinica in Taiwan. Again, Chern influenced my career. When Wang came to New York City, he invited my wife and me to dinner. He said he was very pleased that I would be able to go back. He would like me to serve as Director of the Institute of Mathematics for a year, as the current director had just decided to leave Academia Sinica.

My sabbatical leave turned into something far more involved and gratifying than I had anticipated. After my arrival in Taiwan, Siliang Chien became the President of Academia Sinica and he wanted to renew my appointment. I said I would consider it if he doubled the budget for the Institute. He agreed. It was a small amount, but it made the Institute of Mathematics relatively rich for a long time since the budget of individual institutes increased by a percentage every year. I served six more years as the director, from 1971 to 1976, and traveled frequently between New York and Taiwan.

In 1968, President Chiang Kai-shek consulted with physicist Luke C. L. Yuan about ways Taiwan could attract overseas Chinese scientists. Yuan suggested that the government should allow people to be in charge of scientific institutions on a part-time basis, letting them keep their permanent positions abroad, instead of demanding they choose between Taiwan and the U.S. Yuan said that such appointees would be more objective and feel free to serve to the best interest of their units. Chiang Kai-shek approved. I was one of the first to serve under this new policy.

Query: Could you tell us a bit about the formation of the Institute of Statistical Science in Academia Sinica?

Chow: My main goal as Director of the Institute of Mathematics was to promote computer science and statistics in Taiwan. In those years, Taiwan urgently needed to improve living standards. The per capita income in Taiwan was $450, and the monthly income for a professor was $100, barely enough to support a family. Professors did not have enough money to buy books, let alone to conduct research. Within the science, computer science and statistics are fields that can lead directly to greater prosperity. During my tenure at Academia Sinica, I pushed hard for these two fields. We formed research groups within the Institute of Mathematics to attract people, so that they could eventually split out to form new institutes. We tried to find ways to use Taiwanese parts to build smaller computers like the DEC PDP-8 instead of the bigger IBM

FIG. 5. *Y. S. Chow in 2003.*

type. Many well-known scientists helped our institute in this project, including some who later played important roles in the development of the computer industry in Taiwan. In 1977, shortly after I left, Academia Sinica formed the Preparatory Office for the Institute of Information Science, which formally became an institute in 1982.

From 1982 to 1987, I served as Chairman of the Board of the Preparatory Office for the Institute of Statistics, and Min-Te Chao served as Director. George C. Tiao, Gregory Chow and C. C. Li also played important roles. We invited C. R. Rao over to Taiwan so we could learn from the great success of the Indian Statistical Institute. We realized that it was not wise to copy its model exactly, since we would like to play a more active role in promoting industry and commerce. We therefore invited Moriguti from the Imperial University, Tokyo, who was instrumental in the development of the computer industry in Japan in the 1960s. He served as President of the International Statistical Institute and was very talented in administration. Moreover, I wanted the institute to maintain close ties to the government Statistical Bureau. For this, I solicited help from many people outside of statistics. The institute was formally established in 1987.

Query: Would you tell us something about the development of statistics in Taiwan?

Chow: Statistical sciences cover a wide range of areas, but we must focus on applications, on real problems. To develop statistics in Taiwan, we must rely on our own efforts and do what others are not willing to do and what others have never done. Now applied statistics is thriving in Taiwan. For example, we conduct all kinds of public polls, including election polls. Recently, Taiwan started to focus on medical science and public health, and decided to establish its National Health Research Institute, modeled after the NIH in the U.S. There are many talented biostatisticians in NIH, but they belong to different groups in various research institutes. Before the formation of Taiwan's NIH, I had a long talk with Gordon Lan. I said we must seize this opportunity to establish an independent institute of biostatistics. Taiwan's National Health Research Institute now has five divisions, including the Division of Biostatistics and Bioinformatics. Chao Agnes Hsiung is serving as Director of the division, which will eventually become an institute. It will be the first such statistical institute in the world.

We cultivated a very close relationship between the statistical community and the government in Taiwan. The Statistical Society in Taiwan is a semi-official organization, whose members are mainly from the National Statistical Bureau. The Chairman of its Board of Directors is always from the Bureau, but the majority of its directors are academics. One of the most powerful committees in the National Statistical Bureau is the Supervisory Committee on National Income, which convenes every year to approve the national income figure announced by the government. Its Chair is the General Director of Budget, Accounting and Statistics, and its members are composed of deputy secretaries of various departments (Financial Ministry, Central Bank, etc.) of the Executive Yuan (the Cabinet) and academics, in about a two-to-one ratio. The committee has to reach a unanimous decision. Some people say this is just a formality, but we have to be in the government to get to know the right people, to express our opinion and to get things done. In Taiwan, the government provides tremendous support to statistics and effective communication with it is essential. This is by no means easy to achieve, but Taiwan has been quite successful at it.

CHINESE STUDENTS IN THE UNITED STATES

Query: Many Chinese students come to the U.S. to pursue their postgraduate education. You are one of the pioneers. Could you share with us your thoughts on this?

Chow: I would first like to share with you some background information. I belong to the first generation of students who came to the United States with financial support from American universities. Most students coming here before the revolution were supported by the Chinese government, somewhat like visiting scholars from Mainland China in the eighties. Whatever they did, whatever they said, they had to take into consideration the reaction from the government. We did not have such a close relationship with the Taiwan government, so we did not feel we had any obligation. In short, we had a different mentality.

It was extremely difficult to get scholarships from American universities, and the amount was often quite small. We had to use borrowed money to certify our financial ability when we applied for visas, and we had to find part-time jobs to support ourselves here. School authorities were well aware of this and allowed us to work on campus. The situation improved in the late fifties, partly because of the excellent academic performance of Chinese students. I remember that newspapers in Taiwan used to carry articles saying that Chinese universities had reached the world level because their graduates were directly admitted into postgraduate schools in the United States. Most students before us started from senior undergraduate courses in the U.S., including graduates from Zhejiang and Taiwan Universities, since college degrees from China were not recognized at that time.

After I joined Columbia University in 1962, I found that the Statistics Department had never admitted a Chinese student. I suggested to Robbins that Columbia offer two scholarships, one to the Mathematics Department at National Taiwan University and the other to Taiwan Normal University. They were the only two mathematics departments in Taiwan at that time, and there was no statistics department. Mathematics majors did not know that they could naturally take statistics as their specialty for postgraduate study. Robbins offered each university a four-year fellowship. Taiwan Normal sent Pi-Erh Lin, who later joined the faculty at the Florida State University, but the student nominated by National Taiwan University failed an English test at the U.S. Embassy. It turned out that National Taiwan University recommended neither the one who wanted to come most nor its best. The one who wanted to come most, Min-Te Chao, was also the best mathematics senior at National Taiwan University for that year. He spent one more year in graduate school in Taiwan before he went to Berkeley. Although he did not come to Columbia, Robbins' letter helped him to choose statistics. These two were the earliest students. From then on, many mathematics majors in Taiwan followed their path.

In 1972, Richard Nixon went to China. I saw an opportunity and suggested to Robbins that he write to Caohao Gu, my contemporary at Zhejiang University and a well-known mathematician at Fudan University. Robbins did, but because of the Cultural Revolution, Gu never received the letter. In 1979, Zhejiang University sent a delegation to the U.S. It was composed of its president and seven vice presidents, the first delegation at such a high level from a Chinese university. Since I am an alumnus of Zhejiang University, I received the delegation when it visited Columbia. We went to see Robbins after their campus tour, and Robbins offered the Mathematics Department at Zhejiang University a fellowship in the meeting upon my suggestion.

Caohao Gu was then visiting SUNY at Stony Brook. When he learned this, he asked me if the Mathematics Department at Fudan University could also have a fellowship, since it was formed by faculty from Zhejiang University. Robbins agreed to that too. Gu is a smart person—he recommended two students without ranking them, and we had no choice but to accept both. That was the year Zukang Zheng and Cun-Hui Zhang came to Columbia.

The Statistics Department at Purdue was also a pioneer in recruiting Chinese students. In 1963, when I went to Purdue, I was surprised to see eight students from Taiwan, all mathematics majors from National Taiwan University. Since I was also from there, we became good friends. Seven of them returned to Taiwan after completing their Ph.D.'s. Although I did not help them come over, I did aid their return. I take credit

FIG. 6. *Zhiliang Ying, Y. S. Chow and Cun-Hui Zhang.*

for the fact that the statistical community in Taiwan has a high percentage of Columbia and Purdue alumni. Sometimes I feel a little bit guilty, though, because I have remained in the States while encouraging others to return.

ADVISING STUDENTS

Query: Would you share with us your experience in advising students?

Chow: Some people teach students how to appreciate mathematics. They describe the beauty of mathematical ideas. This is analogous to teaching swimming by asking students to watch how others swim. This is not going to work. The more you watch, the more afraid you become; you have to jump into the water. Appreciating mathematics is one thing, but actually doing it is an entirely different matter. Many influential papers are often incomplete and imperfect. For example, the proof in Birkhoff's 1932 paper on ergodic theory was not correct and the original proof of optimality of the sequential probability ratio test was also problematic. Referees nowadays may reject these papers, but they are among the best ones. It is impossible to start something with a perfect paper. We should allow our students to explore, to make a few mistakes and eventually correct them. Do not show your students too much. That will spoil them. Let them practice first and broaden their vision gradually. It is just like rolling a snowball. You have to have something small to start with. It is important to start doing research early.

We should credit students for all the good stuff in their theses and credit their advisors for the rest. Good students are made by themselves, not by their professors. A good friend of mine once said that an advisor would give students only two types of problems: those he could not solve and those he did not care to solve. I wonder why we give student problems.

When I first came to Illinois, Brahana had a big influence on me. Although my college grades were not great, he said I had a strong background and forced me to take all the hard courses. I became a good student all of a sudden. This changed my life. I realized that you could become a good student when your teacher treats you as one; if your teacher treats you as a poor student, you simply get worse and worse.

RETIREMENT

Query: You retired from Columbia about ten years ago. Could you tell us something about your retirement?

Chow: I grew up in China and received all my early education there, for which I never paid anything. After college, I went to Taiwan. I have done something for the Chinese people in Taiwan, but not enough for those in the Mainland, to whom I owe quite a lot. I am retired, but have energy, so I have been contributing both my time and my money to education.

We, along with our friends, are subsidizing some high school students in my hometown, which is poverty-stricken. The tuition for senior high school is about the same as the average annual family income there. Some families are allotted two *mus* (about 1/3 acre) of land to farm. If their children pass the

FIG. 7. *Steve Kou, Y. S. Chow and Burnett Sams.*

entrance exam for high school, they will have difficulties finding the money. In the past ten years, we have supported about 300 such poor students for their high school education.

We also funded two scholarship programs for college education. In the first program, we award a scholarship every year to anyone in my home county who is admitted to one of the five most selective universities in China: Peking, Tsinghua, University of Science and Technology, Fudan and Nankai. Up to now, 26 students from my hometown have been enrolled at the above universities. In the second program, we selected 20 poor students from the key high school in my hometown, and if they succeed in entering one of a group of 15 good universities, we lend a hand. Since I started the scholarship program, the number of students from this high school admitted to college has increased quite dramatically.

We also set up a math competition within the key high school in 1989, which the county government insisted on naming the "Chow Yuan Shih Math Award," despite my objections. The top five scorers in the competition each receive a prize, a small amount in the U.S., but a nontrivial figure in my hometown. Besides the prizes, I also give a banquet, so I have to go back there every year. The county magistrate, the deputy magistrate and the principal of the high school are all invited, along with the parents of the winners. The county government sends a car to pick up winners and their parents for the dinner. In addition, the local TV station always reports the results of the competition. It is a big event each year. In this way, I spend very little for very good results. In recent years, students from my hometown have always taken the top two spots in the Xiangfan Regional math competition.

Another thing in my retirement plan was to bring American-style home architecture to my hometown. Xiangfan is an ancient city, but wars stripped off all its historical landmarks. The Communists and the Nationalists fought their last battle in the city, with 25,000 casualties on each side. After the revolution, the city became desolate and required rebuilding. When I went back in 1989, I saw houses along a street 5 kilometers long, all of the same height (about 7 meters), all of the same grey color and none with a bathroom. I could not believe my eyes! Those who need the toilet have to go down the street to the public latrine. It was impossible to explain to the locals what could be done, so I decided to build a house myself. I modeled my house after Illinois' Arlington Park and brought architects to modify my design. I built the house using Chinese products no matter how much I had to spend, and hired local builders no matter how inefficient they were. Nevertheless, the house was finished in three years. It is beautiful, but it needs repairing often. When I pay for replacement materials, I complain loudly and always ask merchants to stock better stuff. Now, when the locals build their own houses, they include bathrooms too. What's more, I used a light blue color, and it has become a fashionable color for houses in my hometown. I also used parquet hardwood floors, which was new to the locals, but has become very popular now.

Publications of Yuan Shih Chow

1954:

1. On the Cesaro summability of double Fourier series. *Tohoku Math. J.* **5**, 277-283.

1960:

2. A martingale inequality and the law of large numbers. *Proc. Amer. Soc.* **11**, 107-111.

3. Martingales in a σ-finite measure space indexed by directed sets. *Trans. Amer. Math. Soc.* **97**, 254-285.

1961:

4. On the sums of independent random variables with infinite moments and fair games. *Proc. Nat. Acad. Sci. U.S.A.* **47**, 330-335 (with H. Robbins).

5. Iterates on conditional expectation operators. *Proc. Amer. Math. Soc.* **12**, 490-495 (with D.L. Burkholder).

6. A martingale system theorem and applications. *Proc. Fourth Berkeley Symp. on Math. Statist. and Probab.* **1**, 93-104 (with H. Robbins).

1962:

7. Convergence theorems of martingales. *Zeit. Wahrscheinlichkeitstheorie* **1**, 340-346.

1963:

8. On optimal stopping rules. *Zeit. Wahrscheinlichkeitstheorie* **2**, 33-49 (with H. Robbins).

9. A renewal theorem for random variables which are dependent or non-identically distributed. *Ann. Math. Statist.* **34**, 390-396 (with H. Robbins).

1964:

10. On optimal selection based on relative rank. *Israel J. Math.* **2**, 81-90 (with S. Moriguti, H. Robbins, and S.M. Samuels).

1965:

11. Local convergence of martingales and law of large numbers. *Ann. Math. Statist.* **36**, 552-558.

12. On the asymptotic theory of fixed-width sequential intervals for the mean. *Ann. Math. Statist.* **36**, 457-462 (with H. Robbins).

13. Moments of randomly stopped sums. *Ann. Math. Statist.* **36**, 789-799 (with H. Robbins and H. Teicher).

14. On optimal stopping rules for S_n/n. *Ill. J. Math.* **9**, 444-454 (with H. Robbins).

15. A martingale convergence theorem of Ward's type. *Ill. J. Math.* **9**, 569-576.

1966:

16. On the moments of some one-sided stopping rules. *Ann. Math. Statist.* **37**, 382-387.

17. On the second moments of some two-sided stopping rules. *Ann. Math. Statist.* **37**, 388-392 (with H. Teicher).

18. Some convergence theorems for independent random variables. *Ann. Math. Statist.* **37**, 1482-1493.

1967:

19. On the values of stochastic sequence. *Proc. Fifth Berkeley Symp. on Math. Statist. and Probab.* **1**, 427-440 (with H. Robbins).

20. A class of optimal stopping problems. *Proc. Fifth Berkeley Symp. on Math. Statist. and Probab.* **1**, 419-426 (with H. Robbins).

21. On the expected value of a stopped submartingale. *Ann. Math. Statist.* **38**, 608-609.

22. On a strong law of large numbers for martingales. *Ann. Math. Statist.* **38**, 610.

1968:

23. Convergence of sums of squares of martingales differences. *Ann. Math. Statist.* **39**, 123-133.

24. On the monotonicity of $E_p(S_t/t)$. *Ann. Math. Statist.* **39**, 1755 (with W.J. Studden).

1969:

25. Martingale extensions of a theorem of Marcinkiewicz and Zygmund. *Ann. Math. Statist.* **40**, 427-433.

26. On the expected value of a stopped stochastic sequence. *Ann. Math. Statist.* **40**, 456-461 (with W.F. Stout).

27. Monotonicity of the variance under truncation and variations of Jensen's inequality. *Ann. Math. Statist.* **40**, 1106-1108 (with W.J. Studden).

28. Stopping rules for X_n/n and related problems. *Israel J. Math.* **7**, 240-248 (with A. Dvoretzky).

1970:

29. The variance of some one-sided stopping rules, studies and essays (Presented to Yu-why Chen on his 60th Birthday), 299-307. *Math. Res. Center*, Nat. Taiwan University.

1971:

30. Almost certain summability of independent, identically distributed random variables. *Ann. Math. Statist.* **42**, 401-404 (with H. Teicher).

31. The L_p-convergence for $n^{-1/p}S_n$, $0 < p < 2$. *Ann. Math. Statist.* **42**, 392-394.

32. *Great Expectations: The Theory of Optimal Stopping.* Houghton Mifflin, Boston (with H. Robbins and D. Siegmund).

1973:

33. Iterated logarithm laws for weighted averages. *Zeit. Wahrscheinlichkeitstheorie* **26**, 87-94 (with H. Teicher).

34. Limiting behavior of weighted sums of independent random variables. *Ann. Probab.* **1**, 810-824 (with T.L. Lai).

35. Some necessary conditions for a.s. convergence of sums of independent random variables. *Bull. Inst. Math. Acad. Sinica* **1**, 1-7 (with J.H. Abbott).

36. Delayed sums and Borel summability for independent, identically distributed random variables. *Bull. Inst. Math. Acad. Sinica* **1**, 207-220.

1974:

37. Optimal stopping – a new development in mathematical statistics. *J. Chinese Statist. Assoc.* **9**. (In Chinese).

1975:

38. *Optimal Stopping Rules for X_n/n and S_n/n, Statistical Inference and Related Topics*, Vol. 2. Academic Press, New York, 159-177 (with K.K. Lan).

39. Some one-sided theorems on the tail distributions of sample sums with applications to the last time and largest excess of boundary crossings. *Trans. Amer. Math. Soc.* **208**, 51-72 (with T.L. Lai).

1976:

40. Limiting behavior of $\max_{j<n} S_j j^{-\alpha}$ and the first passage times in a random walk with positive drift. *Bull. Inst. Math. Acad. Sinica* **4**, 35-44 (with Agnes Hsiung).

1977:

41. On the existence of optimal stopping rules for the reward sequence S_n/n. *Sankhya, Ser. A* **39**, 278-289 (with A.K. Basu).

1978:

42. *Probability Theory.* Springer-Verlag, New York (with H. Teicher).

43. Paley-type inequalities and convergence rates related to the law of large numbers and extended renewal theory. *Z. Wahrscheinlichkeitstheorie verw. Gebiete* **45**, 1-19.

1979:

44. Extended renewal theory and moment convergence in Anscombe's theorem. *Ann. Probability* **7**, 304-318 (with Chao A. Hsiung and T.L. Lai).

45. On the maximal excess in boundary crossings of random walks related to fluctuation theory and laws of large numbers. *Bull. Inst. Math. Acad. Sinica* **7**, 271-289 (with T.L. Lai).

46. Conditions for the existence and nonexistence of optimal stopping rules for S_n/n. *Proc. Amer. Math. Soc.* **75**, 300-307 (with J. Cuzick).

47. Moments of ladder variables for driftless random ralks. *Z. Wahrscheinlichkeitstheorie verw. Gebiete* **48**, 253-257 (with T.L. Lai).

1980:

48. Limit theorems for a positively drifting process and iots related first passage times. *Bull. Inst. Math. Acad. Sinica* **8**, 141-172 (with C.A. Hsiung and K.F. Yu).

1981:

49. The performance of a sequential procedure for the estimation of the mean. *Ann. Statist.* **9**, 184-189 (with K.F. Yu).

50. Iterated logarithm laws with random subsequences. *Z. Wahrscheinlichkeitstheorie verw. Gebiete* **57**, 235-251 (with H. Teicher, C.Z. Wei and K.F. Yu).

1982:

51. Bounded regret of a sequential procedure for estimation of the mean. *Ann. Statist.* **10**, 909-914 (with A.T. Martinsek).

1983:

52. A renewal theorem and its applications to some sequential procedures. *Z. Wahrscheinlichkeitstheorie verw. Gebiete* **64**, 241-250 (with C.A. Hsiung and K.F. Yu).

1984:

53. Some limit theorems for a subcritical branching process with immigration. *J. Appl. Probab.* **21**, 50-57 (with K.F. Yu).

1986:

54. A note on Feller's strong law of large numbers. *Ann. Probab.* **14**, 1088-1094 (with C.-H. Zhang).

55. On moments of ladder height variables. *Adv. Appl. Math.* **7**, 46-54.

1988:

56. *Probability Theory*, 2nd ed., Springer-Verlag, New York (with H. Teicher).

57. On the rate of moment convergence of sample sums and extremes. *Bull. Inst. Math. Acad. Sinica* **16**, 219-243.

1991:

58. Moment convergence of reciprocals of some first passage times. *Amer. J. Math. and Management Sci.* **11**, 135-146.

1992:

59. Wald's equation for a class of de-normalized U-statistics. *Ann. Probab.* **21**, 1151-1158 (with V.H. de la Pena and H. Teicher).

60. On the law of iterated logarithm and the central limit theorem for the number of runs. *Chinese J. Math. (Taiwan)* **20**, 217-221.

61. On a Poisson convergence for the number of runs. *J. Chinese Statist. Assoc.* **30**, 241-244. (In Chinese).

1993:

62. A moment inequality for Hsu and Robbins series. *Statistica Sinica* **3**, 1-7.

1995:

63. On a moment inequality. *J. Chinese Statist. Assoc.* **33** 117-121. (In Chinese).

64. A moment inequality for products. *J. Chinese Statist. Assoc.* **33** 429-436. (In Chinese).

1997:

65. *Probability Theory*, 3rd ed., Springer-Verlag, New York (with H. Teicher).

66. On Spitzer's formula for the moment of ladder variables. *Statistica Sinica* **7**, 149-156.

67. A central limit theorem for the number of success runs: an example of regenerative processes. *Statistica Sinica* **7**, 157-166 (with S.G. Kou).

2002:

68. On stable convergence in the central limit theorems. *Statist. Probab. Letters* **57**, 307-313 (with Tsung-Lin Cheng).

PREFACE

Yuan-Shih Chow, Professor Emeritus of Statistics, Columbia University and Academician of Academia Sinica, Taiwan, celebrated his 80th birthday on September 1, 2004. In honor of this occasion and to pay tribute to his influential work in probability theory and mathematical statistics as well as his contributions to mathematics education and the development of statistical research and application in Taiwan and Mainland China, the International Conference on Random Walks, Sequential Analysis and Related Topics was held at Fudan University, Shanghai, China during July 18 and 19, 2004.

This volume includes papers presented at the Fudan conference and contributions from a number of authors who were not able to make the trip to Shanghai for the occasion. The 22 papers in the volume cover a wide range of topics which reflect the broad scope of areas where Professor Chow has made major contributions as well as recent advances in probability and statistics. The authors were invited to participate in this endeavor to honor Professor Chow because of their association with Professor Chow or because of their activities in the areas of research the conference focused on. In addition, the volume includes "A conversation with Yuan Shih Chow," originally published in the February, 2006 issue of *Statistical Science*, which contains some fascinating stories of Professor Chow's life experience as a student, teacher, scholar, colleague, friend, and intellectual in general. It is hoped that this volume will capture some of the spirit of the Shanghai conference, the important advances in random walk, sequential analysis and related topics, and the inspiration of Professor Chow.

The Editors would like to thank Professors Tze Leung Lai and Zukang Zheng who were instrumental in the organization of the conference and publication of this volume, members of the advisory and organizing committees of the conference, Chaohao Gu, Tze-Leung Lai, Jia-Ding Chen, Xi-Ru Chen, Chun-Tu Lin, Shaw-Hwa Lo, Henry Teicher, Jia-Gang Wang, Ching-Zong Wei, Rong Wu, Minggao Gu, Gordon Lan, Zhaohai Li, Kai F. Yu, Ming Zheng, Jinjin Hu, Jia Shen, Qinfeng Xu, and Xinshend Zhang, and all the participants of the conference and contributors to the volume. The Editors would also like to thank about 20 people who served as anonymous referees. Finally, the Editors would like to thank Fudan University for providing financial support for the conference, and Fudan's staff who helped create a wonderful and pleasant atmosphere for the conference.

Agnes Chao Hsiung, Zhiliang Ying, and Cun-Hui Zhang
April 3, 2006

CONTENTS

Sequential Analysis

MAXIMIZING EXPECTED VALUE WITH TWO STAGE STOPPING RULES

DAVID ASSAF*

Department of Statistics, Hebrew University of Jerusalem, Israel

LARRY GOLDSTEIN

Department of Mathematics, University of Southern California, U.S.A.

ESTER SAMUEL-CAHN* †

Department of Statistics, Hebrew University of Jerusalem, Israel

Let $X_n, \ldots, X_1$ be i.i.d. random variables with distribution function F and finite expectation. A statistician, knowing F, observes the X values sequentially and is given two chances to choose X's using stopping rules. The statistician's goal is to select a value of X as large as possible. Let V_n^2 equal the expectation of the larger of the two values chosen by the statistician when proceeding optimally. We obtain the asymptotic behavior of the sequence V_n^2 for a large class of F's belonging to the domain of attraction (for the maximum) $\mathcal{D}(G_{II}^{\alpha})$, where $G_{II}^{\alpha}(x) = \exp(-x^{-\alpha})\mathbf{I}(x > 0)$ with $\alpha > 1$. The results are compared with those for the asymptotic behavior of the classical one choice value sequence V_n^1, as well as with the "prophet value" sequence $E(\max\{X_n, \ldots, X_1\})$, and indicate that substantial improvement is obtained when given two chances to stop, rather than one.

Some key words: Optimal Stopping; Prophet value; Two-choice; Domains of attraction; Asymptotic value.

1. Introduction

Kennedy and Kertz (1991) study the asymptotic behavior of the value sequence of a one choice optimal stopping problem, as $n \to \infty$, where one observes $X_n, \ldots, X_1$ independent, identically distributed random variables with known distribution F, and the payoff is the random variable at which

*Research supported by the Israel Science Foundation Grant 879/01

†Research supported by the Israel Science Foundation Grant 467/04.

one stops. The goal is to maximize the expected payoff. The value of such a sequence is therefore $\sup EX_t = V_n^1$, where the supremum is over all stopping times t, which stop after at most n observations, with probability one. They show that the asymptotic behavior depends on which of the three well-known domains of attraction (for the maximum) F belongs to.

Recently Assaf and Samuel-Cahn (2000) and Assaf, Goldstein and Samuel-Cahn (2002) study stopping problems in which the optimal stopper is given more than one choice. There "Prophet Inequalities" are derived for general finite sequences of independent, non-negative, not necessarily identically distributed, random variables.

The present paper focuses on a different aspect of multiple choices. We consider the situation where the observations are i.i.d. random variables, and the stopper is given two choices. His payoff is the larger of the two values chosen, and the goal is to maximize the expected return. As an example, the situation we consider here may correspond to a situation in which you put your first selected item (perhaps a house or a job offer) "on hold" as a guaranteed fallback value. You then proceed sequentially to select a second item (which should be of greater value than the first, unless it is the last one) and finish by taking the better of the two items selected.

Let $\mathcal{D}(G_{II}^{\alpha})$ be the domain of attraction for the maximum to

$$G_{II}^{\alpha}(x) = \exp(-x^{-\alpha})\mathbf{I}(x > 0),$$

(see notation in Kennedy and Kertz (1991), or p.4 of Leadbetter, Lindgren and Rootzen, (1983)), and $V_n^2 = \sup_{1 \le t_1 \le t_2 \le n} E[\max\{X_{t_1}, X_{t_2}\}]$. We study the asymptotic behavior of V_n^2 for $F \in \mathcal{D}(G_{II}^{\alpha})$. The case $F \in \mathcal{D}(G_{III}^{\alpha})$ has recently been considered in Assaf, Goldstein and Samuel-Cahn (2003), henceforth referred to as AGS. Because the distributions whose maxima are attracted to G_{III}^{α} are bounded above, it was more natural there to consider the goal to be the minimization of the expected value of the minimum of the two values chosen. The results obtained there about the minimum were then translated back to results about the maximum; see Remark 7.4 of AGS.

A necessary and sufficient condition for $F \in \mathcal{D}(G_{II}^{\alpha})$ (see Theorem 1.6.2 of Leadbetter, Lindgren and Rootzen, 1983) is $x_F = \sup\{x : F(x) < 1\} = \infty$, and that for some $\alpha > 0$,

$$\lim_{t\to\infty} [1 - F(tx)]/[1 - F(t)] = x^{-\alpha} \quad \text{for all } x > 0,$$

which can also be written as

$$1 - F(x) = x^{-\alpha}L(x) \quad \text{as } x \to \infty \tag{1}$$

where $L(x)$ is slowly varying at ∞. We will treat the case where

$$\lim_{x\to\infty} L(x) = \mathcal{L} \in (0,\infty). \tag{2}$$

Throughout we will consider the case $\alpha > 1$; the case $0 < \alpha \leq 1$ is uninteresting, as then the expectation of X is infinite. It is known that for $X_n, \ldots, X_1$ i.i.d. with distribution $F \in \mathcal{D}(G_{II}^{\alpha})$ with finite expectation, for $M_n = \max(X_n, \ldots X_1)$ we have

$$\lim_{n\to\infty} n[1 - F(EM_n)] = [\Gamma\left(1 - \frac{1}{\alpha}\right)]^{-\alpha}. \tag{3}$$

(See proposition 2.1 of Resnick, 1987). Let V_n^1 be the optimal one-choice value, when the goal is to maximize the expected value of the item chosen. Then by Kennedy and Kertz (1991),

$$\lim_{n\to\infty} n[1 - F(V_n^1)] = \frac{\alpha - 1}{\alpha}. \tag{4}$$

When (2) is satisfied, then taking $\mathcal{L} = 1$ for convenience, the limits in (3) and (4) can be rewritten, respectively, as

$$\lim_{n\to\infty} n^{-1/\alpha} EM_n = \Gamma(1 - 1/\alpha)$$

$$\text{and} \quad \lim_{n\to\infty} n^{-1/\alpha} V_n^1 = \left(\frac{\alpha}{\alpha - 1}\right)^{1/\alpha}. \tag{5}$$

Thus if we denote by $V_n = V_n^2$ the optimal two choice value, it is reasonable to expect that $\lim_{n\to\infty} n^{-1/\alpha} V_n = d_\alpha$, where $\left(\frac{\alpha}{\alpha-1}\right)^{1/\alpha} < d_\alpha < \Gamma(1 - 1/\alpha)$. We prove the following

Theorem 1.1. *Let $X_n, X_{n-1}, \ldots, X_1$ be i.i.d. random variables with finite expectation, having distribution function F satisfying $x_F = \infty$, $1 - F(x) = x^{-\alpha} L(x)$ where $\lim_{x\to\infty} L(x) = \mathcal{L} \in (0,\infty)$. Define*

$$h(u) = \left(\frac{\alpha}{\alpha - 1} + \frac{1}{u}\right)^{1/\alpha} \quad \text{for } 0 < u < \infty$$

and let β_α be the unique solution y to

$$\int_0^y h(u)du - \left(\frac{1}{\alpha} + y\right) h(y) = 0. \tag{6}$$

Then

$$\lim_{n\to\infty} n[1 - F(V_n^2)] = [h(\beta_\alpha)]^{-\alpha},$$

so in particular, for $\mathcal{L} = 1$

$$\lim_{n\to\infty} n^{-1/\alpha} V_n^2 = h(\beta_\alpha) = d_\alpha.$$

The proof of Theorem 1.1 is given in Section 6.

The paper is organized as follows. In Section 2 we derive the optimality equations and give a heuristic argument leading to Theorem 1.1. Section 3 contains some lemmas. In Section 4 we derive a general theorem for the convergence of recursions. These are applied in Section 5 to the family of Pareto distributions, and are then, in Section 6, generalized to all distributions considered in Theorem 1.1. Section 7 contains a table of numerical evaluations and comparisons of the limiting scaled values, as n tends to infinity, of the optimal one and two stop values, and EM_n. It is seen that considerable improvement is obtained, through the possibility of a second choice. For example, numerical computations indicate that the limit, as n tends to infinity, of the ratio $[V_n^2 - V_n^1]/[EM_n - V_n^1]$ is never less than 78%.

After the present work was accepted, the paper of Kühne and Rüschendorf (2002) came to our attention. That paper treats the same problem as the one here, using Poisson-approximations, and in full generality, basing their results on their earlier detailed paper Kühne and Rüschendorf (2000). Detailed results are given for the domain of attraction $\mathcal{D}(e^{-e^{-x}})$, only. Our approach in the present paper is more direct, and our results for the present domain of attraction are more explicit than theirs, when (2) holds.

2. Preliminaries and Heuristics

For X an integrable random variable with distribution function F, let

$$g(x) = g_1(x) = E[X \vee x] \quad \text{and} \quad g_{n+1}(x) = g(g_n(x)), \quad n \geq 1.$$

The function $g_n(x)$ equals the optimal one-choice value when there are n observations to be made, and one is guaranteed the value x; for this reason when stressing this interpretation we denote $g_n(x)$ by $V_n^1(x)$. On the other hand, for notational clarity, we will drop the superscript on V_n^2, denoting it V_n. Clearly $g_n(x)$ is increasing in x for all n. The optimality equations for the two choice values are

$$\begin{aligned} V_2 &= E[X_1 \vee X_2] \\ V_{n+1} &= E[V_n \vee V_n^1(X_{n+1})] \quad n \geq 2. \end{aligned} \tag{7}$$

Note that we 'reverse' index our sequence of variables $X_n, \ldots, X_1$ for convenience, so that X_k denotes the kth variable from the end of the horizon. The first term in the expectation (7) corresponds to passing up X_{n+1} and keeping two choices on the remaining n variables; the second term corresponds to choosing X_{n+1} and continuing with one choice on the remaining variables with the value X_{n+1} guaranteed.

We now present an outline of the argument which yields our result and an accompanying simple heuristic. From (7), letting b_n be defined through

$$V_n = g_n(b_n), \tag{8}$$

we see that if $X_{n+1} > b_n$ then it should be chosen and otherwise passed up. Write (7) in the form

$$V_{n+1} = V_n F(b_n) + \int_{b_n}^{\infty} g_n(x) dF(x). \tag{9}$$

As a representative of $F \in \mathcal{D}(G_{II}^{\alpha})$ we first consider, for $\alpha > 1$ fixed, the Pareto distribution with

$$F_\alpha(x) = [1 - x^{-\alpha}]\mathbf{I}(x \geq 1) \quad \text{and density} \quad \alpha x^{-(\alpha+1)}\mathbf{I}(x \geq 1). \tag{10}$$

For this family

$$g(x) = x + \frac{1}{\alpha - 1} x^{-(\alpha-1)} \qquad \text{for } x \geq 1 \tag{11}$$

$$\text{so} \quad g_{n+1}(x) = g_n(x) + \frac{1}{\alpha - 1} g_n(x)^{-(\alpha-1)} \qquad \text{for } x \geq 1. \tag{12}$$

Since $P(X \geq 1) = 1$ we have $V_n^1 = g_n(1)$.

In particular, for the Pareto family, (9) becomes

$$V_{n+1} = V_n[1 - b_n^{-\alpha}] + \alpha \int_{b_n}^{\infty} g_n(x) x^{-(\alpha+1)} dx. \tag{13}$$

For $y \geq 1/n$ let

$$f_n(y) = n^{-1/\alpha} g_n((ny)^{1/\alpha}), \quad \text{that is,} \quad g_n(x) = n^{1/\alpha} f_n(x^\alpha/n) \tag{14}$$

and

$$W_n^1 = n^{-1/\alpha} V_n, \quad \text{and} \quad B_n = n^{-1} b_n^\alpha,$$

thus, from (8),

$$W_n = f_n(B_n). \tag{15}$$

Note that $g_n(x)$ is strictly increasing for $x \geq 1$, and hence $f_n(x)$ is strictly increasing for $x \geq 1/n$. Rewriting (13) in this notation and multiplying by $n^{-1/\alpha}$ yields

$$\left(\frac{n+1}{n}\right)^{1/\alpha} W_{n+1} = W_n \left[1 - \frac{1}{nB_n}\right] + \alpha \int_{b_n}^{\infty} f_n\left(\frac{x^\alpha}{n}\right) x^{-(\alpha+1)} dx,$$

and making the change of variable $y = x^\alpha/n$, we obtain

$$\left(\frac{n+1}{n}\right)^{1/\alpha} W_{n+1} = W_n\left[1 - \frac{1}{nB_n}\right] + n^{-1}\int_{B_n}^{\infty} f_n(y)y^{-2}dy. \qquad (16)$$

When the change of variable is done directly in (7), this expression can also be written as

$$\left(\frac{n+1}{n}\right)^{1/\alpha} W_{n+1} = n^{-1}\int_{1/n}^{\infty} [W_n \vee f_n(y)]y^{-2}dy. \qquad (17)$$

Since $((n+1)/n)^{1/\alpha} = 1 + 1/(\alpha n) + O(n^{-2})$, if we multiply (16) by n we have

$$n(W_{n+1} - W_n) = -\frac{1}{\alpha}W_{n+1} - \frac{1}{B_n}W_n + \int_{B_n}^{\infty} f_n(y)y^{-2}dy + O(n^{-1}). \quad (18)$$

If $W_n \to W$ such that $W_n = W + a/n + O(n^{-2})$, then $n(W_{n+1} - W_n) \to 0$ and if also $B_n \to B_\alpha$ and $f_n(y) \to f(y)$ then taking limits in (18) yields

$$0 = -\left(\frac{1}{\alpha} + \frac{1}{B_\alpha}\right)W + \int_{B_\alpha}^{\infty} f(y)y^{-2}dy \qquad (19)$$

and by (15)

$$W = f(B_\alpha). \qquad (20)$$

To find f we use the following heusistics: set, for short $\rho = \alpha/(\alpha - 1)$. By (5)

$$V_n^1 = g_n(1) \approx (n\rho)^{1/\alpha}.$$

Suppose for a given x we can find k such that

$$g_n(x) \approx g_{n+k}(1) \approx ((n+k)\rho)^{1/\alpha}.$$

But

$$g_{n+k}(1) = g_n(g_k(1)), \text{ thus } x \approx g_k(1) \approx (k\rho)^{1/\alpha},$$

i.e.

$$k \approx x^\alpha/\rho,$$

and thus

$$g_n(x) \approx ((n + x^\alpha/\rho)\rho)^{1/\alpha} = (n\rho + x^\alpha)^{1/\alpha}.$$

Set $y = x^\alpha/n$ to get

$$g_n((ny)^{1/\alpha}) \approx n^{1/\alpha}(\rho + y)^{1/\alpha}.$$

Thus by (14)

$$f_n(y) \approx \left(\frac{\alpha}{\alpha-1}+y\right)^{1/\alpha}.$$

Let

$$f(y) = \left(\frac{\alpha}{\alpha-1}+y\right)^{1/\alpha}. \tag{21}$$

Note that by (5) and (21)

$$\lim_{y\to 0} f(y) = \left(\frac{\alpha}{\alpha-1}\right)^{1/\alpha} = \lim_{n\to\infty} n^{-1/\alpha} V_n^1.$$

Also, (20) translates to

$$W = \left(\frac{\alpha}{\alpha-1}+B_\alpha\right)^{1/\alpha}. \tag{22}$$

Set $y = 1/u$ and $h(u) = f(1/u)$ in (22) and (21), and write $\beta_\alpha = 1/B_\alpha$ to obtain

$$h(u) = \left(\frac{\alpha}{\alpha-1}+\frac{1}{u}\right)^{1/\alpha}, \tag{23}$$

and

$$W = h(\beta_\alpha).$$

Now substituting into (19)

$$\int_0^{\beta_\alpha} h(u)du - \left(\frac{1}{\alpha}+\beta_\alpha\right) h(\beta_\alpha) = 0$$

which explains the theorem.

3. Some Lemmas

In this section we determine the limiting behavior of the sequence of functions $f_n, n = 1, 2, \ldots$ given in (14), for determining the asymptotic behavior of the two stop value for an i.i.d. sequence with distribution (10), a representative of $F \in \mathcal{D}(G_{II}^\alpha)$.

Lemma 3.1. *Let $f_n(y)$ and $f(y)$ be given in (14) and (21), respectively. Then*

$$f_n(y) > f(y) \text{ for all } y \geq 1/n. \tag{24}$$

Proof. We prove (24) by induction. For $n = 1$ by (11)

$$f_1(y) = g_1(y^{1/\alpha}) = y^{1/\alpha} + \frac{1}{\alpha - 1} y^{-1+1/\alpha} = y^{1/\alpha}\left(1 + \frac{1}{(\alpha-1)y}\right), y \geq 1$$

while

$$f(y) = y^{1/\alpha}\left(1 + \frac{\alpha}{(\alpha-1)y}\right)^{1/\alpha}.$$

But $1/\alpha < 1$, therefore $(1+x)^{1/\alpha} < 1 + x/\alpha$, thus $f(y) < y^{1/\alpha}\left(1 + \frac{1}{(\alpha-1)y}\right) = f_1(y)$, and (24) holds for $n = 1$. Now assume (24) holds for some $n \geq 1$. Then by (14), for $x \geq 1$,

$$g_n(x) = n^{1/\alpha} f_n\left(\frac{x^\alpha}{n}\right) > n^{1/\alpha} f\left(\frac{x^\alpha}{n}\right) = x\left(\frac{\alpha n}{(\alpha-1)x^\alpha} + 1\right)^{1/\alpha} \tag{25}$$

and thus, since g is increasing

$$\begin{aligned} g_{n+1}(x) &= g(g_n(x)) \\ &> x\left(\frac{\alpha n}{(\alpha-1)x^\alpha} + 1\right)^{1/\alpha}\left\{1 + \frac{1}{\alpha-1}\left[x^\alpha\left(\frac{\alpha n}{(\alpha-1)x^\alpha} + 1\right)\right]^{-1}\right\} \\ &= x\left(\frac{\alpha n}{(\alpha-1)x^\alpha} + 1\right)^{1/\alpha}\left[1 + \frac{1}{\alpha n + (\alpha-1)x^\alpha}\right]. \end{aligned} \tag{26}$$

Thus, similar to (25), it suffices to show that for $x \geq 1$ the right hand side of (26) is greater than

$$x\left(\frac{\alpha(n+1)}{(\alpha-1)x^\alpha} + 1\right)^{1/\alpha},$$

i.e. that

$$\frac{\alpha n + (\alpha-1)x^\alpha}{(\alpha-1)x^\alpha}\left(1 + \frac{1}{\alpha n + (\alpha-1)x^\alpha}\right)^\alpha > \frac{\alpha(n+1) + (\alpha-1)x^\alpha}{(\alpha-1)x^\alpha}.$$

Put over the common factor $(\alpha-1)x^\alpha$, the numerator on the left hand side is greater than $(\alpha n + (\alpha-1)x^\alpha)\left(1 + \frac{\alpha}{\alpha n + (\alpha-1)x^\alpha}\right) = \alpha(n+1) + (\alpha-1)x^\alpha$ which is the numerator of the right hand side. ∎

Lemma 3.2. *Let* $\varepsilon_n(y) = f_n(y) - f(y)$. *Then for* $n \geq 1$

$$\varepsilon_n(y) < \frac{y^{1/\alpha - 2}}{2(\alpha-1)n} \quad \textit{for } y \geq 1/n. \tag{27}$$

Proof: Consider $(1+t)^\delta$ where $0<\delta<1,\ t>0$. Then by Taylor expansion, for some $0<\theta<1$

$$(1+t)^\delta = 1+\delta t - \frac{\delta(1-\delta)t^2}{2(1+\theta t)^{2-\delta}} > 1+\delta t - \frac{\delta(1-\delta)t^2}{2}.$$

Hence for $t=\alpha/((\alpha-1)y)$ and $\delta=1/\alpha$,

$$f(y) = y^{1/\alpha}\left(1+\frac{\alpha}{(\alpha-1)y}\right)^{1/\alpha} > y^{1/\alpha}\left[1+\frac{1}{(\alpha-1)y}-\frac{1}{2(\alpha-1)y^2}\right]. \quad (28)$$

We prove (27) by induction. For $n=1$ we have $f_1(y)=y^{1/\alpha}(1+\frac{1}{(\alpha-1)y})$, for $y\geq 1$. Thus (27) follows immediately for $n=1$ by (28). Now suppose (27) holds for some $n\geq 1$. Set $a_n=n/(n+1)$. Then by (14) and (12) it follows, after setting $y=x^\alpha/n$, that

$$a_n^{-1/\alpha} f_{n+1}(a_n y) = f_n(y) + \frac{1}{(\alpha-1)n} f_n(y)^{-(\alpha-1)}. \quad (29)$$

We want to show that $\varepsilon_{n+1}(y) < \frac{y^{1/\alpha-2}}{2(\alpha-1)(n+1)}$ for $y\geq 1/(n+1)$. This is equivalent to

$$\varepsilon_{n+1}(a_n y) < \frac{a_n^{1/\alpha-2} y^{1/\alpha-2}}{2(\alpha-1)(n+1)} = \frac{a_n^{1/\alpha} y^{1/\alpha-2}(n+1)}{2(\alpha-1)n^2} \quad \text{for } y\geq 1/n. \quad (30)$$

Now, by (29)

$$\begin{aligned}\varepsilon_{n+1}(a_n y) &= f_{n+1}(a_n y) - f(a_n y) \quad (31)\\ &= a_n^{1/\alpha}\left\{f_n(y)+\frac{1}{(\alpha-1)n}f_n(y)^{-(\alpha-1)} - a_n^{-1/\alpha} f(a_n y)\right\}.\end{aligned}$$

Note that

$$f'(y) = \frac{1}{\alpha} f(y)^{-(\alpha-1)} > 0 \text{ and } f''(y) = \frac{-(\alpha-1)}{\alpha^2} f(y)^{-(2\alpha-1)} < 0.$$

Thus by Taylor expansion we can write, for some $0<\theta<1$,

$$f(x+\Delta) = f(x) + \frac{\Delta}{\alpha} f(x)^{-(\alpha-1)} - \frac{\Delta^2(\alpha-1)}{2\alpha^2} f(x+\theta\Delta)^{-(2\alpha-1)}.$$

Thus with $\Delta=\frac{\alpha}{n(\alpha-1)}$, we have

$$\begin{aligned}a_n^{-1/\alpha} f(a_n y) &= \left(\frac{n+1}{n}\frac{\alpha}{\alpha-1}+y\right)^{1/\alpha} = f\left(y+\frac{\alpha}{n(\alpha-1)}\right) \quad (32)\\ &= f(y)+\frac{1}{n(\alpha-1)}f(y)^{-(\alpha-1)} - \frac{1}{2n^2(\alpha-1)} f(y+\frac{\theta\alpha}{n(\alpha-1)})^{-(2\alpha-1)}.\end{aligned}$$

Substituting (32) into (31) we obtain $\varepsilon_{n+1}(a_n y)$ is $a_n^{1/\alpha}$ times

$$\varepsilon_n(y) + \frac{1}{n(\alpha-1)} \ (f_n(y)^{-(\alpha-1)} - f(y)^{-(\alpha-1)}) \\ + \ \frac{1}{2n^2(\alpha-1)} f(y + \frac{\theta\alpha}{n(\alpha-1)})^{-(2\alpha-1)}. \qquad (33)$$

Now by Lemma 3.1, since $\alpha > 1$, we have $f_n(y)^{-(\alpha-1)} < f(y)^{-(\alpha-1)}$. Thus the second term in (33) is negative. Also $f(y)$ is increasing, thus $f(y)^{-(2\alpha-1)}$ is decreasing, and

$$f^{-(2\alpha-1)}\left(y + \frac{\theta\alpha}{n(\alpha-1)}\right) < f^{-(2\alpha-1)}(y) = \frac{1}{\left(\frac{\alpha}{\alpha-1} + y\right)^{2-1/\alpha}} < y^{1/\alpha-2}.$$

Thus from (33) and the induction hypothesis, for $y \geq 1/n$, and therefore for $a_n y \geq 1/(n+1)$, we have

$$\begin{aligned}\varepsilon_{n+1}(a_n y) &< a_n^{1/\alpha}\{\frac{y^{1/\alpha-2}}{2(\alpha-1)n} + \frac{1}{2n^2(\alpha-1)} y^{1/\alpha-2}\} \\ &= a_n^{1/\alpha} y^{1/\alpha-2}(n+1)/[2(\alpha-1)n^2]\end{aligned}$$

which shows (30). ∎

The following Corollary is an immediate consequence of Lemmas 3.1 and 3.2.

Corollary 3.1. *Let*

$$h_n(u) = f_n\left(\frac{1}{u}\right) \qquad 0 < u \leq n, \qquad (34)$$

where f_n is defined in (14). Then for all $y > 0$, as $n \to \infty$,

$$f_n(y) \to f(y) = \left(\frac{\alpha}{\alpha-1} + y\right)^{1/\alpha} \qquad \textit{and}$$

$$h_n(y) \to h(y) = \left(\frac{\alpha}{\alpha-1} + \frac{1}{y}\right)^{1/\alpha}.$$

Note that

$$\lim_{y\to\infty} h(y) = \left(\frac{\alpha}{\alpha-1}\right)^{1/\alpha} = \lim_{n\to\infty} n^{1/\alpha} V_n^1,$$

and that (2) holds for (10) with $\mathcal{L} = 1$.

With the notation (34) we can rewrite (17) as

$$\left(\frac{n+1}{n}\right)^{1/\alpha} W_{n+1} = n^{-1} \int_0^n [h_n(u) \vee W_n] du \tag{35}$$

and if we let $\beta_n = 1/B_n$, then, from (15)

$$W_n = h_n(\beta_n).$$

4. Convergence of Recursions

We consider functions q and Q satisfying the following condition. We allow $q(0)$ to take the value infinity.

Condition 4.1. There exist positive numbers A and a such that $q(y)$ is nonnegative for $y \geq 0$, strictly monotone decreasing and differentiable for $0 < y < A$, non-increasing for $y \geq A$ and $\int_0^a q(u)du < \infty$. Further, $Q(A) > 0$ where $Q(y)$ is defined by

$$Q(y) = \int_0^y q(u)du - (1/\alpha + y)\, q(y). \tag{36}$$

Lemma 4.1. *Under Condition 4.1, the function $Q(y)$ is strictly increasing in $(0, A)$, non-decreasing for $y > A$ and there exists a unique value b for which $Q(b) = 0$ and $b < A$.*

Proof. Let $0 \leq y_1 < y_2$. Elementary calculations yield

$$Q(y_2) - Q(y_1) \geq \left(\frac{1}{\alpha} + y_1\right)(q(y_1) - q(y_2)),$$

and monotonicity now follows from noting the latter expression is positive if $y_1 < A$, and non-negative otherwise. Since $Q(0) < 0$ and $Q(A) > 0$ the claim on the root b follows. ■

Our aim is to prove

Theorem 4.1. *Let Condition 4.1 be satisfied, and let $m \geq 1$ and c be arbitrary. Define*

$$Z_m = c \text{ and } \left(\frac{n+1}{n}\right)^{1/\alpha} Z_{n+1} = \frac{1}{n} \int_0^n (q(u) \vee Z_n) du \quad \text{for } n \geq m. \tag{37}$$

Then the limit of Z_n exists and satisfies

$$\lim_{n\to\infty} Z_n = d = q(b)$$

where b is the unique root of $Q(y) = 0$.

The crux of the proof of Theorem 4.1 is the following Lemma.

Lemma 4.2. *Let Condition 4.1 be satisfied, and let $m \geq 1$ be any integer and c any constant, and suppose that Z_n for $n \geq m$ is defined by (37). Then for every $\delta \in (0, \min\{q(0) - d, d - q(A)\})$ there exist $\Delta > 0$ and n_0 such that for all $n \geq n_0$,*

$$\text{if } Z_n > d + \delta \text{ then } Z_{n+1} \leq (1 - \Delta/n)Z_n, \tag{38}$$

$$\text{if } Z_n < d - \delta \text{ then } Z_{n+1} \geq (1 + \Delta/n)Z_n, \tag{39}$$

$$\text{if } Z_n > d \text{ then } Z_{n+1} > d, \text{ and} \tag{40}$$

$$\text{if } |Z_n - d| \leq \delta \text{ then } |Z_{n+1} - d| \leq \delta. \tag{41}$$

Proof: We have

$$(\frac{n}{n+1})^{1/\alpha} = 1 - \frac{1}{\alpha n} + \frac{1}{\alpha}(\frac{1}{\alpha} + 1)\frac{1}{2n^2} + O_\alpha(n^{-3}),$$

and hence, for any $\rho \neq 0$

$$(\frac{n}{n+1})^{1/\alpha}(1+\frac{1}{\rho n}) = 1+\frac{1}{n}(\frac{1}{\rho} - \frac{1}{\alpha})+\frac{1}{n^2}\left(\frac{1}{2\alpha}(\frac{1}{\alpha} + 1) - \frac{1}{\alpha\rho}\right)+O_{\alpha,\rho}(n^{-3}), \tag{42}$$

where we write $O_\lambda(c_n)$ to indicate a sequence bounded in absolute value by c_n times a constant depending only on λ, a collection of parameters.

Define

$$M(t) = \int_0^{q^{-1}(t)} \left(\frac{q(y)}{t} - 1\right) dy \quad \text{for } q(A) < t \leq q(0).$$

From (36), $Q(b) = 0$ and $d = q(b)$, we have

$$M(d) = 1/\alpha.$$

It is not hard to see that $M(t)$ is strictly decreasing over its range. Hence, setting $\Delta_2 = (M(d - \delta) - 1/\alpha)/2$ and $\Delta_1 = (1/\alpha - M(d + \delta))/2$ we have $\Delta = \min\{\Delta_1, \Delta_2\} > 0$.

Consider the function

$$r_n(t) = \frac{1}{n}\int_0^n \left(\frac{q(y)}{t} \vee 1\right) dy.$$

We first show that $Z_n < q(0)$ for all n sufficiently large. This is obvious when $q(0) = \infty$. If $q(0) < \infty$ then note first that $Z_n < q(0)$ implies $Z_{n+1} <$

$q(0)$. Now assuming that $Z_n > q(0)$ for all n sufficiently large gives $Z_{n+1} \leq (n/(n+1))^{1/\alpha} Z_n$ and the contradiction that $Z_n \to 0$ by $\prod_k (k/(k+1))^{1/\alpha} = 0$.

Since $Z_{m+1} > 0$ we have $Z_n > 0$ for all $n \geq m+1$, and now by (37)

$$Z_{n+1}/Z_n = \left(\frac{n}{n+1}\right)^{1/\alpha} r_n(Z_n). \tag{43}$$

By definition, for $q(A) < t \leq q(0)$

$$r_n(t) = 1 + \frac{1}{n}\int_0^{q^{-1}(t)\wedge n} \left(\frac{q(y)}{t} - 1\right) dy = 1 + \frac{1}{n} M(t) \quad \text{when } n \geq q^{-1}(t).$$

To prove (38), assume $Z_n > d + \delta$. Since r_n is decreasing, using (43) and (42), we have for all $n > q^{-1}(d+\delta)$,

$$\begin{aligned} Z_{n+1} &\leq Z_n(\frac{n}{n+1})^{1/\alpha} r_n(d+\delta) \\ &= Z_n(\frac{n}{n+1})^{1/\alpha}(1 + \frac{1}{n}M(d+\delta)) \\ &= \left[1 + \frac{1}{n}(M(d+\delta) - \frac{1}{\alpha}) + O_{\alpha,d-\delta}(n^{-2})\right] Z_n \\ &\leq (1 - \frac{\Delta_1}{n})Z_n \leq (1 - \frac{\Delta}{n})Z_n \end{aligned}$$

for all n sufficiently large, showing (38).

Next we prove (39). When $Z_n < d - \delta$, we have similarly that for $n > q^{-1}(d-\delta)$,

$$\begin{aligned} Z_{n+1} &\geq Z_n(\frac{n}{n+1})^{1/\alpha} r_n(d-\delta) \\ &= Z_n(\frac{n}{n+1})^{1/\alpha}(1 + \frac{1}{n}M(d-\delta)) \\ &= \left[1 + \frac{1}{n}(M(d-\delta) - \frac{1}{\alpha}) + O_{\alpha,d-\delta}(n^{-2})\right] Z_n \\ &\geq (1 + \frac{\Delta_2}{n})Z_n \geq (1 + \frac{\Delta}{n})Z_n \end{aligned}$$

for all n sufficiently large.

Turning now to (40) and (41), for $Z_n \geq d-\delta$, since $d-\delta > q(A)$, $\beta_n < A$ is well defined by

$$q(\beta_n) = Z_n.$$

Now by (37) and (36), for $n \geq A$,

$$\left(\frac{n+1}{n}\right)^{1/\alpha} Z_{n+1} = \frac{1}{n}\left(\int_0^{\beta_n} q(y)dy + (n - \beta_n)q(\beta_n)\right)$$

$$= \frac{1}{n}Q(\beta_n) + (1 + \frac{1}{\alpha n})Z_n;$$

thus

$$Z_{n+1} = \left(\frac{n}{n+1}\right)^{1/\alpha} \frac{1}{n}Q(\beta_n) + R_n Z_n \tag{44}$$

where

$$R_n = (\frac{n}{n+1})^{1/\alpha}(1 + \frac{1}{\alpha n}). \tag{45}$$

For $u > q(A)$ consider

$$Q(q^{-1}(u)) = \int_0^{q^{-1}(u)} q(y)dy - (1/\alpha + q^{-1}(u))u.$$

Since $q^{-1}(u)$ is differentiable for $u > q(A)$, for such u

$$\frac{d}{du}Q(q^{-1}(u)) = -\left(1/\alpha + q^{-1}(u)\right).$$

Hence, evaluating $Q(q^{-1}(u))$ by a Taylor expansion around d, and using $Q(b) = Q(q^{-1}(d)) = 0$, we obtain that there exists some ξ_{Z_n} between d and Z_n such that

$$Q(\beta_n) = Q(q^{-1}(Z_n)) = -(Z_n - d)(1/\alpha + q^{-1}(\xi_{Z_n})). \tag{46}$$

Subtracting d from both sides of (44) and using (46) we obtain

$$Z_{n+1} - d = \left\{1 - (\frac{n}{n+1})^{1/\alpha}\frac{1}{n}(\frac{1}{\alpha} + q^{-1}(\xi_{Z_n}))\right\}(Z_n - d) + [R_n - 1]Z_n. \tag{47}$$

Take n_1 such that for all $n \geq n_1$

$$(\frac{n}{n+1})^{1/\alpha}\frac{1}{n}(\frac{1}{\alpha} + q^{-1}(d-\delta)) < 1 \quad \text{and} \quad 0 < R_n - 1,$$

where we use (45) and (42) with $\rho = \alpha > 1$ for the second inequality.

Take now $Z_n > d$. Then $\xi_{Z_n} > d$, $q^{-1}(\xi_{Z_n}) < q^{-1}(d) < q^{-1}(d - \delta)$, and for $n \geq n_1$

$$0 < \left\{1 - (\frac{n}{n+1})^{1/\alpha}\frac{1}{n}(\frac{1}{\alpha} + q^{-1}(\xi_{Z_n}))\right\},$$

so that the first term on the right hand side of (47) is strictly positive. For $n \geq n_1$ the second term on the right hand side is also positive, and the sum of these two terms is therefore positive. This proves (40).

Turning to (41), suppose that $|Z_n - d| \leq \delta$. Then $|\xi_{Z_n} - d| \leq \delta$, and therefore

$$q^{-1}(d + \delta) \leq q^{-1}(\xi_{Z_n}) \leq q^{-1}(d - \delta).$$

Hence, letting $\Delta_3 = (1/\alpha + q^{-1}(d+\delta))/2 > 0$ we have $(1/\alpha + q^{-1}(\xi_{Z_n}))/2 \geq \Delta_3$ and

$$0 \leq \left\{1 - (\frac{n}{n+1})^{1/\alpha}\frac{1}{n}(\frac{1}{\alpha} + q^{-1}(\xi_{Z_n}))\right\} \leq 1 - \frac{\Delta_3}{n}. \tag{48}$$

Further, from (45), again using (42) with $\rho = \alpha$, there exists K_α such that

$$|R_n - 1| \leq \frac{K_\alpha}{n^2}.$$

Then for all n so large that

$$\frac{K_\alpha}{n}(d+\delta) \leq \Delta_3\delta$$

we have, using (47) and (48),

$$\begin{aligned}|Z_{n+1} - d| &\leq (1 - \frac{\Delta_3}{n})|Z_n - d| + |R_n - 1|Z_n \\ &\leq (1 - \frac{\Delta_3}{n})\delta + \frac{K_\alpha}{n^2}(d+\delta) \\ &\leq \delta.\end{aligned}$$

This proves (41). ■

Proof of Theorem 4.1: Let $0 < \delta < d - q(A)$, and $n \geq n_0$.

Case I: $Z_{n_0} < d - \delta$. If $Z_n < d - \delta$ for all $n \geq n_0$ then by (39) we would have

$$Z_{n+1} \geq \prod_{j=n_0}^{n} (1 + \frac{\Delta}{j})Z_{n_0} \to \infty,$$

a contradiction. Hence for some $n_1 \geq n_0$ we have $Z_{n_1} \geq d - \delta$, and we would therefore be in Case II or Case III.

Case II: $Z_{n_1} > d + \delta$ for some $n_1 \geq n_0$. If $Z_n > d + \delta$ for all $n \geq n_1$ we would have, by (38), that

$$Z_{n+1} \leq \prod_{j=n_1}^{n} (1 - \frac{\Delta}{j})Z_{n_1} \to 0,$$

again a contradiction. Hence there exists $n_2 \geq n_1$ such that $Z_{n_2} \leq d + \delta$. By (40), $Z_{n_2} > d$, reducing to Case III.

Case III: $|Z_{n_1} - d| \leq \delta$ for some $n_1 \geq n_0$. In this case $|Z_n - d| \leq \delta$ for all $n \geq n_1$ by (41). Since δ can be taken arbitrarily small, the Theorem is complete. ■

5. The Pareto Family

Let H be to h in (23) as Q is to q in (36). We first show

Lemma 5.1. *There exists a unique value β_α such that $H(\beta_\alpha) = 0$.*

Proof. Note that $h(y)$ is strictly decreasing and differentiable for $0 < y < \infty$ and that $\int_0^a h(u) < \infty$ for all $a > 0$ since $\alpha > 1$. This Lemma therefore follows from Lemma 4.1 if we show that $H(y)$ is positive for some y.

Now

$$H'(y) = -\left(\frac{1}{\alpha} + y\right) h'(y) = \frac{1}{\alpha}\left(\frac{1}{\alpha} + y\right) y^{-2} h(y)^{-(\alpha-1)}.$$

Since $\lim_{y\to\infty} h(y) = (\alpha/(\alpha-1))^{\frac{1}{\alpha}} < \infty$ it follows that $\int_a^\infty H'(y)dy = \infty$, thus $\lim_{y\to\infty} H(y) = \infty$ and the Lemma follows. ■

By Lemmas 3.1, 3.2 and (34), for $j = 1, 2, \ldots$ and $0 < y \leq j$

$$h(y) < h_j(y) < h(y) + \frac{y^{2-\frac{1}{\alpha}}}{2(\alpha-1)j} := \tilde{h}_j(y). \tag{49}$$

Recall that h and h_j are strictly monotone decreasing in their respective ranges.

The derivative of $\tilde{h}_j$ as defined in (49) is

$$\frac{d\tilde{h}_j(y)}{dy} = -\frac{1}{\alpha} y^{-2} h(y)^{-(\alpha-1)} + \frac{2\alpha - 1}{2\alpha(\alpha-1)j} y^{1-\frac{1}{\alpha}}.$$

Fix $A > \beta_\alpha$. It follows that for all $j > j_0(A)$ the function $\tilde{h}_j(y)$ is strictly monotone decreasing in $y \in (0, A]$. For $j > j_0(A)$ let

$$\tilde{h}_j^A(y) = \begin{cases} \tilde{h}_j(y) & \text{for } 0 < y \leq A \\ \tilde{h}_j(A) & \text{for } A < y < \infty. \end{cases}$$

Since h_j is strictly decreasing it follows from (49) that for $0 < y < j$

$$h_j(y) \leq \tilde{h}_j^A(y). \tag{50}$$

Note also that the sequence $\tilde{h}_j^A(y)$ is monotone decreasing in j, that is, if $j < n$

$$\tilde{h}_n^A(y) < \tilde{h}_j^A(y) \text{ for all } 0 < y < \infty. \tag{51}$$

Lemma 5.2. *Let $\tilde{H}_j^A(y)$ be defined for $\tilde{h}_j^A(y)$ through (36). Then for all $j > j_1(A)$ there exists a value $\beta_{j,\alpha}$ such that $\tilde{H}_j^A(\beta_{j,\alpha}) = 0$, and with $d_\alpha = h(\beta_\alpha)$ and $d_{j,\alpha} = \tilde{h}_j^A(\beta_{j,\alpha})$,*

$$\lim_{j\to\infty} \beta_{j,\alpha} = \beta_\alpha \quad \text{and} \quad \lim_{j\to\infty} d_{j,\alpha} = d_\alpha. \tag{52}$$

Proof: Since $\tilde{H}_j^A(y) \to H(y)$ uniformly on $[0, A]$ as $j \to \infty$ it follows that $\lim \tilde{H}_j^A(A) = H(A) > H(\beta_\alpha) = 0$. Hence for all $j > j_1(A)$ the value $\beta_{j,\alpha}$ exists. Now (52) follows from the uniform convergence of $\tilde{h}_j^A$ and $\tilde{H}_j^A$ to h and H, respectively. ■

It will become convenient to consider value and scaled value sequences arising from stopping on the independent variables $X_n, \ldots, X_{m+1}, Y_m, \ldots, Y_1$. The scaled value sequence W_n for this problem satisfies (35) for $n \geq m$ with starting value $W_m = m^{-1/\alpha} V_m(Y_m, \ldots, Y_1)$. Note that for any m and c there exists $Y_m, \ldots, Y_1$ such that $c = m^{-1/\alpha} V_m(Y_m, \ldots, Y_1)$; the simplest construction is obtained by letting $Y_j = cm^{1/\alpha}$ for $1 \leq j \leq m$. Our suppression of the dependence of W_n on m and c is justified by Theorem 5.1, which states that the limiting value of W_n is the same for all such sequences.

Lemma 5.3. *Let $m \geq 1$ be any integer and c be any constant. Let $W_m = c$ and for $n > m$ let W_n be determined by the recursion (35). Let*

$$Z_m^- = c \text{ and } \left(\frac{n+1}{n}\right)^{1/\alpha} Z_{n+1}^- = \frac{1}{n}\int_0^n (h(y) \vee Z_n^-)dy \text{ for } n \geq m. \tag{53}$$

For $j > j_1(A)$ fixed let $m_j = \max\{m, j\}$ and define the sequence $Z_{j,n}^+$ through

$$Z_{j,m_j}^+ = W_{m_j} \text{ and}$$
$$\left(\frac{n+1}{n}\right)^{1/\alpha} Z_{j,n+1}^+ = \frac{1}{n}\int_0^n (\tilde{h}_j^A(y) \vee Z_{j,n}^+)dy, \quad n \geq m_j. \tag{54}$$

Then for all $n \geq m_j$

$$Z_n^- \leq W_n \leq Z_{j,n}^+ \tag{55}$$

and

$$\lim_{n\to\infty} Z_n^- = d_\alpha = h(\beta_\alpha) \text{ and } \lim_{n\to\infty} Z_{j,n}^+ = d_{j,\alpha} = \tilde{h}_j^A(\beta_{j,\alpha}) \tag{56}$$

Proof. Since by (49), (51) and (50) for all $n > j > j_1(A)$

$$h(y) < h_n(y) \leq \tilde{h}_n^A(y) < \tilde{h}_j^A(y), \quad 0 < y \leq n,$$

we obtain (55) directly by a comparison of the definitions in (35), (53) and (54). The conclusions in (56) are immediate from Theorem 4.1. ■

Theorem 5.1. *Let $m \geq 1$ be any integer, let $X_n, \ldots, X_{m+1}, Y_m, \ldots, Y_1$ be independent random variables, where $X_i \sim F_\alpha$ of (10) and $Y_m, \ldots, Y_1$ have finite expectation. For $n > m$, let*

$$V_{n,m} = V_n(X_n, \ldots, X_{m+1}, Y_m, \ldots, Y_1)$$

be the optimal two choice value. Then $W_n = n^{-1/\alpha} V_{n,m}$ satisfies

$$\lim_{n\to\infty} W_n = h(\beta_\alpha) = d_\alpha \tag{57}$$

where β_α is defined through (6). In particular, the optimal two stop value V_n for the sequence of i.i.d. r.v.'s with distribution function F_α of (10) satisfies

$$\lim_{n\to\infty} n^{-1} V_n^\alpha = h(\beta_\alpha)^\alpha,$$

that is, Theorem 1.1 holds for the family of distributions F_α.

Proof. Applying Lemma 5.3 with $c = m^{-1/\alpha} V_m(Y_m, \ldots, Y_1)$, for all $j > j_1(A)$

$$d_\alpha \leq \lim_{n\to\infty} \inf W_n \leq \limsup_{n\to\infty} W_n \leq d_{j,\alpha}.$$

Now let $j \to \infty$ and use (52) to get (57). Clearly the values W_n for the i.i.d. sequence with distribution F_α are generated by recursion (35), $m = 2$ and $c = 2^{-1/\alpha} E[X_1 \vee X_2]$. ■

6. Extension to General Distributions

Let $F \in \mathcal{D}(G_{II}^\alpha)$. By Proposition 2.1 of Resnick (1987),

$$\text{if for some integer } 0 < k < \alpha, \int_{-\infty}^{0} |x|^k dF(x) < \infty, \tag{58}$$

$$\text{then} \quad \lim_{n\to\infty} E[n^{-1/\alpha} M_n]^k = \Gamma(1 - \alpha^{-1} k).$$

Since we are considering random variables with finite expectation, it follows that F satisfies (58) with $k = 1$.

It suffices to prove Theorem 1.1 for positive random variables. Indeed, let X be a random variable with finite mean but otherwise arbitrary, X^+ be the positive part of X, and V_n and V_n^+ the corresponding two stop values. Clearly we have

$$V_n \le V_n^+. \tag{59}$$

For an inequality in the other direction, note that when we apply the optimal rules on the X^+ sequence, if the first variable selected is at time $t_1 < n$ it was because the positive threshold value b_n was exceeded, so that

$$X_{t_1}^+ \quad \text{is positive on the event} \quad \{t_1 < n\}.$$

Hence, applying the optimal X^+ rules on the X sequence, which may not be optimal for it, we obtain

$$X_{t_1} = X_{t_1}^+ \quad \text{on the event} \quad \{t_1 < n\},$$

and moreover that

$$X_{t_1} \vee X_{t_2} = X_{t_1}^+ \vee X_{t_2}^+ \quad \text{on the event} \quad \{t_1 < n\},$$

yielding

$$V_n \ge V_n^+ P(t_1 < n) - E[\max(0, -X)]P(t_1 = n).$$

Since $E|X| < \infty$ and $P(t_1 < n) \to 1$ as $n \to \infty$, we have

$$\liminf_{n\to\infty} n^{-1/\alpha} V_n \ge \lim_{n\to\infty} n^{-1/\alpha} V_n^+$$

which combined with (59) gives $\lim_{n\to\infty} n^{-1/\alpha} V_n = \lim_{n\to\infty} n^{-1/\alpha} V_n^+$. Thus, without loss of generality, we henceforth assume that $X \ge 0$.

We consider F satisfying (1), and (2). First note that without loss of generality we may assume that (2) holds with $\mathcal{L} = 1$, and prove Theorem 1.1 for this case only. This follows since if X is such that (1) and (2) hold, then for $Y = X/\mathcal{L}^{1/\alpha}$ we have $1 - F_Y(y) = y^{-\alpha}\hat{L}(y)$ for $\hat{L}(y) \to 1$ as $y \to \infty$, and $1 - F_Y(V_n^Y) = 1 - F_X(V_n^X)$, where V_n^X and V_n^Y are the optimal two choice values of iid sequences of length n from the F_X and F_Y distributions, respectively.

Lemma 6.1. *Let $X_\alpha \sim F_\alpha$, where F_α is given in (10) and let $X \ge 0$ with $X \sim F$ where $1 - F(x) = x^{-\alpha}L(x)$ and $\lim_{x\to\infty} L(x) = 1$. Then there exists a bounded function $L^*(x)$ satisfying $\lim_{x\to\infty} L^*(x) = 1$ such that*

$$X =_d X_\alpha L^*(X_\alpha). \tag{60}$$

Proof: Let

$$F^{-1}(u) = \sup\{x : F(x) < u\} \quad \text{for } u \in (0,1],$$

and

$$L^*(y) = F^{-1}(1 - y^{-\alpha})/y \quad \text{for } y > 1.$$

It is well known (see e.g. Lemma 6.4 of AGS) that for $U \sim \mathcal{U}[0,1]$,

$$X =_d F^{-1}(U);$$

since $F_\alpha(X_\alpha) = 1 - X_\alpha^{-\alpha} \sim \mathcal{U}[0,1]$, (60) follows. Now writing

$$F^{-1}(u) = \sup\{x : 1 - x^{-\alpha}L(x) < u\},$$

we have

$$F^{-1}(1 - y^{-\alpha}) = \sup\{x : 1 - x^{-\alpha}L(x) < 1 - y^{-\alpha}\} = \sup\{x : xL^{-1/\alpha}(x) < y\}.$$

Hence $\lim_{y\to\infty} L^*(y) = 1$ is equivalent to

$$\lim_{y\to\infty} \sup\{\frac{x}{y} : \frac{x}{y} < L^{1/\alpha}(x)\} = 1. \tag{61}$$

Since for every fixed x, $\lim_{y\to\infty} x/y = 0$, it follows that $x \to \infty$ as $y \to \infty$. But $\lim_{x\to\infty} L^{1/\alpha}(x) = 1$, and (61) follows. Let B be such that for $y \geq B$ we have $L^*(y) \leq 2$, say. Using $X \geq 0$, on $y \in (1, B]$ we have $0 \leq F^{-1}(0^+) \leq F^{-1}(1 - y^{-\alpha}) \leq F^{-1}(1 - B^{-\alpha})$. Hence the function $L^*(y)$ is bounded on its domain $(1, \infty)$. ∎

Proof of Theorem 1.1 By (60) we write

$$X_i = X_{\alpha,i}L^*(X_{\alpha,i}) \quad \text{a.s.}$$

where $X_n, \ldots, X_1$ are i.i.d. with distribution satisfying the conditions of the Theorem with $\mathcal{L} = 1$, and $X_{\alpha,i}$ are distributed with distribution F_α of (10). Let X_{t_n} and $X_{\alpha,t_n(\alpha)}$ be the optimally stopped two stop random variables on the i.i.d. sequences $X_n, \ldots, X_1$ and $X_{n,\alpha}, \ldots, X_{1,\alpha}$, respectively, where t_n and $t_n(\alpha)$ denote the respective times corresponding to the optimal values. Let $\epsilon > 0$ be given and let c^+ be such that $L^*(x) < 1 + \epsilon$ for all $x \geq c^+$.

Then

$$
\begin{aligned}
&\limsup_{n\to\infty} E[n^{-1/\alpha} X_{t_n}] = \limsup_{n\to\infty} E[n^{-1/\alpha} X_{\alpha,t_n} L^*(X_{\alpha,t_n})] \\
&= \limsup_{n\to\infty} \Big(E[n^{-1/\alpha} X_{\alpha,t_n} L^*(X_{\alpha,t_n}) \mathbf{I}(X_{\alpha,t_n} \geq c^+)] \\
&\quad + E[n^{-1/\alpha} X_{\alpha,t_n} L^*(X_{\alpha,t_n}) \mathbf{I}(X_{\alpha,t_n} < c^+)] \Big) \\
&= \limsup_{n\to\infty} E[n^{-1/\alpha} X_{\alpha,t_n} L^*(X_{\alpha,t_n}) \mathbf{I}(X_{\alpha,t_n} \geq c^+)] && (62) \\
&\leq \limsup_{n\to\infty} E[n^{-1/\alpha} X_{\alpha,t_n} (1+\epsilon) \mathbf{I}(X_{\alpha,t_n} \geq c^+)] && (63) \\
&= (1+\epsilon) \limsup_{n\to\infty} E[n^{-1/\alpha} X_{\alpha,t_n} \mathbf{I}(X_{\alpha,t_n} \geq c^+)] \\
&= (1+\epsilon) \limsup_{n\to\infty} E[n^{-1/\alpha} X_{\alpha,t_n}] \\
&\leq (1+\epsilon) \limsup_{n\to\infty} E[n^{-1/\alpha} X_{\alpha,t_n(\alpha)}] && (64) \\
&= (1+\epsilon) h(\beta_\alpha),
\end{aligned}
$$

where to obtain (62) we have used Lemma 6.1 to conclude that $X_{\alpha,t_n} L^*(X_{\alpha,t_n}) \mathbf{I}(X_{\alpha,t_n} < c^+)$ is bounded, and therefore the second expectation on the line above (62) has limit zero as $n \to \infty$; the last equality follows from Theorem 5.1.

Since (63) holds for any $\epsilon > 0$ we have

$$\limsup_{n\to\infty} n^{-1/\alpha} V_n(X_n, \ldots, X_1) \leq h(\beta_\alpha). \tag{65}$$

Now let c^- be such that $L^*(x) > 1 - \epsilon$ for all $x \geq c^-$. Consider using the rule $t_n(\alpha)$ on the sequence $X_n, \ldots, X_1$. Since this rule may not be optimal for that sequence, we have

$$
\begin{aligned}
&\liminf_{n\to\infty} [n^{-1/\alpha} V_n(X_n, \ldots, X_1)] \geq \liminf_{n\to\infty} E[n^{-1/\alpha} X_{t_n(\alpha)}] \\
&= \liminf_{n\to\infty} E[n^{-1/\alpha} X_{\alpha,t_n(\alpha)} L^*(X_{\alpha,t_n(\alpha)})] \\
&= \liminf_{n\to\infty} E[n^{-1/\alpha} X_{\alpha,t_n(\alpha)} L^*(X_{\alpha,t_n(\alpha)}) \mathbf{I}(X_{\alpha,t_n(\alpha)} \geq c^-)] \\
&\geq \liminf_{n\to\infty} E[n^{-1/\alpha} X_{\alpha,t_n(\alpha)} (1-\epsilon) \mathbf{I}(X_{\alpha,t_n(\alpha)} \geq c^-)] && (66) \\
&= (1-\epsilon) \liminf_{n\to\infty} E[n^{-1/\alpha} X_{\alpha,t_n(\alpha)} \mathbf{I}(X_{\alpha,t_n(\alpha)} \geq c^-)] \\
&= (1-\epsilon) \liminf_{n\to\infty} E[n^{-1/\alpha} X_{\alpha,t_n(\alpha)}] = (1-\epsilon) h(\beta_\alpha).
\end{aligned}
$$

Since (66) is true for every $\epsilon > 0$ we get, by (65),

$$h(\beta_\alpha) \geq \limsup_{n\to\infty} n^{-1/\alpha} V_n(X_n, \ldots, X_1)$$

$$\geq \liminf_{n\to\infty} n^{-1/\alpha} V_n(X_n, \ldots, X_1) \geq h(\beta_\alpha),$$

and Theorem 1.1 follows. ∎

7. Numerical Evaluations and Remarks

In Table 1, for $\alpha = 1.1, 1.2, \ldots, 2, 3, \ldots, 10$, the values in column (1), we tabulate for $\mathcal{L} = 1$ the following quantities in the columns indicated

(2) β_α
(3) $\lim_{n\to\infty} n^{-1/\alpha}\, V_n^1$
(4) $\lim_{n\to\infty} n^{-1/\alpha}\, V_n^2$
(5) $\lim_{n\to\infty} n^{-1/\alpha}\, EM_n$

In columns (6), (7) and (8) we tabulate the ratios (4)/(3), (5)/(4) and (5)/(3) respectively. The final column, column (9), of Table 1 represents the relative (limiting) improvement attained by using two stops rather than one, as compared to the reference value of the prophet, i.e.

$$\lim_{n\to\infty} (V_n^2 - V_n^1)/(EM_n - V_n^1). \tag{67}$$

The ratio in (67) has a minimum value of 0.788041 attained for $\alpha \approx 2.32$. Thus the limiting improvement when using two choices rather than one is never below 78.8%.

The following limiting statements can be shown to hold.

(i) For $\alpha \to \infty$,

$$\lim_{\alpha\to\infty} \beta_\alpha = e - 1$$

$$\lim_{\alpha\to\infty} \lim_{n\to\infty} n(1 - F(V_n^1)) = 1$$

$$\lim_{\alpha\to\infty} \lim_{n\to\infty} n(1 - F(V_n^2)) = 1 - e^{-1}$$

$$\lim_{\alpha\to\infty} \lim_{n\to\infty} n(1 - F(EM_n)) = e^{-\gamma}$$

and $$\lim_{\alpha\to\infty} \lim_{n\to\infty} \frac{V_n^2 - V_n^1}{EM_n - V_n^1} = [1 - \log(e-1)]/\gamma = .7946\ldots,$$

where $\gamma = .5772\ldots$ is Euler's constant.

(ii) For $\alpha \to 1$,

$$\lim_{\alpha\to1}\beta_\alpha = 0$$

$$\lim_{\alpha\to1}\lim_{n\to\infty} n(1-F(c_n)) = 0 \quad \text{for } c_n = V_n^1, V_n^2 \text{ and } EM_n,$$

$$\text{but} \quad \lim_{\alpha\to1}\lim_{n\to\infty}(V_n^2 - V_n^1)/(EM_n - V_n^1) = 1,$$

so the limiting rclative improvement for this case is 100%.

Remark 7.1. The present approach can easily be applied to obtain the asymptotic behavior for the one-choice value (obtained in Kennedy and Kertz (1991) by a different method) when $F(x)$ satisfies (1) and $\lim_{x\to\infty} L(x) = \mathcal{L} \in (0,\infty)$. First assume $X \sim F_\alpha$ of (10). Then for the one choice value V_n^1 we have $V_n^1 = EX$ and

$$V_{n+1}^1 = E[X_{n+1} \vee V_n^1] = \alpha \int_1^\infty [x \vee V_n^1] x^{-(\alpha+1)} dx. \tag{68}$$

Set $W_n^1 = n^{-1/\alpha} V_n^1$. Multiply (68) by $n^{-1/\alpha}$ to obtain

$$\left(\frac{n+1}{n}\right)^{1/\alpha} W_{n+1}^1 = \alpha \int_1^\infty [n^{-1/\alpha} x \vee W_n^1] x^{-(\alpha+1)} dx. \tag{69}$$

Substituting $u = nx^{-\alpha}$, (69) can be rewritten as

$$\left(\frac{n+1}{n}\right)^{1/\alpha} W_{n+1}^1 = \frac{1}{n}\int_0^n [u^{-1/\alpha} \vee W_n^1] du.$$

Now $q(u) = u^{-1/\alpha}$ satisfies Condition 4.1 and one can thus apply Theorem 4.1 directly to show that

$$\lim_{n\to\infty} W_n^1 = d = b^{-1/\alpha},$$

where b solves $Q(y) = 0$ with

$$Q(y) = \int_0^y u^{-1/\alpha} du - \left(\frac{1}{\alpha} + y\right) y^{-1/\alpha}$$

that is, that

$$\frac{\alpha}{\alpha-1} y^{-1/\alpha+1} - \left(\frac{1}{\alpha} + y\right) y^{-1/\alpha} = 0,$$

from which it follows immediately that $b = (\alpha - 1)/\alpha$ and $\lim_{n\to\infty} W_n^1 = \left(\frac{\alpha}{\alpha-1}\right)^{1/\alpha}$. The general result for the wider class of distributions satisfying (1) with $\lim L(x) = \mathcal{L} \in (0, \infty)$ now follows much in the same way as the arguments in Section 6.

Remark 7.2. Hill and Kertz (1982) and Kertz (1986) study one-choice prophet inequalities for non-negative i.i.d. random variables. They show that these prophet inequalities are n-dependent, and for each fixed n the extremal ratio can be obtained by a random variable with $n + 1$ atoms, (thus these variables do not belong to any domain of attraction for the maximum). As $n \to \infty$ the ratio tends to 1.34.... It may therefore be of interest to note, by comparison, that the extremal ratio of $\lim[EM_n/V_1^n]$ as $n \to \infty$, for the family studied here is 1.2882..., attained for $\alpha = 1.4628\ldots$.

(1)	(2)	(3)	(4)	(5)	(6)	(7)	(8)	(9)
α	β_α	$\lim \frac{V_n^1}{n^{1/\alpha}}$	$\lim \frac{V_n^2}{n^{1/\alpha}}$	$\lim \frac{M_n}{n^{1/\alpha}}$	(4)/(3)	(5)/(4)	(5)/(3)	$\frac{[(4)-(3)]}{[(5)-(3)]}$
1.1	0.54315	8.84546	10.18170	10.50590	1.15107	1.03184	1.18771	0.80477
1.2	0.68106	4.45102	5.34177	5.56632	1.20012	1.04204	1.25057	0.79867
1.3	0.78076	3.08932	3.77026	3.94584	1.22042	1.04657	1.27725	0.79501
1.4	0.85971	2.44692	3.00352	3.14912	1.22747	1.04848	1.28697	0.79266
1.5	0.92489	2.08008	2.55383	2.67894	1.22775	1.04899	1.28790	0.79109
1.6	0.98008	1.84599	2.26032	2.37044	1.22445	1.04872	1.28410	0.79003
1.7	1.02764	1.68530	2.05476	2.15338	1.21922	1.04800	1.27774	0.78931
1.8	1.06915	1.56912	1.90340	1.99289	1.21303	1.04702	1.27007	0.78881
1.9	1.10577	1.48182	1.78769	1.86974	1.20641	1.04590	1.26179	0.78848
2	1.13836	1.41421	1.69660	1.77245	1.19968	1.04471	1.25331	0.78827
3	1.33839	1.14471	1.30982	1.35412	1.14423	1.03382	1.18293	0.78846
4	1.43534	1.07457	1.19365	1.22542	1.11081	1.02662	1.14038	0.78939
5	1.49277	1.04564	1.13935	1.16423	1.08962	1.02184	1.11341	0.79017
6	1.53080	1.03085	1.10830	1.12879	1.07512	1.01849	1.09500	0.79077
7	1.55784	1.02227	1.08833	1.10577	1.06463	1.01602	1.08168	0.79124
8	1.57806	1.01683	1.07448	1.08965	1.05669	1.01412	1.07162	0.79161
9	1.59375	1.01317	1.06432	1.07776	1.05048	1.01263	1.06375	0.79191
10	1.60628	1.01059	1.05657	1.06863	1.04549	1.01142	1.05743	0.79215

8. Final Remarks

The last two authors are very saddened to announce that our invaluable colleague and friend David Assaf passed away most suddenly on December 23rd 2003. On that very day, in a last email from Prof. Assaf to us he wrote that he had some ideas and 'I will say more on this in a few days.' We regret on many levels that our work on two stage stopping can now only remain more or less in its current form, without the benefit of those further comments, now forever lost, which would have certainly greatly improved it.

References

1. ASSAF, D. & SAMUEL-CAHN, E. (2000) Simple ratio prophet inequalities for a mortal with multiple choices. *J. Appl. Prob.*, **37**, pp. 1084-1091.
2. ASSAF, D., GOLDSTEIN, L.& SAMUEL-CAHN, E. (2002). Ratio prophet inequalities when the mortal has several choices. *Ann. Appl. Prob.*, **12**, pp. 972-984.
3. ASSAF, D., GOLDSTEIN, L.& SAMUEL-CAHN, E. (2004). Two Choice Optimal Stopping. *Adv. Appl. Prob.* **36**, pp. 1116-1147
4. HILL, T.P. & KERTZ, R.P. (1982). Comparisons of stop rule and supremum expectations of i.i.d. random variables. *Ann. of Prob.*, **10**, pp. 336-345.
5. KENNEDY, D. P. & KERTZ, R.P. (1991). The asymptotic behavior of the reward sequence in the optimal stopping of i.i.d random variables. *Ann. Prob.* **19**, pp. 329-341.
6. KERTZ, R.P. (1986). Stop rule and supremum expectations of i.i.d. random variables: a complete comparison by conjugate duality. *J. Multivariate Analysis*, **19**, pp. 88-112.
7. KÜHNE, R. & RÜSCHENDORF, L. (2002) On optimal two-stopping problems. *Limit theorems in probability and statistics*, Vol. II, eds. Berkes, I, Csáki, E. & Csörgö, M, pp. 261-271, János Bolyai Math. Soc., Budapest.
8. KÜHNE, R. & RÜSCHENDORF, L. (2000) Approximation of optimal stopping problems. *Stochastic Process. Appl.*, **90**, pp. 301-325.
9. LEADBETTER, M. R., LINDGREN, G. & ROOTZÉN, H.(1983). *Extremes and Related Properties of Random Sequences and Processes.* Springer-Verlag, New York, Heidelberg, Berlin
10. RESNICK, S. I. (1987). *Extreme Values, Regular variation, and Point Processes.* Springer. N.Y.

NORMAL APPROXIMATION FOR TWO-STAGE WINNER DESIGN

K.K. GORDON LAN

J&J Pharmaceutical Research & Development, Raritan, New Jersey, U.S.A.

YUHWEN SOO and ZHENMING SHUN

Sanofi-Aventis, Bridgewater, New Jersey, U.S.A.

Some key words: Adaptive design; Dose selection; Interim Analysis; Multiplicity; Normal approximation; Sample size; Skewed normal distribution; Power; Type I error adjustment; Winner design.

1. Introduction

We consider a study where several potential new treatments are compared to control (standard) treatment C. In stage I of the study, a winner W among the new treatments is determined based on the treatment effect compared to C. In stage II, the losers in stage I are dropped and additional patients are allocated only to treatment arms W and C. At the end of the study, the data of W and C patients in stages I and II combined are compared to determine whether W is better than C. This design is called "Two-Stage Winner Design". The concept of selecting a treatment for further evaluation using the maximum test statistic was proposed by Whitehead (1986), followed by Thall, Simon, and Ellenberg (1988) and Simon (1989). These papers dealt with the situation of a binary endpoint, where exact calculations of the tail probabilities were provided. For normal endpoints, Sampson and Sill (2005) provided a test procedure that controls the overall type I error rates strongly.

It is well understood that the procedure of interim treatment arm selection would alter the rejection region and therefore the type I error rate would be inflated due to the nature of multiplicity and the interim adaptability if no correction is made. One may consider there is a certain "% of multiplicity" depending upon the time of interim selection is performed. If the interim selection is performed at early stage of the study, there will be less multiplicity issue compared to the situation where the interim selection is performed in a later time of the study. If there is no interim selection, then the winner design becomes the conventional multiple comparisons with a "many to one" mechanism: each treatment group is compared to a common control at the end of the study. In this case, two-sample tests comparing new treatments with the same control are correlated and the rejection region for testing the winner being better than the control was studied by Dunnett (1955). In general, the earlier the interim selection is performed, the less inflation in type I error rate there will be due to the multiple comparisons.

In this paper, we start with the special case where there are only two new treatments, A and B. To our pleasant surprise, the distribution of the test statistic comparing W and C is extremely close to normal. Thus, we propose a normal approximation approach in dealing with the type I error rate and power associated with the Winner design for normally distributed data. An almost exact method is also proposed. The normal approximation is extremely convenient, especially when an integration program is not handy. For almost all practical situations, it provides a simple way to understand the insights in designing such studies when the two treatment effects are similar. In more general settings when the effects are different, we provide a simple table in Section 4 to facilitate the evaluation of sample size for desired power levels. The case of unknown variance and the extension to more than two treatments are also discussed.

The paper is organized as follows: In Section 2, the Winner design and its test statistic are introduced; in Section 3, we consider tail probabilities using exact calculation, normal approximation, and almost exact approach; in Section 4, the power is evaluated; in Section 5, a small sample case with unknown variance is considered; and finally we complete this article with the discussions in Section 6.

2. Design and Test Statistic

Consider a clinical trial with two new treatment arms A, B and, a control arm C. The goal is to show either A or B is superior to C. The two new treatment arms could be two dose-levels of the same compound or even two different compounds. In stage I of the trial, n patients will be randomly assigned to each of the three arms. At the end of stage I, an interim look is planned for the selection of a winning treatment W between A and B for continual development. In stage II, an additional k patients will be allocated only into each of W and C, i.e., the loser drops out. At the end of the trial the effects of W and C will be compared based on $N = n + k$ patients, the combined patients across two stages in each group. The total sample size at the end of study would be $n + 2N$. The ratio $\tau = n/N$ will be called the information time.

Assume that efficacy of the treatments is captured in continuous variables denoted as $Y_i^j, i = 1, 2, ..., N, j = A, B$, and C, that are independently normally distributed with mean μ_j and common variance σ^2. For the loser L between A and B, only $Y_i^L, i = 1, 2, ..., n, n < N$, are observable. Let $\Delta_j = (\mu_j - \mu_C)/\sigma$, $j = A$ or B. We consider the following one-sided statistical hypotheses,

$$H_0 : \Delta_A = \Delta_B = 0 \text{ versus } H_a : \Delta_A > 0 \text{ or } \Delta_B > 0.$$

That is, larger observations are better outcomes. At the end of the trial, the test statistics for Δ_j, $j = A$ or B, is

$$\begin{aligned} Z_N^j &= \frac{\widehat{\Delta}_j}{\sqrt{\frac{2}{N}}} = \frac{\left(\overline{Y}_N^j - \overline{Y}_N^C\right)}{\sqrt{\frac{2\sigma^2}{N}}} = \sqrt{\tfrac{n}{N}} Z_n^j + \sqrt{\tfrac{k}{N}} Z_{(k)}^j \\ &= \sqrt{\tau} Z_n^j + \sqrt{1-\tau} Z_{(k)}^j \end{aligned}$$

Under the null hypothesis H_0, Z_N^j, Z_n^j, and $Z_{(k)}^j$ are all normally distributed with mean 0 and standard deviation 1.

An interim analysis is planned to select treatment based on the following "Select the Winner" procedure:

$$W : \begin{Bmatrix} \text{if } \overline{Y}_n^A > \overline{Y}_n^B, \text{ keep } A \text{ and } C \text{ to continue the study;} \\ \text{if } \overline{Y}_n^A < \overline{Y}_n^B, \text{ keep } B \text{ and } C \text{ to continue the study.} \end{Bmatrix}$$

There will be no early termination due to a larger $\overline{Y}_n^C$. The case of early termination due to futility will be further discussed (see Section 6). The test statistic based on Winner design is then

$$\begin{aligned} Z^W &= \sqrt{\tau} \max\left(Z_n^A, Z_n^B\right) + \sqrt{1-\tau} Z_{(k)} \\ &= \max\left(\sqrt{\tau} Z_n^A + \sqrt{1-\tau} Z_{(k)}, \sqrt{\tau} Z_n^B + \sqrt{1-\tau} Z_{(k)}\right), \end{aligned}$$

where $Corr\left(\sqrt{\tau} Z_n^A + \sqrt{1-\tau} Z_{(k)}, \sqrt{\tau} Z_n^B + \sqrt{1-\tau} Z_{(k)}\right) = 1 - \tau/2$. Under H_0, Z^W is the maximum of two correlated standard normal variables. The distribution of this maximum is called the skewed-normal distribution, and a thorough summary was given by Azzalini (1985). Under the null hypothesis, the distribution of Z^W can be derived and is expressed as

$$F_{Z^W}(y \mid \tau) = 2 \int_{-\infty}^{y} \Phi\left(\frac{\sqrt{\tau}}{\sqrt{4-\tau}} w\right) \phi(w)\, dw \qquad (1)$$

This gives the density function

$$f_{Z^W}(y \mid \tau) = 2\Phi\left(\frac{\sqrt{\tau}}{\sqrt{4-\tau}}y\right)\phi(y), \quad -\infty < y < \infty.$$

For given τ and under H_0, the mean and variance of Z^W are

$$\mu_\tau = \sqrt{\frac{\tau}{2\pi}} \text{ and } \sigma_\tau^2 = 1 - \frac{\tau}{2\pi}$$

respectively.

3. The Tail Probability

3.1. *Exact method*

Assume that, at the final analysis, an one-sided hypothesis is tested at the $100\alpha\%$ level. The regular critical value is expressed as $z_\alpha = \Phi^{-1}(1-\alpha)$. The exact type I error rate of Winner design can be calculated by

$$\begin{aligned}\alpha_E^W(\tau) &= \Pr\left(Z^W > z_\alpha \mid \tau\right)\\ &= \int_{z_\alpha}^{\infty} f_{Z^W}(w \mid \tau)dw = 2\int_{z_\alpha}^{\infty} \Phi\left(\frac{\sqrt{\tau}}{\sqrt{4-\tau}}w\right)\phi(w)\,dw, 0 \le \tau \le 1.\end{aligned}$$

Because of the distribution shift due to the interim winner selection, the type I error rate is inflated. In fact, it can be shown that $\alpha_{\mathrm{E}}^W(\tau) \ge \alpha$ for any given τ. When $\tau = 0$, it is the same as that only one treatment is considered. When $\tau = 1$, it is the same as comparing 2 treatments with control where the Dunnett's test applies.

Sampson and Sill (2005) have provided a detailed and more general discussion with any number of treatment groups in the study. In their paper they proposed an uniformly most powerful conditionally unbiased (UMPCU) approach based on the exact distribution of the Winner statistic.

The ultimate goal of the Winner design is to demonstrate that the selected winner is superior to the control. That is, the statistical hypothesis considered is one-sided. It is interesting to note that for any given two standard normal random variables Z_1 and Z_2 with correlation coefficient η and $Z_M = \max(Z_1, Z_2)$, the two-sided type I error, $Pr(|Z_M| > z_{\alpha/2}) = \alpha$, which indicates a two-sided test would not demonstrate type I error inflation.

3.2. *Normal approximation*

It can be proved that the standardized Z^W,

$$Z^{WS} = \frac{Z^W - \mu_\tau}{\sigma_\tau}$$

with a density function

$$f_{Z^{WS}}(y \mid \tau) = \sigma_\tau \cdot f_{Z^W}\left(\sigma_\tau y + \mu_\tau \mid \tau\right),$$

is approximated standard normal. In fact, the dth order cumulant $\kappa_{(d)}$ of Z^{WS} has $\mathbf{O}(\mu_\tau/\sigma_\tau)^d$ for $d > 2$. This demonstrates that the cumulants of Z^{WS} of order 2 or higher are close to zero based on the fact that $\mu_\tau/\sigma_\tau = \sqrt{\tau}/\sqrt{2\pi - \tau} \leq 0.435$, for $0 \leq \tau \leq 1$. The mathematical details can be found in Shun, Soo and Lan (2003). A closer look of the difference between the density curve of Z^{WS} and the standard normal curve shows, approximately, a maximum magnitude of about 0.001 (see Figure 1).

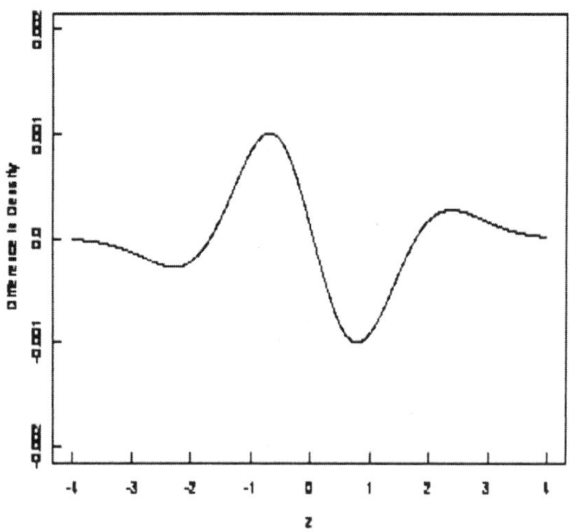

Fig. 1. Difference in Density Functions of Z^{WS} and $N(0,1)$ for $\tau = 0.5$

Therefore, we propose to use a standard normal distribution to approximate Z^{WS}. The type I error rate of the Winner design can be then estimated from the normal approximation as follows:

$$\begin{aligned} \alpha_A^W(\tau) &= \Pr\left(Z^W > z_\alpha \mid \tau\right) = \Pr\left(Z^{WS} > \tfrac{z_\alpha - \mu_\tau}{\sigma_\tau} \mid \tau\right) \\ &\cong 1 - \Phi\left(\tfrac{z_\alpha - \mu_\tau}{\sigma_\tau}\right) \end{aligned} \tag{2}$$

To control the overall type I error at the nominal α level, when an interim treatment selection is performed, statistical inference at the end of study may be done with an adjusted rejection region

$$Z^W > z_\alpha \sigma_\tau + \mu_\tau \equiv z_\alpha^{\text{adj}}.$$

or $Z^{WS} > z_\alpha$. The exact adjusted type I error rate is calculated using the normal approximation as follow:

$$\alpha_{E\_adj}^W(\tau) = \Pr\left(Z^W > z_\alpha \sigma_\tau + \mu_\tau \mid \tau\right) = 2 \int_{z_\alpha \sigma_\tau + \mu_\tau}^{\infty} \Phi\left(\frac{\sqrt{\tau}}{\sqrt{4-\tau}} w\right) \phi(w)\, dw.$$

For $\alpha = 0.025$ or $z_\alpha = 1.959964$, and $\tau = 0.1, 0.2, \cdots, 1$, the exact adjusted type I errors are listed in Table 1. Note that the adjusted critical values perform well, but the results are slightly anti-conservative. A correction is proposed in the next section.

Table 1. Exact Adjusted α Level, for the Nominal $\alpha = 0.025$

τ	0.1	0.2	0.3	0.4	0.5
$\alpha_{E\_adj}^W$	0.02502472	0.02507199	0.02513603	0.02521535	0.02530943
τ	0.6	0.7	0.8	0.9	1.0
$\alpha_{E\_adj}^W$	0.02541824	0.02554197	0.02568106	0.02583605	0.02600762

At the end of the study,

$$E\left[\left(Z^W - \sqrt{\frac{N}{2}}\Delta\right) - \mu_\tau \mid \tau\right] = 0.$$

Naturally, an unbiased estimate for the treatment effect is

$$\widehat{\Delta} = \sqrt{\frac{2}{N}} Z^W - \sqrt{\frac{2}{N}} \mu_\tau.$$

Using the normal approximation, the corresponding approximated $100(1-\alpha)\%$ confidence interval can be easily constructed as

$$\widehat{\Delta} - \sqrt{\frac{2}{N}} \sigma_\tau z_{\alpha/2} < \Delta < \widehat{\Delta} + \sqrt{\frac{2}{N}} \sigma_\tau z_{\alpha/2}. \tag{3}$$

The treatment effect could also be estimated by a biased estimate

$$\tilde{\Delta} = \sqrt{\frac{2}{N}} Z^W$$

with the same confidence interval as in (3).

3.3. *Almost exact approach*

It can be shown that the difference between $\alpha_E^W(\tau)$ and $\alpha_A^W(\tau)$ is maximized at the critical value of around 2, with the difference less than 0.001 for $0 < \tau \leq 1$. These maximum differences were fit as a function of τ, and a fitted function, called $\varepsilon(\tau)$, is given below.

$$\varepsilon(\tau) = 0.0003130571 \times 4.444461^{\tau} - 0.0003334429. \qquad (4)$$

The $\varepsilon(\tau)$ is used to correct the anti-conservative problem associated with the normal approximation, i.e., the "almost exact" type I error rate is given as

$$\alpha_{AE}^W(\tau) = \alpha_A^W(\tau) + \varepsilon(\tau).$$

For $z_\alpha = 2.326348$ and 1.959964, $\alpha_E^W(\tau)$, $\alpha_A^W(\tau)$, and $\alpha_{AE}^W(\tau)$ are calculated for $\tau = 0, 0.2, \cdots, 1$, and are listed in Table 2. In almost all cases, the anti-conservative issue associated with the normal approximation is corrected by the almost exact approach. There are few cases where the issue remains, however its magnitude is so small and can be neglected.

Table 2. Exact, Approximate, and Almost Exact α Level

τ	α_E^W	α_A^W	α_{AE}^W
z_α=2.326348			
0.0	0.010	0.010	0.010
0.2	0.01458029	0.01451940	0.01460784
0.4	0.01623103	0.01604120	0.01627628
0.6	0.01733160	0.01695474	0.01738745
0.8	0.01812354	0.01750208	0.01820111
1.0	0.01870608	0.01778037	0.01883829
z_α=1.959964			
0.0	0.025	0.025	0.025
0.2	0.03517700	0.03510130	0.03518974
0.4	0.03902900	0.03880306	0.03903814
0.6	0.04172584	0.04129011	0.04172282
0.8	0.04377500	0.04307111	0.04377014
1.0	0.04537772	0.04434412	0.04540205

The exact, approximate, and almost exact p-values can be calculated similarly. To control the "almost exact" type I error at the nominal level, α,

$$\alpha_A^W(\tau) = \alpha - \varepsilon(\tau)$$

This implies the almost exact rejection region of

$$Z^{WS} > z_{\alpha-\varepsilon(\tau)} = -\Phi(\alpha - \varepsilon(\tau))$$

or

$$Z^W > z_{\alpha-\varepsilon(\tau)}\sigma_\tau + \mu_\tau$$

For $\alpha = 0.025$ and $\tau = 0.2, 0.4, \cdots, 1$, the almost exact critical values are given in Table 3.

Table 3. Almost Exact Critical Values Based on Z^{WS} for $\alpha = 0.025$

τ	0.2	0.4	0.6	0.8	1.0
$z_{\alpha-\varepsilon(\tau)}$	1.961479	1.964002	1.967422	1.972067	1.978395

For hypothesis testing, the approximate critical value, z_α^{adj}, and the approximate p-values are sufficient in practice most of the time. When in doubt (borderline significance), use the almost exact method, which provides a slightly more accurate and conservative estimate.

4. Power

First we consider, under the alternative hypothesis, the means of the two test treatment arms are identical with constant variance, i.e., $\Delta_A = \Delta_B$. Let the constant **standardized** effect size be δ. For given planned type I error rate α and type II error rate β, it can be shown that the power of the Winner design, with the adjusted critical value $z_\alpha^{\text{adj}} = z_\alpha\sigma_\tau + \mu_\tau$, can be approximated (see (2)) in terms of $z_\alpha = \Phi^{-1}(1-\alpha)$ and $z_\beta = \Phi^{-1}(1-\beta)$ as follows:

$$\sqrt{N}\delta = \sqrt{2}\sigma_\tau (z_\alpha + z_\beta). \tag{5}$$

That is, the power is approximated with

$$1 - \Phi\left(z_\alpha - \frac{\sqrt{N}\delta}{\sqrt{2}\sigma_\tau}\right).$$

Note that the exact power of the Winner design can be calculated with

$$2\int_l^\infty \Phi\left(\frac{\sqrt{\tau}}{\sqrt{4-\tau}}x\right)\phi(x)\,dx,$$

where

$$l = z_\alpha \sigma_\tau + \mu_\tau - \frac{\sqrt{N}\delta}{\sqrt{2}\sigma_\tau}.$$

If sample sizes during interim are n_A, n_B and n_C, and final sample sizes are N_W and N_C, then

$$\tau \approx \left(\frac{1}{n_W} + \frac{1}{n_C}\right)^{-1} / \left(\frac{1}{N_W} + \frac{1}{N_C}\right)^{-1}$$

(Lan and Zucker, 1993).

When design a study using the Winner design, the sample size can be easily calculated using (5) for any giving α and β. Table 4 shows the sample size per group, calculated using normal approximation and expressed as $\sqrt{N}\delta$, and the corresponding exact power for various τ.

Table 4. Sample Size and Exact Power of the Winner Design, $\alpha = .025$

τ	$\sqrt{N}\delta$	Approx. Power	Exact Power
0	4.584195	0.9	0.9
0.25	4.492070	0.9	0.9000742
0.50	4.398015	0.9	0.9002322
0.75	4.301905	0.9	0.9004700
1	4.203597	0.9	0.9007977
0	4.237546	0.85	0.85
0.25	4.152387	0.85	0.8500198
0.50	4.065444	0.85	0.8500716
0.75	3.976602	0.85	0.8501596
1	3.885728	0.85	0.8502911

The normal approximate not only provides an extremely convenient way for calculating the sample size for the Winner design but also the corresponding exact power is maintained at its nominal level. It is worthy to note that when the interim winner selection is done at $\tau = 0.5$, the sample size required to have 90% power to detect a treatment effect $\delta = 0.5$ at the final test is $(4.398015/0.5)^2 \approx 78$, and for 85% is $(4.065444/0.5)^2 \approx 67$. Note that the total sample size for the Winner design is $2N + n = (2 + \tau)N$.

Secondly, we consider the case where $\Delta_A \neq \Delta_B$ under the alternative hypothesis. No simple mathematical way is derived yet to deal with unequal means. When **standardized** effect size, under the alternative hypothesis, $\delta_A = \delta \neq \delta_B = \lambda\delta$, $0 \leq \lambda < 1$, for the planned type I error rate α and sample size per arm, N, it can be shown that the exact power of the Winner design is calculated with

$$\Pr\left(Z_1 > l,\ Z_2 > -\sqrt{\tau}(1-\lambda)\frac{\sqrt{N}\delta}{\sqrt{2}}\right)$$
$$+\Pr\left(Z_1 > l + (1-\lambda)\frac{\sqrt{N}\delta}{\sqrt{2}},\ Z_2 > \sqrt{\tau}(1-\lambda)\frac{\sqrt{N}\delta}{\sqrt{2}}\right)$$

where Z_1 and Z_2 are standard normal random variables with correlation coefficient $\sqrt{\tau}/2$, and

$$l = z_\alpha \sigma_\tau + \mu_\tau - \frac{\sqrt{N}\delta}{\sqrt{2}}.$$

The table below shows the sample size per arm needed, expressed in term of $N/N_{\lambda=1}$, for the Winner design at the exact power of 85% or 90%, and $\lambda = 0.1, 0.2, \cdots, 0.9$ and $\tau = 0.25, 0.75$, and 1, where $N_{\lambda=1}$ is the number of sample size per arm in the case of equal treatment effect.

Table 5. Sample Size as $N/N_{\lambda=1}$ of the Winner Design for $\alpha = 0.025$

Power = 0.90									
	λ								
τ	0.1	0.2	0.3	0.4	0.5	0.6	0.7	0.8	0.9
0.25	1.325	1.385	1.444	1.479	1.468	1.410	1.319	1.212	1.104
0.50	1.256	1.273	1.295	1.313	1.321	1.305	1.260	1.188	1.098
0.75	1.304	1.307	1.310	1.313	1.307	1.288	1.247	1.182	1.097
1	1.372	1.371	1.367	1.360	1.342	1.310	1.258	1.186	1.098
Power = 0.85									
	λ								
τ	0.1	0.2	0.3	0.4	0.5	0.6	0.7	0.8	0.9
0.25	1.319	1.367	1.412	1.439	1.432	1.384	1.304	1.206	1.103
0.50	1.268	1.284	1.302	1.317	1.320	1.302	1.257	1.187	1.098
0.75	1.317	1.319	1.322	1.322	1.314	1.292	1.248	1.182	1.097
1	1.384	1.383	1.378	1.369	1.349	1.315	1.261	1.187	1.098

Assuming a Winner design is planned with interim look to be performed at 50% information time. i.e., $\tau = 0.5$, if the standardized treatment effect for one treatment is assumed to be $\delta = 0.5$ and the standardized effect for another treatment is 0.4, that is, $\lambda = 0.8$, then to detect the treatment effect at 90% power it would require $1.1883 \times 78 \approx 93$ subjects per arm. For 85% power, it would require $1.1871 \times 67 \approx 80$ subjects per arm.

5. The Small Sample Case

In the small sample case where the variance σ^2 at the final analysis is estimated by the sample variance s^2, it is natural to consider the random variable

$$T^W = \frac{Z^W}{s/\sigma},$$

for hypothesis testing. The density function of T^W, $f_{T^W}(t \mid \nu, \eta)$ with ν being the degrees of freedom, can be easily derived (Shun, Soo, and Lan, 2003) and shown to be

$$f_{T^W}(t \mid \tau) = 2\int_0^\infty \nu\sqrt{x}\Phi\left(\frac{\sqrt{\tau}}{\sqrt{4-\tau}}t\sqrt{x}\right)\phi\left(t\sqrt{x}\right)f_{\chi^2_\lambda}(\nu x)\,dx, \ -\infty < t < \infty,$$

where $f_{\chi^2_\lambda}$ is the density function of a χ^2 random variable with ν degrees of freedom. The exact tail probability under null hypothesis, $p_E^W(t;\tau)$, can be calculated upon numerical integration. Since Z^W/σ_τ is approximately normally distributed with mean μ_τ/σ_τ and variance 1, the following transformed random variable

$$T^{WS} = \frac{Z^W/\sigma_\tau}{s/\sigma}$$

could be approximated by a non-central t distribution with degrees of freedom ν and non-centrality parameter

$$\varsigma = \frac{\mu_\tau}{\sigma_\tau} = \sqrt{\frac{\tau}{2\pi - \tau}}$$

The non-central t approximated p-value, $p_A^W(t,\tau)$, at the final analysis in the Winner design can be calculated using the similar approaches as in the large sample case as follows:

$$p_A^W(t;\tau) = \Pr\left(T^W > t|\tau\right) = \Pr\left(T^{WS} > \frac{t}{\sigma_\tau} \mid \tau\right) \approx \Pr\left(t(\nu,\varsigma) > \frac{t}{\sigma_\tau} \mid \tau\right).$$

The degrees of freedom ν depends on how s is calculated. Two approaches are proposed as follow:

[1] Use only the data from W and C, i.e., $\nu = 2N - 2$;

[2] Use all available data from A, B, and C, i.e., $\nu = 2N + n - 3$.

The difference between the approximate and exact p-values is given in Figure 2 for $N = 10$ (with $\nu = 18$) and $\tau = 0, 0.2, 0.4, 0.6, 0.8, 1$, which again shows that at the tail the non-central t approximation is slightly anti-conservative.

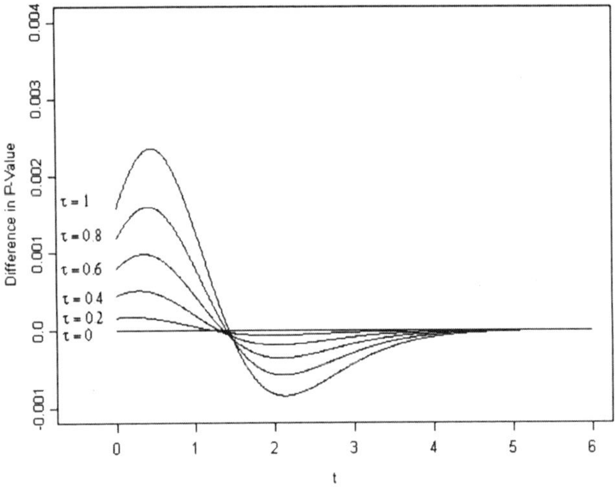

Fig. 2. Difference Between Approximate and Exact P-Values for $\nu = 18$

Table 6. Exact, Approximate, and Almost Exact p-Values for Small Sample ($\tau = 0.5, N = 10, \lambda = 18$)

t^W	p^W_{E}	p^W_{A}	p^W_{AE}
1.97	0.05070	0.05044	0.05076
2.33	0.02561	0.02536	0.02569
2.40	0.02231	0.02207	0.02240
2.60	0.01491	0.01471	0.01504
2.80	0.00987	0.00970	0.01003
3.00	0.00647	0.00634	0.00667

To correct the anti-conservative problem, the same adjustment, $\varepsilon(\tau)$ (see (4)), as used in the large sample case is used, and the "almost exact" p-value is given as $p^W_{AE}(\tau) = p^W_A(\tau) + \varepsilon(\tau)$. The differences between the almost exact and exact p-values are given in Figure 3 for $N = 10$ (with $\nu = 18$) and $\tau = 0, 0.2, 0.4, 0.6, 0.8, 1$, which shown that the adjustment works well for small sample case as well. For $\tau = 0.5$ and $N = 10$ (with $\lambda = 18$), p^W_E, p^W_A, and p^W_{AE} are listed in Table 6.

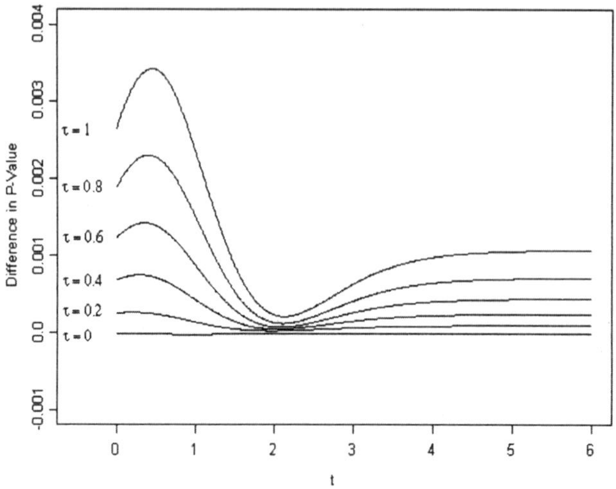

Fig. 3. Difference Between Almost Exact and Exact P-Values for $\nu = 18$

6. Discussions

Some fundamental ideas behind the Winner design can easily be extended to the case where a winner is to be selected from K ($K > 2$) new treatment arms.

Lemma 6.1. *Let $W_1, W_2, \cdots, W_K$ be i.i.d. $N(0,1)$ and $W_M = \max(W_1, W_2, \cdots, W_K)$. Denote $\mu_M = E(W_M)$ and $\sigma_M^2 = Var(W_M)$. When $K = 2$, $\mu_M = \sqrt{1/\pi}$ and $\sigma_M^2 = 1 - 1/\pi$. The exact mean and variance of W_M, calculated by numerical integration, for various K are given in the following table.*

K	μ_M	σ_M^2
2	0.5641896	0.6816901
3	0.8462844	0.5594672
4	1.0293754	0.4917152
5	1.1629645	0.4475341

(6)

Theorem 1 *Let* $Z_1, Z_2, \cdots, Z_K$ *be* $N(0,1)$, $Cov(Z_i, Z_j) = \eta$, *for all* $i < j$. *Let also* $Z_{M\eta} = \max(Z_1, Z_2, \cdots, Z_K)$. *Then*

$$E(Z_{M\eta}) = \mu_M \sqrt{1-\eta}, \tag{7}$$

$$Var(Z_{M\eta}) = (1-\eta)\sigma_M^2 + \eta. \tag{8}$$

Proof. Let $W, W_1, W_2, \cdots, W_K$ be i.i.d. $N(0,1)$. Z_i can be rewritten as $Z_i = \sqrt{1-\eta}W_i + \sqrt{\eta}W, i = 1, 2, \cdots, K$, such that $Z_1, Z_2, \cdots, Z_K$ are i.i.d. $N(0,1)$ and $Cov(Z_i, Z_j) = \eta$. Then

$$Z_{M\eta} = \sqrt{1-\eta}\max(W_1, W_2, \cdots, W_K) + \sqrt{\eta}W$$

The mean and the variance of $Z_{M\eta}$ as shown in (7) and (8) are then obtained. Note that when $K = 2$

$$\begin{aligned} E(Z_{M\eta}) &= \sqrt{(1-\eta)/\pi}, \\ Var(Z_{M\eta}) &= (1-\eta)(1-1/\pi) + \eta \\ &= 1 - (1-\eta)/\pi. \end{aligned}$$

□

The distribution of the test statistics Z^W, under the null hypothesis, is given below.

$$F_{Z^W}(y \mid \tau) = \int_{-\infty}^{\infty} \Phi^K\left(\frac{\sqrt{2}y - \sqrt{2-\tau}w}{\sqrt{\tau}}\right)\phi(w)\,dw$$

The mean and the variance of Z^W can be calculated using (6), (7), and (8), with $\eta = 1 - \tau/2$. For $z_\alpha = 2.326348$ and 1.959964, $\alpha_{\mathrm{E}}^W(\tau)$, $\alpha_{\mathrm{A}}^W(\tau)$, and $\alpha_{\mathrm{AE}}^W(\tau)$ are calculated for $\tau = 0, 0.2, 0.4, 0.6, 0.8, 1$, and $K = 5$, and listed in Table 7. The results indicate that the proposed normal approximation and almost exact adjustment approaches work well even when $K > 2$. It should note that since the adjustment $\varepsilon(\tau)$ is determined based on two new treatment case, the anti-conservation is not completely corrected (but only within a very small magnitude) when applied to more than 2 treatment cases.

Table 7. Exact, Approximate, and Almost Exact α Level for $K = 5$

τ	α_E^W	α_A^W	α_{AE}^W
z_α= 2.326348			
0.0	0.010	0.010	0.010
0.2	0.02203550	0.02195028	0.02203872
0.4	0.02801475	0.02773627	0.02797136
0.6	0.03275939	0.03218710	0.03261981
0.8	0.03671706	0.03574171	0.03644074
1.0	0.04004779	0.03854120	0.03959912
z_α= 1.959964			
0.0	0.025	0.025	0.025
0.2	0.05079088	0.05070096	0.05078940
0.4	0.06367338	0.06341997	0.06365505
0.6	0.07419953	0.07374576	0.07417847
0.8	0.08334917	0.08267767	0.08337670
1.0	0.09146926	0.09057720	0.09163513

It is not unusual that the data used at interim for treatment selection are from a surrogate endpoint, $X_i^j, i = 1, 2, ..., N; j = A, B, C$, which are independently normally distributed with mean ν_j and common variance σ_X^2. The correlation between X_i^j and Y_i^j is expressed in terms of correlation coefficient ρ. The statistical inference discussed in this paper is primarily about $\Delta_j = \mu_j - \mu_C$, $j = A$ or B, although there is another set of parameters ν_j's involved. It is understood that under the null hypothesis that $\Delta_A = \Delta_B = 0$, it is not necessary $\nu_A = \nu_B = \nu_C$ when $\rho \neq 1$. However, it is reasonable to assume under the null hypothesis that there is no treatment effect on the surrogate endpoint either. The distribution function of Z^W given in (1) can also be expressed as

$$\Pr(Z^W < y) = 2\Pr\left(\sqrt{\tau}Z_n^A + \sqrt{1-\tau}Z_{(k)} < y, Z_n^A - Z_n^B > 0\right)$$

with $Corr\left(\sqrt{\tau}Z_n^A + \sqrt{1-\tau}Z_{(k)}, Z_n^A - Z_n^B\right) = \sqrt{\tau}/2$. When a surrogate endpoint is used for the interim treatment selection, the distribution of Z^W becomes

$$\Pr(Z^W < y) = 2\Pr\left(\sqrt{\tau}Z_n^A + \sqrt{1-\tau}Z_{(k)} < y, \frac{\overline{X}_n^A - \overline{X}_n^B}{\sigma_X\sqrt{\frac{2}{n}}} > 0\right)$$

with $Corr\left(\sqrt{\tau}Z_n^A + \sqrt{1-\tau}Z_{(k)}, \frac{\overline{X}_n^A - \overline{X}_n^B}{\sigma_X\sqrt{\frac{2}{n}}}\right) = \sqrt{\tau}\rho/2$. That is, the proposed procedure still holds by replacing τ with $\tau\rho^2$.

In general, the correlation coefficient ρ depends on the nature of the data, and is difficult to evaluate in cases other than normality. One approach that can be adopted, at least for the purpose of determining the overall type I error rate, is to estimate ρ from the observed data at the study completion. In real life situations, however, it is often that the correlation coefficient estimation is a challenge. We propose the following approach for practical purpose:

(1) A re-sampling method can be used for the estimation;
(2) When the estimation is not feasible, one can always choose $\rho = 1$, although very conservative.

From the mathematical derivation one can see that the type I error rates actually only depend on the interim and final statistics, not on the individual observations. Therefore, all large sample results provided in this paper are not limited to normal distribution as long as the statistics used for the interim and the final analyses are asymptotically normal.

In case of early termination for futility is considered when the control treatment C is better than A and B at the end of Stage I, the procedures presented in this paper would be pretty conservative. That is, the type I error rate becomes much smaller than the nominal α. However, numerical integration can be used to evaluate the exact p-values.

References

1. Azzalini, A. (1985). A Class of Distributions Which Includes the Normal Ones, *Scand. J. Statistics*, 12, 171-178.
2. Dunnett, C. (1955). A multiple comparison procedure for comparing several treatments with a control. *Journal of American Statistical Association,* 50 (272), 1096-1121.
3. Lan, K.K.G. and Zucker, D.M. (1993). Sequential Monitoring of Clinical Trials: the Role of Information and Brownian Motion, *Statistics in Medicine*, 12, 753-765.
4. Sampson, A.R. and Sill, M.W. (2005). Drop-the-Losers Design: Normal Case. *Biometrical Journal*, 47, 257-268.
5. Shun, Z., Soo, Y., and Lan, K.K.G. (2003). Maximum of Two Normal Variables and Interim Dose Selection in Clinical Trials, *Technical Report,* Aventis Pharmaceuticals, Inc.
6. Simon, R. (1989). Optimal Two-Stage Designs for Phase II Clinical Trials. *Controlled Clinical Trials*, 10, 1-10.
7. Thall, P.F., Simon, R., and Ellenberg, S.S. (1988). Two-Stage Selection and Testing Designs for Comparative Clinical Trials. *Biometrika*, 75, 303-310.
8. Whitehead, J. (1986). Sample Sizes for Phase II and Phase III Clinical Trials: An Integrated Approach. *Statistics in Medicine*, 5, 459-464.

A RESTRICTED MINIMAX DETERMINATION OF THE INITIAL SAMPLE SIZE IN STEIN'S AND RELATED TWO-STAGE PROCEDURES

JOON SANG LEE and MICHAEL WOODROOFE

Statistics Department, University of Michigan, Ann Arbor, MI 48109, U.S.A.

When the variance is known, a level $1 - \alpha$ confidence interval of specified width $2h > 0$ for the mean of a normal distribution requires a sample of size at least $\eta = c^2\sigma^2/h^2$, where c is the upper $(1 - \frac{1}{2}\alpha)^{th}$ quantile of the standard normal distribution. If the variance is unknown, then such an interval may be constructed using Stein's double sampling procedure in which an initial sample of size $m \geq 2$ is drawn and used to estimate η. Here is it shown that if the experimenter can specify a prior guess, η_0 say, for η, then $\sqrt{\frac{1}{2}(1 + c^2)\eta_0}$ is an approximately minimax choice for the initial sample size. The formulation is, in fact, more general and includes point estimation with equivariant loss as well as interval estimation.

Some key words: Decision problem; Minimax solution; Sequential point and interval estimation.

1. Introduction

Stein's (1945) two-stage procedure for setting a confidence interval of prescribed width $2h > 0$ and confidence $1 - \alpha$ for the mean of a normal distribution may be described as follows: Let $X_1, X_2, \cdots$ be independent normal variables with unknown mean μ and variance σ^2; and let $\bar{X}_n$ and S_n^2 denote the sample mean and variance of $X_1, \cdots, X_n$, so that $n\bar{X}_n = X_1 + \cdots + X_n$ and $(n-1)S_n^2 = (X_1 - \bar{X}_n)^2 + \cdots + (X_n - \bar{X}_n)^2$. First take an initial sample of size $m \geq 2$ and compute S_m^2; next let

$$T_m = \max\{m, \lceil \frac{c_{m-1}^2 S_m^2}{h^2} \rceil\},$$

where c_m is the upper $1-\alpha/2$ quantile of the t-distribution with m degrees of freedom and $\lceil x \rceil$ denotes the least integer that is greater than or equal to x; then take an additional $T_m - m$ observations and declare $[\bar{X}_{T_m} - h, \bar{X}_{T_m} + h]$

to be the interval. Stein showed that $P_{\mu,\sigma^2}[|\bar{X}_{T_m} - \mu| \leq h] \geq 1 - \alpha$ for all μ and σ^2 for any $m \geq 2$. Of course, if σ^2 were known, then a confidence interval of width $2h$ and confidence $1 - \alpha$ could be constructed by taking a sample of size at least

$$\eta = \frac{c^2\sigma^2}{h^2}, \tag{1}$$

where c is the upper $1 - \alpha/2$ quantile of the standard normal distribution. Stein's procedure effectively estimates η by T_m.

The present paper centers on the choice of m in Stein's procedure and related ones. The choice of m has to be subjective at some level, because there is no data when it is chosen. It is required here that the experimenter specify a prior expectation, σ_0^2 say, for σ^2. That is, it is assumed that the experimenter has a prior distribution ξ for which

$$\int_0^\infty \sigma^2 \xi\{d\sigma\} = \sigma_0^2. \tag{2}$$

Equivalently, the experimenter must specify the prior expectation $\eta_0 = c^2\sigma_0^2/h^2$ for η. No other details of the prior are required.

Let Ξ_0 be the class of prior distributions for which (2) holds. Then a corollary to the main result asserts that

$$m_h = \lceil \sqrt{(\frac{1+c^2}{2})\eta_0}\, \rceil \tag{3}$$

is an asymptotically minimax choice for m within the class Ξ_0 in the following sense:

$$\inf_{m \geq 2} \sup_{\xi \in \Xi_0} \int_0^\infty E_{\sigma^2}(T_m)\xi\{d\sigma\} = \eta_0 + 2\sqrt{(\frac{1+c^2}{2})\eta_0} + o(\sqrt{\eta_0}) \tag{4}$$

$$= \sup_{\xi \in \Xi_0} \int_0^\infty E_{\sigma^2}(T_{m_h})\xi\{d\sigma\}$$

as $h \downarrow 0$. For example, if $\alpha = .05$, $h = .1$, and $\sigma_0 = 1$, then $\eta_0 = 384.16$ and the approximately minimax choice of m is $m_h = 31$. While only a prior guess for σ^2 is required, the minimax choice (3) is sensitive to that choice, since m_h depends linearly on σ_0. This is unfortunate, but in some sense unavoidable, since only prior knowledge/opinion is available when m is chosen.

There has been continuing interest in two-stage procedures for estimating means since [7], but few explicit recommendations for the initial sample size. On one hand there has been interest in asymptotic properties of two-stage procedures. Much of this work is described in [6] and does include some order of magnitude restrictions on the initial sample size. If a lower bound for σ is known, say $\sigma \geq \sigma_* > 0$, then m must be at least $m_* = c^2\sigma_*^2/h^2$ and $\lceil m_* \rceil$ is suggested as an initial sample size. This remark can be useful when $\sigma_* \gg h$. See [5] for an example. There has also been some recent interest in Bayesian solutions, [3], [4] and [8], when the prior distribution can be specified completely. A special case of our main result appears in [12], the case of point estimation with squared error loss, as described in the next section. For that case, it is shown that a statistician who can specify the prior distribution completely may be led to take a larger initial sample size than the minimax choice.

The proof of (4) requires finding an appropriate decision problem, one for which Stein's procedure is the (asymptotic minimax) solution. The decision problem is identified in Section 2. The main result is stated there, and (4) proved. Section 3 contains some preliminary lemmas. The first two of these, at least, are intuitive. The main result is then established in Section 4. The proof is substantially more involved than the special case in [12], where simple explicit expressions were used extensively.

2. The Main Result

The proof of Stein's result uses the observation: If $t = t(S_m^2) \geq m$ is an (measurable) integer valued function of S_m^2, then the conditional distribution of $\bar{X}_t$ given S_m^2 is normal with mean μ and variance σ^2/t. It then follows that

$$P_{\mu,\sigma^2}\big[|\bar{X}_t - \mu| \leq h\big] = E_{\sigma^2}\big[2\Phi\big(\frac{h\sqrt{t}}{\sigma}\big) - 1\big], \tag{5}$$

where Φ denotes the standard normal distribution function. Using extensions of (5) to fully sequential procedures, Woodroofe [11] showed a close connection between the problem of setting a fixed width confidence interval and an optimal stopping problem in which the statistician must find a stopping time t with respect to $S_2^2, S_3^2, \cdots$ to minimize $\eta E_{\sigma^2}[K(t/\eta)]$, where

$$K(x) = \frac{2[1 - \Phi(c\sqrt{x})]}{c\varphi(c)} + x,$$

η is as in (1) with $\Phi(c) = 1 - \alpha/2$, and φ the standard normal density. This connection is elaborated in the proof of Corollary 1 at the end of this section.

A similar form arises in point estimation. To see how, suppose that it is required to estimate μ with a loss of the form $A|\bar{X}_n - \mu|^{2p}$ for estimation error, where $p > 0$, and unit cost for each observation. Here A determines the importance of estimation error relative to the cost of sampling and is assumed to be large. If $m \geq 2$, $t = t(S_m^2) \geq m$, and a sample of size t is taken then, as in (5), the expected loss plus sampling cost is

$$E\big[A|\bar{X}_t - \mu|^{2p} + t\big] = E_{\sigma^2}\big[\frac{A\gamma_p\sigma^{2p}}{t^p} + t\big],$$

where $\gamma_p = 2^p\Gamma(\frac{1}{2} + p)/\sqrt{\pi}$, and this is of the form $\eta E_{\sigma^2}[K(t/\eta)]$ with $\eta = (Ap\gamma_p)^{\frac{1}{p+1}}\sigma^{\frac{2p}{p+1}}$ and $K(x) = 1/(px^p) + x$.

Observe that for both the interval and point estimation problems, K is of the form

$$K(x) = K_0(x) + x, \tag{6}$$

where K_0 is a non-negative decreasing strictly convex function for which $K_0'(1) = -1$ (so that $K(x)$ is minimized when $x = 1$), and

$$\eta = a\sigma^{2q}, \tag{7}$$

where $0 < a, q < \infty$. In the point estimation problem, $K_0(x) = 1/(px^p)$, $q = p/(p+1)$, and $a = (Ap\gamma_p)^{1/(p+1)}$. For interval estimation, $K_0(x) = 2[1 - \Phi(c\sqrt{x})]/[c\varphi(c)]$, $q = 1$, and $a = c^2/h^2$.

In the statement of the main result, the statistician must specify a pair $\delta = (m, t)$, where $m \geq 2$ is an integer, the initial sample size, and $t = t(S_m^2)$ is a measurable integer valued function for which $t \geq m$. The loss incurred when δ is used is taken to be $\eta K(t/\eta)$, where K and η are as in (6) and (7), and the risk is then

$$R_a(\delta; \sigma^2) = \eta E_{\sigma^2}\big[K(\frac{t}{\eta})\big]. \tag{8}$$

The function K_0 is required to be a twice continuously differentiable convex function for which

$$K_0(x) = O(x^{-\ell})$$

as $x \downarrow 0$ for some $\ell > 0$.

For a given σ_0, let Ξ_0 be the class of prior distributions ξ for which

$$\int_0^\infty \sigma^{2q}\xi\{d\sigma\} = \sigma_0^{2q}.$$

Further, let

$$r_a(\delta; \sigma^2) = R_a(\delta; \sigma^2) - K(1)\eta,$$

$$\bar{R}_a(\delta; \xi) = \int_0^\infty R_a(\delta; \sigma^2)\xi\{d\sigma\},$$

and

$$\bar{r}_a(\delta;\xi) = \int_0^\infty r_a(\delta;\sigma^2)\xi\{d\sigma\}$$

for $\xi \in \Xi_0$. Finally, let $\eta_0 = a\sigma_0^{2q}$, the prior expectation of η, and let δ_a be the procedure defined by

$$m_a = \lceil q\sqrt{K''(1)\eta_0}\rceil \tag{9}$$

and

$$t_a = \max\{m_a, \lceil aS_m^{2q}\rceil\}. \tag{10}$$

Theorem 2.1. *With the notation and assumptions of the previous two paragraphs,*

$$\inf_\delta \sup_{\xi\in\Xi_0} \bar{r}_a(\delta;\xi) = 2q\sqrt{K''(1)\eta_0} + o(\sqrt{\eta_0}) = \sup_{\xi\in\Xi_0} \bar{r}_a(\delta_a;\xi) \tag{11}$$

as $a \to \infty$.

The proof of Theorem 2.1) is presented in Section 4. The following corollary contains (4) as a special case. In its statement $0 < \alpha < 1$ and $0 < \sigma_0^2 < \infty$ are fixed, $a = c^2/h^2$, $\eta_0 = c^2\sigma_0^2/h^2$, and $h \downarrow 0$.

Corollary 2.1. *If* $\delta_h = (m_h, t_h)$ *are any procedures for which*

$$P_{\mu,\sigma^2}\left[|\bar{X}_{t_h} - \mu| \le h\right] \ge 1 - \alpha \tag{12}$$

for all μ *and* σ^2, *then*

$$\sup_{\xi\in\Xi_0} \int_0^\infty E_{\sigma^2}(t_h)\xi\{d\sigma\} \ge \eta_0 + 2\sqrt{(\frac{1+c^2}{2})\eta_0} + o(\sqrt{\eta_0}) \tag{13}$$

as $h \downarrow 0$. *Moreover, there is equality in (13) if* m_h *is as in (3) and* $t_h = T_{m_h}$.

Proof. If δ_h satisfies (12), then $E_{\sigma^2}[K_0(t/\eta)] \le \alpha/[c\varphi(c)]$,

$$\bar{R}_a(\delta_h;\xi) \le \frac{\alpha\eta_0}{c\varphi(c)} + E_\xi(t_h) = K(1)\eta_0 + E_\xi(t_h - \eta_0),$$

and, therefore,

$$\int_0^\infty E_{\sigma^2}(t_h)\xi\{d\sigma\} - \eta_0 \ge \bar{r}_a(\delta_h;\xi)$$

for all $\xi \in \Xi_0$. The inequality asserted in (13) now follows directly from the theorem, since $q = 1$ and $K''(1) = (1 + c^2)/2$ for interval estimation.

The equality asserted in the corollary follows from a direct calculation. Clearly, $T_m \le c_{m-1}^2 S_m^2/h^2 + m + 1$ for any m and, therefore,

$$E_\xi(T_m - \eta_0) \le (\frac{c_{m-1}^2 - c^2}{h^2})\sigma_0^2 + m + 1$$

for any $\xi \in \Xi_0$. From [1], p. 949, or first principles, $c_m = c + (c + c^3)/4m + O(1/m^2)$ as $m \to \infty$ and, therefore,

$$\left(\frac{c_{m-1}^2 - c^2}{h^2}\right)\sigma_0^2 = \left(\frac{c^2 + c^4}{2mh^2}\right)\sigma_0^2 + O(\frac{1}{m^2h^2}) = (\frac{1 + c^2}{2m})\eta_0 + O(\frac{1}{m^2h^2}).$$

Letting $m = m_h$ and combining the last two expressions now shows that there is equality in (13) when m_h is as in (3). ◊

3. Preliminary Lemmas

In the statements of lemmas below q is fixed,

$$g_{\alpha,\beta}(y) = \frac{\beta^{\frac{1}{2}\alpha} y^{\frac{1}{2}\alpha - 1} e^{-\frac{1}{2}\beta y}}{2^{\frac{1}{2}\alpha}\Gamma(\frac{1}{2}\alpha)}$$

$$\nu_\alpha = \int_0^\infty y^q g_{\alpha,1}(y)dy = \frac{2^q\Gamma(\frac{1}{2}\alpha + q)}{\Gamma(\frac{1}{2}\alpha)}.$$

and

$$\bar{K}_\alpha(x) = \int_0^\infty K(\frac{xy^q}{\nu_\alpha})g_{\alpha,1}(y)dy, \tag{14}$$

where Γ denotes the gamma function. Here $\bar{K}_\alpha(x)$ is defined for all $0 < \alpha, x < \infty$, though possibly infinite. Properties of $\bar{K}_\alpha$ are central to the proof.

In the proofs of the following Lemmas, use is made of Stirling's Formula,

$$\log\Gamma(z) = (z - \frac{1}{2})\log(z) - z + \frac{1}{2}\log(2\pi) + \frac{1}{12z} + O(\frac{1}{z^3}) \tag{15}$$

as $z \to \infty$. See, for example, [1], p. 267. In particular, it follows from Stirling's Formula that

$$\nu_\alpha = \alpha^q + O(1)$$

as $\alpha \to \infty$.

Lemma 3.1. *i) The integral in (14) converges for all* $\alpha > 2\ell q$.

ii) If $\bar{K}_\alpha(x) < \infty$ *for some* $0 < \alpha, x < \infty$*, then* $\bar{K}_{\alpha'}(x') < \infty$ *for all* $\alpha' \ge \alpha$ *all* x'.

iii) In this case, $\bar{K}_\alpha$ is a twice continuously differentiable, strictly convex function, and $\bar{K}_\alpha(x) > K(x)$ for all x.

iv) Finally,

$$\bar{K}_\alpha(x) = K(x) + \frac{q^2x^2K''(x)}{\alpha} + o(\frac{1}{\alpha}) \tag{16}$$

uniformly in x on compact subintervals of $(0,\infty)$ as $\alpha \to \infty$.

Proof. Clearly, $\bar{K}_\alpha(x) < \infty$ iff $\int_0^1 K_0(xy^q/\nu_\alpha)g_{\alpha,1}(y)dy < \infty$. The first assertion follows directly. For ii) and iii), write

$$\bar{K}_\alpha = \int_0^\infty K(\frac{y^q}{\nu_\alpha})g_{\alpha,x^{-\frac{1}{q}}}(y)dy. \tag{17}$$

The second assertion follows since $g_{\alpha',x'^{-\frac{1}{q}}}(y)/g_{\alpha,x^{-\frac{1}{q}}}(y)$ remains bounded as $y \to 0$ when $\alpha' \geq \alpha$ and $0 < x, x' < \infty$. The continuous differentiability asserted in iii) also follows from (17), since $g_{\alpha,\beta}$ is an exponential family in β; $\bar{K}_\alpha$ is, in fact, analytic in $x > 0$. See Brown [2], pp. 34-36. That $\bar{K}_\alpha$ is strictly convex and $\bar{K}_\alpha(x) > K(x)$, then follow from the convexity of K, the positivity of $g_{\alpha,1}$, and Jensen's Inequality. Relation (16) may be formally obtained by applying the delta method and rigorously established along the lines of Chapter 3 in [9]. ◇

Lemma 3.2. *For all $\alpha > 0$,*

$$\inf_{0<x<\infty} \bar{K}_\alpha(x) > K(1). \tag{18}$$

If $\bar{K}_\alpha$ is finite, then $\bar{K}_\alpha(x)$ attains its minimum at a unique value x_α. As $\alpha \to \infty$, $x_\alpha \to 1$ and

$$\inf_{x>0} \bar{K}_\alpha(x) = \bar{K}_\alpha(1) + o(\frac{1}{\alpha}). \tag{19}$$

Proof. Since (18) is clear when $\bar{K}_\alpha \equiv \infty$, it suffices to consider α for which $\bar{K}_\alpha$ is finite. For such α, $\lim_{x\to\infty} \bar{K}_\alpha(x) = \infty$ and

$$\lim_{x\downarrow 0} \bar{K}'_\alpha(x) = \lim_{x\downarrow 0} K'(x) < 0,$$

so that the infimum cannot be attained as $x \downarrow 0$ or $x \to \infty$. That the infimum is attained at a unique point then follows from the strict convexity, and $\bar{K}_\alpha(x_\alpha) > K(x_\alpha) \geq K(1)$, establishing (18). Next $\inf_{|x-1|\geq\epsilon} \bar{K}_\alpha(x) \geq \inf_{|x-1|\geq\epsilon} K(x) > \bar{K}_\alpha(1)$ for all sufficiently large α for any $\epsilon > 0$ and, therefore, that $x_\alpha \to 1$ as $\alpha \to \infty$. Relation (19) then follows from (16), since

$$\begin{aligned}\bar{K}_\alpha(x_\alpha) &= K(x_\alpha) + \frac{q^2 x_\alpha^2 K''(x_\alpha)}{\alpha} + o(\frac{1}{\alpha}) \\ &\geq K(1) + \frac{q^2 K''(1)}{\alpha} + o(\frac{1}{\alpha}) \\ &= \bar{K}_\alpha(1) + o(\frac{1}{\alpha}),\end{aligned}$$

where the inequality uses $K(x) \geq K(1)$ and $x_\alpha \to 1$. ◇

Let $\xi_{\alpha,\beta}$ be the prior distribution for σ under which $\theta = 1/\sigma^2$ has density $g_{\alpha,\beta}$. Also, write $E_{\alpha,\beta}$ for unconditional expectation in the Bayesian model and $E^m_{\alpha,\beta}$ for conditional expectation given S^2_m, when σ has prior $\xi_{\alpha,\beta}$. Then

$$U_0 := \int_0^\infty \sigma^{2q} \xi_{\alpha,\beta}\{d\sigma\} = \int_0^\infty \theta^{-q} g_{\alpha,\beta}(\theta) d\theta = \frac{\beta^q}{\nu_{\alpha-2q}}$$

for $\alpha > 2q$, and $\xi \in \Xi_0$ iff $\beta = \nu_{\alpha-2q}^{1/q}\sigma_0^2$, in which case $\eta_0 = aU_0$. Similarly,

$$\int_0^\infty \eta K(\frac{x}{\eta})\xi_{\alpha,\beta}\{d\sigma\} = a\int_0^\infty \theta^{-q} K(\frac{x\theta^q}{a}) g_{\alpha,\beta}(\theta) d\theta = \eta_0 \bar{K}_{\alpha-2q}(\frac{x}{\eta_0}).$$

Inverted gamma priors $\xi_{\alpha,\beta}$ are conjugate to scaled chi-squared distributions, and

$$U_m := E^m_{\alpha,\beta}(\sigma^{2q}) = \frac{\beta_m^q}{\nu_{\alpha_m - 2q}},$$

where

$$\begin{aligned}\alpha_m &= \alpha + m - 1, \\ \beta_m &= \beta + (m-1)S^2_m.\end{aligned}$$

Lemma 3.3. *If* $\alpha > 2(\ell+1)q$ *and* $x > 0$, *then*

$$E_{\alpha,\beta}\big[\eta K(\frac{x}{\eta})\big] = \bar{\eta}\bar{K}_{\alpha-2q}(\frac{x}{\bar{\eta}}),$$

where $\bar{\eta} = aE_{\alpha,\beta}(\sigma^{2q})$.

Proof. The proof depends on the simple relation

$$\theta^{-q} g_{\alpha,\beta}(\theta) = \frac{\beta^q}{\nu_{\alpha-2q}} g_{\alpha-2q,\beta}(\theta).$$

It follows that $\bar{\eta} = a\beta^q/\nu_{\alpha-2q}$ and

$$\begin{aligned} E_{\alpha,\beta}[\eta K(\frac{x}{\eta})] &= \int_0^\infty a\theta^{-q}K(\frac{x\theta^q}{a})g_{\alpha,\beta}(\theta)d\theta \\ &= \frac{a\beta^q}{\nu_{\alpha-2q}}\int_0^\infty K(\frac{x\theta^q}{a})g_{\alpha-2q,\beta}(\theta)d\theta \\ &= \bar{\eta}\int_0^\infty K(\frac{x\theta^q}{\bar{\eta}\nu_{\alpha-2q}})g_{\alpha-2q,1}(\theta)d\theta \\ &= \bar{\eta}\bar{K}_{\alpha-2q}(\frac{x}{\bar{\eta}}), \end{aligned}$$

as asserted. ◇

Lemma 3.4. *There is a constant C for which*

$$P_{\alpha,\beta}[U_m > u] \le C(\beta u^{-\frac{1}{q}})^{\frac{1}{2}\alpha}$$

for all $u > 0$, provided that $m \ge 2$ and $\alpha + m - 1 \ge 2q + 1$.

Proof. Let $Z_m = \beta_m/\beta$. Then the marginal density of Z_m is

$$\frac{\Gamma(\frac{\alpha+m-1}{2})}{\Gamma(\frac{\alpha}{2})\Gamma(\frac{m-1}{2})}\frac{(z-1)^{\frac{m-3}{2}}}{z^{\frac{\alpha+m-1}{2}}}$$

for $1 \le z < \infty$. This is at most $C_0 m^{\frac{1}{2}\alpha}/z^{1+\frac{1}{2}\alpha}$ for all $\alpha > 2q$ and all $z \ge 1$ for some constant C_0. So,

$$P_{\alpha,\beta}[Z_m > z] \le \frac{C_0 m^{\frac{1}{2}\alpha}}{qz^{\frac{1}{2}\alpha}}$$

for all $\alpha > 2q$ and $z \ge 1$. The Lemma is an easy consequence of this and (15), since

$$P_{\alpha,\beta}[U_m > u] = P_{\alpha,\beta}\big[Z_m > \frac{\nu_{\alpha_m-q}^{\frac{1}{q}}u^{\frac{1}{q}}}{\beta}\big].$$

4. Proof of the Main Result

It suffices to show that

$$\sup_{\xi\in\Xi_0}\inf_{\delta}\ \bar{r}_a(\delta;\xi) \ge 2q\sqrt{K''(1)\eta_0} + o(\sqrt{\eta_0}) \tag{20}$$

and

$$\sup_{\xi\in\Xi_0}\ \bar{r}_a(\delta_a;\xi) \le 2q\sqrt{K''(1)\eta_0} + o(\sqrt{\eta_0}) \tag{21}$$

The Upper Bound. Relation (21) is established first. With m_a and t_a as in (9) and (10), let a be so large that $m_a \geq 2\ell q$. Then it is clear from the monotonicity of K_0 that

$$K(\frac{t_a}{\eta}) \leq K[(\frac{S^2_{m_a}}{\sigma^2})^q] + \frac{m_a+1}{\eta}.$$

So,

$$R_a(\delta_a;\sigma^2) \leq \eta \bar{K}_{m_a-1}(y_a) + m_a + 1,$$

where $y_a = \nu_{m_a-1}/(m_a-1)^q$ and, therefore,

$$\sup_{\xi\in\Xi_0} \bar{r}_a(\delta_a;\xi) \leq \eta_0[\bar{K}_{m_a-1}(y_a) - K(1)] + m_a + 1,$$

The right side is easily analyized. From Stirling's Formula, $y_a = 1 + O(1/m_a)$. Then $\bar{K}_{m_a-1}(y_a) = K(y_a) + q^2y_a^2K''(y_a)/m_a + o(1/m_a)$, by Lemma 3.1, and $K(y_a)+q^2y_a^2K''(y_a)/m_a = K(1)+q^2K''(1)/m_a+o(1/m_a)$ by the continuity of K''. This leaves

$$\begin{aligned}\sup_{\xi\in\Xi_0} \bar{r}_a(\delta_a;\xi) &\leq \frac{q^2K''(1)\eta_0}{m_a} + m_a + 1 + o(\frac{\eta_0}{m_a}) \\ &\leq 2q\sqrt{K''(1)\eta_0} + o(\sqrt{\eta_0})\end{aligned}$$

where the last step uses the definition (9) of m_a. This establishes (21).

The Lower Bound. Let $\beta = \beta_a = a^{-2/q}$, and let $\alpha = \alpha_a$ be determined by $\beta^q/\nu_{\alpha-2q} = \sigma_0^{2q}$. Then the inverted gamma priors $\xi_{\alpha,\beta}$ are in Ξ_0, $\beta \to 0$, $\alpha \downarrow 2q$ as $a \to \infty$, and

$$P_{\alpha,\beta}[U_m > \frac{\epsilon m}{a}] \leq C\big[\beta(\frac{a}{\epsilon m})^{\frac{1}{q}}\big]^{\frac{1}{2}\alpha} \leq \frac{C}{\epsilon\sqrt{a}} \tag{22}$$

for all $m \geq 2$, $0 < \epsilon < 1$, and large a. It will be shown that

$$\inf_{\delta} \bar{r}(\delta;\xi_{\alpha,\beta}) \geq 2q\sqrt{K''(1)\eta_0} + o(\sqrt{\eta_0}). \tag{23}$$

To establish this, first observe that

$$\inf_{\delta} \bar{R}(\delta;\xi_{\alpha,\beta}) = \inf_{m\geq 2} E_{\alpha,\beta}\Big\{\inf_{t\geq m} E^m_{\alpha,\beta}\big[\eta K(\frac{t}{\eta})\big]\Big\} \tag{24}$$

and

$$E^m_{\alpha,\beta}\big[\eta K(\frac{t}{\eta})\big] = \bar{\eta}_m \bar{K}_{\alpha_m-2q}(\frac{t}{\bar{\eta}_m})$$

for fixed $m \geq 2$ and $t \geq m$, where $\bar{\eta}_m = aU_m$. So, for fixed m,

$$\inf_{t\geq m} E^m_{\alpha,\beta}\big[\eta K(\frac{t}{\eta})\big] \geq I_m + I\!I_m,$$

where

$$I_m = \inf_{x>0} \bar{\eta}_m \bar{K}_{\alpha_m - 2q}(x)$$

$$I\!I_m = \bar{\eta}_m \big[\bar{K}_{\alpha_m-2q}(\frac{m}{\bar{\eta}_m}) - \inf_{x>0} \bar{K}_{\alpha_m-2q}(x)\big] \mathbf{1}_{\{x_m^* < m/\bar{\eta}_m\}},$$

and x_m^* is the value at which $\bar{K}_{\alpha_m-2q}(x)$ is minimized. Clearly, $E_{\alpha,\beta}(U_m) = U_0$ and, therefore, $E_{\alpha,\beta}(\bar{\eta}_m) = \eta_0$. So,

$$E_{\alpha,\beta}(I_m) = \eta_0 \inf_{x>0} \bar{K}_{\alpha_m-2q}(x).$$

Now, $\inf_{\alpha \le \alpha^o} \inf_{x>0} \bar{K}_\alpha(x) > K(1)$ for any α^o by Lemma 3.2. Combining this with (21), it follows that the infimum in (24) can be restricted of $m \ge m_0$ for any given m_0 for all sufficiently large a.

Considering only large a and m,

$$E_{\alpha,\beta}(I_m) = \eta_0 \big[K(1) + \frac{q^2 K''(1)}{\alpha + m - 1} + o(\frac{1}{m})\big]$$

Next,

$$E_{\alpha,\beta}(I\!I_m) = I\!I_{1,m} - I\!I_{2,m},$$

where

$$I\!I_{1,m} = \int_{x_m^* < m/\bar{\eta}_m} \bar{\eta}_m \big[\bar{K}_{\alpha_m-2q}(\frac{m}{\bar{\eta}_m}) - K(1)\big] dP_{\alpha,\beta}$$

and

$$I\!I_{2,m} = \int_{x_m^* < m/\bar{\eta}_m} \bar{\eta}_m \big[\inf_{x>0} \bar{K}_{\alpha_m-2q}(x) - K(1)\big] dP_{\alpha,\beta}.$$

Here

$$\begin{aligned} |I\!I_{2,m}| &\le \int_{x_m^* < m/\bar{\eta}_m} \bar{\eta}_m \big[\frac{q^2 K''(1)}{m} + o(\frac{1}{m})\big] dP_{\alpha,\beta} \\ &\le \frac{m}{x_m^*} \big[\frac{q^2 K''(1)}{m} + o(\frac{1}{m})\big] \\ &= O(1). \end{aligned}$$

For $I\!I_{1,m}$ observe that $x_m^* \le m/\bar{\eta}_m$ iff $\bar{\eta}_m \le m/x_m^*$ and recall that $x_m^* \to 1$. So, for any $0 < \epsilon < 1$,

$$I\!I_{1,m} \ge (1-\epsilon) m P_{\alpha,\beta}[\bar{\eta}_m \le \frac{\epsilon m}{K(1)}] \ge (1-\epsilon)^2 m$$

for all $m \geq m_0$ for all sufficiently large a, by (22). This leaves

$$\inf_{\delta} \bar{r}(\delta; \xi_{\alpha,\beta}) \geq \inf_{m \geq m_0} \left\{ [\frac{q^2 K''(1)}{m} + o(\frac{1}{m})]\eta_0 + (1-\epsilon)^2 m + O(1) \right\}$$
$$= 2(1-\epsilon)q\sqrt{K''(1)\eta_0} + o(\sqrt{\eta_0}),$$

establishing (23) since ϵ was arbitrary.

5. Acknowledgements

Thanks to Anirban DasGupta, Bob Keener, and Jiayang Sun for helpful discussions. This research was supported by the National Science Foundation and U.S. Army Research Office.

References

1. Abromowitz, Milton and Irene Stegun (1964). *Handbook of Mathematical Functions.* National Bureau of Standards.
2. Brown, Larry (1986). *Fundamentals of Statistical Exponential Families with Applications in Statistical Decision Theory.* IMS Lecture Notes - Monograph Series, **9**.
3. Cohen, Arthur and Harold Sackrowitz (1985). Bayes double sampling procedures. *Ann. Statist*, **12**, 1035-1049.
4. Cohen, Arthur and Harold Sackrowitz (1985). Double sampling estimation when cost depends on parameters. *Statistical Decision Theory and Related Topics*, **V**, 253-266.
5. Denne, Jonathan and Christopher Jennision (1999). Estimation the sample size for a *t*-test using an internal pilot. *Stat. Med.*, **18**, 1575-1585.
6. Ghosh, Malay, Nitus Mukhopadhyay, and Pranab Kumar Sen (1997). *Sequential Estimation.* Wiley.
7. Stein, Charles (1945). A two-stage test for a linear hypothesis whose power is independent of the variance. *Ann. Math. Statist.*, **16**, 243-258.
8. Treder, R.P. and J. Sedransk (1996). Bayesian Sequential Two-Phase Sampling. *J. Amer. Statist. Assn.*, **91**, 782-790.
9. Van der Vaart, Aad (1998). *Asymptotic Statistics.* Cambridge University Press
10. Woodroofe, Michael (1982). *Non-linear Renewal Theory in Sequential Analysis.* S.I.A.M.
11. Woodroofe, Michael (1986). Asymptotic optimality in sequential interval estimation. *Adv. Appl. Math.*, **7**, 70-79.
12. Woodroofe, Michael (2004). An asymptotic minimax determination of the initial sample size in a two-stage procedure. *IMS Lecture Notes - Monograph Series*, **45**, 228-236

INCORPORATING OVERRUNNING DATA INTO THE ANALYSIS OF BOTH PRIMARY AND SECONDARY ENDPOINTS IN A SEQUENTIAL TRIAL

AIYI LIU, CHENGQING WU* and KAI F. YU

Biometry and Mathematical Statistics Branch
National Institute of Child Health and Human Development
Department of Health and Human Services
6100 Executive Blvd., Rockville, MD, U. S. A.
liua@mail.nih.gov, wuch@mail.nih.gov, yukf@mail.nih.gov

We propose methods for incorporating overrunning data, either balanced or unbalanced, into the final analysis of a sequential clinical trial comparing an experimental arm with a control arm. We consider inference on the primary endpoint for which the sequential test is designed and a correlated secondary endpoint. By separating the monitoring process into arm-specific processes, we derive the sufficient statistics and show how the independent overrunning data can be combined with the trial data at stopping and thus allow the likelihood-based inference to be conducted.

Some key words: Additional data; Arm-specific analysis; Clinical trials; Interim analysis; Patients allocation; Secondary endpoints; Stopping boundaries.

1. Introduction

Group sequential testing procedures are frequently used in the conduct of clinical trials for ethical and economic concerns. Using a group sequential testing scheme, the null hypothesis of no treatment difference is being tested at a number of time points and the trial is stopped (and a decision with respect to rejection of the null hypothesis is made) if at any stage the stopping criterion is met.

After a sequential trial is stopped, additional data concerning the treatment effect is often available, but mechanisms that generate these data vary.

---

*Also at Department of Statistics and Finance, Univeristy of Science and Technology of China, Hefei, Anhui, China

Hall and Liu (2002) presented the overrunning problem in the context of a Brownian motion continuing to be observed after it reaches a stopping boundary. Sooriyarachchi *et al.* (2003) discussed implementation and comparison of several methods to incorporate such overrunning data. These settings are applicable in clinical trials when an independent overrunning path of the monitoring process that summarizes the treatment difference can be constructed from the additional data. A good illustrative example of overrunning is that after a sequential clinical trial comparing survival rates of two arms stops, patients survival is updated and additional information becomes available later due to lagged data.

We deal with two possible scenarios of overrunning, balanced and unbalanced, and propose a unified approach to construct the independent overrunning path of the monitoring process. Throughout, balanced overrunning refers to data generated when patient allocation ratio remains unchanged. In this case no knowledge about data from each treatment arm upon stopping is needed to conduct the likelihood-based inference. Unbalanced overrunning data occur when the allocation ratio changes after stopping. For example, after a sequential clinical trial is terminated, investigators decide to assign all future patients due to continued accrual to the experimental arm shown to be superior to the control arm, and thus additional data are available only from the experimental arm; no additional data are observed from the control arm. In this case, these additional data, though need to be incorporated into the final analysis, can not be used to construct an independent estimate of the treatment difference. In order to allow such construction, data from each treatment arm need to be known so that overrunning data can be combined with the already observed data from each treatment arm. Another example that bears the same nature is a phase II cancer clinical trial which shows satisfactory response rate and thus warrants a phase III trial to compare with the standard treatment. If patients in the phase II trial are considered to be a randomly selected part of the targeted phase III patient population, then patients' survival data observed in the phase II trial follow-up should also be incorporated into the analysis of phase III trial data after the latter is stopped. In either case, however, special treatments are needed so that the methods presented in Hall and Liu (2002) and Sooriyarachchi *et al.* (2003) can be used for the updated analysis.

We propose a separation-recombination approach to incorporate the overrunning data, balanced or unbalanced, in order to make a final analysis of the trial data. Such approach separates the monitoring process into two arm-specific processes, corresponding to each treatment arm. Overrunning data from each arm are then combined with the corresponding process to make inference on arm-specific parameters and then the treatment difference parameter. A mathematical model is presented in section 2 following a brief introduction to group sequential testing. Several methods are proposed in Section 3 to estimate the treatment difference and other parameters. Section 4 develops inferential methods for parameters associated with a secondary endpoint. Section 5 presents some simulation results. The paper ends with some discussions in Section 6.

2. Separation of the Monitoring Process

2.1. *The group sequential tests*

Consider a clinical trial comparing two treatment arms. Let X_{Ei} and X_{Ci}, $i = 1, \ldots,$ be the patients' response outcomes (primary endpoint) in the experimental arm and the control arm, respectively. Without loss of generality (asymptotically) we assume that outcomes are normally distributed with $X_{Ei} \sim N(\mu_E, 1)$ and $X_{Ci} \sim N(\mu_C, 1)$. Denote by $\delta = \mu_E - \mu_C$ the treatment difference, referred to as the primary parameter. The null hypothesis of no treatment difference is thus $H_0 : \delta = 0$. Throughout we consider two sided test with power $1 - \beta$ at $\delta = \delta_1$.

Suppose that a $K-$stage group sequential procedure is used to test the null hypothesis H_0. For simplicity we assume that at each stage subjects are equally allocated to each treatment arm and that the kth group has g_k observations in each treatment arm; thus $n_k = g_1 + g_2 + \ldots + g_k$ is the cumulative sample size per arm at stage k. Put $X_i = X_{Ei} - X_{Ci}, i = 1, 2, \cdots,$ the outcome difference between the ith patient in treatment arm and the ith patient in the control arm. Then X_i is normally distributed with mean δ and variance 2. Let $S_n = X_1 + \cdots + X_n, n = 1, 2, \cdots$, be the difference between the cumulative sample sums. The stopping boundaries are determined based on the random sequence $S_{n_1}, S_{n_2}, \cdots$, which throughout will be referred to as the monitoring process. The null hypothesis H_0 is rejected at the kth stage if

$$|S_{n_j}| < c_j,\ 1 \leq j \leq k-1,\ |S_{n_k} \geq c_k, \tag{1}$$

and no further sample takes place thereafter. Popular methods for setting up the critical values c_k include that of, among others, O'Brien-Fleming (1979), Pocock (1977) and Lan and DeMets (1983).

Terminal inference on the treatment difference parameter δ upon stopping of the test can be solely based on M, the number of looks conducted upon stopping, and S_{n_M}, the observed sum difference. Their joint density function (Emerson and Fleming, 1990) at $(M, S_{n_M}) = (k, s)$ can be computed numerically using the following recursive formula:

$$f_\delta(k,s) = h_0(k,s)\phi_{2n_k}(s - n_k\delta), \tag{2}$$

where $\phi_v(.)$ and $\Phi_v(.)$ are respectively the density and cumulative distribution function of a normal random variable with zero mean and variance v (omitting the subscription if $v = 1$), and $h_0(1,s) = 1$,

$$h_0(k,s) = \int_{-c_{k-1}}^{c_{k-1}} h_0(k-1,t)\phi_{\frac{2n_{k-1}g_k}{n_k}}(t - \frac{n_{k-1}}{n_k}s)dt \ .$$

Theory and methods have been developed extensively (based on a single normal distribution or a Brownian motion applicable asymptotically to the above two-arm comparison model) to construct point estimates, confidence intervals and P-values according to various ordering of the sample space; see Whitehead (1997) and Jennison and Turnbull (2000) for a summary.

After the stopping criterion for the sequential clinical trials has been met, data on additional patients may become available due to continued accrual or lagged data-reporting. These additional data should be considered as an integrated part of the trial and need to be included in a final and complete analysis. Suppose that the trial stops at the kth stage and that observations following stopping are taken from a number of m_k and l_k additional patients assigned to the experimental arm and the control arm respectively. If the allocation ratio remains the same (implying $l_k = m_k$), then an additional path of the monitoring process, independent of the one observed before stopping, can be constructed using the S_n-statistic based on these overrunning data, and hence methods described in Hall and Liu (2002) and Sooriyarachchi et al. (2003) can be employed. However, if the allocation ratio changes (e.g. all additional patients are assigned to the experimental arm), and thus $l_k \neq m_k$, resulting in the so called unbalanced overrunning, then these methods are not directly applicable since construction of the additional path of the monitoring process is not obvious. (The one-arm-only overrunning data do not provide information on the treatment difference parameter δ; see more details in Section 6.) We propose below a separation-recombination method that allows to incorporate the unbalanced overrunning data into the final analysis of the trial; the method readily applies to balanced overrunning data.

2.2. *The arm-specific processes and the sufficient statistics*

In order to make inference on the treatment difference parameter δ by including the unbalanced overrunning data, we consider, instead of the monitoring process, the two arm-specific processes at stopping stage k, i.e. $S_E = \sum_{i=1}^{n_k} X_{Ei}$ and $S_C = \sum_{i=1}^{n_k} X_{Ci}$, each having an independent overrunning path, $S_{Eo} = \sum_{i=n_k+1}^{n_k+m_k} X_{Ei}$ and $S_{Co} = \sum_{i=n_k+1}^{n_k+l_k} X_{Ci}$, the accumulative sample sum of observations from the experimental arm and the control arm, respectively. Such separation allows the pre-post stopping paths of an arm-specific process to be combined. To obtain a minimal sufficient statistic based on $(M, S_E, S_C, S_{Eo}, S_{Co})$, write $W_i = X_{Ei} + X_{Ci}, i = 1, 2, \cdots$, and $U_n = W_1 + \cdots + W_n, n = 1, 2, \cdots$. Then W_i is normally distributed with mean θ and variance 2, and is independent of X_i, where $\theta = \mu_E + \mu_C$. Thus the random variable U_n is normally distributed with mean $n\theta$ and variance $2n$, and for each n is independent of sequence S_n based on which the stopping boundaries are determined. Therefore the distribution of U_n shall not be affected by the sequential test due to independence. Thus, from (2), the joint density function of (M, S_{n_M}, U_{n_M}) at (k, s, u) is

$$f_{\delta,\theta}(k, s, u) = h_0(k, s)\phi_{2n_k}(s - n_k\delta)\phi_{2n_k}(u - n_k\theta) \ . \tag{3}$$

Note that $S_{n_M} = S_E - S_C$ and $U_{n_M} = S_E + S_C$. From (3), the joint density function of (M, S_E, S_C) at (k, s_E, s_C) is

$$f_{\mu_E,\mu_C}(k, s_E, s_C) = h_0(k, s_E - s_C)\phi_{n_k}(s_E - n_k\mu_E)\phi_{n_k}(s_C - n_k\mu_C) \ . \tag{4}$$

Conditioning on $M = k$, the distribution of S_{Eo} and S_{Co} are $N(m_k\mu_E, m_k)$ and $N(l_k\mu_C, l_k)$, respectively. Then the jointly conditional density function of S_{Eo} and S_{Co} is

$$f_{\mu_E,\mu_C}(s_{Eo}, s_{Co}|M = k) = \phi_{m_k}(s_{Eo} - m_k\mu_E)\phi_{l_k}(s_{Co} - l_k\mu_C) \ . \tag{5}$$

Multiplication of (4) with (5) thus yields the density of $(M, S_E, S_C, S_{Eo}, S_{Co})$ as

$$f_{\mu_E,\mu_C}(k, s_E, s_C, s_{Eo}, s_{Co}) = f_{0,0}(k, s_E, s_C, s_{Eo}, s_{Co}) \exp\Big\{(s_E + s_{Eo})\mu_E$$
$$+(s_C + s_{Co})\mu_C - \frac{1}{2}(n_k + m_k)\mu_E^2 - \frac{1}{2}(n_k + l_k)\mu_C^2\Big\}, \tag{6}$$

where

$$f_{0,0}(k, s_E, s_C, s_{Eo}, s_{Co}) = h_0(k, s_E - s_C)\phi_{n_k}(s_E)\phi_{n_k}(s_C)\phi_{m_k}(s_{Eo})\phi_{l_k}(s_{Co}) \ .$$

Therefore, the triplet (M, S_E^o, S_C^o) is a minimal sufficient statistic for (μ_E, μ_C) and thus for δ, where

$$S_E^o = \sum_{i=1}^{n_M+m_M} X_{Ei}, \quad \text{and } S_C^o = \sum_{i=1}^{n_M+l_M} X_{Ci},$$

the final observed sample sum of the experimental arm and the control arm respectively. From (6), with the fact that $\phi_u(s)\phi_v(t-s) = \phi_{u+v}(t)\phi_{uv/(u+v)}(s - tu/(u+v))$, its density is given by

$$f_{\mu_E,\mu_C}(k, s_E^o, s_C^o) = f_{0,0}(k, s_E^o, s_C^o)\phi_{n_k+m_k}(s_E^o - (n_k + m_k)\mu_E) \times \phi_{n_k+l_k}(s_C^o - (n_k + l_k)\mu_C), \tag{7}$$

with

$$f_{0,0}(k, s_E^o, s_C^o) = \int_{s_E - s_C \in B_k} \left\{ h_0(k, s_E - s_C)\phi_{v_{k1}}(s_E - \frac{n_k}{n_k + m_k} s_E^o) \times \phi_{v_{k2}}(s_C - \frac{n_k}{n_k + l_k} s_C^o) \right\} ds_E\, ds_C,$$

where $v_{k1} = n_k m_k/(n_k + m_k)$, $v_{k_2} = n_k l_k/(n_k + l_k)$, and B_k is the stop region at stage k, i.e. $B_k = \{s : |s| \geq c_k\}$ for $k = 1, \cdots, K-1$ and $B_K = (-\infty, \infty)$. Notice that $\phi_u(t)\phi_v(t-s) = \phi_{u+v}(s)\phi_{uv/(u+v)}(t-su/(u+v))$,with variable changes, one can show that

$$f_{0,0}(k, s_E^o, s_C^o) = \int_{s\in B_k} h_0(k, s)\phi_{v_{k1}+v_{k2}}\left(s - (\frac{n_k}{n_k + m_k} s_E^o - \frac{n_k}{n_k + l_k} s_C^o)\right) ds. \tag{8}$$

3. Statistical Inference on the Primary Endpoint

The separation-recombination of the monitoring process allows us to make inference on the three parameters related to the primary endpoint, the treatment difference parameter δ and the two arm-specific mean parameters μ_E and μ_C. We consider inference based on likelihood orderings. Write $\bar{X} = S_{n_M}/n_M$, , $\bar{X}_E^o = S_E^o/(n_M + m_M)$ and $\bar{X}_C^o = S_C^o/(n_M + l_M)$. By maximizing the likelihood function (7) with respect to μ_E and μ_C, we obtain the maximum likelihood estimator of μ_E and μ_C to be $\bar{X}_E^o$ and $\bar{X}_C^o$ respectively, hence the maximum estimation of treatment differencc δ is $\hat{\delta} - \bar{X}_E^o - \bar{X}_C^o$.

The maximum likelihood estimators following a sequential sampling scheme are known to be biased. From (7), the joint density function of

$\bar{X}_E^o$ and $\bar{X}_C^o$ is

$$g_{\mu_E,\mu_C}(x_E,x_C)=\sum_{k=1}^{K} g_{0,0}(k,x_E,x_C)\phi_{\frac{1}{n_k+m_k}}(x_E-\mu_E)\phi_{\frac{1}{n_k+l_k}}(x_C-\mu_C), \tag{9}$$

where $g_{0,0}(k,x_E,x_C)=f_{0,0}(k,(n_k+m_k)x_E,(n_k+l_k)x_C)$.

From (8) and (9), it can be shown that the bias of $\bar{X}_E^o$ and $\bar{X}_C^o$ only depends on the primary parameter δ. Indeed, let $b_E(\delta)$ and $b_C(\delta)$ be respectively the bias function of the corresponding maximum likelihood estimators. Then

$$b_E(\delta)=E\left(\frac{n_M}{2(n_M+m_M)}(\bar{X}-\delta)\right), \tag{10}$$

$$b_C(\delta)=-E\left(\frac{n_M}{2(n_M+l_M)}(\bar{X}-\delta)\right). \tag{11}$$

(For the proof, see the Appendix.) Hence the bias of $\hat{\delta}$ is

$$b(\delta)=E\left(\frac{n_M(2n_M+m_M+l_M)}{2(n_M+m_M)(n_M+l_M)}(\bar{X}-\delta)\right). \tag{12}$$

To reduce the bias, the method of Whitehead (1986a,b) can be used. The bias adjusted estimate, $\tilde{\delta}$ can be obtained by solving the equation

$$b(\tilde{\delta})+\tilde{\delta}=\hat{\delta}\ , \tag{13}$$

that is, the expectation of the maximum likelihood estimate evaluated at $\tilde{\delta}$ is equal to $\hat{\delta}$. The solution can be found numerically. From Proposition A1 in the Appendix of Hall and Liu (2002), $\hat{\delta}$ is stochastically increasing in δ. Hence the expectation of $\hat{\delta}$, therefore the left side of (13), is increasing in $\tilde{\delta}$. This monotonicity property ensures the unique solution to the equation (13); a bisection procedure is a suitable and robust numerical method to obtain the solution.

To reduce the bias of $\bar{X}_E^o$ and $\bar{X}_C^o$, a multidimensional extension of the Whitehead approach can be used. With (10) and (11), it is easy to show that the bias-adjusted estimate, $\tilde{\mu}_E$ and $\tilde{\mu}_C$, are

$$\tilde{\mu}_E=\bar{X}_E-b_E(\tilde{\delta})\ \text{ and }\ \tilde{\mu}_C=\bar{X}_C-b_C(\tilde{\delta})\ , \tag{14}$$

where $\tilde{\delta}$ is the solution of equation (13).

The p-value of the test, confidence intervals and median unbiased estimate of the parameter δ can be obtained from $\bar{H}_\delta(x)=P(\hat{\delta}>x)$, the upper tail probability function of the maximum likelihood estimate. With observed value $\hat{\delta}_{obs}$ of $\hat{\delta}$, a corresponding p-value is given by $\bar{H}_\delta(\hat{\delta}_{obs})$. In turn, a p-value for the null hypothesis against the alternative $\delta > 0$ is $\bar{H}_0(\hat{\delta}_{obs})$. For a two side test, with the symmetry of $\bar{H}_0(x)$, p-value is double of $\bar{H}_0(|\hat{\delta}_{obs}|)$.

4. Secondary Endpoints

Trial monitoring activities toward possible early stopping focus exclusively on the designated primary outcome. In practice, there are other clinical endpoints that provide additional characterization of the treatment effect which is not monitored directly in the design, and analyzing these secondary endpoints are often desirable. Since in general these secondary endpoints are correlated with the primary endpoint, potential bias introduced by sequential sampling needs also to be reduced.

Let Y_{Ei} and $Y_{Ci}, i = 1, 2, \cdots$, be patients' responses of a secondary endpoint from the experimental arm and the control arm, respectively. We assume that these secondary outcomes are normally distributed with unit variance and mean λ_E for the experimental arm λ_C for the control arm. Further, (X_{Ei}, Y_{Ei}) and (X_{Ci}, Y_{Ci}) follow bivariate normal distributions with a correlation coefficient parameter ρ_E and ρ_C respectively, hereafter assumed to be known. The secondary treatment effect parameter $\lambda = \lambda_E - \lambda_C$ is the focus of inference.

Define $Z_{Ei} = Y_{Ei} - \rho_E X_{Ei}$ and $Z_{Ci} = Y_{Ci} - \rho_C X_{Ci}$, $i = 1, 2, \cdots$. Then Z_{Ei} and Z_{Ci} are normally distributed with mean $\lambda_E - \rho_E \mu_E$ and $\lambda_C - \rho_C \mu_C$, and variance $1 - \rho_E^2$ and $1 - \rho_C^2$, and uncorrelated with X_{Ei}, X_{Ci} and X_i. Hence the distributions of their sum $Z_{E1} + \cdots + Z_{En}$ and $Z_{C1} + \cdots + Z_{Cn}$ are not affected by possible early stopping. In particular we have,

$$E\left(\bar{Z}_E\right) = \lambda_E - \rho_E \mu_E \quad \text{and} \quad E\left(\bar{Z}_C\right) = \lambda_C - \rho_C \mu_C \ , \tag{15}$$

where $\bar{Z}_E$ and $\bar{Z}_C$ are the sample mean of the sequence $Z_{E1}, \cdots, Z_{En}$ and $Z_{C1}, \cdots, Z_{Cn}$ respectively.

Let $\bar{Y}_E$, $\bar{Y}_C$ denote the overall sample mean of the patients' response outcomes in the experimental arm and the control arm, respectively, of the secondary endpoint; these are the maximum likelihood estimates of λ_E and λ_C, respectively. Then $\hat{\lambda}$, the maximum likelihood estimate of λ, is $\bar{Y}_E - \bar{Y}_C$. With (15), it can be shown that

$$\mathrm{BIAS}_\delta(\bar{Y}_E) = \rho_E b_E(\delta) \quad \text{and} \quad \mathrm{BIAS}_\delta(\bar{Y}_C) = \rho_C b_C(\delta) \ . \tag{16}$$

Then

$$\mathrm{BIAS}_\delta(\hat{\lambda}) = \rho_E b_E(\delta) - \rho_C b_C(\delta) \ . \tag{17}$$

From (16) and (17) along with the definition of Whitehead's bias adjusted estimation, it is easy to show that the bias adjusted estimates of λ_E, λ_C and λ are, respectively,

$$\tilde{\lambda}_E = \bar{Y}_E - \rho_E b_E(\tilde{\delta}),\ \tilde{\lambda}_C = \bar{Y}_C - \rho_C b_C(\tilde{\delta}) \text{ and } \tilde{\lambda} = \hat{\lambda} - \left(\rho_E b_E(\tilde{\delta}) - \rho_C b_C(\tilde{\delta})\right). \tag{18}$$

If $\rho_E = \rho_E = \rho$, then the bias of $\hat{\lambda}$ is

$$\text{BIAS}(\hat{\lambda}) = \rho\text{BIAS}(\hat{\delta}) , \tag{19}$$

that is, the bias of the maximum likelihood estimator of the secondary parameter is proportional to the bias of the maximum likelihood estimator of the primary parameter. (Similar results were obtained in Whitehead, 1986b, and Liu and Hall, 2001, for terminal inference upon stopping.) Then the bias adjusted estimator is

$$\tilde{\lambda} = \hat{\lambda} - \rho(\hat{\delta} - \tilde{\delta})$$

and its bias is given by

$$\text{BIAS}(\tilde{\lambda}) = \rho\text{BIAS}(\tilde{\delta}) , \tag{20}$$

following the same proportionality as the maximum likelihood estimates.

5. Numerical Results

Simulation studies were conducted to investigate the performance of the proposed point estimation and confidence interval of the parameter δ, for a O'Brien and Fleming (1979) type group sequential boundaries. At stage k, the critical value of the standardized test statistic $S_{n_k}/\sqrt{2n_k}$ is $C_B(K,\alpha)\sqrt{K/k}$, where $C_B(K,\alpha)$ is chosen to give overall type I error α. Hence the critical value $c_k = C_B(K,\alpha)\sqrt{2n_k K/k}$. For simplicity, we chose a two stage design with equal group size 20 at each stage per arm. Table 2.3 of Jennison and Turnbull (2001, pp.29) gives $C_B(K,\alpha) = 1.977$ for $\alpha = 0.05$. Then the boundary of the first stage sample sum is 17.683. Also, we arbitrarily set $m_1 = m_2 = 10$ and $l_1 = l_2 = 5$.

For this simple case of a two-stage design, bias functions (10) and (11) can be expressed analytically. Let

$$a(\delta) = \phi_{n_1}(c_1 - n_1\delta) - \phi_{n_1}(c_1 - n_1\delta) ,$$

then $b_E(\delta) = d_E a(\delta)$ and $b_C(\delta) = d_C a(\delta)$, where

$$d_E = \frac{n_1(n_2 + m_2 - n_1 - m_1)}{2(n_1 + m_1)(n_2 + m_2)} \quad \text{and} \quad d_C = -\frac{n_1(n_2 + l_2 - n_1 - l_1)}{2(n_1 + l_1)(n_2 + l_2)} .$$

Table 1 presents the results obtained from 1000 replicates using the methods described in this paper applied to the O'Brian and Fleming test. The left hand side presents the means and standard deviations of the maximum likelihood estimate $\hat{\delta}$, the bias adjusted estimate $\tilde{\delta}$ and the median unbiased estimate. Compared to the maximum likelihood estimate, the median unbiased estimate has smaller bias and standard deviation. However the improvement is quite negligible. The bias adjusted estimate, on the other hand, has the smallest bias and standard deviation among the three. It reduces almost half of the bias of the maximum likelihood estimation.

Table 1. Point estimation of δ with standard deviation, together with coverage probabilities and symmetry of 95% confidence intervals for an O'Brein & Fleming test with unbalanced overrunning data.

δ	MLE Mean(std)	Bias-Adjusted Mean(std)	Median-Unbiased Mean(std)	$p(\delta_L)$	$p(\delta_U)$	Coverage
0.0	−0.0036(0.2083)	−0.0032(0.2028)	−0.0033(0.2071)	0.023	0.977	0.954
0.1	0.1113(0.2105)	0.1085(0.2045)	0.1108(0.2088)	0.023	0.976	0.953
0.3	0.3152(0.2176)	0.3055(0.2092)	0.3124(0.2137)	0.024	0.977	0.953
0.5	0.5254(0.2261)	0.5061(0.2160)	0.5167(0.2185)	0.025	0.0977	0.952
0.8	0.8459(0.2346)	0.8147(0.2325)	0.8265(0.2326)	0.024	0.975	0.951
1.0	1.0445(0.2396)	1.0144(0.2330)	1.0264(0.2394)	0.023	0.975	0.952

The three right columns in Table 1 show coverage probabilities, the proportion of intervals which include the true treatment improvement δ, and the probability that the lower and upper confidence limits exceed true δ. It can be seen that the intervals obtained are accurate. For 1000 replicates, the standard error of the estimated coverage probability is 0.007. The coverage probabilities are all within one standard deviation of the target value of 0.95. The proportions for the lower and upper limits are close to their target values of 0.025 and 0.975, respectively.

6. Discussion

Suppose the sequential test stops at the $M = k$th look. In absence of overrunning data, inference on the primary parameter δ can be solely based on the statistic (M, S_{n_k}), with $S_{n_k} = S_E - S_C$.

In the case of balanced overrunning data, i.e. $m_k = l_k = m$, data at stopping from each treatment arm are not needed for the likelihood-based inference; only the monitoring process needs to be observed. This is so because, with $M = k$,

$$\hat{\delta} = \bar{X}_E^o - \bar{X}_C^o = \frac{1}{n_k + m}(S_{n_k} + S_o),$$

with $S_o = S_{Eo} - S_{Co}$. Thus methods of Hall and Liu (2002) and Sooriyarachchi *et al.* (2003) can then be applied to the two independent paths of the monitoring process, $S_{n_k}/2$ and $S_o/2$, the latter being an overrunning path, both resemble a Brownian motion with a common drift parameter δ and time $n_k/2$ and $m/2$, respectively.

With unbalanced overrunning data ($m_k \neq l_k$), arm-specific data collected at the time of stopping are needed to combined with the overrunning data from each treatment arm in order to make likelihood-based inference. This can be seen from the likelihood function derived in Section 2.2.

When only the monitoring process is observed upon stopping and data from each treatment arm are not available, an independent overrunning path S_o of the monitoring process can still be constructed as follows, using only the overrunning data. Continue to use the notations in Section 2.2 and assume that $m_k, l_k > 0$, and define $S_o = (l_k S_{Eo} - m_k S_{Co})/(l_k + m_k)$. Then $S_{n_k}/2$ and S_o are independent, both resemble a Brownian motion with a common drift parameter δ and time $n_k/2$ and $m_k l_k/(m_k + l_k)$, respectively. Hence S_o is an independent overrunning path of the monitoring process after stopping.

Therefore the methods of Hall and Liu (2002) and Sooriyarachchi *et al.* (2003) continue to be applicable. However, it should be pointed out that inference procedures based on (M, S_{n_M}, S_o) constructed above are not functions of the sufficient statistics and thus may suffer from loss of efficiency. Furthermore, such construction procedure is not applicable to the extreme overrunning case when only one treatment arm has additional data ($m_k = 0$ or $l_k = 0$); the separation-recombination appears to be the only feasible approach to incorporate such overrunning data into the final analysis.

Acknowledgments

This research was supported by the Intramural Research Program of the National Institute of Child Health and Human Development, National Institutes of Health. The opinions expressed are those of the authors and not necessarily of the National Institutes of Health.

Appendix

Proof of (10) and (11) We only show (10) here since (11) follows similarly. From (8) and (9), we have

$$
\begin{aligned}
E\bar{X}_E &= \int_{-\infty}^{\infty} x_E g_{\mu_E,\mu_C}(x_E, x_C) dx_E \\
&= \sum_{k=1}^{K} \int_{B_k} h(k, n_k x) \phi_{\frac{2}{n_k}}(x-\delta) dx \\
&\quad \times \int_{-\infty}^{\infty} x_E \phi_{v_{k3}}\left(x_E - \mu_E - \frac{n_k}{2(n_k+m_k)}(x-\delta)\right) dx_E \\
&\quad \times \int_{-\infty}^{\infty} \phi_{v_{k4}}\left(x_C - \mu_C - a_k(x - \delta - (x_E - \mu_E))\right) dx_C \\
&= \sum_{k=1}^{K} \int_{B_k} \left(\mu_E + \frac{n_k}{2(n_k+m_k)}(x-\delta)\right) h(k, n_k x) \phi_{\frac{2}{n_k}}(x-\delta) dx \\
&= \mu_E + E\left(\frac{n_M}{2(n_M+m_M)}(\bar{X}-\delta))\right),
\end{aligned}
$$

where

$$
\begin{aligned}
a_k &= -\frac{n_k(n_k+m_k)}{(n_k+l_k)(n_k+2m_k)}, \\
v_{k3} &= \frac{n_k+2m_k}{2(n_k+m_k)^2}, \\
v_{k4} &= \frac{n_k(m_k+l_k)+2m_k l_k}{(n_k+2m_k)(n_k+l_k)^2}.
\end{aligned}
$$

References

1. Emerson, S. S. and Fleming, T. R. (1990). Parameter estimation following sequential hypothesis testing. *Biometrika* **77**, 875-92.
2. Hall, W. J. and Liu, A. (2003). Sequential tests and estimators after overrunning based on maximum likelihood orderring. *Biometrica* **89**, 699-707.
3. Jennison, C. and Turnbull, B. W. (2001). *Group Sequential Methods with Applications to Clinical Trials.* Boca Raton, FL: Chapman & Hall/CRC.
4. Lan, K. K. G. and Demets, D. L. (1983). Discrete sequential boundaries for clinical trials. *Biometrika* **70**, 659-663.
5. Liu, A. and Hall, W. J. (2001). Unbiased estimation of secondary parameters following a sequential test. *Biometika* **88**, 895-900.
6. O'Brien, P. C. and Fleming, T. R. (1979). A multiple testing procedure for clinical trials. *Biometrics* **35**, 549-556.
7. Pocock, S. J. (1977). Group sequential methods in the design and analysis of clinical trials. *Biometrika* **64**, 191-199.
8. Sooriyarachchi, M. R., Whitehead, J., Matsushita, T., Bolland, K. and Whitehead, A. (2003). Incorporating data received after a sequential trial has

stopped into the final analysis:implementation and acomparison of methods. *Biometrics*, **59**, 701-709.

9. Whitehead, J. (1986a). On the bias of maximum likelihood estimation following a sequential test. *Biometrika* **73**, 573-581.
10. Whitehead, J. (1986b). Supplementary analysis at the conclusion of a sequential clinical trial. *Biometrics* **42**, 461-471.
11. Whitehead, J. (1997). *The Design and Analysis of Sequential Clinical Trials, Revised 2nd ed.* New York: Wiley.

Probability Theory

CENTRAL LIMIT THEOREMS FOR INSTANTANEOUS FILTERS OF LINEAR RANDOM FIELDS ON Z^2

TSUNG-LIN CHENG*

National Changhua University of Education, Taiwan, R.O.C.

HWAI-CHUNG HO

Academia Sinica, Taiwan, R.O.C.

This note considers the stationary sequence generated by applying an instantaneous filter to a linear random field in Z^2. The class of filters under consideration includes polynomials and indicator functions. Instead of imposing certain mixing conditions on the random fields, it is assumed that the weights of the innovations satisfy a summability property. By building a martingale decomposition based on a suitable filtration, asymptotic normality is proved for the partial sums of the stationary sequence.

Some key words: Linear random fields; Central limit theorem; Martingale decomposition; ℓ-approximation

1. Introduction

The theory of statistical inference procedure on random fields has developed significantly over the last two decades (see, e.g., Doukhan (1994) for a review of early development). One of the central issues is to establish the central limit theorem for the cumulative distribution function of a sample generated from the underlying random field. The traditional approach usually requires that the random field either satisfy a certain type of mixing condition or have some structures that eventually imply the condition. Some recent results along this direction can be found in Lahiri (1999), Lahiri et al. (1999), Zhu et al. (2002), among others. The first drawback of this approach is that mixing conditions are in general hard to verify. Secondly the structure that ensures a mixing condition typically translates to stringent

*He is indebted to Prof. Y.S. Chow for his guidance and encouragement.

restrictions such as the normality assumption together with a continuous and positive spectral density (Rosenblatt (1972)) or some regularity conditions on the weights and the probability density function of the innovations if the random field is linear (Bradley (1986) and Pham and Tran (1985)). The present note is focused on a class of random fields from which the similar central limit theorem can be derived without relying on the assumption of mixing conditions. Specifically, the model under consideration is the stationary linear random field on Z^2 defined by $X_{(s,t)} = \sum_{(i,j)\in Z^2} a_{i,j}\varepsilon_{s-i,t-j}$, where $(s,t) \in Z^2$, $\sum_{(i,j)\in Z^2} a_{i,j}^2 < \infty$ and the innovations $\varepsilon_{s,t}$, $(s,t) \in Z^2$, are mean-zero i.i.d. random variables having at least finite second moments. We consider linear random fields because it is one of the most popular and useful models in practice. We choose Z^2 instead of the more general case Z^d, $d \geq 2$, merely for the ease of presentation. The central limit theorems stated later on can be extended to the case of Z^d, $d \geq 2$, with only straightforward but tedious modifications. As described in the following the framework we consider not only includes the empirical distribution as a special case but also covers other common statistics such as moment estimators. Let $K(\cdot)$ be a Borel measurable function and A is a non-empty subset of Z^2. We aim to show that under a certain summability condition on the weights $\{a_{i,j}\}$ of $X_{(s,t)}$, the normalized partial sum

$$\frac{1}{|A_k|^{1/2}} S_A = \frac{1}{|A|^{1/2}} \sum_{(s,t)\in A} (K(X_{(s,t)}) - EK(X_{(s,t)})),$$

for an increasing sequence of squares A_k is asymptotically normal as the cardinality of A_k goes to infinity. To achieve the goal, the two technical issues we need to deal with are: the $X_{(s,t)}$ are non-causal and the function K is highly non-linear. The latter was resolved in Ho and Hsing (1997) but only under the setting that the linear process of interest is causal. Another factor that needs to be taken into account is whether the underlying random field is short or long range dependent. Techniques work for one case may not for the other especially in the presence of non-causality. The distinction can be seen by comparing Doukhan and Surgalis (1998) with Cheng and Ho (2005). For Z^d with $d = 1$, Giraitis and Surgailis (1999) proved a functional central limit theorem for the empirical process of a non-causal moving average with long-range dependence, i.e.,the weights of innovations are regularly varying and summed to infinity. Doukhan et al. (2002) later extended the result to a natural non-causal model of random fields on $Z^d, d \geq 2$, that is also long range dependent. In both papers, because of long-range dependence, the normalizing constant is $N^{d-\delta}$ with $\delta < d/2$ instead of $N^{d/2}$ which appears in the standard central limit theorem. So far it is still unanswered whether the method employed by Giratis and Surgalis (1999) or Doukhan et al. (2002) can be applied to the framework discussed in this paper where the non-linear functionals are more general than indicator functions and the linear random field is of short-range dependence for which the standard central limit theorem is expected to hold. Using an approach different from those by Giraitis and Surgalis (1999) and Doukhan et al. (2002), Cheng and Ho (2005) recently obtain the asymptotic normality for $n^{-1/2}S_n = n^{1/2}\sum_{j=1}^{n}(K(X_j) - EK(X_j))$, where K belongs to a fairly general class of functions which includes indicator functions and polynomials, and $X_t = \sum_{j=-\infty}^{\infty} a_j\varepsilon_{t-j}, \quad t \geq 1$, is a non-causal linear process with the weights $\{a_i\}$ satisfying that $\sum_{j=-\infty}^{\infty} |a_j|^{1/2} < \infty$. In this paper, we modify the method by Cheng and Ho (2005) to extend their results to the linear random field on Z^2, which is a more natural and useful non-causal model in applications than the non-causal linear process considered in Cheng and Ho (2005).

Our main results consist of two central limit theorems which are stated in Theorem 1 and Corollary 1 in the next section. Their proofs are given in Section 3.

2. Main Results

Throughout this paper the $\varepsilon_{i,j}, (i,j) \in Z^2$, are i.i.d. random variables having zero mean and finite second moments, and the weights of the linear random field $X_{(s,t)} = \sum_{(i,j)\in Z^2} a_{i,j}\varepsilon_{s-i,t-j}$ of interest is assumed to satisfy

$$\sum_{(i,j)\in Z^2} |a_{i,j}|^{1/2} < \infty. \tag{1}$$

Let $K(\cdot)$ be a Borel measurable function satisfying $EK^2(X_{(s,t)}) < \infty$. For each subset A of Z^2 let $\mathcal{F}_A$ be the σ-algebra generated by $\{\varepsilon_{s,t}, (s,t) \in A\}$, and define $K_A(x) = \int K(x+u)dF_A(u)$ whenever the integral exists. Set

$$\varepsilon_{0,0} \overset{d}{\sim} G\ , \qquad a_{i,j}\varepsilon_{0,0} \overset{d}{\sim} G_{i,j},$$

$$X_{(s,t)} = \sum_{(i,j)\in Z^2} a_{i,j}\varepsilon_{s-i,t-j} \overset{d}{\sim} F\ , \qquad X_{(s,t),A} = \sum_{(i,j)\in A} a_{(i,j)}\varepsilon_{(s-i,t-j)} \overset{d}{\sim} F_A,$$

$$\tilde{X}_{(s,t),A} = \sum_{(i,j)\notin A} a_{i,j}\varepsilon_{s-i,t-j} \overset{d}{\sim} F_{A^c},$$

where the symbol $\overset{d}{\sim}$ signifies the equality in distribution. We adopt the convention that for the empty set $\emptyset$, $X_{(s,t),\emptyset} \equiv 0$ and $F_\emptyset(\cdot)$ denotes the Dirac point mass measure at zero. Hence, $K_\emptyset(x) = K(x)$ and

$$K_{A_1 \cup A_2}(x) = \int K(x+u+v)dF_{A_1}(u)dF_{A_2}(v)$$

for disjoint subsets A_1 and A_2 of Z^2.

In order to study the asymptotic behavior of the partial sum S_A, some regularity conditions are needed. Let H be a Borel measurable function from $\mathcal{R}$ to $\mathcal{R}$. The triplet $\{H, X_{(0,0)}, \varepsilon_{0,0}\}$ is said to be in $\mathcal{L}$ if it satisfies

(C)$EH(X_{(0,0)})^2 < \infty$, the third derivative $H^{(3)}_{\{(0,0)\}}(x)$ of $H_{\{(0,0)\}}((x))$ is continuous, and either of the following two conditions holds:

(i) $H^{(i)}_{\{(0,0)\}}(x)$ is bounded, $0 \le i \le 3$, and $E\varepsilon^8_{0,0} < \infty$.

(ii) There is a polynomial function $U(x)$ of degree M such that $|H^{(i)}_{\{(0,0)\}}(x)| \le |U(x)|$ for all $0 \le i \le 3$ and $x \in \mathcal{R}$, and $E\varepsilon_{0,0}^{\max\{8,4M\}} < \infty$

Condition (C), which describes a concrete class of functionals K, is not to its full generality as given in Ho and Hsing (1997), yet it covers many interesting cases considered in the literature such as the sample moment and the empirical distribution. We choose to use (C) merely for presentation simplicity since our main purpose is to introduce a new method for analyzing the asymptotic behavior of the statistics of generated by linear random fields rather than seeking a class of functionals K as general as possible. Note in part (i) of (C) that indicator functions are included if the distribution function G of $\varepsilon_{0,0}$ is sufficiently smooth. Condition (C) assures the following useful property needed later for proving the main theorem: for every nonempty $A \in Z^2$ and for each $(i,j) \in A^c$, $K^{(t)}_{A\cup\{(i,j)\}}(x)$ is continuous and satisfies

$$K^{(t)}_{A\cup\{(i,j)\}}(x) = EK^{(t)}_A(x + a_{i,j}\varepsilon_{0,0}) = \int K^{(t)}_A(x+u)dG_{i,j}(u) \qquad (2)$$

for all $0 \le t \le 3$. Equation (2) can be shown by a similar argument used in proving Lemma 2.1 of Ho and Hsing(1997). In condition (C) the two moment conditions, $EH(X_{(0,0)})^2 < \infty$ and $E\varepsilon_{0,0} < \infty$ may not be mutually exclusive when K is a polynomial function as covered in case (ii) of (C).

Let $\Gamma_n, n \ge 0$, denote loops in Z^2 defined by

$$\Gamma_0 = \{(0,0)\} \quad \text{and } \Gamma_n = \{(i,j) \in Z^2 : max(|i|,|j|) = n\}, n \ge 1.$$

Geometrically speaking, for each $n \geq 1$, Γ_n consists of the lattice points on the edges of a square centered at the origin with length $2n$. For example,

$$\Gamma_1 = \{(-1,1),(0,-1),(1,-1),(1,0),(1,1),(0,1),(-1,1),(-1,0)\}.$$

Set $U_n = \cup_{j=0}^{n}\Gamma_j$, $n = 0,1,2,\ldots$. For each $A \subset Z^2$ let $|A|$ denote the number of elements of A. Then $|\Gamma_0| = 1$ and $|\Gamma_n| = 8n$ for all $n \geq 1$. Set $a_{i,j}^{\ell}$ to be $a_{i,j}$ if $(i,j) \in U_\ell$ and 0 otherwise. Define $X_{(s,t),A}^{\ell} = \sum_{(i,j)\in A} a_{i,j}^{\ell}\varepsilon_{s-i,t-j}$, and for simplicity, $X_{(s,t)}^{\ell} = \sum_{(i,j)\in Z^2} a_{i,j}^{\ell}\varepsilon_{s-i,t-j}$ if $A = Z^2$. For any positive integer ℓ and any subset $A \subset Z^2$, denote by

$$S_A^{\ell} = \sum_{(s,t)\in A} (K(X_{(s,t)}^{\ell}) - EK(X_{(s,t)}^{\ell}))$$

the truncated version of the partial sum

$$S_A = \sum_{(s,t)\in A} (K(X_{(s,t)}) - EK(X_{(s,t)})). \tag{3}$$

Because of (1), it is clear that $\sum_{(i,j)\in Z^2\setminus U_\ell} |a_{i,j}| \to 0$, as $\ell \to \infty$, which implies

$$E[(X_{(i,j)} - X_{(i,j)}^{\ell})^2] \to 0 \text{ as } \ell \to \infty, \qquad \text{for each } (i,j) \in Z^2.$$

Since we are dealing with the asymptotics in Z^2, it is necessary to specify how the index set A, over which the partial sum in (3) is taken, grows when $|A|$ increases to infinity. Toward this we define a complete order relation symbolized by "$\prec$" among the elements in Z^2.

Definition. For any two distinct latice points (i_1,j_1) and (i_2,j_2) in Z^2, denote $(i_1,j_1) \prec (i_2,j_2)$ if one of the following two conditions is fulfilled:

(i) There exist two positive integers $n_1 < n_2$ such that $(i_1,j_1) \in \Gamma_{n_1}$, and $(i_2,j_2) \in \Gamma_{n_2}$.

(ii) Both of (i_1,j_1) and (i_2,j_2) are located at the same Γ_n for some $n \geq 0$ and, if traveling in counterclockwise direction along the contour Γ_n initiated at $(-n,-n)$, one reaches (i_1,j_1) earlier than (i_2,j_2).

For each pair $(i,j) \in Z^2$, in the partial order "$\prec$", we denote its predecessor by $(i,j)^-$ and its successor by $(i,j)^+$. For example, $(2,1)^- = (2,0)$ and $(2,1)^+ = (2,2)$.

In this note we focus on the case where the set A specifying the indices of the summands in (3) is a square. Although this is the simplest case, the approach we employ can also be used to deal with index sets which are in the form of general rectangles. This extension, however, requires only straightforward modifications and introduces no new ideas, and is therefore omitted. For a sequence of non-empty subsets in Z^2, $\{A_k, k \geq 1\}$, we say $\{A_k, k \geq 1\} \in \mathcal{H}$, if $A_0 = (0,0)$, A_k's are all squares, $A_k \subset A_{k+1}$ and $|A_k| \to \infty$ as $k \to \infty$.

Theorem 1. *Assume (1) and that for any non-empty* $A \in Z^2$, $\{K_A, X_{(0,0)}, \varepsilon_{0,0}\} \in \mathcal{L}$. *In addition, assume that for every* $x \in \Re$,

$$\sup_{A \subset Z^2} EK^2(x + X_{(0,0),A}) < \infty, \tag{4}$$

and that for any two independent random variables X *and* Y *with* $E(K^2(X) + K^2(Y) + K^2(X+Y)) \leq M < \infty$,

$$E[K(X+Y) - K(X)]^2 \leq C(EY^2)^\alpha \quad \text{for some } \alpha \geq 1/2, \tag{5}$$

where the constant C *only depends on the bound* M. *Let* $\{A_k, k \geq 1\} \in \mathcal{H}$. *Then, as* $k \to \infty$,

$$\frac{1}{|A_k|^{\frac{1}{2}}} S_{A_k} \xrightarrow{d} N(0, \sigma^2),$$

where

$$\sigma^2 = \lim_{k \to \infty} \frac{1}{|A_k|} Var(S_{A_k}) = \lim_{\ell \to \infty} \lim_{k \to \infty} \frac{1}{|A_k|} Var(S^{\ell}_{A_k}).$$

If K is an indicator function and satisfies (C), then the following central limit theorem for the empirical distribution function holds provided that the distribution function $G(\cdot)$ of $\varepsilon_{0,0}$ satisfies certain regularity conditions.

Corollary 1. *Assume (1),* $\{K, X_{(0,0)}, \varepsilon_{0,0}\} \in \mathcal{L}$, *and that the distribution function* $G(x)$ *of* $\varepsilon_{0,0}$ *has bounded first and integrable second derivatives. Then for each* x *in the support of* $X_{(0,0)}$ *and any sequence* $\{A_k\}$ *of squares that is in* $\mathcal{H}$,

$$\sqrt{|A_N|}(F_{A_N}(x) - F(x)) \xrightarrow{d} W(x) \quad \text{as } N \to \infty, \tag{6}$$

where $F_{A_N}(x) = |A_N|^{-1} \sum_{(i,j) \in A_N} I(X_{(i,j)} \leq x)$, *and* $W(x)$ *is Gaussian random variable with zero mean and covariance function*

$$Var(W(x)) = \sum_{(i,j) \in Z^2} Cov(I(X_{(0,0)} \leq x), I(X_{(i,j)} \leq x)). \tag{7}$$

3. Proofs

To prove the theorem, three auxiliary lemmas stated below are needed. The first two of them, Lemmas 1 and 2, can be obtained by an argument similar to that in Ho and Hsing (1997) (see also Lemma 2 of Cheng and Ho (2005)); and Lemma 3 was shown in Lemma 4 of Cheng and Ho (2005).

Lemma 1. *Suppose* $\{K, X_{(0,0)}, \varepsilon_{0,0}\} \in \mathcal{L}$. *If* $(\hat{\imath}, \hat{\jmath}) \in Z^2$ *and* $\hat{\ell}$ *is a positive integer, then for every finite subset* A *of* Z^2

$$E\Big\{[K_A(\tilde{X}_{(s,t),A}) - K_{A\cup\{(\hat{\imath},\hat{\jmath})\}}(\tilde{X}_{(s,t),A\cup\{(\hat{\imath},\hat{\jmath})\}})] - [K_A(\tilde{X}^{\hat{\ell}}_{(s,t),A}) - K_{A\cup\{(\hat{\imath},\hat{\jmath})\}}(\tilde{X}^{\hat{\ell}}_{(s,t),A\cup\{(\hat{\imath},\hat{\jmath})\}})]\Big\}^2 \leq \begin{cases} C(\sum_{(m,n)\in Z^2\setminus U_{\hat{\ell}}} a_{m,n})a^2_{\hat{\imath},\hat{\jmath}}, & \text{if } a^* \in U_{\hat{\ell}}, \\ Ca^2_{\hat{\imath},\hat{\jmath}}, & \text{otherwise}, \end{cases} \quad (8)$$

where $C > 0$ *is a constant independent of* $A, (s,t), (\hat{\imath}, \hat{\jmath})$ *and* $\hat{\ell}$, *and* a^* *stands for the maximal element of set* A *with respect to the order relation* $\prec$.

Lemma 2. *Suppose that the conditions assumed in Theorem 1 hold. Then*

$$\sup_{(s,t)\in Z^2} E\{[K(X_{s,t}) - K(X^{\ell}_{(s,t)})]^2\} = o(1) \text{ as } \ell \to \infty. \quad (9)$$

Lemma 3. *Let* X *and* Y *be two independent, square integrable random variables. If* $Y \overset{d}{\sim} F_Y(y), X \overset{d}{\sim} F_X(x)$ *with* $|F_X(u) - F_X(v)| \leq C|u-v|$ *for all* $u, v \in \Re$ *and some constant* $0 < C < \infty$ *which is independent of* u, v. *Then, for any* $z \in \Re$,

$$E\Big(|I(X+Y \leq z) - I(X \leq z)|^2\Big) \leq C(EY^2)^{\frac{1}{2}}.$$

Proof of Theorem 1. The proof has two important ingredients: an ℓ-approximation technique and an orthogonal expansion for $K(X_{(s,t)})$. The main step of the latter is to build a martingale decomposition on a suitable filtration. Since for fixed ℓ, $S^{\ell}_{A_k}/|A_k|$ is asymptotically normal, it suffices to show that

$$\lim_{\ell\to\infty} \limsup_{|A|_k\to\infty} \frac{1}{|A|_k} Var(S_A - S^{\ell}_A) = 0. \quad (10)$$

Because of the stationarity of $X_{s,t}$, if A_k consists of all lattice points in a rectangle, we may assume without loss of generality that $A_k = \{(i,j) : i = 1,2,\cdots,m_k; j = 1,2,\cdots,n_k\}$ with $m_k n_k \to \infty$. Then

$$S_{A_k} - S^{\ell}_{A_k} = \sum_{s=1}^{m_k}\sum_{t=1}^{n_k} \Lambda^{\ell}_{s,t},$$

where $\Lambda^{\ell}_{s,t} = [K(X_{(s,t)}) - EK(X_{(s,t)})] - [K(X^{\ell}_{(s,t)}) - EK(X^{\ell}_{(s,t)})]$.

Note that

$$Var(S_{A_k} - S^{\ell}_{A_k}) \leq \sum_{s=1}^{m_k}\sum_{t=1}^{n_k} E(\Lambda^{\ell}_{s,t})^2 + 2\sum_{s=1}^{m_k}\sum_{t=1}^{n_k}\sum_{p=s-m_k}^{m_k-s}\sum_{q=0}^{n_k-t} \left|Cov(\Lambda^{\ell}_{s,t}, \Lambda^{\ell}_{s+p,t+q})\right|.$$

By (9),

$$\lim_{\ell\to\infty} \frac{1}{|A|_k}\sum_{s=1}^{m_k}\sum_{t=1}^{n_k} E(\Lambda^{\ell}_{s,t})^2 = 0.$$

To prove (10), it remains to show that

$$\frac{1}{|A|_k}\sum_{s=1}^{m_k}\sum_{t=1}^{n_k}\sum_{p=s-m_k}^{m_k-s}\sum_{q=0}^{n_k-t} |Cov(\Lambda^{\ell}_{s,t}, \Lambda^{\ell}_{s+p,t+q})| = o(1), \qquad (11)$$

as k and $\ell \to \infty$.

Set $\Phi_{(i,j)} = \{(u,v) \in Z^2 : (u,v) \prec (i,j)\}$ for each $(i,j) \in Z^2$, and, for each fixed $(s,t) \in Z^2$, we define $\Phi^{(s,t)}_{(i,j)} = \{(u+s, v+t) : (u,v) \in \Phi_{(i,j)}\}$. Note that $\Phi^{(s,t)}_{(i,j)}$ is nothing more than the set $\Phi_{(i,j)}$ shifted by (s,t). Let $f(i,j)$ be any function on Z^2. The notation $\sum_{(i,j)\in\overrightarrow{A}} f(i,j)$ denotes the sum of $f(i,j)'s$ over the lattice points of $A \subset Z^2$ following the partial order " $\prec$ " defined in the previous section. For example,

$$\begin{aligned}\sum_{(i,j)\in\overrightarrow{\Phi}_{(-2,-1)}} f(i,j) &= f(0,0) + f(-1,-1) + f(0,-1)\\ &\quad + f(1,-1) + f(1,0) + f(1,1) + f(0,1)\\ &\quad + f(-1,1) + f(-1,0) + f(-2,-2) + f(-1,-2)\\ &\quad + \cdots + f(-2,0),\end{aligned}$$

and

$$\sum_{(i,j)\in\vec{Z^2}} f(i,j) = \sum_{n=0}^{\infty} \sum_{(i,j)\in\vec{\Gamma}_n} f(i,j),$$

if the sum on both sides are well-defined. A technically crucial step is the observation that $\Lambda^{\ell}_{s,t}$ can be expressed as the following telescoping identity.

$$\begin{aligned}
\Lambda^{\ell}_{s,t} &= [K(X_{(s,t)}) - EK(X_{(s,t)})] - [K(X^{\ell}_{(s,t)}) - EK(X^{\ell}_{(s,t)})] \\
&= \sum_{(i,j)\in\vec{Z^2}} \Big\{[E(K(X_{(s,t)})|\mathcal{F}_{Z^2\setminus\Phi_{(i,j)}}) - E(K(X_{(s,t)})|\mathcal{F}_{Z^2\setminus\Phi_{(i,j)+}})] \\
&- [E(K(X^{\ell}_{(s,t)})|\mathcal{F}_{Z^2\setminus\Phi_{(i,j)}}) - E(K(X^{\ell}_{(s,t)})|\mathcal{F}_{Z^2\setminus\Phi_{(i,j)+}})]\Big\} \\
&= \sum_{(i,j)\in\vec{Z^2}} \Big\{[K_{\Phi^{(s,t)}_{(i,j)}}(\tilde{X}_{(s,t),\Phi^{(s,t)}_{(i,j)}}) - K_{\Phi^{(s,t)}_{(i,j)+}}(\tilde{X}_{(s,t),\Phi^{(s,t)}_{(i,j)+}})] \\
&- [K_{\Phi^{(s,t)}_{(i,j)}}(\tilde{X}^{\ell}_{(s,t),\Phi^{(s,t)}_{(i,j)}}) - K_{\Phi^{(s,t)}_{(i,j)+}}(\tilde{X}^{\ell}_{(s,t),\Phi^{(s,t)}_{(i,j)+}})]\Big\} \\
&\equiv \sum_{(i,j)\in\vec{Z^2}} \Psi^{\ell}_{s,t,i,j}.
\end{aligned} \tag{12}$$

By the argument of iterative conditioning, it is not difficult to show that for different pairs $(s,t) \in A_k$ and $(s',t') \in A_k$, $\Psi^{\ell}_{s,t,i,j}$ and $\Psi^{\ell}_{s',t',i',j'}$ are orthogonal if $(i,j) \neq (i',j')$. Hence

$$Cov(\Lambda^{\ell}_{s,t}, \Lambda^{\ell}_{s',t'}) = \sum_{(i,j)\in\vec{Z^2}} Cov(\Psi^{\ell}_{s,t,i,j}, \Psi^{\ell}_{s',t',i,j}), \tag{13}$$

and, by stationarity,

$$Cov(\Lambda^{\ell}_{s,t}, \Lambda^{\ell}_{s',t'}) = \sum_{(i,j)\in\vec{Z^2}} Cov(\Psi^{\ell}_{0,0,i,j}, \Psi^{\ell}_{s'-s,t'-t,i,j}).$$

We shall show (11) for two separate cases that the function K is smooth and non-smooth. First, K is assumed smooth so that triplet $\{K, X_{(0,0)}, \varepsilon_{0,0})\} \in$

$\mathcal{L}$. Using (13), we have

$$|\sum_{s=1}^{m_k}\sum_{t=1}^{n_k}\sum_{p=1}^{m_k-s}\sum_{q=1}^{n_k-t} Cov(\Lambda^{\ell}_{s,t},\Lambda^{\ell}_{s+p,t+q})|$$

$$\leq \sum_{s=1}^{m_k}\sum_{t=1}^{n_k} \sum_{\substack{(p,q)\in U_\ell \\ s-m_k\le p\le m_k-s \\ 0\le q\le n_k-q}} \sum_{\substack{(s+i,t+j)\in U_\ell \text{ or} \\ (s+p+i,t+j+q)\in U_\ell}} |Cov(\Psi^{\ell}_{s,t,i,j},\Psi^{\ell}_{s+p,t+q,i,j})|$$

$$+\sum_{s=1}^{m_k}\sum_{t=1}^{n_k} \sum_{\substack{(p,q)\in U_\ell \\ s-m_k\le p\le m_k-s \\ 0\le q\le n_k-q}} \sum_{\substack{(s+i,t+j)\notin U_\ell \text{ and} \\ (s+p+i,t+j+q)\notin U_\ell}} |Cov(\Psi^{\ell}_{s,t,i,j},\Psi^{\ell}_{s+p,t+q,i,j})|$$

$$+\sum_{s=1}^{m_k}\sum_{t=1}^{n_k} \sum_{\substack{(p,q)\notin U_\ell \\ s-m_k\le p\le m_k-s \\ 0\le q\le n_k-q}} \sum_{\substack{(s+i,t+j)\in U_\ell \text{ or} \\ (s+p+i,t+j+q)\in U_\ell}} |Cov(\Psi^{\ell}_{s,t,i,j},\Psi^{\ell}_{s+p,t+q,i,j})|$$

$$+\sum_{s=1}^{m_k}\sum_{t=1}^{n_k} \sum_{\substack{(p,q)\notin U_\ell \\ s-m_k\le p\le m_k-s \\ 0\le q\le n_k-q}} \sum_{\substack{(s+i,t+j)\notin U_\ell \text{ and} \\ (s+p+i,t+j+q)\notin U_\ell}} |Cov(\Psi^{\ell}_{s,t,i,j},\Psi^{\ell}_{s+p,t+q,i,j})|$$

$$\equiv I_1+I_2+I_3+I_4.$$

By Cauchy-Schwartz inequality and the upper bounds obtained in (8), we have

$$I_1+I_2+I_3+I_4\le C|A|_k\Big[\big(\sum_{(i,j)\notin U_\ell} a^2_{i,j}\big)^{1/2}+\sum_{(i,j)\notin U_\ell}|a_{i,j}|\Big].$$

Therefore, (11) holds and so does (10).We now show (11) for the second case that K is not smooth, but $\{K_A, X_{(0,0)}, \varepsilon_{0,0}\}$ is in $\mathcal{L}$. By stationarity, $Cov(\Lambda^{\ell}_{s,t},\Lambda^{\ell}_{s+p,t+q}) = Cov(\Lambda^{\ell}_{0,0},\Lambda^{\ell}_{p,q})$, for all $(s,t)\in Z^2$ and $(p,q)\in Z^2$

with $s - m_k \le p \le m_k - s$ and $0 \le q \le n_k - t$. For fixed p and q, we can rewrite $\Lambda^{\ell}_{0,0}$ and $\Lambda^{\ell}_{p,q}$ as

$$\Lambda^{\ell}_{0,0} = I^{\ell}_1 + I^{\ell}_2 + I^{\ell}_3 \text{ and } \Lambda^{\ell}_{p,q} = J^{\ell}_1 + J^{\ell}_2 + J^{\ell}_3$$

where

$$\begin{aligned}
I^{\ell}_1 &= \Big[K(X_{(0,0)}) - K_{\{(0,0)\}}(\tilde{X}_{(0,0),\{(0,0)\}})\Big] \\
&\quad -\Big[K(X^{\ell}_{(0,0)}) - K_{\{(0,0)\}}(\tilde{X}^{\ell}_{(0,0),\{(0,0)\}})\Big], \\
I^{\ell}_2 &= \Big[K_{\{(0,0)\}}(\tilde{X}_{(0,0),\{(0,0)\}}) - K_{\{(0,0),(-p,-q)\}}(\tilde{X}_{(0,0),\{(0,0),(-p,-q)\}})\Big] \\
&\quad -\Big[K_{\{(0,0)\}}(\tilde{X}_{(0,0),\{(0,0)\}}) - K_{\{(0,0),(-p,-q)\}}(\tilde{X}_{(0,0),\{(0,0),(-p,-q)\}})\Big], \\
I^{\ell}_3 &= \Lambda^{\ell}_{0,0} - I^{\ell}_1 - I^{\ell}_2,
\end{aligned}$$

and

$$\begin{aligned}
J^{\ell}_1 &= \Big[K(X_{(p,q)}) - K_{\{(p,q)\}}(\tilde{X}_{(p,q),\{(p,q)\}})\Big] \\
&\quad -\Big[K(X^{\ell}_{\{(p,q)\}}) - K_{\{(p,q)\}}(\tilde{X}^{\ell}_{(p,q),\{(p,q)\}})\Big], \\
J^{\ell}_2 &= \Big[K_{\{(p,q)\}}(\tilde{X}_{(p,q),\{(p,q)\}}) - K_{\{(p,q),(0,0)\}}(\tilde{X}_{(p,q),\{(p,q),(0,0)\}})\Big] \\
&\quad -\Big[K_{\{(p,q)\}}(\tilde{X}^{\ell}_{(p,q),\{(p,q)\}}) - K_{\{(p,q),(0,0)\}}(\tilde{X}^{\ell}_{(p,q),\{(p,q),(0,0)\}})\Big], \\
J^{\ell}_3 &= \Lambda^{\ell}_{p,q} - J^{\ell}_1 - J^{\ell}_2.
\end{aligned}$$

Again by the argument of iterative conditioning, we have

$$Cov(I^{\ell}_i, J^{\ell}_j) = 0, \quad \text{for all } i, j, 1 \le i \ne j \le 3.$$

Therefore,

$$|Cov(\Lambda^{\ell}_{0,0}, \Lambda^{\ell}_{p,q})| \le \sum_{i-1}^{3} |Cov(I^{\ell}_i, J^{\ell}_i)|.$$

We first find upper bounds for $|Cov(I^{\ell}_i, J^{\ell}_i)|, i = 1, 2$, which, by Cauchy-Schwartz inequality, is bounded by $\Big(E(I^{\ell}_i)^2\Big)^{1/2}\Big(E(J^{\ell}_i)^2\Big)^{1/2}$. Since

$$I^{\ell}_1 = [K(X_{(0,0)}) - K(X^{\ell}_{(0,0)})] - E[K(X_{(0,0)}) - K(X^{\ell}_{(0,0)})|\mathcal{F}_{Z^2\setminus\Phi_{0,0}}],$$

we have, by Jensen's inequality and (4),

$$\begin{aligned} E(I_1^\ell)^2 &\le C\Big\{E[K(X_{(0,0)}) - K(X_{(0,0)}^\ell)]^2 \\ &\quad + E\Big(E[(K(X_{(0,0)}) - K(X_{(0,0)}^\ell))^2|\mathcal{F}_{Z^2\setminus\Phi_{0,0}}]\Big)^2\Big\} \\ &\le C\bigg(\sum_{(i,j)\notin U_\ell} a_{i,j}^2\bigg)^\alpha. \end{aligned} \tag{14}$$

By Lemma 1, we can bound $E(I_2^\ell)$ by

$$E(I_2^\ell)^2 \le \begin{cases} C\Big(\sum_{(i,j)\notin U_\ell} a_{i,j}^2\Big)a_{-p,-q}^2\,, & \text{if } (-p,-q)\in U_\ell \\ Ca_{(-p,-q)}^2 \,, & \text{otherwise} \end{cases}. \tag{15}$$

Rewrite J_1^ℓ and J_2^ℓ as

$$\begin{aligned} J_1^\ell &= \int [K(X_{(p,q)}) - K(\tilde{X}_{(p,q),\{(p,q)\}})]dG(x) \\ &\quad + \int [K(\tilde{X}_{(p,q),\{(p,q)\}}) - K(a_{p,q}x + \tilde{X}_{(p,q),\{(p,q)\}})]dG(x) \\ &\quad - \int [K(X_{(p,q)}^\ell) - K(\tilde{X}_{(p,q),\{(p,q)\}}^\ell)]dG(x) \\ &\quad - \int [K(\tilde{X}_{(p,q),\{(p,q)\}}^\ell) - K(a_{p,q}x + \tilde{X}_{(p,q),\{(p,q)\}}^\ell)]dG(x), \end{aligned}$$

$$\begin{aligned} J_2^\ell &= \int [K(a_{p,q}x + \tilde{X}_{(p,q),\{(p,q)\}}) - K(a_{p,q}x + \tilde{X}_{(p,q),\{(p,q)\}}^\ell)]dG(x) \\ &\quad + \int [K(a_{0,0}x + a_{p,q}y + \tilde{X}_{(p,q),\{(p,q)\}}) \\ &\qquad\qquad - K(a_{0,0}x + a_{p,q}y + \tilde{X}_{(p,q),\{(p,q)\}}^\ell)]dG(x)dG(y). \end{aligned}$$

Applying Jensen's inequality and (4) again to the righthand side of the preceding two equations, we have

$$E(J_1^\ell)^2 \le Ca_{p,q}^{2\alpha} \quad \text{and } E(J_2^\ell)^2 \le C\Big(\sum_{(i,j)\notin U_\ell} a_{i,j}^2\Big)^\alpha. \tag{16}$$

Combining (14)-(16) yields

$$\sum_{i=1}^{2} |Cov(I_i^\ell, J_i^\ell)| \leq C\Bigg[\Big(\sum_{(i,j)\notin U_\ell} a_{i,j}^2 \Big)^{\alpha/2} |a_{p,q}|^\alpha + \begin{cases} \Big(\sum_{(i,j)\notin U_\ell} a_{i,j}^2\Big)^{1/2} \Big(\sum_{(i,j)\notin U_\ell} a_{i,j}^2\Big)^{\alpha/2} |a_{-p,-q}|\,, & \text{if } (-p,-q) \in U_\ell \\ \Big(\sum_{(i,j)\notin U_\ell} a_{i,j}^2\Big)^{\alpha/2} |a_{-p,-q}|\,, & \text{otherwise} \end{cases} \Bigg]. \tag{17}$$

To bound $|Cov(I_3^\ell, J_3^\ell)|$, we first observe that both I_3^ℓ and J_3^ℓ are formed by smooth functionals $K_{\{(0,0),(-p,-q)\}}$ and $K_{\{(0,0),(p,q)\}}$, respectively. Therefore we can use the same technique as in (12) to expand I_3^ℓ and J_3^ℓ, and obtain the following bound

$$|Cov(I_3^\ell, J_3^\ell)| \leq C\Bigg[\sum_{i=-\infty}^{\infty} \sum_{j=-\infty}^{\infty} |a_{i,j} a_{i-p,j-q}| \times \Big[\Big(\sum_{(i,j)\notin U_\ell} a_{i,j}^2 \Big)^{1/2} + \sum_{(i,j)\notin U_\ell} |a_{i,j}| \Big] \Bigg]. \tag{18}$$

From (17) and (18), it follows that

$$\begin{aligned} &\Big| \sum_{s=1}^{m_k} \sum_{t=1}^{n_k} \sum_{p=s-m_k}^{m_k-s} \sum_{q=0}^{n_k-t} Cov(\Lambda_{s,t}^\ell, \Lambda_{s+p,t+q}^\ell) \Big| \\ &\leq C|A| \Bigg\{ \sum_{(i,j)\in Z^2} \sum_{(p,q)\in Z^2} |a_{i,j} a_{i-p,j-q}| \Big[\Big(\sum_{(i,j)\notin U_\ell} a_{i,j}^2 \Big)^{1/2} + \sum_{(i,j)\notin U_\ell} |a_{i,j}| \Big] \\ &\quad + \sum_{(p,q)\in Z^2} \Big(\sum_{(i,j)\notin U_\ell} a_{i,j}^2 \Big)^{\alpha/2} \big(|a_{p,q}|^\alpha + |a_{-p,-q}| \big) \Bigg\}, \end{aligned}$$

which together with the summability condition specified in (1) and the fact that $\alpha \geq 1/2$ imply (11). This completes the proof.

PROOF OF COROLLARY 1. First of all, condition (4) holds since an indicator function is bounded. By using Lemma 3 and the regularity conditions imposed on the distribution function G of $\varepsilon_{0,0}$, it is immediate that condition (5) is satisfied with $\alpha = 1/2$. Therefore, by the same argument as employed in proving the non-smooth case of Theorem 1, the asymptotic normality of (6) and (7) follows.

References

1. Bradley, R. C. (1986). Basic properties of strong mixing conditions. In *Dependence in Probability and Statistics* (E. Eberlein and M. S. Taqqu, eds.) 162-192. Birkhäuser, Boston.
2. Cheng, T.L. and Ho, H.C. (2005) Asymptotic normality for non-linear functionals of non-causal linear processes with summable weights. *J. Theoretical Probab.* **18**, 345-358.
3. Doukhan, P. (1994). *Mixing: Properties and Examples.* Lecture Notes in Statistics. No. 85, Springer.
4. Doukhan, P. and Surgailis, D. (1998). Functional central limit theorem for the empirical process of short memory linear processes. *C.R. Acad. Sci. Paris, t. 326, Série I,* 87-92.
5. Doukhan, P., Lang, G. and Surgailis, D. (2002). Asymptotics of weighted empirical processes of linear fields with long-range dependence. *Ann. I. H. Poincaré*-PR **38**, 879-896.
6. Giraitis, L. and Surgailis, D. (1999). Central limit theorem for the empirical process of a linear sequence with long memory. *J. Statist. Plan. Infer.* **80**, 81-93.
7. Ho, H.-C. and Hsing, T. (1997). Limit theorems for functional of moving averages. *Ann. Probab.* **25**, 1636-1669.
8. Lahiri, S.N. (1999). Asymptotic distribution of the empirical spatial cumulative distribution function predictor and prediction bands based on a subsampling method. *Probability Theory and Related Fields.* **114**, 55-84.
9. Lahiri, S.N., Kaiser, M.S., Cressie, N., and Hsu, N.J. (1999). Prediction of spatial cumulative distribution functions using subsampling. *Journal of the American Statistical Association.* **94**, 86-110.
10. Pham, T. D. and Tran, T. T. (1985). Some mixing properties of time series models. *Stochastic Process. Appl.* **19** 297-303.
11. Rosenblatt, M. (1972). Central limit theorem for stationary processes. Proc. Sixth Berkeley Symp., **2**, pp. 551-561.
12. Zhu, J., Lahiri, S.N., and Cressie, N. (2002) Asymptotic inference for spatial CDFs over time. *Statistica Sinica* **12**, 843-861.

A STUDY OF INVERSES OF THINNED RENEWAL PROCESSES

WEN-JANG HUANG* and CHUEN-DOW HUANG

Department of Applied Mathematics, National University of Kaohsiung Kaohsiung, Taiwan, 811, R.O.C.

We study properties of thinning and Markow chain thinning of renewal processes. Among others, for some special renewal processes we investigate whether these processes can be obtained through Markov chain thinning.

Some key words: Completely monotone; Gamma-c distribution; Laplace transform; Markov chain; Negative binomial distribution; Poisson process; Renewal process; Thinned point process.

1. Introduction

In this work we will investigate the problem of inverses of thinned renewal processes. Let $\mathcal{N} \equiv \{N(t),\ t \geq 0\}$ be a point process, and denote by $\mathcal{N}_p \equiv \{N_p(t),\ t \geq 0\}$ the point process obtained by retaining (in the same location) every point of $\mathcal{N}$ with a constant probability p and deleting it with probability $1-p$, independent of all other points and independent of the point process $\mathcal{N}$. $\mathcal{N}_p$ is called the p-thinning of $\mathcal{N}$, and $\mathcal{N}$ is called the p-inverse of $\mathcal{N}_p$. As mentioned in Yannaros (1988a), the p-inverse of any thinned point process is unique in distributional meaning, and it is also called the original process.

Now let $\mathcal{N}$ be a renewal process, and $\{\chi_i,\ i \geq 1\}$, independent of $\mathcal{N}$, be a sequence of binary variables which form a stationary Markov chain with marginal distribution

$$P(\chi_i = 1) = p = 1 - P(\chi_i = 0), \quad 0 < p \leq 1,$$

*Support for this research was provided in part by the National Science Council of the Republic of China, Grant No. NSC 93-2118-M-390-001 and transition probabilities

$$P(\chi_{i+1} = 1|\chi_i = 1) = \alpha_1 = 1 - P(\chi_{i+1} = 0|\chi_i = 1),$$

$$P(\chi_{i+1} = 1|\chi_i = 0) = \alpha_0 = 1 - P(\chi_{i+1} = 0|\chi_i = 0),$$

where $0 \le \alpha_0,\ \alpha_1 \le 1,\ i \ge 1$. The stationarity of the chain imposes that α_0, α_1 and p satisfy the following constraint

$$p = \alpha_1 p + \alpha_0(1-p)\ . \tag{1}$$

Then a thinned point process $\mathcal{A} \equiv \{A(t),\ t \ge 0\}$ can be obtained by retaining the i-th point of $\mathcal{N}$ if $\chi_i = 1$ and deleting it if $\chi_i = 0$. $\mathcal{A}$ is called the Markov chain thinning of $\mathcal{N}$. Generating this way, it can be proved easily that $\mathcal{A}$ is a delayed renewal process. Conversely, we are interested in knowing that given a stationary Markov chain $\{\chi_i,\ i \ge 1\}$ and a delayed renewal process $\mathcal{A}$, under what conditions there exists an ordinary renewal process $\mathcal{N}$, say the M-inverse of $\mathcal{A}$, such that $\mathcal{A}$ can be obtained through the Markov chain thinning of $\mathcal{N}$.

When the sequence $\{\chi_i,\ i \ge 1\}$ are independent and identically distributed (i.i.d.), namely $\alpha_0 = \alpha_1 = p$, the Markov chain thinning becomes p-thinning, and the above inverse problem has been studied by many authors. Yannaros (1988a) proved that the p-inverse of a thinned renewal process is unique and is also a renewal process. Next in (1988b), Yannaros characterized when an ordinary gamma renewal process has a p-inverse. He also gave necessary and sufficient conditions for a delayed gamma renewal process can be obtained through p-thinning. Later, in (1991), Yannaros extended the above model to the thinned random walks, and gave the limit behaviour of p-thinned random walks, as $p \to 0$. Yannaros (1994) investigated the class of renewal processes with Weibull lifetime distribution from the point of view of the general theory of point processes. On the other hand, Isham (1980), Chandramohan et al. (1985) discussed Markov chain thinning in various problems. That is the motivation in this note we will study properties of thinning and Markov chain thinning of renewal processes. Also we will investigate whether some special renewal processes can be obtained through Markov chain thinning.

In Section 2, we present some properties of completely monotone functions and Laplace transforms. In Section 3, we give some simple properties related to the Markov chain thinning. In Sections 4 and 5, when $\mathcal{A}$ is a delayed renewal process, we give conditions such that the M-inverse exists, with interarrival times being gamma-c or negative binomial distributed, respectively. Here a random variable X is said to be $\Gamma_c(a, \lambda)$ distributed, if it has the Laplace transform $(1+\lambda s^c)^{-a}$, for some $0 < c \le 1$, $\lambda > 0$, $a > 0$, for every $s \ge 0$, and the delayed gamma-c renewal process can be defined similarly. Gamma-c distribution was studied by Huang and Chen (1989) and (1991). Let Z be a nonnegative random variable having the distribution function H so that H has support in $[0,\infty)$, namely $H(0-) = 0$. The Laplace transform of Z or H is the function $\hat{h}$ on $[0,\infty)$ given by

$$\hat{h}(s) = E(e^{-sZ}) = \int_0^\infty e^{-sx} dH(x) \,.$$

In Section 6, we give an example of delayed renewal process which does not belong to any of the two classes discussed in Sections 4 and 5, and a stationary Markov chain such that the M-inverse exists. Finally, in Section 7, we discuss some unsolved problems of inverses of thinned renewal processes.

When $\mathcal{N}$ is a delayed renewal process, let $\{X_i,\ i \ge 1\}$ be the sequence of interarrival times with G being the distribution function of X_1 and F being the distribution function of $\{X_i,\ i \ge 2\}$, where $F(0) = G(0) = 0$. Also let $\hat{g}(s)$ and $\hat{f}(s)$ denote the Laplace transforms of G and F, respectively. Given the stationary Markov chain $\{\chi_i,\ i \ge 1\}$, it can be derived easily (see, e.g., Isham (1980)) that

$$\hat{\xi}(s) = \frac{p\hat{g}(s) + (\alpha_0 - p)\hat{f}(s)\hat{g}(s)}{1-(1-\alpha_0)\hat{f}(s)}, \tag{2}$$

$$\hat{\eta}(s) = \frac{\alpha_1\hat{f}(s) + (\alpha_0 - \alpha_1)\hat{f}^2(s)}{1-(1-\alpha_0)\hat{f}(s)}, \tag{3}$$

where $\hat{\xi}(s) = E(e^{-sY_1})$, $\hat{\eta}(s) = E(e^{-sY_2})$ and $\{Y_i,\ i \ge 1\}$ is the sequence of interarrival times of the thinned point process $\mathcal{A}$ which is a delayed renewal process also. Note that the delayed renewal process $\mathcal{N}$ is stationary if and only if

$$G(x) = \frac{\int_0^x (1-F(y))dy}{E(X_2)}, \quad x \ge 0, \tag{4}$$

or equivalently

$$\hat{g}(s) = \frac{1}{E(X_2)s}(1-\hat{f}(s)), \quad s > 0 \,. \tag{5}$$

2. Preliminaries

First a function ψ on $(0,\infty)$ is called completely monotone if it possesses derivatives $\psi^{(n)}$ of all orders and

$$(-1)^n \psi^{(n)}(s) \ge 0, \tag{6}$$

for each $n \geq 0$ and each s in $(0, \infty)$ .The following is a useful characterization of Laplace transforms of measures on $(0, \infty)$ due to Chung (1974).

Theorem 1. A function ψ on $(0, \infty)$ is the Laplace transform of a distribution function B, namely

$$\psi(s) = \int_0^\infty e^{-sx} dB(x),$$

if and only if it is completely monotone in $(0, \infty)$ with $\psi(0+) = 1$.

In the following we give two criteria of completely monotone functions which can be found in books such as Feller (1971).

Criterion 1. If ψ and ϕ are completely monotone so is their product $\psi\phi$.

Criterion 2. If ψ is completely monotone and ϕ a positive function with a completely monotone derivative then $\psi(\phi)$ is completely monotone.

The next result was proved by Kolsrud (1986), which gives a simple consequence of Bernstein functions. Here a function ϕ on $(0, \infty)$ is said to be a Bernstein function if it has a completely monotone derivative, i.e. if $(-1)^n \phi^{(n)}(s) \leq 0$, $\forall s > 0$, for $n = 1, 2, \cdots$.

Lemma 1. If ϕ is a Bernstein function with $\phi(0+) = 1$, then for any α in $(0, 1]$, $0 < p < 1$, the function $(p + (1-p)\phi^\alpha)^{-1}$ is the Laplace transform of a probability measure.

As mentioned in Section 1, let $\mathcal{N}$ be a renewal process with interarrival distribution function F, then $\mathcal{N}_p$ is also a renewal process with interarrival distribution function G. Let $\hat{f}$ and $\hat{g}$ be the Laplace transforms of F and G, respectively. Yannaros (1988c) gave the relation of $\hat{f}$ and $\hat{g}$, and proved the following lemma.

Lemma 2. The function $\hat{f} = \hat{g}/(p + (1-p)\hat{g})$ is completely monotone for every $p \in (0, 1]$, if and only if $\hat{g} = 1/(1 + \phi)$, where ϕ is a Bernstein function with $\phi(0+) = 0$.

In Lemma 2, if $\hat{g} = 1/(1 + \phi)$, then $\hat{f} = 1/(1 + p\phi)$, which gives a description of the class of completely monotone functions.

3. Some Basic Properties for Thinning via a Markov Chain

In this section we give some elementary theorems. First we characterize the class of ordinary renewal processes.

Theorem 2. Assume $\mathcal{N}$ is an ordinary renewal process which is thinned by a stationary Markov chain $\{\chi_i,\ i \geq 1\}$ as defined in Section 1. Then the thinned point process $\mathcal{A}$ is an ordinary renewal process if and only if $\{\chi_i,\ i \geq 1\}$ is an independent sequence.

Proof. From (2) and (3) we find that $\mathcal{A}$ is an ordinary renewal process if and only if

$$\alpha_1 \hat{f}(s) + (\alpha_0 - \alpha_1)\hat{f}^2(s) = p\hat{f}(s) + (\alpha_0 - p)\hat{f}^2(s), \quad s > 0, \tag{7}$$

or

$$(p - \alpha_1)\hat{f}(s)(\hat{f}(s) - 1) = 0, \quad s > 0\ . \tag{8}$$

As $0 < \hat{f}(s) < 1,\ s > 0$, (8) is equivalent to $\alpha_1 = p$.

On the other hand, the assumption that $\{\chi_i,\ i \geq 1\}$ is a stationary Markov chain gives $p = \alpha_1 p + \alpha_0(1-p)$. This together with $\alpha_1 = p$ implies $p = 1$ or $\alpha_0 = p$. Obviously, either $\alpha_0 = \alpha_1 = p$ or $p = 1$, implies $\{\chi_i,\ i \geq 1\}$ is an independent sequence.

Conversely, it is easy to see that if $\{\chi_i,\ i \geq 1\}$ forms an independent sequence then $\mathcal{A}$ is an ordinary renewal process. This completes the proof.

The "if" part of the following corollary is well known, the "only if" part can be proved by using Theorem 2 and the fact that Poisson process is also an ordinary renewal process.

Corollary 1. Let $\mathcal{N}$ be a Poisson process and $\{\chi_i,\ i \geq 1\}$ be a stationary Markov chain. Then $\mathcal{A}$ is Poisson if and only if $\{\chi_i,\ i \geq 1\}$ is an independent sequence.

The next theorem is about stationary renewal process.

Theorem 3. Let $\mathcal{N}$ be a delayed renewal process, $\{\chi_i,\ i \geq 1\}$ be a stationary Markov chain. Then $\mathcal{A}$ is a stationary renewal process if and only if $\mathcal{N}$ is a stationary renewal process.

Proof. If $\mathcal{N}$ is stationary, then $\hat{g}(s) = (E(X_2)s)^{-1}(1 - \hat{f}(s))$. Substituting this into (2), yields

$$\hat{\xi}(s) = \frac{(p + (\alpha_0 - p)\hat{f}(s))(1 - \hat{f}(s))}{E(X_2)s(1 - (1 - \alpha_0)\hat{f}(s))} . \tag{9}$$

The stationarity of $\{\chi_i,\ i \geq 1\}$ in turn implies $p = \alpha_1 p + \alpha_0(1 - p)$, or $\alpha_0 - p = (\alpha_0 - \alpha_1)p$. Hence (9) can be rewritten as, by replacing $\alpha_0 - p$ by $(\alpha_0 - \alpha_1)p$,

$$\begin{aligned} \hat{\xi}(s) &= \frac{p}{E(X_2)s}(1 - \frac{\alpha_1\hat{f}(s) + (\alpha_0 - \alpha_1)\hat{f}^2(s)}{1 - (1 - \alpha_0)\hat{f}(s)}) \\ &= \frac{p}{E(X_2)s}(1 - \hat{\eta}(s)) . \end{aligned} \tag{10}$$

This proves $\mathcal{A}$ is a stationary renewal process and the "if" part is obtained.

Conversely, assume $\mathcal{A}$ is a stationary renewal process, then

$$\hat{\xi}(s) = \frac{1}{E(Y_2)s}(1 - \hat{\eta}(s)) . \tag{11}$$

In view of (2) and (3), (11) implies

$$\frac{p\hat{g}(s) + (\alpha_0 - p)\hat{f}(s)\hat{g}(s)}{1 - (1 - \alpha_0)\hat{f}(s)} = \frac{p}{E(X_2)s}(1 - \frac{\alpha_1\hat{f}(s) + (\alpha_0 - \alpha_1)\hat{f}^2(s)}{1 - (1 - \alpha_0)\hat{f}(s)}) . \tag{12}$$

Note that since $\{\chi_i,\ i \geq 1\}$ is stationary,

$$E(Y_2) = \frac{1 - \alpha_1 + \alpha_0}{\alpha_0}E(X_2) = \frac{1}{p}E(X_2) . \tag{13}$$

Again substituting $\alpha_0 - p$ by $(\alpha_0 - \alpha_1)p$ in the left side of (12) and after some simplifications, gives

$$\hat{g}(s) = \frac{1}{E(X_2)s}(1 - \hat{f}(s)) . \tag{14}$$

Therefore $\mathcal{N}$ is a stationary renewal process as required. This completes the proof of this theorem.

We also have the following immediate consequence which can be compared with Corollary 1.

Corollary 2. Let $\mathcal{N}$ be an ordinary renewal process, $\{\chi_i,\ i \geq 1\}$ be a stationary Markov chain. Then $\mathcal{A}$ is a stationary renewal process if and only if $\mathcal{N}$ is Poisson process.

Proof. From Theorem 3 we obtain that $\mathcal{A}$ is a stationary renewal process if and only if $\mathcal{N}$ is a stationary renewal process. Yet the only ordinary stationary renewal process is Poisson. This completes the proof.

4. Delayed Gamma-c Renewal Process

It is desirable to know that given an arbitrarily delayed renewal process $\mathcal{A}$ and a sequence of Markov chain $\{\chi_i,\ i \geq 1\}$, whether there exists an original process $\mathcal{N}$, i.e., the M-inverse of $\mathcal{A}$, such that $\mathcal{A}$ can be obtained through the Markov chain thinning. When $\{\chi_i,\ i \geq 1\}$ is an i.i.d. sequence with $P(\chi_i = 1) = p$, Yannaros (1988a) has shown that a renewal process cannot be obtained through the thinning of a non-renewal process for any $p \leq 1$. That is the class of renewal processes is closed under inverse thinning. Yannaros (1988b) also gave necessary and sufficient condition for a delayed gamma renewal process to be a Cox process. Note that a Cox process can be viewed as a Poisson process with a random intensity.

In the following let $\mathcal{A}$ be a delayed gamma-c renewal process as defined in Section 1 with interarrival distribution functions H and K, which are both gamma-c with shape parameters β and α, respectively, and for simplicity we assume both scale parameters equal to 1 (hence $\hat{\xi}(s) = \int_0^\infty e^{-sx} dH(x) = (1+s^c)^{-\beta}$ and $\hat{\eta}(s) = \int_0^\infty e^{-sx} dK(x) = (1+s^c)^{-\alpha}$); and let $\{\chi_i,\ i \geq 1\}$ be a stationary Markov chain as defined in Section 1. We find conditions for the existence of an ordinary renewal process to be the M-inverse of $\mathcal{A}$. In the special case $c = 1$, $\mathcal{A}$ becomes a delayed gamma renewal process.

Case 1. $\alpha = \beta$.

In this case $\mathcal{A}$ becomes an ordinary renewal process. Hence by Theorem 2, we obtain $\{\chi_i,\ i \geq 1\}$ must be an independent sequence. So this reduces to the problem of determining the p-inverse. The case $p = 1$ is trivial, $\mathcal{A}$ is the inverse of itself for every $\alpha > 0$. For every $0 < p < 1$, from (2) and (3) with $\hat{g}(s) = \hat{f}(s)$, we obtain

$$\hat{f}(s) = \frac{\hat{\xi}(s)}{p + (1-p)\hat{\xi}(s)} .$$

Since $\hat{\xi}(s) = \hat{\eta}(s) = (1 + s^c)^{-\alpha}$, it yields

$$\hat{f}(s) = \frac{1}{p(1+s^c)^\alpha + (1-p)} . \tag{15}$$

As $\hat{f}(0+) = 1$, being a Laplace transform, $\hat{f}(s)$ must be completely monotone. In order to determine the conditions such that $\hat{f}(s)$ is completely monotone, we consider the following three situations: $0 < \alpha \leq 1$, $1 < \alpha \leq 1/c$, and $\alpha > 1/c$, where $0 < c \leq 1$.

Firstly, we study the case $0 < \alpha \leq 1$. Let $\phi(s) = 1 + s^c$, then $\phi(0+) = 1$. Since for $0 < c \leq 1$,

$$(-1)^n \phi^{(n)}(s) \leq 0, \quad s > 0,$$

$\phi(s)$ is a Bernstein function. Consequently, $\hat{f}(s)$ as defined in (15), is a Laplace transform by Lemma 1.

Although for the case $1 < \alpha \leq 1/c$, we are unable to determine whether the function $\hat{f}(s)$ is a Laplace transform, we have some partial result. Let α be an integer, and $\phi(s) = (1 + s^c)^\alpha - 1$, then $\phi(0+) = 0$. Applying the Binomial theorem, we have

$$\phi(s) = (1 + s^c)^\alpha - 1 = \sum_{j=0}^{\alpha} \binom{\alpha}{j} s^{cj} - 1,$$

hence

$$\phi'(s) = \sum_{j=0}^{\alpha} \binom{\alpha}{j} cj s^{cj-1} .$$

It is easy to see that $\phi(s)$ has a completely monotone derivative if $cj \leq 1$ for every $j = 0, 1, \cdots, \alpha$. Thus $\phi(s)$ is a Berstein function if $c\alpha \leq 1$. From Lemma 2, $\hat{f}(s)$ as defined in (15), is a Laplace transform.

Finally, consider the case $\alpha > 1/c$. Again let $\phi(s) = (1 + s^c)^\alpha - 1$, then $\phi(0+) = 0$. It is easy to get

$$\begin{aligned} \phi'(s) &= \alpha(1 + s^c)^{\alpha-1} c s^{c-1}, \\ \phi''(s) &= \alpha(\alpha - 1)(1 + s^c)^{\alpha-2} c^2 s^{2c-2} + \alpha(1 + s^c)^{\alpha-1} c(c-1) s^{c-2} \\ &= c\alpha(1 + s^c)^{\alpha-2}\{(c\alpha - 1)s^c + (c - 1)\} . \end{aligned}$$

Thus, $\phi''(s) \geq 0$ when s is large enough. Hence $\phi(s)$ is not a Bernstein function. From Lemma 2, $\hat{f}(s)$ as defined in (15), is not a Laplace transform.

Case 2. $\alpha > \beta$.

First note that as $\mathcal{A}$ is not an ordinary renewal process, by Theorem 2, $\{\chi_i,\ i \geq 1\}$ cannot be an independent sequence. That is $\alpha_0 \neq \alpha_1$. From (2) and (3) with $\hat{g}(s) = \hat{f}(s)$, we obtain

$$\frac{\hat{\xi}(s)}{\hat{\eta}(s)} = \frac{p + (\alpha_0 - p)\hat{f}(s)}{\alpha_1 + (\alpha_0 - \alpha_1)\hat{f}(s)} . \tag{16}$$

Substituting $\hat{\xi}(s) = (1+s^c)^{-\beta}$, $\hat{\eta}(s) = (1+s^c)^{-\alpha}$ and $\alpha_0 - p = p(\alpha_0 - \alpha_1)$ into (16), then solving for $\hat{f}(s)$, yields

$$\hat{f}(s) = \frac{p(1+s^c)^{-\alpha} - \alpha_1(1+s^c)^{-\beta}}{(\alpha_0-\alpha_1)((1+s^c)^{-\beta} - p(1+s^c)^{-\alpha})} . \tag{17}$$

Being a Laplace transform,

$$\hat{f}'(s) = \frac{p(1-\alpha_1)(\beta-\alpha)}{\alpha_0-\alpha_1} \cdot \frac{(1+s^c)^{\beta-\alpha-1}}{(1-p(1+s^c)^{\beta-\alpha})^2} \le 0 . \tag{18}$$

Thus $\alpha_0 > \alpha_1$. Again (17) can be rewritten as

$$\hat{f}(s) = \frac{1}{\alpha_0-\alpha_1} \cdot \frac{p - \alpha_1(1+s^c)^{\alpha-\beta}}{(1+s^c)^{\alpha-\beta} - p} . \tag{19}$$

For $\alpha_1 \ne 0$, since $\alpha > \beta$, the function $\{(1+s^c)^{\alpha-\beta} - p\}$ is positive for every $s > 0$, and the function $\{p - \alpha_1(1+s^c)^{\alpha-\beta}\}$ is negative for s large enough. Consequently, $\hat{f}(s) < 0$, $\hat{f}(s)$ as defined in (19), is not a Laplace transform for $\alpha_1 \ne 0$.

For the special case $\alpha_1 = 0$, we investigate when the function

$$\hat{f}(s) = \frac{1}{\alpha_0} \cdot \frac{p}{(1+s^c)^{\alpha-\beta} - p} \tag{20}$$

is a Laplace transform. By (1), (20) becomes

$$\hat{f}(s) = \frac{1-p}{(1+s^c)^{\alpha-\beta} - p} . \tag{21}$$

Again, let

$$\phi(s) = \frac{1}{p(1-p)}(1+s^c)^{\alpha-\beta} - \frac{1}{p(1-p)} . \tag{22}$$

Since $\phi(0+) = 0$, the problem becomes to determine when the function $\phi(s)$ is a Bernstein function. It can be seen that as in Case 1 by considering the three situations $0 < \alpha - \beta \le 1$, $1 < \alpha - \beta \le 1/c$, and $\alpha - \beta > 1/c$, parallel results can be obtained.

Case 3. $\alpha < \beta$.

Again (17) can be rewritten as

$$\hat{f}(s) = \frac{1}{\alpha_0-\alpha_1} \cdot \frac{p(1+s^c)^{\beta-\alpha} - \alpha_1}{1 - p(1+s^c)^{\beta-\alpha}} . \tag{23}$$

Since $\alpha_0 < \alpha_1$, if both $\{p(1+s^c)^{\beta-\alpha} - \alpha_1\}$ and $\{1 - p(1+s^c)^{\beta-\alpha}\}$ are positive, then $\hat{f}(s) < 0$. It is easy to obtain the inequality

$$\{(\frac{\alpha_1}{p})^{\frac{1}{\beta-\alpha}} - 1\}^{\frac{1}{c}} < s < \{(\frac{1}{p})^{\frac{1}{\beta-\alpha}} - 1\}^{\frac{1}{c}}, \quad \alpha_1 \ne 1 .$$

Therefore, $\hat{f}(s) < 0$, $\hat{f}(s)$ as defined in (23), is not a Laplace transform for $\alpha_1 \neq 1$.

We now consider the special case $\alpha_1 = 1$. From (1), we have

$$p = \frac{\alpha_0}{1 + \alpha_0 - \alpha_1} = \frac{\alpha_0}{1 + \alpha_0 - 1} = 1 \ .$$

This shows that $\mathcal{A}$ is the inverse of itself. In other words, given a stationary Markov chain, the M-inverse exists if $\alpha_1 = 1$ when $\alpha < \beta$.

5. Delayed Negative Binomial Renewal Process

In the above section, we consider renewal process with continuous interarrival distribution function. In this section we consider the discrete situation. More precisely, we consider a delayed renewal process with interarrival times being $NB(k, \theta)$ and $NB(r, \theta)$ distributed (hence $\hat{\xi}(s) = (\frac{\theta e^{-s}}{1 - (1 - \theta)e^{-s}})^k$ and $\hat{\eta}(s) = (\frac{\theta e^{-s}}{1 - (1 - \theta)e^{-s}})^r$); and let $\{\chi_i,\ i \geq 1\}$ be a stationary Markov chain as defined in Section 1. We find conditions for the existence of an ordinary renewal process to be the M-inverse of $\mathcal{A}$.

Case 1. $k = r$.

In this case $\mathcal{A}$ becomes an ordinary renewal process. Hence by Theorem 2, we obtain $\{\chi_i,\ i \geq 1\}$ is an independent sequence. So this reduces to the problem of determining the p-thinning. The case $p = 1$ is trivial, $\mathcal{A}$ is the inverse of itself for every $\alpha > 0$. For every $0 < p < 1$, from (2) and (3) with $\hat{g}(s) = \hat{f}(s)$, we obtain

$$\hat{f}(s) = \frac{\hat{\xi}(s)}{p + (1 - p)\hat{\xi}(s)} \ .$$

Since $\hat{\xi}(s) = \hat{\eta}(s) = (\frac{\theta e^{-s}}{1 - (1 - \theta)e^{-s}})^r$, it yields

$$\hat{f}(s) = \frac{\theta^r}{p(e^s - (1 - \theta))^r + (1 - p)\theta^r} \ . \tag{24}$$

Let

$$\phi(s) = (\frac{e^s - (1 - \theta)}{\theta})^r - 1 \ . \tag{25}$$

It is easy to obtain

$$\phi'(s) = (\frac{1}{\theta})^r r(e^s - (1-\theta))^{r-1} e^s,$$

$$\phi''(s) = (\frac{1}{\theta})^r r\{(r-1)(e^s - (1-\theta))^{r-2} e^{2s} + (e^s - (1-\theta))^{r-1} e^s\}$$

$$= (\frac{1}{\theta})^r r(e^s - (1-\theta))^{r-2} e^s \{re^s - (1-\theta)\} .$$

Since r is an integer, $\phi''(s)$ is positive for every $s > 0$. Hence $\phi(s)$ is not a Bernstein function. Consequently, by Lemma 2, $\hat{f}(s)$ as defined in (24) is not a Laplace transform.

Case 2. $k < r$.

First note that as $\mathcal{A}$ is not an ordinary renewal process, by Theorem 2 $\{\chi_i,\ i \geq 1\}$ cannot be an independent sequence. Hence $\alpha_0 \neq \alpha_1$. From (2) and (3) with $\hat{g}(s) = \hat{f}(s)$, we obtain

$$\frac{\hat{\xi}(s)}{\hat{\eta}(s)} = \frac{p + (\alpha_0 - p)\hat{f}(s)}{\alpha_1 + (\alpha_0 - \alpha_1)\hat{f}(s)} . \tag{26}$$

Substituting $\hat{\xi}(s) = (\frac{\theta e^{-s}}{1-(1-\theta)e^{-s}})^k$, $\hat{\eta}(s) = (\frac{\theta e^{-s}}{1-(1-\theta)e^{-s}})^r$ and $\alpha_0 - p = p(\alpha_0 - \alpha_1)$ into (26), then solving for $\hat{f}(s)$, yields

$$\hat{f}(s) = \frac{1}{\alpha_0 - \alpha_1} \cdot \frac{\alpha_1(\frac{\theta e^{-s}}{1-(1-\theta)e^{-s}})^k - p(\frac{\theta e^{-s}}{1-(1-\theta)e^{-s}})^r}{p(\frac{\theta e^{-s}}{1-(1-\theta)e^{-s}})^r - (\frac{\theta e^{-s}}{1-(1-\theta)e^{-s}})^k}$$

$$= \frac{1}{\alpha_0 - \alpha_1} \cdot \frac{\alpha_1 \theta^k (e^s - (1-\theta))^r - p\,\theta^r (e^s - (1-\theta))^k}{p\,\theta^r (e^s - (1-\theta))^k - \theta^k (e^s - (1-\theta))^r} . \tag{27}$$

Being a Laplace transform,

$$\hat{f}'(s) = \frac{p(1-p)(k-r)}{\alpha_0 - \alpha_1} \cdot \frac{\theta^{k+r}(e^s - (1-\theta))^{k+r-1}}{(p\,\theta^r (e^s - (1-\theta))^k - \theta^k (e^s - (1-\theta))^r)^2} \leq 0 . \tag{28}$$

Thus $\alpha_0 > \alpha_1$, and from (1) we have $\alpha_0 > p > \alpha_1$.

Again (27) can be rewritten as

$$\hat{f}(s) = \frac{1}{\alpha_0 - \alpha_1} \cdot \frac{\alpha_1 (e^s - (1-\theta))^{r-k} - p\,\theta^{r-k}}{p\,\theta^{r-k} - (e^s - (1-\theta))^{r-k}} . \tag{29}$$

If $\alpha_1 \neq 0$, since $\theta \leq e^s - (1-\theta)$, $\{\alpha_1 (e^s - (1-\theta))^{r-k} - p\,\theta^{r-k}\}$ is positive for s large enough and $\{p\,\theta^{r-k} - (e^s - (1-\theta))^{r-k}\}$ is negative for every

$s > 0$, we obtain $\hat{f}(s) < 0$ for s large enough. Therefore, $\hat{f}(s)$ as defined in (29), is not a Laplace transform when $\alpha_1 \neq 0$.

Now consider the case $\alpha_1 = 0$. In this case

$$\hat{f}(s) = \frac{1}{\alpha_0} \cdot \frac{p\,\theta^{r-k}}{(e^s - (1-\theta))^{r-k} - p\,\theta^{r-k}} .$$

Furthermore, by using (1), we obtain

$$\hat{f}(s) = \frac{1-p}{(\frac{e^s-(1-\theta)}{\theta})^{r-k} - p} . \tag{30}$$

Again, let

$$\phi(s) = \frac{1}{p(1-p)}(\frac{e^s - (1-\theta)}{\theta})^{r-k} - \frac{1}{p(1-p)} .$$

Then

$$\begin{aligned}
\phi'(s) &= \frac{1}{p(1-p)\theta^{r-k}}(r-k)(e^s-(1-\theta))^{r-k-1}e^s, \\
\phi''(s) &= \frac{r-k}{p(1-p)\theta^{r-k}}\Big\{(r-k-1)(e^s-(1-\theta))^{r-k-2}e^{2s} \\
&\qquad + (e^s-(1-\theta))^{r-k-1}e^s\Big\} \\
&= \frac{(r-k)(e^s-(1-\theta))^{r-k-2}e^s}{p(1-p)\theta^{r-k}}\{(r-k)e^s-(1-\theta)\} .
\end{aligned}$$

Since $r-k$ is an integer, $\phi''(s)$ is positive for every $s > 0$. Hence $\phi(s)$ is not a Bernstein function. Consequently, $\hat{f}(s)$ as defined in (30), is not a Laplace transform by Lemma 2. This shows that when $r > k$, for any Markov chain, there does not exist an ordinary renewal process $\mathcal{N}$ such that $\mathcal{A}$ can be obtained through this Markov chain thinning.

Case 3. $k > r$.

Again from (27), by a similar argument as in Case 2, we obtain

$$\hat{f}(s) = \frac{1}{\alpha_0 - \alpha_1} \cdot \frac{\alpha_1\,\theta^{k-r} - p(e^s-(1-\theta))^{k-r}}{p(e^s-(1-\theta))^{k-r} - \theta^{k-r}}, \tag{31}$$

here $\alpha_0 < \alpha_1$, and from (1), we have $\alpha_0 < p < \alpha_1$.

Similarly, if both $\{\alpha_1\,\theta^{k-r} - p(e^s-(1-\theta))^{k-r}\}$ and $\{p(e^s-(1-\theta))^{k-r} - \theta^{k-r}\}$ are negative, then $\hat{f}(s) < 0$. It is easy to obtain the inequality

$$((\frac{\alpha_1}{p})^{\frac{1}{k-r}} - 1)\theta + 1 < e^s < ((\frac{1}{p})^{\frac{1}{k-r}} - 1)\theta + 1, \quad \alpha_1 \neq 1 .$$

Therefore, $\hat{f}(s) < 0$, and $\hat{f}(s)$ is not a Laplace transform when $\alpha_1 \neq 1$.

Now consider the special case $\alpha_1 = 1$. From (1), we have

$$p = \frac{\alpha_0}{1+\alpha_0-\alpha_1} = \frac{\alpha_0}{1+\alpha_0-1} = 1 .$$

This shows that $\mathcal{A}$ is the inverse of itself. In other words when $k > r$, given a Markov chain, the M-inverse exists if and only if $\alpha_1 = 1$.

6. An Example

In Sections 4 and 5, where $\mathcal{A}$ is a delayed renewal process, we give conditions such that the M-inverse exists, with interarrival times being gamma-c or negative binomial distributed, respectively. For some special delayed renewal process $\mathcal{A}$, which does not belong to any of the above two classes, the M-inverse may also exist. We give an example in this section.

Let $\mathcal{A}$ be a delayed renewal process with distribution functions of the interarrival times $\{Y_i,\ i \geq 1\}$ being :

$$\begin{aligned} \mathrm{P}(Y_1 \leq x) &= 1 - \frac{1}{2}(2+x)e^{-2x}, \quad x > 0, \\ \mathrm{P}(Y_k \leq x) &= 1 - (1+x)e^{-2x}, \quad x > 0,\ k \geq 2, \end{aligned}$$

and $P(Y_k \leq x) = 0$, if $x < 0$. Then the Laplace transforms of Y_1 and Y_2 are

$$\hat{\xi}(s) = \mathrm{E}(e^{-sY_1}) = \frac{4+1.5s}{(2+s)^2}, \quad s > 0, \tag{32}$$

and

$$\hat{\eta}(s) = \mathrm{E}(e^{-sY_2}) = \frac{4+s}{(2+s)^2}, \quad s > 0, \tag{33}$$

respectively. In the following we find the conditions that given $\mathcal{A}$, and a stationary Markov chain $\{\chi_i,\ i \geq 1\}$, as defined in Section 1, when the M-inverse of $\mathcal{A}$ exists.

From (2) and (3) with $\hat{g}(s) = \hat{f}(s)$, we obtain

$$\frac{\hat{\xi}(s)}{\hat{\eta}(s)} = \frac{p + (\alpha_0 - p)\hat{f}(s)}{\alpha_1 + (\alpha_0 - \alpha_1)\hat{f}(s)} . \tag{34}$$

Substituting (32), (33) and $\alpha_0 - p = p(\alpha_0 - \alpha_1)$ into (34), then solving for $\hat{f}(s)$, yields

$$\hat{f}(s) = \frac{1}{\alpha_0 - \alpha_1} \cdot \frac{(p - 1.5\alpha_1)s + 4(p - \alpha_1)}{(1.5-p)s + 4(1-p)} . \tag{35}$$

Being a Laplace transform,

$$\hat{f}'(s) = \frac{1}{\alpha_0 - \alpha_1} \cdot \frac{2p(\alpha_1 - 1)}{((1.5-p)s + 4(1-p))^2} \leq 0 . \tag{36}$$

Thus $\alpha_0 > \alpha_1$. Note that from (1) we have $\alpha_0 > p > \alpha_1$.

Since $(-1)^n \hat{f}^{(n)}(s) \geq 0, \forall\, n \geq 1, s > 0$, with $\alpha_0 > \alpha_1$, we only need to find the conditions such that whether $\hat{f}(s) \geq 0, \forall s > 0$. The problem is equivalent to determining when $p - 1.5\alpha_1 \geq 0$. Solving the above inequality with $p = \dfrac{\alpha_0}{1+\alpha_0-\alpha_1}$ and noting that $0 \leq \alpha_0, \alpha_1 \leq 1$, yields

$$1 \geq \alpha_0 \geq \frac{3\alpha_1(1-\alpha_1)}{2-3\alpha_1} \qquad \text{and} \qquad \alpha_1 \leq \frac{3-\sqrt{3}}{3}\,. \tag{37}$$

(37) is then a necessary and sufficient condition for $\hat{f}(s)$ being a Laplace transform. This is also the necessary and sufficient condition for $\mathcal{A}$ having an M-inverse.

7. Discussion

As mentioned in Theorem 1, a function ψ on $(0,\infty)$ is a Laplace transform if and only if it is completely monotone with $\psi(0+) = 0$. Usually, it is difficult to determine whether a function is completely monotone. It is also difficult to determine whether the function $\phi(s) = (1+s^c)^\alpha, 0 < c \leq 1,\ 1 < \alpha \leq 1/c$ and $s > 0$, is a Bernstein function. In this work we have solved the problem for the case that α is an integer. The case that α is not an integer will be investigated in the future work.

References

1. Chandramohan, J. and Liang, L. K. (1985). Bernoulli, multinomial and markov chain thinning of some point processes and some results about the superposition of dependent renewal processes. *J. Appl. Prob.* **22**, 828-835.
2. Chung, K. L. (1974). *A Course in Probability Theory*, 2nd ed. Academic Press, New York.
3. Feller, W. (1971). *An Introduction to Probability Theory and Its Applications*, Vol. 2, 2nd ed. John Wiley & Sons, New York.
4. Huang, W. J. and Chen, L. S. (1989). Note on a characterization of gamma distribution. *Statist. & Prob. Lett.* **8**, 485-487.
5. Huang, W. J. and Chen, L. S. (1991). On a study of certain power mixtures. *Chinese J. Math.* **19(2)**, 95-104.
6. Isham, V. (1980). Dependent thinning of point processes. *J. Appl. Prob.* **17**, 987-995.
7. Kolsrud, T. (1986). Some comments on thinned renewal processes. *Scand. Actuarial. J.*, 236-241.
8. Yannaros, N. (1988a). The inverses of thinned renewal processes. *J. Appl. Prob.* **25**, 822-828.

9. Yannaros, N. (1988b). On Cox processes and gamma renewal processes. *J. Appl. Prob.* **25**, 423-427.
10. Yannaros, N. (1988c). Some comments on the inverse problem for thinned renewal processes. *Scand. Actuarial J.*, 113-116.
11. Yannaros, N. (1991). Randomly observed random walks. *Commun. Statist.-Stochastic Models.* **7(2)**, 219-231.
12. Yannaros, N. (1994). Weibull renewal processes. *Ann. Inst. Statist. Math.* **46**, 641-648.

STOCHASTIC GAMES OF CONTROL AND STOPPING FOR A LINEAR DIFFUSION *

IOANNIS KARATZAS

Departments of Mathematics and Statistics, Columbia University
New York, NY 10027, U.S.A.
ik@math.columbia.edu

WILLIAM SUDDERTH

School of Statistics, University of Minnesota
Minneapolis, MN 55455, U.S.A.
bill@stat.umn.edu

We study three stochastic differential games. In each game, two players control a process $X = \{X_t\,,\ 0 \le t < \infty\}$ which takes values in the interval $I = (0,1)$, is absorbed at the endpoints of I, and satisfies a stochastic differential equation

$$dX_t \;=\; \mu\big(X_t, \alpha(X_t), \beta(X_t)\big)\,dt \;+\; \sigma\big(X_t, \alpha(X_t), \beta(X_t)\big)\,dW_t\,, \quad X_0 = x \in I\,.$$

The *control functions* $\alpha(\cdot)$ and $\beta(\cdot)$ are chosen by players $\mathfrak{A}$ and $\mathfrak{B}$, respectively.

In the first of our games, which is zero-sum, player $\mathfrak{A}$ has a continuous reward function $u : [0,1] \to \mathbb{R}$. In addition to $\alpha(\cdot)$, player $\mathfrak{A}$ chooses a stopping rule τ and seeks to maximize the expectation of $u(X_\tau)$; whereas player $\mathfrak{B}$ chooses $\beta(\cdot)$ and aims to minimize this expectation.

In the second game, players $\mathfrak{A}$ and $\mathfrak{B}$ each have continuous reward functions $u(\cdot)$ and $v(\cdot)$, choose stopping rules τ and ρ, and seek to maximize the expectations of $u(X_\tau)$ and $v(X_\rho)$, respectively.

In the third game the two players again have continuous reward functions $u(\cdot)$ and $v(\cdot)$, now assumed to be unimodal, and choose stopping rules τ and ρ. This game terminates at the minimum $\tau \wedge \rho$ of the stopping rules τ and ρ, and players $\mathfrak{A}$, $\mathfrak{B}$ want to maximize the expectations of $u(X_{\tau\wedge\rho})$ and $v(X_{\tau\wedge\rho})$, respectively.

Under mild technical assumptions we show that the first game has a value, and find a saddle point of optimal strategies for the players. The other two games are not zero-sum, in general, and for them we construct Nash equilibria.

Some key Words: Stochastic games; Optimal control; Optimal stopping; Nash equilibrium; Linear diffusions; Local time; Generalized Itô rule.

*Work supported in part by the National Science Foundation Grant NSF-DMS-06-01774.

1. Preliminaries

The state space for the two-player games studied in this paper, is the interval $I = (0,1)$. For each $x \in [0,1]$, players $\mathfrak{A}$ and $\mathfrak{B}$ are given nonempty action sets $A(x)$ and $B(x)$, respectively. For simplicity, we assume that $A(x)$ and $B(x)$ are Borel sets of the real line $\mathbb{R}$.

Denote by $\mathfrak{Z} = C([0,\infty))$ the space of continuous functions $\xi : [0,\infty) \to \mathbb{R}$ with the topology of uniform convergence on compact sets. A `stopping rule` is a mapping $\tau : \mathfrak{Z} \to [0,\infty]$ such that

$$\{\xi \in \mathfrak{Z} \,:\, \tau(\xi) \le t\} \,\in\, \mathcal{B}_t \,:=\, \varphi_t^{-1}(\mathcal{B})$$

holds for every $t \in [0,\infty)$. Here $\mathcal{B}$ is the Borel σ-algebra generated by the open sets in $\mathfrak{Z}$, and $\varphi_t : \mathfrak{Z} \to \mathfrak{Z}$ is the mapping

$$(\varphi_t\,\xi)(s) \,:=\, \xi(t \wedge s)\,, \qquad 0 \le s < \infty\,.$$

We shall denote by $\mathcal{S}$ the set of all stopping rules.

The two players are allowed to choose measurable functions $\alpha : [0,1] \to \mathbb{R}$ and $\beta : [0,1] \to \mathbb{R}$ with $\alpha(x) \in A(x)$ and $\beta(x) \in B(x)$ for every $x \in [0,1]$. (We assume that the correspondences $A(\cdot)$ and $B(\cdot)$ are sufficiently regular, to guarantee the existence of such 'choice functions'.) For each such pair $\alpha(\cdot)$ and $\beta(\cdot)$, we have the dynamics

$$dX_t \,=\, \mu\big(X_t, \alpha(X_t), \beta(X_t)\big)\,dt \,+\, \sigma\big(X_t, \alpha(X_t), \beta(X_t)\big)\,dW_t\,,\ X_0 = x \in I \tag{1}$$

of a diffusion process $X \equiv X^{x,\alpha,\beta}$ that lives in the interval I and is absorbed the first time it reaches either of the endpoints. Here $W = \{W_t\,,\ 0 \le t < \infty\}$ is a standard, one-dimensional Brownian motion, and the functions $\mu(\,\cdot\,,\,\cdot\,,\,\cdot\,)$, $\sigma(\,\cdot\,,\,\cdot\,,\,\cdot\,)$ are measurable and map the product space $[0,1] \times \mathbb{R} \times \mathbb{R}$ into $\mathbb{R}$ and $(0,\infty)$, respectively.

More precisely, we are assuming that for any pair of functions $\alpha(\cdot)$ and $\beta(\cdot)$ chosen as above, the local drift and variance

$$b(x) \,:=\, \mu\big(x, \alpha(x), \beta(x)\big)\,, \qquad s^2(x) \,:=\, \sigma^2\big(x, \alpha(x), \beta(x)\big)\,, \qquad x \in [0,1]$$

are measurable functions, which satisfy the conditions

$$s^2(x) \,>\, 0\,, \qquad \int_{x-\varepsilon}^{x+\varepsilon} \frac{1 + |b(y)|}{s^2(y)}\,dy \,<\, \infty \qquad \text{for some } \ \varepsilon = \varepsilon(x) > 0$$

for each $x \in I$, as well as

$$b(x) \,=\, s^2(x) \,=\, 0 \qquad \text{for} \qquad x \in \{0,1\}$$

(in other words, both endpoints are absorbing). The first of these conditions implies, in particular, that the resulting diffusion process

$$dX_t = b(X_t)\,dt + s(X_t)\,dW_t\,, \qquad X_0 = x$$

is *regular:* it exits in finite expected time from any open neighborhood $(x-\varepsilon, x+\varepsilon) \subset I$ (cf. Karatzas & Shreve (1991), p.344).

If τ is a stopping rule, we often write X_τ as abbreviation for a random variable that equals $X_{\tau(X)}$ when $\tau(X)$ is finite on the path $\{X_t\,, 0 \le t < \infty\}$. Also, for a bounded and measurable function $u : [0,1] \to \mathbb{R}$, we set

$$u(X_\tau) \equiv \limsup_{t\to\infty} u\left(X_{\tau(X)\wedge t}\right) = \begin{Bmatrix} u\left(X_{\tau(X)}\right) & ; \text{ if } \tau(X) < \infty \\ \limsup_{t\to\infty} u\left(X_t\right) & ; \text{ if } \tau(X) = \infty \end{Bmatrix}.$$

2. A Zero-Sum Game

Assume player $\mathfrak{A}$ selects the control function $\alpha(\cdot)$ and a stopping rule τ, whereas player $\mathfrak{B}$ has only the choice of the control function $\beta(\cdot)$. Let $u : [0,1] \to \mathbb{R}$ be a continuous function, which we regard as a *reward function* for player $\mathfrak{A}$. Then the expected payoff from player $\mathfrak{B}$ to player $\mathfrak{A}$ is

$$\mathbb{E}[\,u(X_\tau)\,]\,,$$

where $X \equiv X^{x,\alpha,\beta}$ and the game begins at $X_0 = x \in I$. Assume further that $u(\cdot)$ attains its maximum value at a unique position $m \in [0,1]$. Call this game $\mathfrak{G}_1(x)$.

Optimal strategies for the players have a simple intuitive description: In the interval $(0,m)$ to the left of the maximum, player $\mathfrak{A}$ chooses a control function $\alpha^*(\cdot)$ that maximizes the "signal-to-noise" (mean/variance) ratio μ/σ^2, whereas player $\mathfrak{B}$ chooses a control function $\beta^*(\cdot)$ that minimizes this ratio. In the interval $(m,1)$ to the right of the maximum, player $\mathfrak{A}$ chooses $\alpha^*(\cdot)$ to minimize μ/σ^2, whereas player $\mathfrak{B}$ chooses $\beta^*(\cdot)$ to maximize it. Finally, player $\mathfrak{A}$ takes the stopping rule τ^* to be optimal for maximizing $\mathbb{E}[\,u(Z_\tau)\,]$ over all stopping rules τ of the diffusion process $Z \equiv X^{x,\alpha^*,\beta^*}$, namely

$$dZ_t = b^*(Z_t)\,dt + s^*(Z_t)\,dW_t\,, \qquad Z_0 = x\,, \tag{2}$$

with drift and dispersion coëfficients given by

$$b^*(z) := \mu\big(z, \alpha^*(z), \beta^*(z)\big)\,, \qquad s^*(z) := \sigma\big(z, \alpha^*(z), \beta^*(z)\big)\,, \tag{3}$$

respectively. More precisely, we impose the following condition.

C.1: Local Saddle-Point Condition. *There exist measurable functions $\alpha^*(\cdot)$, $\beta^*(\cdot)$ such that $(\alpha^*(z), \beta^*(z)) \in A(z) \times B(z)$ holds for all $z \in (0,1)$;*

$$\left(\frac{\mu}{\sigma^2}\right)(z, a, \beta^*(z)) \le \left(\frac{\mu}{\sigma^2}\right)(z, \alpha^*(z), \beta^*(z))$$
$$\le \left(\frac{\mu}{\sigma^2}\right)(z, \alpha^*(z), b), \quad \forall\ (a,b) \in A(z) \times B(z)$$

holds for every $z \in (0, m)$; and

$$\left(\frac{\mu}{\sigma^2}\right)(z, a, \beta^*(z)) \ge \left(\frac{\mu}{\sigma^2}\right)(z, \alpha^*(z), \beta^*(z))$$
$$\ge \left(\frac{\mu}{\sigma^2}\right)(z, \alpha^*(z), b), \quad \forall\ (a,b) \in A(z) \times B(z)$$

holds for every $z \in (m, 1)$.

Remark 1: For fixed z, the existence of $\alpha^*(z)$, $\beta^*(z)$ satisfying the condition $C.1$ corresponds to the existence of optimal strategies for a "local" (one-shot) game with action sets $A(z)$ and $B(z)$, and with payoff given by $(\mu/\sigma^2)(z,a,b)$ for $0 < z < m$ and by $(-\mu/\sigma^2)(z,a,b)$ for $m < z < 1$. Such strategies exist, for example, if $(\mu/\sigma^2)(z, \cdot, \cdot)$ is continuous, and the sets $A(z)$, $B(z)$ are compact for every $z \in I$.

Suppose that the local saddle-point condition $C.1$ holds, and recall the diffusion process $Z \equiv X^{x,\alpha^*,\beta^*}$ of (2), (3) for a fixed initial position $x \in I$. Fix an arbitrary point η of I, and write the *scale function* of Z as

$$p(z) := \int_\eta^z p'(u)\, du, \qquad z \in I,$$

where we have set

$$p'(z) := \exp\left\{-2 \int_\eta^z \frac{b^*(u)}{(s^*(u))^2}\, du\right\} = 1 + \int_\eta^z p''(u)\, du > 0$$

and

$$p''(z) := -2\, p'(z) \cdot \frac{b^*(z)}{(s^*(z))^2}.$$

C.2: Strong Non-degeneracy Condition. *We have*

$$\inf_{z \in I} \left(\inf_{\substack{a \in A(z) \\ b \in B(z)}} \sigma^2(z, a, b) \right) > 0.$$

Theorem 1: *Assume that $\alpha^*(\cdot)$, $\beta^*(\cdot)$ satisfy the local saddle-point condition $C.1$. For each $x \in I$, define the process $Z^x \equiv X^{x,\alpha^*,\beta^*}$ of (2), (3), and assume that the strong non-degeneracy condition $C.2$ holds. Let*

$$U(x) := \sup_{\tau \in \mathcal{S}} \mathbb{E}[\, u(Z^x_\tau)\,], \quad x \in I \tag{4}$$

where τ ranges over the set $\mathcal{S}$ of all stopping rules, and consider the element τ^* of $\mathcal{S}$ given by

$$\tau^*(\xi) := \inf\{t \geq 0 : U(\xi_t) = u(\xi_t)\}, \quad \xi \in \mathfrak{Z}. \tag{5}$$

Then for each x, the game $\mathfrak{G}_1(x)$ has value $U(x)$ and saddle-point $((\alpha^*(\cdot), \tau^*), \beta^*(\cdot))$.

Namely, the pair $(\alpha^*(\cdot), \tau^*)$ is optimal for player $\mathfrak{A}$, when player $\mathfrak{B}$ chooses $\beta^*(\cdot)$; and the function $\beta^*(\cdot)$ is optimal for player $\mathfrak{B}$, when player $\mathfrak{A}$ chooses $(\alpha^*(\cdot), \tau^*)$.

Before broaching the proof of this result, let us recall some well-known facts from the theory of optimal stopping (as developed, for instance, in Snell (1952), Chow et al. (1971), Neveu (1975) or Shiryaev (1978)). The function $U(\cdot)$ of (4) is the *value function* for the optimal stopping problem associated with the random process $u(Z^x) = \{u(Z_t^x), 0 \leq t < \infty\}$; it is obviously bounded, because the reward function $u(\cdot)$ is continuous (therefore bounded) on the compact interval [0,1]. In addition, the function $U(\cdot)$ is known to be continuous, and the stopping rule τ^* to be optimal: i.e., $U(x) = \mathbb{E}[u(Z_{\tau^*}^x)]$ (see, for example, Shiryaev (1978)). Furthermore,

$$U(x) = \widetilde{U}(p(x)), \qquad x \in I,$$

where $\widetilde{U}(\cdot)$ is the least concave majorant of the function $\widetilde{u}(\cdot)$ defined through

$$u(x) = \widetilde{u}(p(x));$$

for instance, see Karatzas & Sudderth (1999) or Dayanik & Karatzas (2003), pp.178-181. The process $Y \equiv Y^y := p(Z^x)$ is easily seen to be a "diffusion in natural scale", that is,

$$dY_t = \widetilde{s}(Y_t)\, dW_t, \qquad Y_0 = y := p(x),$$

where $\widetilde{s}(\cdot)$ is defined via $\widetilde{s}(p(x)) = s(x)\, p'(x)$, $x \in I$. The state-space of this diffusion is the open interval

$$\widetilde{I} = (p(0+), p(1-)).$$

The function $\widetilde{U}(\cdot)$ is the value function of the optimal stopping problem for the process $\widetilde{u}(Y)$ with starting point $Y_0 = y \in \widetilde{I}$; to wit,

$$\widetilde{U}(y) = \sup_{\varrho \in \mathcal{S}} \mathbb{E}[\widetilde{u}(Y_\varrho^y)], \quad Y_0^y = y \in \widetilde{I}.$$

We shall refer to this as the "auxiliary optimal stopping problem".

As the least concave majorant of $\widetilde{u}(\cdot)$, the function $\widetilde{U}(\cdot)$ is clearly increasing on the interval $(p(0+), \widetilde{m})$ and decreasing on the interval $(\widetilde{m}, p(1-))$, where $\widetilde{m} := p(m)$ is the unique point at which the function $\widetilde{u}(\cdot)$ attains its maximum over the interval $\widetilde{I}$. Furthermore, the function $\widetilde{U}(\cdot)$ is affine on the connected components of the open set

$$\widetilde{\mathcal{O}} := \left\{ y \in \widetilde{I} \,:\, \widetilde{U}(y) > \widetilde{u}(y) \right\},$$

the optimal continuation region for the auxiliary stopping problem. Consequently, the positive measure $\widetilde{\nu}$ defined by

$$\widetilde{\nu}\big([c, d) \big) := D^- \widetilde{U}(c) - D^- \widetilde{U}(d), \qquad p(0+) < c < d < p(1-),$$

assigns zero mass to the set $\widetilde{\mathcal{O}}$: $\widetilde{\nu}(\widetilde{\mathcal{O}}) = 0$. Note also that $\widetilde{\mathcal{O}} = p(\mathcal{O})$, where

$$\mathcal{O} := \{ x \in I \,:\, U(x) > u(x) \}$$

is the optimal continuation region for the original stopping problem.

Proof of Theorem 1: Let $\alpha(\cdot)$ and $\beta(\cdot)$ be arbitrary control functions for the two players, and let $\tau \in \mathcal{S}$ be an arbitrary stopping rule for player $\mathfrak{A}$. Consider the processes

$$Z \equiv X^{x,\alpha^*,\beta^*}, \quad H \equiv X^{x,\alpha^*,\beta}, \quad \Theta \equiv X^{x,\alpha,\beta^*}.$$

It suffices to show

$$\mathbb{E}[\,u(\Theta_\tau)\,] \le \mathbb{E}[\,u(Z_{\tau^*})\,] \le \mathbb{E}[\,u(H_{\tau^*})\,]. \tag{6}$$

Indeed, these inequalities mean that the pair $(\alpha^*(\cdot), \tau^*)$ is optimal for player $\mathfrak{A}$, when player $\mathfrak{B}$ chooses $\beta^*(\cdot)$; and that $\beta^*(\cdot)$ is optimal for player $\mathfrak{B}$, when player $\mathfrak{A}$ chooses $(\alpha^*(\cdot), \tau^*)$.

In other words, $\big((\alpha^*(\cdot), \tau^*),\, \beta^*(\cdot)\big)$ is a saddle-point for the game $\mathfrak{G}_1(x)$, and this game has value

$$\mathbb{E}\,[\,u(Z_{\tau^*})\,] = \mathbb{E}\left[u\left(X^{x,\alpha^*,\beta^*}_{\tau^*} \right) \right] = U(x)$$

equal to the middle term in (6). This is the same as the value $U(x)$ of the optimal stopping problem for the diffusion process Z, as in (4).

Furthermore, it suffices to prove (6) when $m = 1$, to wit, when the maximum of $u(\cdot)$ occurs at the right-endpoint of I. This is because it is obviously optimal for player $\mathfrak{A}$ to stop at m, so the intervals $(0, m)$ and $(m, 1)$ can be treated separately and similarly.

• To prove the second inequality of (6), consider the semimartingale $\widetilde{H} := p(H)$ which satisfies

$$\begin{aligned} d\widetilde{H}_t &= p'(H_t)\, dH_t \,+\, \frac{1}{2}\, p''(H_t)\, \sigma_H^2(t)\, dt \\ &= p'(H_t)\, [\, \mu_H(t)\, dt + \sigma_H(t)\, dW_t\,] \,+\, \frac{1}{2}\, p''(H_t)\, \sigma_H^2(t)\, dt \end{aligned}$$

or equivalently

$$d\widetilde{H}_t \;=\; \sigma_H^2(t) \left[\frac{1}{2}\, p''(H_t) + \frac{\mu_H(t)}{\sigma_H^2(t)} \cdot p'(H_t) \right] dt \,+\, p'(H_t)\, \sigma_H(t)\, dW_t\,, \quad (7)$$

where we have set

$$\mu_H(t) \;:=\; \mu\big(H_t, \alpha^*(H_t), \beta(H_t)\big) \quad \text{and} \quad \sigma_H(t) \;:=\; \sigma\big(H_t, \alpha^*(H_t), \beta(H_t)\big)\,.$$

By condition $C.1$ and the positivity of $p'(\cdot)$, we have

$$\begin{aligned} &\frac{1}{2}\, p''(H_t) + \frac{\mu_H(t)}{\sigma_H^2(t)} \cdot p'(H_t) \\ &\geq \frac{1}{2}\, p''(H_t) + \left(\frac{\mu}{\sigma^2}\right) \big(H_t, \alpha^*(H_t), \beta^*(H_t)\big) \cdot p'(H_t) \\ &= \frac{1}{2}\, p''(H_t) + \frac{b^*(H_t)}{\big(s^*(H_t)\big)^2} \cdot p'(H_t) \;=\; 0 \end{aligned}$$

from (3). Therefore, the process $\widetilde{H} = p(H)$ is a local submartingale.

Now let us look at the semimartingale $U(H_t) = \widetilde{U}(\widetilde{H}_t)$, $0 \leq t < \infty$. By the generalized Itô rule for concave functions (cf. Karatzas & Shreve (1991), section 3.7), this process satisfies

$$\begin{aligned} U(H_T) \;=\; \widetilde{U}(\widetilde{H}_T) \;=\; U(x) + \int_0^T D^- \widetilde{U}(\widetilde{H}_t) \cdot d\widetilde{H}_t \;-\; \int_{\widetilde{I}} L_T^{\widetilde{H}}(\zeta)\, \widetilde{\nu}(d\zeta) \\ = U(x) + \int_0^T D^- \widetilde{U}(p(H_t)) \cdot \sigma_H^2(t) \left[\frac{1}{2}\, p''(H_t) + \frac{\mu_H(t)}{\sigma_H^2(t)} \cdot p'(H_t) \right] dt \\ + \int_0^T D^- \widetilde{U}(p(H_t)) \cdot p'(H_t)\, \sigma_H(t)\, dW_t \;-\; \int_{\widetilde{I}} L_T^{\widetilde{H}}(\zeta)\, \widetilde{\nu}(d\zeta)\,. \quad (8) \end{aligned}$$

We are using the notation $L_T^{\Upsilon}(\zeta)$ for the local time of a continuous semimartingale Υ, accumulated by calendar time T at the site ζ. The last integral of (8) is equal to

$$\int_{\widetilde{I} \setminus \widetilde{\mathcal{O}}} L_T^{\widetilde{H}}(\zeta)\, \widetilde{\nu}(d\zeta)\,,$$

and therefore vanishes for $T = \tau^*(H)$; just recall that $\widetilde{\nu}(\widetilde{\mathcal{O}}) = 0$, $\widetilde{\mathcal{O}} = p(\mathcal{O})$, and observe that the stopping rule $\tau^*(H)$ of (5) can be written as

$$\tau^*(H) \;=\; \inf\{t \geq 0 \,:\, \widetilde{U}(\widetilde{H}_t) = \widetilde{u}(\widetilde{H}_t)\} \;=\; \inf\{t \geq 0 \,:\, \widetilde{H}_t \notin \widetilde{\mathcal{O}}\,\}\,. \quad (9)$$

Thus, the nonnegativity of $D^-\widetilde{U}(\cdot)$ and of the term in brackets on the second line of (8) guarantee that $U(H_{\cdot\wedge\tau^*})$ is a bounded local submartingale, hence a (bounded, and genuine) submartingale. Consequently, the optional sampling theorem gives

$$U(x) \leq \mathbb{E}[\,U(H_{\tau^*})\,] = \mathbb{E}[\,u(H_{\tau^*})\,].$$

For this last equality, we have used (9) and the property

$$\tau^*(X) < \infty \quad \text{a.s., for every process } X \equiv X^{x,\alpha,\beta} \quad \text{as in (1);}$$

this is a consequence of the strong non-degeneracy condition C.2. Since $U(x) = \mathbb{E}[\,u(Z_{\tau^*})\,]$, the proof of the second inequality in (6) is now complete.

- To verify the first inequality of (6), namely

$$U(x) = \mathbb{E}[\,u(Z_{\tau^*})\,] \geq \mathbb{E}[\,u(\Theta_\tau)\,],$$

we observe that an analysis of the processes $\widetilde{\Theta} := p(\Theta)$ and $\widetilde{U}(\widetilde{\Theta}) \equiv U(\Theta)$, similar to that carried out above, shows they are local supermartingales, and that $U(\Theta)$ is a genuine supermartingale because it is bounded. By the optional sampling theorem we obtain then

$$U(x) \geq \mathbb{E}[\,U(\Theta_\tau)\,] \geq \mathbb{E}[\,u(\Theta_\tau)\,]$$

for an arbitrary stopping rule $\tau \in \mathcal{S}$. □

Remark 2: Before proceeding to the next section it is useful to recall that, in the special case where player $\mathfrak{B}$ is a dummy, we have solved a one-person game control-and-stopping problem for player $\mathfrak{A}$. That is, against any fixed control function $\beta(\cdot)$ for player $\mathfrak{B}$, it is optimal for player $\mathfrak{A}$ to choose an $\alpha(\cdot)$ that maximizes (μ/σ^2) to the left of m and minimizes (μ/σ^2) to the right of m; and then to choose a rule τ which is optimal for stopping the process

$$u\big(X^{x,\alpha,\beta}\big).$$

This essentially recovers, and slightly extends, the main result of Karatzas & Sudderth (1999).

General results about the existence of values for zero-sum stochastic differential games have been established in the literature, using both analytical and probabilistic tools; see, for instance, Friedman (1972), Elliott (1976), Nisio (1988) and Fleming & Souganidis (1989).

3. A Non-Zero-Sum Game of Control and Stopping

Suppose now that *both* players $\mathfrak{A}$ and $\mathfrak{B}$ have continuous reward functions, $u(\cdot)$ and $v(\cdot)$ respectively, which map the unit interval [0,1] into the real line. Will shall assume that $u(\cdot)$ and $v(\cdot)$ attain their maximum values over the interval [0,1] at unique points m and ℓ, respectively; and without loss of generality, we shall take $\ell \le m$.

As in the previous section, the two players choose control functions $\alpha(\cdot)$ and $\beta(\cdot)$, but now we assume that *both* players can select stopping rules, say τ and ρ respectively. The dynamics of the process $X \equiv X^{x,\alpha,\beta}$ are given by (1), just as before. The expected rewards to the two players resulting from their choices of $(\alpha(\cdot),\tau)$ and $(\beta(\cdot),\rho)$ at the initial state $x \in I$ are defined to be

$$\mathbb{E}[\, u(X_\tau)\,] \qquad \text{and} \qquad \mathbb{E}[\, v(X_\rho)\,]\,,$$

respectively. Let $\mathfrak{G}_2(x)$ denote this game.

As this game is typically *not* zero-sum, we seek a *Nash equilibrium:* namely, pairs $(\alpha^*(\cdot),\tau^*)$ and $(\beta^*(\cdot),\rho^*)$ such that

$$\mathbb{E}\left[u\left(X_\tau^{x,\alpha,\beta^*}\right)\right] \le \mathbb{E}\left[u\left(X_{\tau^*}^{x,\alpha^*,\beta^*}\right)\right], \qquad \forall\ (\alpha(\cdot),\tau)$$

and

$$\mathbb{E}\left[v\left(X_\rho^{x,\alpha^*,\beta}\right)\right] \le \mathbb{E}\left[v\left(X_{\rho^*}^{x,\alpha^*,\beta^*}\right)\right], \qquad \forall\ (\beta(\cdot),\rho)\,.$$

This simply says that the choice $(\alpha^*(\cdot),\tau^*)$ is optimal for player $\mathfrak{A}$ when player $\mathfrak{B}$ uses $(\beta^*(\cdot),\rho^*)$, and vice-versa.

Speaking intuitively: in positions to the left of the point ℓ, both players want the process X to move to the right; in the interval (ℓ, m), player $\mathfrak{A}$ wants the process to move to the right, and player $\mathfrak{B}$ wants it to move to the left; and at positions to the right of m, both players prefer X to move to the left. Here, "moving to the right" (resp., to the left) is a euphemism for "trying to maximize (resp., to minimize) the signal-to-noise ratio (μ/σ^2)"; recall the discussion in Remark 4.1 of Karatzas & Sudderth (1999), in connection with the problem of that paper and with the *goal problem* of Pestien & Sudderth (1985).

These considerations suggest the following condition.

C.3: Local Nash Equilibrium Condition. *There exist measurable functions $\alpha^*(\cdot)$, $\beta^*(\cdot)$ such that $(\alpha^*(z), \beta^*(z)) \in A(z) \times B(z)$ holds for all $z \in (0,1)$;*

$$\left(\frac{\mu}{\sigma^2}\right)(z,a,b) \le \left(\frac{\mu}{\sigma^2}\right)(z,\alpha^*(z),\beta^*(z)), \quad \forall\ (a,b) \in A(z) \times B(z)$$

holds for every $z \in (0, \ell)$ *;*

$$\left(\frac{\mu}{\sigma^2}\right)(z, a, \beta^*(z)) \le \left(\frac{\mu}{\sigma^2}\right)(z, \alpha^*(z), \beta^*(z))$$
$$\le \left(\frac{\mu}{\sigma^2}\right)(z, \alpha^*(z), b)\,, \quad \forall\ (a,b) \in A(z) \times B(z)$$

holds for every $z \in (\ell, m)$ *; and*

$$\left(\frac{\mu}{\sigma^2}\right)(z, \alpha^*(z), \beta^*(z)) \le \left(\frac{\mu}{\sigma^2}\right)(z, a, b)\,, \quad \forall\ (a,b) \in A(z) \times B(z)$$

holds for every $z \in (m, 1)$ *.*

Suppose there exist measurable functions $\alpha^*(\cdot)$ and $\beta^*(\cdot)$ that satisfy condition $C.3$, and construct, for each fixed $x \in I$, the diffusion process $Z^x \equiv X^{x,\alpha^*,\beta^*}$ of (2), (3). Next, consider the optimal stopping problems for these two players with value functions

$$U(x) := \sup_{\tau \in \mathcal{S}} \mathbb{E}[\, u(Z^x_\tau)\,] \quad \text{and} \qquad V(x) := \sup_{\rho \in \mathcal{S}} \mathbb{E}[\, v(Z^x_\rho)\,]\,, \qquad x \in I\,,$$

respectively. Define also the stopping rules

$$\tau^*(\xi) := \inf\{t \ge 0 \,:\, U(\xi_t) = u(\xi_t)\}$$

and

$$\rho^*(\xi) := \inf\{t \ge 0 \,:\, V(\xi_t) = v(\xi_t)\}\,, \quad \xi \in \mathfrak{Z}\,.$$

Theorem 2: *Assume that the strong non-degeneracy condition C.2 holds, and that there exist measurable functions* $\alpha^*(\cdot)$, $\beta^*(\cdot)$ *satisfying the local Nash equilibrium condition C.3.*

For each $x \in I$, *construct the process* $Z^x \equiv X^{x,\alpha^*,\beta^*}$ *of (2) and (3). Then the pairs* $(\alpha^*(\cdot), \tau^*)$ *and* $(\beta^*(\cdot), \rho^*)$ *form a Nash equilibrium for the game* $\mathfrak{G}_2(x)$ *.*

Proof : Consider the control-and-stopping problem faced by player $\mathfrak{A}$, when player $\mathfrak{B}$ chooses the pair $(\beta^*(\cdot), \rho^*)$. By condition $C.3$, the control function $\alpha^*(\cdot)$ maximizes the ratio (μ/σ^2) to the left of m, and minimizes it to the right of m. Also, the rule τ^* is optimal for stopping the process $u(Z^x)$ (player $\mathfrak{A}$). Thus, by Remark 2 at the end of section 2, we see that the pair $(\alpha^*(\cdot), \tau^*)$ is optimal for player $\mathfrak{A}$ when player $\mathfrak{B}$ chooses the pair $(\beta^*(\cdot), \rho^*)$.

Similar considerations show that the pair $(\beta^*(\cdot), \rho^*)$ is optimal for player $\mathfrak{B}$ when player $\mathfrak{A}$ chooses the pair $(\alpha^*(\cdot), \tau^*)$. □

It is not difficult to extend Theorem 2 to any finite number of players. General existence results for Nash equilibria in non-zero-sum stochastic differential games have been established in the literature; see, for instance, Uchida (1978, 1979), Bensoussan & Frehse (2000), Buckdhan et al. (2004). For examples of special games with explicit solutions, see Mazumdar & Radner (1991), Nilakantan (1993) and Brown (2000).

4. A General Game of Stopping

In this section we address a game of stopping, which will be crucial to our study of the control-and-stopping game in the next section. This stopping game is a cousin of the so-called *Dynkin (1967) Games,* which have been studied by a number of authors; see, for instance, Friedman (1973), Bensoussan & Friedman (1974, 1977), Neveu (1975), Bismut (1977), Stettner (1982), Alario-Nazaret et al. (1982), Morimoto (1984), Lepeltier & Maingueneau (1984), Ohtsubo (1987), Hamadène & Lepeltier (1995), Karatzas (1996), Cvitanić & Karatzas (1996), Karatzas & Wang (2001), Touzi & Vieille (2002), Fukushima & Taksar (2002) and Boetius (2005), among others.

Let $\mathfrak{X}$ be a locally compact metric space and suppose that, for each $x \in \mathfrak{X}$, the process $Z^x = \{Z^x_t,\, 0 \le t < \infty\}$ is a continuous strong Markov process with $Z^x_0 = x$ and values in the space $\mathfrak{X}$. We shall assume that Z^x is *standard* in the sense of Shiryaev (1978), p.19. (For the application to the next section, Z^x will be a diffusion process in the interval (0,1) with absorption at the endpoints, just as it was in sections 2 and 3.)

Consider now, for each $x \in \mathfrak{X}$, a two-person game $\mathfrak{G}_3(x)$ in which players $\mathfrak{A}$ and $\mathfrak{B}$ have bounded, continuous reward functions $u(\cdot)$ and $v(\cdot)$, respectively, that map $\mathfrak{X}$ into the real line. Player $\mathfrak{A}$ chooses a stopping rule τ, player $\mathfrak{B}$ chooses a stopping rule ρ, and their respective expected rewards are

$$\mathbb{E}[\,u(Z^x_{\tau\wedge\rho})\,] \qquad \text{and} \qquad \mathbb{E}[\,v(Z^x_{\tau\wedge\rho})\,]\,.$$

The resulting non-zero-sum game of timing has a trivial Nash equilibrium $\check{\tau} = \check{\rho} = 0$, which leaves player $\mathfrak{A}$ with reward $u(x)$ and player $\mathfrak{B}$ with reward $v(x)$. A somewhat less trivial Nash equilibrium is described in Theorem 3 and its proof below.

Theorem 3: *There exist stopping rules τ^* and ρ^* such that*

$$\mathbb{E}[\,u(Z^x_{\tau^*\wedge\rho^*})\,] \ge \mathbb{E}[\,u(Z^x_{\tau\wedge\rho^*})\,], \qquad \forall\ \tau \in \mathcal{S}$$

and

$$\mathbb{E}[\,v(Z^x_{\tau^*\wedge\rho^*})\,] \ge \mathbb{E}[\,v(Z^x_{\tau^*\wedge\rho})\,], \qquad \forall\ \rho \in \mathcal{S}$$

hold for each $x \in \mathfrak{X}$, to wit, the game $\mathfrak{G}_3(x)$ has (τ^*, ρ^*) as a Nash equilibrium; in this equilibrium the expected rewards of players $\mathfrak{A}$ and $\mathfrak{B}$ satisfy

$$\mathbb{E}[\,u(Z^x_{\tau^* \wedge \rho^*})\,] \geq u(x) \qquad \textit{and} \qquad \mathbb{E}[\,v(Z^x_{\tau^* \wedge \rho^*})\,] \geq v(x)\,, \qquad \forall\ x \in \mathfrak{X}\,, \tag{10}$$

respectively; and the inequalities of (10) can typically be strict.

The stopping rules τ^*, ρ^* of the theorem will be constructed as limits of two decreasing sequences $\{\tau_n\}$ and $\{\rho_n\}$ of stopping rules. The specific structure of these sequences will be important for the application we undertake in the next section.

These stopping rules will be defined inductively, in the order τ_1, ρ_1, τ_2, ρ_2, $\cdots$, as solutions to a sequence of optimal stopping problems. First, let us define

$$\begin{aligned} U_1(x) &:= \sup_{\tau \in \mathcal{S}} \mathbb{E}[\,u(Z^x_\tau)\,] \\ \tau_1(\xi) &:= \inf\{t \geq 0 \,:\, \xi_t \in F_1\,\}\,, \quad \xi \in \mathfrak{Z}\,, \end{aligned} \tag{11}$$

where

$$F_1 := \{x \in \mathfrak{X} \,:\, U_1(x) = u(x)\}\,.$$

Next, let

$$\begin{aligned} V_1(x) &:= \sup_{\rho \in \mathcal{S}} \mathbb{E}[\,v(Z^x_{\tau_1 \wedge \rho})\,]\,, \\ \rho_1(\xi) &:= \inf\{t \geq 0 \,:\, \xi_t \in G_1\,\}\,, \quad \xi \in \mathfrak{Z}\,, \end{aligned}$$

where

$$G_1 := \{x \in \mathfrak{X} \,:\, V_1(x) = v(x)\}\,.$$

It is well known (e.g. Shiryaev (1978)) that the stopping rules τ_1 and ρ_1 are optimal in their respective problems, in the sense that $U_1(x) = \mathbb{E}[\,u(Z^x_{\tau_1})\,]$ and $V_1(x) = \mathbb{E}[\,v(Z^x_{\tau_1 \wedge \rho_1})\,]$; and that the functions $U_1(\cdot)$, $V_1(\cdot)$ are continuous. Thus the optimal stopping regions for these two problems, namely, F_1 and G_1, are closed sets.

Suppose now that we have already constructed: the stopping rules τ_1, $\rho_1, \cdots, \tau_n$, ρ_n; the continuous functions $U_1(\cdot)$, $V_1(\cdot), \cdots, U_n(\cdot)$, $V_n(\cdot)$; and the closed sets F_1, G_1, $\cdots$, F_n, G_n. Let

$$\begin{aligned} U_{n+1}(x) &:= \sup_{\tau \in \mathcal{S}} \mathbb{E}[\,u(Z^x_{\tau \wedge \rho_n})\,]\,, \\ F_{n+1} &:= \{x \in \mathfrak{X} \,:\, U_{n+1}(x) = u(x)\}\,, \\ \tau_{n+1}(\xi) &:= \inf\{t \geq 0 \,:\, \xi_t \in F_{n+1}\,\}\,, \quad \xi \in \mathfrak{Z}\,; \end{aligned}$$

as well as

$$V_{n+1}(x) := \sup_{\rho \in \mathcal{S}} \mathbb{E}[\, v(Z^x_{\tau_{n+1} \wedge \rho})\,]\,,$$
$$G_{n+1} := \{x \in \mathfrak{X} \,:\, V_{n+1}(x) = v(x)\}$$

and

$$\rho_{n+1}(\xi) := \inf\{t \geq 0 \,:\, \xi_t \in G_{n+1}\}\,, \quad \xi \in \mathfrak{Z}\,.$$

Observe that

$$u(x) \leq U_2(x) = \sup_{\tau \in \mathcal{S}} \mathbb{E}[\, u(Z^x_{\tau \wedge \rho_1})\,] \leq U_1(x)\,, \qquad x \in \mathfrak{X}\,.$$

Thus $F_1 \subseteq F_2$ and so $\tau_1 \geq \tau_2$. Similarly, $G_1 \subseteq G_2$ and $\rho_1 \geq \rho_2$. An argument by induction shows that $\tau_n \geq \tau_{n+1}$ and $\rho_n \geq \rho_{n+1}$, as well as

$$u(x) \leq U_{n+1}(x) \leq U_n(x)\,, \qquad v(x) \leq V_{n+1}(x) \leq V_n(x)\,,$$

hold for all $n \in \mathbb{N}$.

Also by construction, we have

$$\mathbb{E}[\, u(Z^x_{\tau_{n+1} \wedge \rho_n})\,] \geq \mathbb{E}[\, u(Z^x_{\tau \wedge \rho_n})\,] \tag{12}$$

and

$$\mathbb{E}[\, v(Z^x_{\tau_n \wedge \rho_n})\,] \geq \mathbb{E}[\, v(Z^x_{\tau_n \wedge \rho})\,]\,, \tag{13}$$

for all integers n and stopping rules τ, ρ. Define the decreasing limits

$$\tau^* := \lim_{n \to \infty} \downarrow \tau_n\,, \qquad \rho^* := \lim_{n \to \infty} \downarrow \rho_n\,,$$

and let $n \to \infty$ in (12), (13).

The above construction shows also that, for every integer $n \in \mathbb{N}$, the functions $U_n(\cdot)$, $V_n(\cdot)$ are continuous and the sets F_n, G_n closed.

On the other hand, it is clear that the inequalities in (10) can be strict. For instance, take $u(\cdot) \equiv v(\cdot)$ and suppose that the optimal stopping problem of (11) is not trivial, that is, has a non-empty optimal continuation region $\mathcal{O}_1 := \mathfrak{X} \setminus F_1 = \{x \in \mathfrak{X} \,:\, U_1(x) > u(x)\}$. Then $\rho_1 = \tau_1$, $V_1(\cdot) \equiv U_1(\cdot)$ as well as $\rho_2 = \rho_1 = \tau_1$, $U_2(\cdot) \equiv U_1(\cdot)$, and by induction: $V_n(\cdot) \equiv U_n \equiv U_1(\cdot)$, $\tau_n = \rho_n = \tau_1$ for all n. For every $x \in \mathcal{O}_1$, the inequalities in (10) are then strict, since

$$\mathbb{E}[\, u(Z^x_{\tau^* \wedge \rho^*})\,] = \mathbb{E}[\, u(Z^x_{\tau_1})\,] = U_1(x) > u(x)\,. \qquad \square$$

Remark 3: It is straightforward to generalize Theorem 3 to the case of $K \geq 2$ players with reward functions $u_1(\cdot), \cdots, u_K(\cdot)$, who choose stopping rules $\tau_1, \cdots, \tau_K$ and receive payoffs

$$\mathbb{E}[\, u_i(Z^x_{\tau_1 \wedge \cdots \wedge \tau_K})\,]\,, \qquad i = 1, \cdots, K\,.$$

5. Another Non-Zero-Sum Game of Control and Stopping

For each point x in the interval $I = (0,1)$, let $\mathfrak{G}_4(x)$ be a two-player game which is the same as the game $\mathfrak{G}_2(x)$ of section 3, except that the payoffs to the two players, resulting from their choices $(\alpha(\cdot), \tau)$ and $(\beta(\cdot), \rho)$, are now, respectively,

$$\mathbb{E}\left[u\left(X^{x,\alpha,\beta}_{\tau\wedge\rho}\right)\right] \quad \text{and} \quad \mathbb{E}\left[v\left(X^{x,\alpha,\beta}_{\tau\wedge\rho}\right)\right].$$

Just like the game of the previous section, this one has the trivial Nash equilibrium $(\breve{\alpha}(\cdot), \breve{\tau})$ and $(\breve{\beta}(\cdot), \breve{\rho})$, with arbitrary $\breve{\alpha}(\cdot)$, $\breve{\beta}(\cdot)$ and with $\breve{\tau} = \breve{\rho} = 0$.

A somewhat less trivial equilibrium for this game is constructed in Theorem 4 below. It uses the additional assumption that the reward functions $u(\cdot)$ and $v(\cdot)$ are *unimodal* with unique maxima at m and ℓ, respectively, and with $0 \le \ell \le m \le 1$. To say that $u(\cdot)$, for example, is unimodal, means that $u(\cdot)$ is increasing on the interval $(0, m)$ and decreasing on the interval $(m, 1)$.

Theorem 4: *Assume that the strong non-degeneracy condition C.2 holds; that there exist measurable functions $\alpha^*(\cdot)$, $\beta^*(\cdot)$ satisfying the local Nash equilibrium condition C.3; and that the reward functions $u(\cdot)$, $v(\cdot)$ are unimodal. For any given $x \in I$, define the process $Z^x \equiv X^{x,\alpha^*,\beta^*}$ as in (2) and (3).*

Let τ^*, ρ^* be the Nash equilibrium of Theorem 3 for the stopping game $\mathfrak{G}_3(x)$ of section 4. Then the pairs $(\alpha^*(\cdot), \tau^*)$ and $(\beta^*(\cdot), \rho^*)$ form a Nash equilibrium for the game $\mathfrak{G}_4(x)$.

Proof : We must establish the comparisons

$$\mathbb{E}\left[u\left(X^{x,\alpha,\beta^*}_{\tau\wedge\rho^*}\right)\right] \le \mathbb{E}\left[u\left(X^{x,\alpha^*,\beta^*}_{\tau^*\wedge\rho^*}\right)\right], \qquad \forall\ (\alpha(\cdot), \tau) \tag{14}$$

and

$$\mathbb{E}\left[v\left(X^{x,\alpha^*,\beta}_{\tau^*\wedge\rho}\right)\right] \le \mathbb{E}\left[v\left(X^{x,\alpha^*,\beta^*}_{\tau^*\wedge\rho^*}\right)\right], \qquad \forall\ (\beta(\cdot), \rho). \tag{15}$$

It suffices to show that for all $n \in \mathbb{N}$, we have

$$\mathbb{E}\left[u\left(X^{x,\alpha,\beta^*}_{\tau\wedge\rho_n}\right)\right] \le \mathbb{E}\left[u\left(X^{x,\alpha^*,\beta^*}_{\tau_{n+1}\wedge\rho_n}\right)\right], \qquad \forall\ (\alpha(\cdot), \tau), \tag{16}$$

and

$$\mathbb{E}\left[v\left(X^{x,\alpha^*,\beta}_{\tau_n\wedge\rho}\right)\right] \le \mathbb{E}\left[v\left(X^{x,\alpha^*,\beta^*}_{\tau_n\wedge\rho_n}\right)\right], \qquad \forall\ (\beta(\cdot), \rho), \tag{17}$$

where the stopping rules τ_1, ρ_1, τ_2, ρ_2, $\cdots$ are as defined in section 4. The latter inequalities are sufficient, because we can obtain (14) and (15) from them by passing to the limit as $n \to \infty$.

Let us prove the first of these latter inequalities – the second, (17), will follow then by symmetry. This inequality (16) says that $(\alpha^*(\cdot), \tau_{n+1})$ is optimal for player $\mathfrak{A}$ in the one-person problem of control-and-stopping that occurs when player $\mathfrak{B}$ fixes the pair $(\beta^*(\cdot), \rho_n)$. (See Remark 2, at the end of section 2.)

Consider two cases: namely, when the initial state x belongs to G_n, the stopping region for the rule ρ_n, and when x is not an element of G_n.

- If $x \in G_n$, then ρ_n stops the process immediately, and every pair $(\alpha(\cdot), \tau)$ that player $\mathfrak{A}$ may choose is optimal.
- If x belongs to the open set $\mathcal{O}_n := [0,1] \setminus G_n$, then there exists an interval (q,r) such that $x \in (q,r) \subseteq \mathcal{O}_n$ and the process $X^{x,\alpha,\beta^*}_{\cdot\wedge\tau\wedge\rho_n}$ is absorbed at the endpoints of this interval, for all $(\alpha(\cdot), \tau)$.

It is clearly optimal for player $\mathfrak{A}$ to stop at the point m where $u(\cdot)$ attains its maximum, so we can assume that the interval (q,r) lies to one side of m. Suppose, to be precise, that $x \in (q,r) \subseteq \mathcal{O}_n \cap (0,m)$. Because it is unimodal, the function $u(\cdot)$ increases on (q,r) and attains its maximum over the interval $[q,r]$ at the point r. By Remark 2 and the condition $C.3$, it is optimal for player $\mathfrak{A}$ to use the control function $\alpha^*(\cdot)$, and then to choose a rule that is optimal for stopping the process $X^{x,\alpha^*,\beta^*}_{\cdot\wedge\rho_n} \equiv Z^{x}_{\cdot\wedge\rho_n}$, namely, the stopping rule τ_{n+1}. This completes the proof. $\square$

As in previous sections, we conjecture that Theorem 4 can be extended to a $K-$person game in a straightforward manner. However, it may be more difficult to generalize to reward functions that are not unimodal.

6. Generalizations

The results of the paper can easily be extended to much more general information structures on the part of the two players, than we have indicated so far.

Consider, for example, the zero-sum game $\mathfrak{G}_1(x)$ of section 2. Suppose that on some filtered probability space $(\Omega, \mathcal{F}, \mathbb{P})$, $\mathfrak{F} = \{\mathcal{F}_t\}_{0\le t<\infty}$ we can construct, not just the diffusion process Z of (2), but also progressively measurable processes H and Θ satisfying the stochastic differential equations with random coëfficients

$$dH_t = \mu\,(H_t, \alpha^*(H_t), \mathfrak{b}_t)\; dt + \sigma\,(H_t, \alpha^*(H_t), \mathfrak{b}_t)\; dW_t\,, \qquad H_0 = x$$

and

$$d\Theta_t = \mu\,(\Theta_t, \mathfrak{a}_t, \beta^*(\Theta_t))\; dt + \sigma\,(\Theta_t, \mathfrak{a}_t, \beta^*(\Theta_t))\; dW_t\,, \qquad \Theta_0 = x\,,$$

respectively. Here $\mathfrak{a}$ and $\mathfrak{b}$ are progressively measurable processes, with

$$\mathfrak{a}_t \in A(\Theta_t) \qquad \text{and} \qquad \mathfrak{b}_t \in B(H_t)\,, \qquad \forall\, 0 \le t < \infty\,.$$

Then the same analysis as in the proof of Theorem 1, leads to the comparisons

$$\mathbb{E}[\,u(\Theta_{\mathfrak{t}})\,] \le \mathbb{E}[\,u(Z_{\tau^*})\,] \le \mathbb{E}[\,u(H_{\tau^*})\,]$$

as in (6), for an arbitrary stopping time $\mathfrak{t}$ of the filtration $\mathfrak{F}$.

In other words: the pair $(\alpha^*(\cdot), \tau^*)$ is optimal for player $\mathfrak{A}$ against all such pairs $(\mathfrak{a}, \mathfrak{t})$, and the function $\beta^*(\cdot)$ is optimal for player $\mathfrak{B}$ against all such $\mathfrak{b}$.

7. Acknowledgements

We express our thanks to Dr. Constantinos Kardaras for pointing out and helping us remove an unnecessary condition in an earlier version of this paper, and to Dr. Mihai Sirbu for his interest in the problems discussed in this paper, his careful reading of the manuscript and his many suggestions.

References

1. ALARIO-NAZARET, M., LEPELTIER, J.P. & MARCHAL, B. (1982) Dynkin Games. *Lecture Notes in Control & Information Sciences* **43**, 23-32. Springer-Verlag, Berlin.
2. BENSOUSSAN, A. & FREHSE, J. (2000) Stochastic games for n players. *J. Optimiz. Theory Appl.* **105**, 543-565.
3. BENSOUSSAN, A. & FRIEDMAN, A. (1974) Nonlinear variational inequalities and differential games with stopping times. *Journal of Functional Analysis* **16**, 305-352.
4. BENSOUSSAN, A. & FRIEDMAN, A. (1977) Non-zero sum stochastic differential games with stopping times and free-boundary problems. *Trans. Amer. Math. Society* **231**, 275-327.
5. BOETIUS, F. (2005) Bounded variation singular control and Dynkin game. *SIAM Journal on Control & Optimization* **44**, 1289-1321.
6. BROWN, S. (2000) Stochastic differential portfolio games. *J. Appl. Probab.* **37**, 126-147.
7. BISMUT, J.M. (1977) Sur un problème de Dynkin. *Z. Wahrsch. verw. Gebiete* **39**, 31-53.
8. BUCKDHAN, R., CARDALIAGUET, P. & RAINER, Ch. (2004) Nash equilibirum payoffs for stochastic dufferential games. *SIAM Journal on Control & Optimization* **43**, 624-642.
9. CHOW, Y.S., ROBBINS, H.E. & SIEGMUND, D. (1971) *Great Expectations: The Theory of Optimal Stopping.* Houghton Mifflin, Boston.

10. CVITANIĆ, J. & KARATZAS, I. (1996) Backward stochastic differential equations with reflection and Dynkin games. *Annals of Probability* **24**, 2024-2056.
11. DAYANIK, S. & KARATZAS, I. (2003) On the optimal stopping problem for one-dimensional diffusions. *Stochastic Processes & Applications* **107**, 173-212.
12. DYNKIN, E.B. (1967) Game variant of a problem on optimal stopping. *Soviet Mathematics Doklady* **10**, 270-274.
13. DYNKIN, E.B. & YUSHKEVICH, A.A. (1969) *Markov Processes: Theorems and Problems.* Plenum Press, New York.
14. ELLIOTT, R. J. (1976) The existence of value in stochastic dufferential games. *SIAM Journal on Control & Optimization* **14**, 85-94.
15. FLEMING, W.H. & SOUGANIDIS, P. (1989) On the existence of value functions in two-player, zero-sum stochastic differential games. *Indiana Univ. Math. Journal* **38**, 293-314.
16. FRIEDMAN, A. (1972) Stochastic differential games. *J. Differential Equations* **11**, 79-108.
17. FRIEDMAN, A. (1973) Stochastic games and variational inequalities. *Arch. Rational Mech. Anal.* **51**, 321-346.
18. FUKUSHIMA, M. & TAKSAR, M. (2002) Dynkin games via Dirichlet forms, and singular control of one-dimensional diffusions. *SIAM Journal on Control & Optimization* **41**, 682-699.
19. HAMADÈNE, S. & LEPELTIER, J.P. (1995) Zero-sum stochastic differential games and backward equations. *Systems & Control Letters* **24**, 259-263.
20. KARATZAS, I. (1996) A pathwise approach to Dynkin games. In *Statistics, Probability and Game Theory: Papers in Honor of David Blackwell* (T.S. Ferguson, L.S. Shapley & J.B. MacQueen, Editors.) *IMS Lecture Notes-Monograph Series* **30**, 115-125.
21. KARATZAS, I. & SHREVE, S.E. (1991) *Brownian Motion and Stochastic Calculus.* Springer-Verlag, New York.
22. KARATZAS, I. & SUDDERTH, W.D. (1999) Control and stopping of a diffusion process on an interval. *Annals of Applied Probability* **9**, 188-196.
23. KARATZAS, I. & SUDDERTH, W.D. (2001) The controller and stopper game for a linear diffusion. *Annals of Probability* **29**, 1111-1127.
24. KARATZAS, I. & WANG, H. (2001) Connections between bounded variation controller and Dynkin games. In *Optimal Control and Partial Differential Equations*, Volume in Honor of A. Bensoussan. IOS Press, Amsterdam, 363-373.
25. LEPELTIER, J.M. & MAINGUENEAU, M.A. (1984) Le jeu de Dynkin en théorie générale sans l' hypothèse de Mokobodski. *Stochastics* **13**, 25-44.
26. MAITRA, A.P. & SUDDERTH, W.D. (1996) The gambler and the stopper. In *Statistics, Probability and Game Theory: Papers in Honor of David Blackwell* (T.S. Ferguson, L.S. Shapley & J.B. MacQueen, Editors.) *IMS Lecture Notes-Monograph Series* **30**, 191-208.
27. MORIMOTO, H. (1984) Dynkin games and martingale methods. *Stochastics* **13**, 213-228.

28. NEVEU, J. (1975) *Discrete-Parameter Martingales.* North-Holland, Amsterdam.
29. NILAKANTAN, L. (1993) *Continuous-Time Stochastic Games.* Doctoral Dissertation, University of Minnesota.
30. NISIO, M. (1988) Stochastic dufferential games and viscosity solutions of the Isaacs equation. *Nagoya Mathematical Journal* **110**, 163-184.
31. OHTSUBO, Y. (1987) A non-zero-sum extension of Dynkin's stopping problem. *Math. Oper. Research* **12**, 277-296.
32. PESTIEN, V.C. & SUDDERTH, W.D. (1985) Continuous-time red-and-black: how to control a diffusion to a goal. *Mathematics of Operations Research* **10**, 599-611.
33. SHIRYAEV, A.N. (1978) *Optimal Stopping Rules.* Springer-Verlag, New York.
34. SNELL, J.L. (1952) Applications of martingale systems theorems. *Trans. Amer. Math. Society* **73**, 293-312.
35. STETTNER, L. (1982) Zero-sum Markov games with stopping and impulsive strategies. *Applied Mathematics & Optimization* **9**, 1-24.
36. TOUZI, N. & VIEILLE, N. (2002) Continuous-time Dynkin games with mixed strategies. *SIAM Journal on Control & Optimization* **41**, 1073-1088.
37. UCHIDA, K. (1978) On existence of a Nash equilibrium point in n-person, non-zero-sum stochastic dufferential games. *SIAM Journal on Control & Optimization* **16**, 142-149.
38. UCHIDA, K. (1979) A note on the existence of a Nash equilibrium point in stochastic differential games. *SIAM Journal on Control & Optimization* **17**, 1-4.

ON DIRICHLET MULTINOMIAL DISTRIBUTIONS

ROBERT W. KEENER

Department of Statistics, University of Michigan, Ann Arbor, MI 48109, U.S.A.
keener@umich.edu

WEI BIAO WU*

Department of Statistics, University of Chicago, Chicago, IL 60637, U.S.A.
wbwu@galton.uchicago.edu

Dedicated to Professor Y. S. Chow on the Occasion of his 80th Birthday

Let Y have a symmetric Dirichlet multinomial distributions in $\mathbb{R}^m$, and let $S_m = h(Y_1) + \cdots + h(Y_m)$. We derive a central limit theorem for S_m as the sample size n and the number of cells m tend to infinity at the same rate. The rate of convergence is shown to be of order $m^{1/6}$. The approach is based on approximation of marginal distributions for the Dirichlet multinomial distribution by negative binomial distributions, and a blocking technique similar to that used to study renormalization groups in statistical physics. These theorems generalize and refine results for the classical occupancy problem.

Some key words: Occupancy problems; Central limit theorem; Exchangeable distributions.

1. Introduction and Main Results.

Let Y have a multinomial distribution $\mathcal{M}(n, p)$ with n trials and success probabilities $p = (p_1, \ldots, p_m)$. Classical occupancy problems concern counts $\ell_k = \#\{j : Y_j = k\}$ and coverage $m - \ell_0$. If m and n tend to infinity at the same rate and the multinomial distribution is symmetric, $p_1 = \cdots = p_m = 1/m$, Weiss (1958) gives a central limit theorem for the number of cells covered or ℓ_0. This result has been extended in various directions. Rényi (1962) gives proofs extending Weiss' result in more general limits, Kopocinska and Kopocinski (1992) prove joint asymptotic normality for a collection of the ℓ_k, and Englund (1981) gives a Berry-Esseen

*This work was supported by National Science Foundation Grant DMS-0448704

bound for the error of normal approximation. In asymmetric cases where the cell probabilities p_j vary, Esty (1983) gives a central limit theorem for the coverage and Quine and Robinson (1982) obtain a Berry-Esseen bound. Most relevant to the results presented here, Chen (1980) introduces mixture models, described below, in which the multinomial cell probabilities are sampled from a Dirichlet distribution. Extensions are presented in Chen (1981a, 1981b).

The models and results here are of some interest in statistics, particularly in situations where multinomial data divide up a sample, but cells are discovered as the experiment is performed. Then quantities like ℓ_0, the number of unobserved cells, will be unknown parameters, but the other counts ℓ_k, $k \geq 1$, are observed and various proposed estimators are based on these. See Fisher, Corbet and Williams (1943), Good and Toulmin (1956) and Keener, Rothman and Starr (1987) for further details. Related models also arise studying bootstrap procedures, although the distributional questions of interest are not that related to the results here. See Rubin (1981) or Csörgő and Wu (2000).

If $G \sim \prod_{i=1}^{m} \Gamma(A_i, 1)$, then $\mathfrak{p} = G/(G_1 + \cdots + G_m)$ has a Dirichlet distribution,

$$\mathfrak{p} = \frac{G}{G_1 + \cdots + G_m} \sim \mathcal{D}_m(A).$$

If the conditional distribution of Y given $\mathfrak{p}$ is multinomial,

$$Y | \mathfrak{p} = p \sim \mathcal{M}(n, p),$$

then Y has the Dirichlet multinomial distribution,

$$Y \sim \mathcal{DM}(n, A).$$

By smoothing (law of total probability),

$$\begin{aligned} \mathbf{P}(Y = y) &= \mathbf{EP}(Y = y | \mathfrak{p}) \\ &= \binom{n}{y_1, \ldots, y_m} \mathbf{E} \prod_{i=1}^{m} \mathfrak{p}_i^{y_i} \\ &= \frac{n! \Gamma(A_1 + \cdots + A_m)}{\Gamma(n + A_1 + \cdots + A_m)} \prod_{i=1}^{m} \frac{\Gamma(y_i + A_i)}{y_i! \Gamma(A_i)}. \end{aligned}$$

In the sequel, we will be particularly interested in the symmetric case in which $A = (a, \ldots, a) = a\mathbf{1}_m$. Special cases of interest include the Bose-Einstein distribution in which $a = 1$ and $\mathbf{P}(Y = y)$ is independent of y, and the Maxwell-Boltzmann distribution which arises in the limit as

$a \to \infty$. When $m = 2$, $\mathfrak{p}_1$ has the beta distribution, $\mathcal{B}(A_1, A_2)$ and Y_1 has the beta-binomial distribution, $\mathcal{BB}(n, A_1, A_2)$.

In the limit theory developed in this paper, the negative binomial distribution $\mathcal{NB}(a, \eta)$ with mass function

$$\frac{\Gamma(a+y)a^a\eta^y}{y!\Gamma(a)(a+\eta)^{a+y}}, \qquad y = 0, 1, \ldots$$

will play a central role. The shape parameter a here is not restricted to be an integer, and the parameter η is the mean, instead of the usual "success probability" $\eta/(a+\eta)$. The variance with this parameterization is $\eta(a+\eta)/a$.

Let h be a function from $\mathbb{N}^+ = \{0, 1, \ldots\}$ to $\mathbb{R}$, and define

$$S_m = \sum_{i=1}^{m} h(Y_i).$$

The main result below is a central limit theorem for S_m as $m \to \infty$ with Y from a symmetric Dirichlet multinomial distribution. Note that when $h(x) = I\{z = j\}$, S_m equals ℓ_k, showing the connection to the occupancy problems mentioned above.

Theorem 1.1. *Consider a limiting situation in which*

$$m \to \infty \quad \text{and} \quad n \to \infty, \quad \text{with} \quad \eta = \frac{n}{m} \to \eta_\infty \in (0, \infty). \tag{1}$$

Assume

$$Y \sim \mathcal{DM}(n, a\mathbf{1}_m),$$

and that h is a nonlinear function with

$$\sup_{y \in \mathbb{N}^+} h^4(y)e^{-\Lambda y} < \infty,$$

where $\Lambda < \log(1 + a/\eta_\infty)$. Take $Z \sim \mathcal{NB}(a, \eta)$ and define

$$\hat{\mu} = \hat{\mu}(\eta) = E_\eta h(Z).$$

Also, let $\beta = \mathrm{Cov}[h(Z), Z)]/\mathrm{Var}(Z)$, so that $\hat{\mu} + \beta(Z - \eta)$ is the best linear predictor of $h(Z)$, and define

$$\hat{\sigma}^2 = \hat{\sigma}^2(\eta) = \mathrm{Var}\big[h(Z) - \hat{\mu} - \beta(Z - \eta)\big].$$

(Note that since h is nonlinear, $\hat{\sigma}^2 > 0$.) Then

$$\sup_{x \in \mathbb{R}} \left| P(S_m \le x) - \Phi\left(\frac{x - m\hat{\mu}}{\sqrt{m}\hat{\sigma}}\right) \right| = O(1/m^{1/6}).$$

This result remains true if the corresponding moments of $h(Y_1)$ or the mean and standard deviation of S_m are used to center and scale the normal approximation. The version stated seems a bit more convenient for explicit calculation.

The next result complements Theorem 1.1 providing an exponential bound for tail probabilities of S_m.

Theorem 1.2. *Assume*

$$E_{\eta_\infty} e^{\epsilon_0 |h(Z)|} < \infty \tag{2}$$

for some $\epsilon_0 > 0$. *Then for some constant* $c > 0$,

$$P\big[|S_m - m\hat{\mu}| > m\epsilon\big] = O\big(\sqrt{m} e^{-c\epsilon^2 m}\big)$$

in limit (1), uniformly for ϵ *in any bounded subset of* $[0, \infty)$.

2. Marginal Distributions

This section provides approximations for marginal distributions in the limiting situation described in Theorem 1.1. Throughout, Y will have the symmetric Dirichlet multinomial distribution $\mathcal{DM}(n, a\mathbf{1}_m)$, and Z, Z_1, Z_2, ... will be i.i.d. from $\mathcal{NB}(a, \eta)$ with $\eta = n/m$. In limit (1), joint distributions for $(Y_1, \dots, Y_k)$ and $(Z_1, \dots, Z_k)$ converge. We will be interested in approximating moments of $S_k = h(Y_1) + \cdots + h(Y_k)$, and it will be convenient to measure distance between measures using a variant of the total variation norm with exponential weights for large values. Specifically, if ν is a signed measure on $\mathbb{N}^k$, define

$$\|\nu\|_\Lambda = \sum_{y \in \mathbb{N}^k} \big|\nu(\{y\})\big| e^{\Lambda(y_1 + \cdots + y_k)}.$$

If Q and $\hat{Q}$ are finite signed measures on $\mathbb{N}^k$, then

$$\begin{aligned}
&\left|\int f\,dQ - \int f\,d\hat{Q}\right| \\
&\quad = \left|\sum_{y \in \mathbb{N}^k} \Big[f(y) e^{-\Lambda(y_1 + \cdots + y_k)}\Big] e^{\Lambda(y_1 + \cdots + y_k)} \big[Q(\{y\}) - \hat{Q}(\{y\})\big]\right| \\
&\quad \le \|Q - \hat{Q}\|_\Lambda \sup_{y \in \mathbb{N}^k} |f(y)| e^{-\Lambda(y_1 + \cdots + y_k)}.
\end{aligned} \tag{3}$$

The likelihood ratio between the marginal distribution of $(Y_1, \dots, Y_k)$

and $(Z_1, \ldots, Z_k)$ is

$$L(y) = \frac{\mathbf{P}(Y_1 = y_1, \ldots, Y_k = y_k)}{\mathbf{P}(Z = y_1) \cdots \mathbf{P}(Z = y_k)}$$
$$= \frac{\Gamma(n+1)}{\Gamma(n+1-v)n^v} \times \frac{\Gamma(ma)}{\Gamma(ma-ka)(ma)^{ka}}$$
$$\times \frac{\Gamma(n+ma-ka-v)(n+ma)^{ka+v}}{\Gamma(n+ma)},$$

where $v = y_1 + \cdots + y_k$. The three terms here can be approximated using the following lemma, which follows fairly easily from Stirling's approximation for the gamma function.

Lemma 2.1. *If $a = a_x$ and $b = b_x$ are both $o(\sqrt{x})$, then*

$$\frac{\Gamma(x+a)}{x^{a-b}\Gamma(x+b)} = 1 + \frac{(a-b)(a+b-1)}{2x} + O\left(\frac{1+a^4+b^4}{x^2}\right)$$

as $x \to \infty$.

Using this lemma,

$$L(y) = 1 + \frac{1}{m}\left[\frac{v(1-v)}{2\eta} - \frac{k(1+ka)}{2} + \frac{(ka+v)(ka+v+1)}{2(a+\eta)}\right] \tag{4}$$
$$+O\left(\frac{1+v^4}{m^2}\right) \tag{5}$$

in limit (1), provided $v = o(\sqrt{m})$. When v is large, the approximation to L breaks down, and errors from these values will be estimated using Bernstein inequality bounds based on moment generating functions. The moment generating function for the negative binomial distribution is

$$M_Z(u) = \mathbf{E}e^{uZ} = \left(\frac{a}{a+\eta-\eta e^u}\right)^a,$$

finite provided $e^u < 1 + a/\eta$.

Lemma 2.2. *Let $V_k = Y_1 + \cdots + Y_k$. If $e^u < 1 + a/\eta_\infty$, then in limit (1),*

$$\mathbf{E}e^{uV_k} \to \left(\frac{a}{a+\eta_\infty - \eta_\infty e^u}\right)^{ka}.$$

Proof. The moment generating function for the binomial distribution with n trials and success probability p is

$$\left[1 + p(e^u - 1)\right]^n.$$

Since $V_k \sim \mathcal{BB}(n, ka, ma-ka)$, its distribution is a beta mixture of binomial distributions and

$$\begin{aligned}\mathbf{E}e^{uV_k} &= \frac{\Gamma(ma)}{\Gamma(ka)\Gamma(ma-ka)}\int_0^1 \big[1+x(e^u-1)\big]^n x^{ka-1}(1-x)^{ma-ka-1}\,dx \\ &= \frac{\Gamma(ma)}{\Gamma(ka)\Gamma(ma-ka)}\int_0^1 \frac{x^{ka-1}}{(1-x)^{ka-1}}e^{-mf(x)}\,dx,\end{aligned}$$

where

$$\begin{aligned}f(x) &= -\eta\log\big[1+x(e^u-1)\big] - a\log(1-x) \\ &= \big[a-(e^u-1)\big]x + O(x^2)\end{aligned}$$

as $x \to 0$. A change of variables rescaling x by m gives

$$\mathbf{E}e^{u(Y_1+\cdots+Y_k)} = \frac{\Gamma(ma)(ma)^{-ka}}{\Gamma(ma-ka)} \times \frac{a^{ka}}{\Gamma(ka)}\int_0^m \frac{x^{ka-1}e^{-mf(x/m)}}{(1-x/m)^{ka-1}}\,dx.$$

The first factor here tends to one by Lemma 2.1, and the desired result then follows by a dominated convergence argument since $mf(x/m) \to [a+\eta_\infty - \eta_\infty e^u]x$, with the error uniformly bounded for $x \in (0, \sqrt{m})$. To be careful, there should be a separate argument to show that the contribution integrating over $x \in (\sqrt{m}, m)$ is negligible. Since this is fairly routine, the details are omitted. □

Considering the approximation for the likelihood ratio L given in (5), it is natural to approximate the joint distribution Q for $(Y_1, \ldots, Y_k)$ by

$$\hat{Q} = \hat{Q}_0 + \frac{1}{m}\hat{Q}_1$$

where $\hat{Q}_0 = \mathcal{NB}(a,\eta)^k$, the joint distribution of $(Z_1, \ldots, Z_k)$, and

$$\hat{Q}_1(\{z\}) = q_1(z)\hat{Q}_0(\{z\})$$

with

$$\begin{aligned}q_1(z) &= \frac{v(1-v)}{2\eta} - \frac{k(1+ka)}{2} + \frac{(ka+v)(ka+v+1)}{2(a+\eta)} \\ &= \frac{(a+2\eta)\tilde{v}}{2\eta(a+\eta)} - \frac{a\tilde{v}^2}{2\eta(a+\eta)} + \frac{k}{2},\end{aligned}$$

where $v = z_1 + \cdots + z_k$ and $\tilde{v} = v - k\eta$.

Theorem 2.1. *If $\Lambda < \log(1+a/\eta_\infty)$, then in limit (1),*

$$\|Q - \hat{Q}\|_\Lambda = O(1/m^2).$$

Proof. Let $B = B_m = \{y \in \mathbb{N}^k : y_1 + \cdots + y_k \geq n^{1/3}\}$. Then since $Q(\{y\}) = L(y)\hat{Q}_0(\{y\})$,

$$\|Q - \hat{Q}\|_\Lambda \leq \sum_{y \in B^c} |L(y) - 1 - q_1(y)| e^{\Lambda(y_1+\cdots+y_k)} \hat{Q}_0(\{y\})$$
$$+ \sum_{y \in B} \big[Q(\{y\}) + Q_0(\{y\}) + |q_1(y)| Q_0(\{y\})\big] e^{\Lambda(y_1+\cdots+y_k)}.$$

The first term here is $O(1/m^2)$ by the approximation for L in (5), and the second term is of order $e^{-\epsilon n^{1/3}}$ for some ϵ, since moment generating functions in Lemma 2.2 converge for some $u > \Lambda$. □

As a corollary the next result provides approximations to moments of $h(Y_i)$. Let $\mu(\eta) = Eh(Y_i)$ and $\hat{\mu}(\eta) = Eh(Z_i)$.

Corollary 2.1. *Assume* $\Lambda < \log(1 + a/\eta_\infty)$ *and that*

$$\sup_{y \in \mathbb{N}+} |h^4(y) e^{-\Lambda y}| < \infty.$$

Then in limit (1),

$$\mu(\eta) = \hat{\mu}(\eta) + O(1/m),$$

$$\operatorname{Var}\big(h(Y_1)\big) = \operatorname{Var}\big(h(Z)\big) + O(1/m),$$

$$\operatorname{Cov}\big(h(Y_1), h(Y_2)\big) = -\frac{a\big[Eh(Z)(Z-\eta)\big]^2}{m\eta(a+\eta)} + O(1/m^2),$$

$$E\big(h(Y_1) - \mu(\eta)\big)^4 = E\big(h(Z) - \hat{\mu}(\eta)\big)^4 + O(1/m),$$

$$E\big(h(Y_1) - \mu(\eta)\big)^3\big(h(Y_2) - \mu(\eta)\big) = O(1/m),$$

$$E\big(h(Y_1) - \mu(\eta)\big)^2\big(h(Y_2) - \mu(\eta)\big)^2 = \operatorname{Var}^2\big(h(Z)\big) + O(1/m),$$

$$E\big(h(Y_1) - \mu(\eta)\big)^2\big(h(Y_2) - \mu(\eta)\big)\big(h(Y_3) - \mu(\eta)\big) = O(1/m),$$

and

$$E\big(h(Y_1) - \mu(\eta)\big)\big(h(Y_2) - \mu(\eta)\big)\big(h(Y_3) - \mu(\eta)\big)\big(h(Y_4) - \mu(\eta)\big) = O(1/m^2).$$

As a consequence, in this limit

$$ES_m = m\hat{\mu}(\eta) + O(1),$$

$$\begin{aligned}\mathrm{Var}(S_m) &= m\mathrm{Var}\big(h(Z)\big) - \frac{ma\big[Eh(Z)(Z-\eta)\big]^2}{\eta(a+\eta)} + O(1)\\ &= mE\left[h(Z) - \hat{\mu}(\eta) - \frac{\mathrm{Cov}\big(h(Z),Z\big)}{\mathrm{Var}(Z)}(Z-\eta)\right]^2 + O(1),\\ &= m\hat{\sigma}^2(\eta) + O(1),\end{aligned}$$

and

$$E\big(S_m - m\mu(\eta)\big)^4 = O(m^2).$$

Proof. Using (3) the initial assertions all follow from Theorem 2.1. Note that

$$\int f\,d\hat{Q} = Ef(Z_1,\ldots,Z_k) + \frac{1}{m}Eq_1(Z_1,\ldots,Z_k)f(Z_1,\ldots,Z_k),$$

and that q_1 is a quadratic function with

$$Eq_1(Z_1,\ldots,Z_k) = 0.$$

So, for instance, $q_1(Z_1,\ldots,Z_4)$ can be written as a sum of quadratic functions of (Z_i,Z_j), $1 \le i \le j \le 4$ and

$$Eq_1(Z_1,\ldots,Z_4)\prod_{i=1}^{4}\big(h(Z_i) - \mu(\eta)\big) = 0.$$

The results about moments of S_m then follow directly after a bit of combinatorics. □

3. Partial Sums

Given a subset $B = \{b_1,\ldots,b_j\}$ (with $b_1 < \cdots < b_j$) of $\{1,\ldots,m\}$ let Y_B denote the random vector $(Y_{b_1},\ldots,Y_{b_j})$ and let $Y_{+B} = \sum_{i\in B} Y_i$. Define A_B, G_B and G_{+B} similarly, and take $S_{+B} = \sum_{i\in B} h(Y_i)$.

Lemma 3.1. *Let $B_1,\ldots,B_\gamma$ be sets partitioning $\{1,\ldots,m\}$. If $Y \sim \mathcal{DM}_m(A)$, then $Y_{B_1},\ldots,Y_{B_\gamma}$ given $Y_{+B_1} = n_1,\ldots,Y_{+B_\gamma} = n_\gamma$ are conditionally independent with*

$$Y_{B_j}|Y_{+B_1} = n_1,\ldots,Y_{+B_\gamma} = n_\gamma \sim \mathcal{DM}(n_j, A_{B_j}).$$

Proof. Since $Y|G = g \sim \mathcal{M}(n,p)$, the conditional joint mass function

$$\begin{aligned}&P(Y_{B_1} = y_{B_1},\ldots,Y_{B_\gamma} = y_{B_\gamma}|G = g, Y_{+B_1} = n_1,\ldots,Y_{+B_\gamma} = n_\gamma)\\ &\quad= \frac{P(Y = y|G = g)}{P(Y_{+B_1} = n_1,\ldots,Y_{+B_\gamma} = n_\gamma)}\end{aligned}$$

is a ratio of multinomial probabilities. Straightforward algebra then shows that $Y_{B_1}, \dots, Y_{B_\gamma}$ given $G = g$ and $Y_{+B_1} = n_1, \dots, Y_{+B_\gamma} = n_\gamma$ are conditionally independent with

$$Y_{B_i} | G = g, Y_{+B_1} = n_1, \dots, Y_{+B_\gamma} = n_\gamma \sim \mathcal{M}(n_i, g_{B_i}/g_{+B_i}).$$

The stated results now follows integrating against the distribution for G. Conditional independence is preserved because $G_{B_1}, \dots, G_{B_\gamma}$ are independent. □

Given a partition $B_1, \dots, B_\gamma$ of $\{1, \dots, m\}$ we can write $S_m = S_{B_1} + \cdots + S_{B_\gamma}$, and by this lemma the summands are conditionally independent given $Y_{+B_1}, \dots, Y_{+B_\gamma}$. Theorem 1.1 will be established using a Berry-Esseen limit theorem (in an independent but non-identically distributed setting) to argue that the conditional distribution of S_m is approximately normal. The following two technical lemmas will be needed.

Lemma 3.2. *The mean of the beta-binomial distribution* $\mathcal{BB}(n, A_1, A_2)$ *is*

$$\frac{nA_1}{A_1 + A_2},$$

and the variance is

$$\frac{nA_1A_2(n + A_1 + A_2)}{(A_1 + A_2)^2(A_1 + A_2 + 1)}.$$

If $A_1 + A_2 = ma$, *then the variance is at most* $A_1\eta(a + \eta)/a^3$.

This result follows easily from a conditioning argument.

Lemma 3.3. *For* $c > 0$, $x \in \mathbb{R}$ *and* $y \in \mathbb{R}$,

$$|\Phi(cx) - \Phi(y)| \le \frac{\sqrt{e}|\log c| + |x - y|}{\sqrt{2\pi}}.$$

Proof. By ordinary calculus, $|x|\phi(x) \le \phi(1)$. If $c \ge 1$ then

$$|\Phi(cx) - \Phi(x)| = \left| \int_1^c \frac{ux\Phi(ux)}{u}\, du \right| \le \phi(1) \log c.$$

Similarly, $|\Phi(cx) - \Phi(x)| < \phi(1)|\log c|$ when $0 < c < 1$. Also, since $|\Phi'| = \phi \le 1/\sqrt{2\pi}$,

$$|\Phi(x) - \Phi(y)| \le |x - y|/\sqrt{2\pi}.$$

The desired result follows from these bounds because

$$|\Phi(cx) - \Phi(y)| \le |\Phi(cx) - \Phi(x)| + |\Phi(x) - \Phi(y)|.$$ □

Proof. Proof of Theorem 1.1 Let $\gamma = \lfloor m^{1/3} \rfloor$, and let $B_1, \ldots, B_\gamma$ be an even partition of $\{1, \ldots, m\}$, i.e., a partition chosen so that $m_i = |B_i|$ equals $\lfloor \tilde{m} \rfloor$ or $\lceil \tilde{m} \rceil$, where $\tilde{m} = m/\gamma = m^{2/3} + O(m^{1/3})$. Then $|m_i - \tilde{m}| \leq 1$, so $m_i = \tilde{m} + O(1)$. Define

$$n_i = Y_{+B_i} \qquad \text{and} \qquad \eta_i = \frac{n_i}{m_i},$$

and let $\mathcal{F}$ denote the sigma-field generated by $Y_{+B_1}, \ldots, Y_{+B_\gamma}$. Conditional moments of S_{B_i} given $\mathcal{F}$ will be approximated using Corollary 2.1. The approximations will be accurate when the variables η_i are near the limiting value η_∞. Define the event

$$F = \{|\eta_i - \eta| \leq m^{-1/12}, i = 1, \ldots, \gamma\}.$$

Since $Y_{+B_i} \sim \mathcal{BB}(n, m_i a, (m - m_i)a)$, by Lemma 3.2, η_i has mean η and variance at most $m_i \eta(a + \eta)/(m_i^2 a^2) = O(m^{-2/3})$. By Tchebysheff's inequality,

$$P(|\eta_i - \eta| \geq m^{-1/12}) = O(m^{-1/2}).$$

This bound and asymptotic expressions in the sequel for all quantities indexed by i hold uniformly in i. By Boole's inequality,

$$P(F^c) \leq \sum_{i=1}^{\gamma} P(|\eta_i - \eta| \geq m^{-1/12}) = \gamma O(m^{-1/2}) = O(m^{-1/6}).$$

Let $\mu_i = E[S_{B_i}|\mathcal{F}]$, $\sigma_i^2 = \mathrm{Var}(S_{B_i}|\mathcal{F})$, and $\rho_i = E\big[|S_{B_i} - \mu_i|^3 \bigm| \mathcal{F}\big]$. By Corollary 2.1, on F,

$$\mu_i = m_i \hat{\mu}(\eta_i) + O(1) \qquad \text{and} \qquad \sigma_i^2 = m_i \hat{\sigma}^2(\eta_i) + O(1).$$

Also, by the corollary

$$E\big[(S_{B_i} - \mu_i)^4 \bigm| \mathcal{F}\big] = O(m_i^2)$$

on F, and so

$$\rho_i = O(m_i^{3/2})$$

on F. The function $\hat{\mu}(\cdot)$ has a bounded second derivative in some neighborhood of η_∞. Taylor expansion about $\eta_i = \eta$ gives

$$\mu_i = m_i \hat{\mu}(\eta) + (n_i - m_i \eta)\mu'(\eta) + O\big(m_i(\eta_i - \eta)^2\big) + O(1)$$

on F, and summing over i,

$$\sum \mu_i = m\hat{\mu}(\eta) + VO(m^{2/3}) + O(m^{1/3}), \tag{6}$$

on F, where

$$V = \sum(\eta_i - \eta)^2.$$

Similarly, on F,

$$\sum \sigma_i^2 = m\hat{\sigma}^2(\eta) + VO(m^{2/3}) + O(m^{1/3}).$$

Next, by the Berry-Esseen theorem (cf Theorem 16.5.2 in Feller (1971)),

$$\left| P(S_m \le x|\mathcal{F}) - \Phi\left(\frac{x - \sum \mu_i}{\sqrt{\sum \sigma_i^2}}\right) \right| \le 6 \frac{\sum \rho_i}{(\sum \sigma_i^2)^{3/2}}.$$

Now on F, $V = O(m^{1/6})$, and so $\sum \sigma_i^2 \sim m\hat{\sigma}^2(\eta)$. Hence on F,

$$\frac{\sum \rho_i}{(\sum \sigma_i^2)^{3/2}} = O(m^{-1/6}).$$

Since $P(S_m \le x) = EP(S_m \le x|\mathcal{F})$, the theorem will follow from the bounds presented provided

$$E \sup_x \left| \Phi\left(\frac{x - \sum \mu_i}{\sqrt{\sum \sigma_i^2}}\right) - \Phi\left(\frac{x - m\hat{\mu}}{\sqrt{m}\hat{\sigma}}\right) \right| 1_F = O(m^{-1/6}).$$

By Lemma 3.3, the left hand side here is bounded by the sum of

$$\sqrt{\frac{e}{2\pi}} E \left| \log \frac{\sqrt{m}\hat{\sigma}}{\sqrt{\sum \sigma_i^2}} \right| 1_F \qquad \text{and} \qquad \frac{1}{\sqrt{2\pi}} E \left| \frac{\sum \mu_i - m\hat{\mu}}{\sqrt{m}\hat{\sigma}} \right| 1_F.$$

Since $V = O(m^{1/6})$ on F, the argument of the expectation in the first of these expressions is $O(m^{-1/6})$. The second expression, by (6), is

$$O(m^{-1/6}) + O(m^{1/6} EV).$$

But $\mathrm{Var}(\eta_i) = O(m^{-2/3})$ which implies $EV = O(m^{-1/3})$, and so the second expression is also $O(m^{-1/6})$. □

Proof. Proof of Theorem 1.2 Since $Z_1, \ldots, Z_m \sim \mathcal{NB}(ma, m\eta)$,

$$P_\eta(Z_1 + \cdots + Z_m = n) = \frac{\Gamma(ma+n)(ma)^{ma}(m\eta)^n}{n!\Gamma(ma)(ma+m\eta)^{ma+n}}.$$

Using this, it is easy to check that

$$Z_1, \ldots, Z_m | Z_1 + \cdots + Z_n = n \sim \mathcal{DM}(n, a\mathbf{1}_m), \tag{7}$$

noted as Lemma 1 of Chen (1980). Also, by Stirling's formula

$$P_\eta(Z_1 + \cdots + Z_m = n) \sim \sqrt{\frac{a}{2\pi m\eta(a+\eta)}}$$

in limit (1). Using (7),

$$P\big[|S_m - m\hat{\mu}| > m\epsilon\big] \le \frac{P_\eta\left[|\sum_{i=1}^n W_i| > m\epsilon\right]}{P_\eta(Z_1 + \cdots + Z_m = n)},$$

where $W_i = h(Z_i) - \hat{\mu}$ (and $W = h(Z) - \hat{\mu}$ below), and the theorem will follow if

$$P_\eta\left[\left|\sum_{i=1}^n W_i\right| > m\epsilon\right] = O(e^{-c\epsilon^2 m}).$$

This basically follows from Bernstein's inequality, but a bit of care is necessary to make sure the stated uniformity holds. Note that adjusting c, it is sufficient to show that the asymptotic bound holds uniformly for all ϵ sufficiently small. Let $\delta = \epsilon_0/4$. Since $e^x \le 1 + x + x^2 e^{|x|}/2$ and $x^2 \le 4e^{|x|}/e^2$, for $0 \le u \le \delta$,

$$E_\eta e^{uW} \le 1 + \frac{1}{2}u^2 E_\eta W^2 e^{\delta|W|} \le 1 + \frac{2u^2}{\delta^2 e^2} E_\eta e^{2\delta|W|}$$

Introducing a likelihood ratio and using the Schwarz' inequality,

$$\begin{aligned} E_\eta e^{2\delta|W|} &= E_{\eta_\infty}\left(\frac{\eta}{\eta_\infty}\right)^Z \left(\frac{a+\eta_\infty}{a+\eta}\right)^{a+Z} e^{2\delta|W|} \\ &\le \left\{E_{\eta_\infty}\left(\frac{\eta}{\eta_\infty}\right)^{2Z}\left(\frac{a+\eta_\infty}{a+\eta}\right)^{2a+2Z}\right\}^{1/2} \left\{E_{\eta_\infty} e^{\epsilon_0|h(Z)-\hat{\mu}|}\right\}^{1/2}. \end{aligned}$$

The first factor here converges to one by dominated convergence as $\eta \to \eta_\infty$, and the second factor remains bounded for m and n sufficiently large by (2). So there is a constant c_0 such that

$$E_\eta e^{uW} \le 1 + c_0 u^2 \le e^{c_0 u^2}, \qquad 0 \le u \le \delta,$$

for m and n sufficiently large in limit (1). By Bernstein's inequality,

$$P_\eta\left[\sum_{i=1}^n W_i > m\epsilon\right] \le e^{mc_0u^2 - m\epsilon}$$

for m and n sufficiently large in limit (1). If $\epsilon \le 2c_0\delta$, taking $u = \epsilon/(2c_0)$ this bound becomes $e^{-m\epsilon^2/(4c_0)}$. The theorem then follows from this bound and a corresponding bound for $P_\eta\left[\sum_{i=1}^n W_i < -m\epsilon\right]$. □

Acknowledgment. We thank the referee for a careful reading of our manuscript.

References

1. Chen, Wen-Chen (1980) On the weak form of Zipf's law, *Journal of Applied Probability*, **17**, 611-622
2. Chen, Wen-Chen (1981a) Limit theorems for general size distributions, *Journal of Applied Probability*, **18**, 139–147
3. Chen, Wen-Chen (1981b) Some local limit theorems in the symmetric Dirichlet-multinomial urn models, *Annals of the Institute of Statistical Mathematics*, **33**, 405–415
4. Csörgő, S. and Wu, W. B. (2000) Random graphs and the strong convergence of bootstrap means, *Combinatorics, Probability and Computing*, **10** 315–347.
5. Englund, Gunnar, (1981) A remainder term estimate for the normal approximation in classical occupancy, *Annals of Probability*, **9**, 684–692
6. Esty, Warren W. (1983) A normal limit law for a nonparametric estimator of the coverage of a random sample, *Annals of Statistics*, **11** 905–912
7. Feller, W. (1971) *An Introduction to Probability Theory and its Applications. Wiley, New York*
8. Fisher, R. A., Corbet, A. S. and Williams, C. B., (1943) The relation between the number of species and the number of individuals in a random sample of an animal population, *J. Anial Ecol.*, **12**, 42–58
9. Good, I. J. and Toulmin, G. H., (1956) The number of new species, and the increase in population coverage, when a sample is increased, *Biometrika*, **43**, 45–63
10. Keener, R., Rothman, E. and Starr, N. (1987) Distributions on partitions, *Annals of Statistics*, **15**, 1466-1481
11. Kopocinska, I. and Kopocinski, B. (1992) A new proof of generalized theorem of Irving Weiss, *Periodica Mathematica Hungarica*, **25**, 27–30
12. Quine, M. P., (1979) A functional central limit theorem for a generalized occupancy problem, *Stochastic Processes and their Applications*, **9**, 109–115
13. Quine, M. P. and Robinson, J. (1982) A Berry-Esseen bound for an occupancy problem. *Annals of Probability*, **10**, 663–671
14. Rényi, A., (1962) Three new proofs and a generalization of a theorem of Irving Weiss. *Magyar Tud. Akad. Mat. Kutato Int. Kozl.*, **7**, 203–214
15. Rubin, D. B. (1981) The Bayesian bootstrap, *Annals of Statistics*, **9**, 130–134
16. Weiss, Irving (1958) Limiting distributions in some occupancy problems. *Annals of Mathematical Statistics*, **29** 878–884

THE OPTIMAL STOPPING PROBLEM FOR S_n/n AND ITS RAMIFICATIONS

TZE LEUNG LAI

Department of Statistics, Stanford University, Stanford, CA 94305, U.S.A.

YI-CHING YAO

Institute of Statistical Science, Academia Sinica, Taipei 115, Taiwan, ROC

Optimal stopping has been one of Y.S. Chow's major research areas in probability. This paper reviews the seminal work of Chow and Robbins on the optimal stopping problem for S_n/n and subsequent developments and intercrosses with other areas of probability theory. It also uses certain ideas and techniques from these developments to devise relatively simple but accurate approximations to the optimal stopping boundary.

Some key words: Brownian motion; Optimal stopping; Random walks.

1. Introduction

In an expository article on optimal stopping rules, Breiman (1964) presented a number of examples, one of which was referred to as the E.S.P. (extrasensory perception) problem whose objective is to show that a fair coin is actually biased towards heads by stopping appropriately in repeated tosses of the coin. Specifically, let $X_n = 1$ if the n-th toss results in heads and $X_n = -1$ otherwise, so that $X_1, X_2, \ldots$ are i.i.d. with $P(X_1 = 1) = P(X_1 = -1) = 1/2$. Observing $X_1, X_2, \ldots$ sequentially, the objective is to find a finite-valued stopping rule that maximizes $E(S_\nu/\nu)$ over all stopping rules ν where $S_n = X_1 + \cdots + X_n$. While Breiman made no attempt to solve the problem, Chow and Robbins (1965) coined the term "the S_n/n problem" and proved the existence of an optimal stopping rule. More precisely, they showed that there exists a nondecreasing sequence of positive integers $0 < k_1 \leq k_2 \leq \cdots$ with $k_n \leq 13\sqrt{n}$ for large n such that the stopping rule $\tau = \inf\{n \geq 1 : S_n \geq k_n\}$ is optimal. Note that by the law of the iterated logarithm this stopping rule stops with probability 1.

Earlier Chow and Robbins (1961, 1963) embarked on the development of a "reasonably general theory of the existence and computation of optimal stopping rules".

A number of extensions of the seminal work of Chow and Robbins (1965) and several variations of the S_n/n problem have appeared in the literature. A review is given of the S_n/n problem and its variants in Section 2 when $X, X_1, X_2, \ldots$ are i.i.d. random variables with $E[X^2] < \infty$, and in Section 3 for the case $E[X^2] = \infty$. Section 4 first describes some recent work on corrected random walk approximations to Brownian optimal stopping problems, and then applies it to the S_n/n problem to derive a simple second-order approximation to the optimal stopping boundary for the S_n/n problem. It also presents numerical results for the Bernoulli case and compares them with the corresponding second-order approximation. Furthermore, to improve the (analytic) second-order approximation to the optimal stopping boundary for small n, a relatively simple but accurate hybrid method is proposed that takes advantage of the available explicit solution to the continuous-time version of the S_n/n problem. Section 5 concludes the paper.

2. The Case of Finite Second Moment

Let T (or T_+, resp.) denote the collection of all nonnegative (or positive, resp.) stopping rules that are finite a.s., i.e. $P(\nu < \infty) = 1$. [Following Chow and Teicher (1997, p.138), a stopping time may take the value $+\infty$ with positive probability, and is called a stopping rule if it is finite a.s.] Suppose $E(X^2) < \infty$. Then $E(\sup_n |S_n|/n) < \infty$, so that S_ν/ν is integrable for every $\nu \in T_+$. It will be assumed that $E[X] = 0$ in this section, unless stated otherwise.

Dvoretzky (1967) generalized the Chow-Robbins (1965) result and proved that the stopping rule $\tau = \inf\{n \geq 1 : S_n \geq b_n\}$ is optimal for the S_n/n problem, i.e. $E(S_\tau/\tau) = \sup_{\nu \in T_+} E(S_\nu/\nu)$, where $0 < b_1 < b_2 < \cdots$ is a strictly increasing sequence with b_n uniquely determined by the equation

$$\frac{b_n}{n} = \sup_{\nu \in T_+} E\left(\frac{b_n + S_\nu}{n + \nu}\right). \tag{2.1}$$

(In what follows, $b_1, b_2, \ldots$ will be referred to as the optimal stopping boundary points.) He also derived that

$$0.32 < \liminf_{n\to\infty} b_n/\sqrt{n} \leq \limsup_{n\to\infty} b_n/\sqrt{n} < 4.07, \tag{2.2}$$

and conjectured that $\lim_{n\to\infty} b_n/\sqrt{n}$ exists.

Dvoretzky's conjecture was confirmed independently by Shepp (1969) and Walker (1969). By solving the continuous-time version of the S_n/n problem, they found that $r = \lim_{n\to\infty} b_n/\sqrt{n} = 0.83992...$ is the unique positive root of the equation

$$r\phi(r) = (1 - r^2)\Phi(r) \tag{2.3}$$

where ϕ and Φ are the standard normal density and distribution functions, respectively. Walker also considered the more general maximization problem $\sup_{\nu\in T_+} E(S_\nu/\nu^\alpha)$ with $\alpha > 1/2$, for which the (finite-valued) optimal stopping rule is given by $\inf\{n \geq 1 : S_n \geq b_{n,\alpha}\}$. Here the optimal stopping boundary points $b_{n,\alpha}$ are determined by

$$\frac{b_{n,\alpha}}{n^\alpha} = \sup_{\nu\in T_+} E\left[\frac{b_{n,\alpha} + S_\nu}{(n+\nu)^\alpha}\right]. \tag{2.4}$$

Moreover, $r(\alpha) = \lim_{n\to\infty} b_{n,\alpha}/\sqrt{n}$ exists and is the unique positive root of the equation

$$\int_0^\infty x^{2(\alpha-1)} e^{r(\alpha)x - x^2/2} dx = r(\alpha) \int_0^\infty x^{2\alpha-1} e^{r(\alpha)x-x^2/2} dx. \tag{2.5}$$

Further details of the continuous-time version of the S_n/n problem are given in the last paragraph of this section.

More general payoff functions of the form S_n^β/n^α or $|S_n|^\beta/n^\alpha$ were also considered in the literature. For $\beta = 1$ or 2 and for any positive sequence $\{c_n\}$ satisfying

$$c_{n+1}^2 \leq c_n c_{n+2}, \quad (n+1)^\beta c_{n+1} \leq n^\beta c_n, \tag{2.6}$$

Teicher and Wolfowitz (1966) proved that there exists a stopping rule that maximizes $E[c_\nu S_\nu^\beta]$ over all $\nu \in T_+$. Note that $c_n = n^{-\alpha}$ satisfies (2.6) if $\alpha \geq \beta$. Siegmund, Simons and Feder (1968) generalized the above results to the maximization problem $\sup_{\nu\in T_+} E[|S_\nu|^\beta/\nu^\alpha]$ with $0 < \beta < 2\alpha$, and established the existence of an optimal stopping rule provided $E[|X|^{\max\{2,\beta\}}] < \infty$. This result together with a simple comparison principle implies the existence of an optimal stopping rule for each of the following maximization problems:

(1) $\sup_{\nu\in T_+} E[\nu^{-\alpha}\log|S_\nu|]$ (assuming $\alpha > 0$ and $E[X^2] < \infty$);
(2) $\sup_{\nu\in T_+} E[c_\nu (S_\nu^+)^\beta]$ (assuming $\limsup_{n\to\infty} n^\alpha c_n < \infty$ and $E[|X|^{\max\{2,\beta\}}] < \infty$);
(3) $\sup_{\nu\in T_+} E[c_\nu P_k(S_\nu)]$, where $c_n \geq 0, \limsup_{n\to\infty} n^\alpha c_n < \infty, P_k$ is a polynomial of degree $k < 2\alpha$ with positive leading coefficient, and $E[|X|^{\max\{2,k\}}] < \infty$.

Basu and Chow (1977) subsequently considered the more general setting in which $\{X_n\}$ is a martingale difference with respect to a filtration $\{\mathcal{F}_n\}$ and $E[|X_n|^{\max\{2,\beta\}} \mid \mathcal{F}_{n-1}] \le C$ a.s. for some positive constant C. For $0 < \beta < 2\alpha$, they proved the existence of a stopping rule that maximizes $E[|S_\nu|^\beta/\nu^\alpha]$ over all $\nu \in T_+$ under the additional assumption that the functional central limit theorem holds for $\{S_n/n\}$.

The continuous-time version of the S_n/n problem relates to optimal stopping for $W(t)/t$ where $\{W(t), t \ge 0\}$ is standard Brownian motion. It was solved independently by Taylor (1968), Shepp (1969) and Walker (1969), who showed that starting from any given positive time t_0, it is optimal to stop at $\tau = \inf\{t \ge t_0 : W(t) \ge r\sqrt{t}\}$, where r is the unique root of (2.3); i.e. $\sup_{\nu \in T_+, \nu \ge t_0} E[W(\nu)/\nu] = E[W(\tau)/\tau]$. Simons and Yao (1989) considered optimal stopping for Brownian motion $Z(t) = W(t) + \theta t$ with a random drift rate θ which is independent of $\{W(t), t \ge 0\}$ and which is normally distributed with mean μ and variance σ^2. This corresponds to the continuous-time version of the S_n/n problem, in which the X_i are normal with unknown mean θ that has a prior $N(\mu, \sigma^2)$ distribution. They showed that the optimal stopping problem can be transformed into the finite-horizon optimal stopping problem for $W(t)/t$ considered earlier by Chernoff and Petkau (1984). Assuming a finite horizon at $t = 1$ (i.e. stopping is enforced at $t = 1$), Chernoff and Petkau (1984) showed that the optimal stopping boundary $b(t)$ (yielding the optimal stopping rule $\{t \ge t_0 : W(t) \ge b(t)\} \wedge 1$ for every $0 < t_0 < 1$) satisfies $0 < b(t) \le rt^{1/2}, 0 < t \le 1$, and

$$b(t) = rt^{1/2} + o(t^{1/2}) \text{ as } t \to 0+, \tag{2.7}$$

$$b(t) = \tilde{r}(1-t)^{1/2} + O((1-t)^{3/2}) \text{ as } t \to 1-, \tag{2.8}$$

where $\tilde{r} = 0.63883...$ is the root of the equation

$$(1 - \tilde{r}^2)\phi(\tilde{r}) = \tilde{r}^3\Phi(\tilde{r}). \tag{2.9}$$

For the infinite-horizon optimal stopping for $Z(t)/t$, the posterior mean of θ is $[Z(t)+\sigma^{-2}\mu]/(t+\sigma^{-2})$. Let $u(t) = t/(t+\sigma^{-2})$ so that $u(t)/t$ is the posterior variance. Simons and Yao (1989) showed that the optimal stopping time $\tau^* \ge t_0 (> 0)$ is of the form $\tau^* = \inf\{t \ge t_0 : Z(t)/t \ge \sigma b(u(t))/u(t) + \mu\}$ ($\inf \emptyset = \infty$), with the payoff defined to be $\theta (= \lim_{t\to\infty} Z(t)/t)$ on the event $\{\tau^* = \infty\}$, which has positive probability of occurrence in view of (2.8) (and therefore $\tau^* \notin T_+$).

3. The Case of Infinite Second Moment

When $E[X] = 0$ and $E[X^2] = \infty$, there may exist a stopping rule $\nu \in T_+$ such that $E[S_\nu/\nu] = \infty$. McCabe and Shepp (1970) found that $E[X^+ \log^+ X] < \infty$ is a necessary and sufficient condition for $\sup_{\nu \in T_+} E[S_\nu/\nu] < \infty$, where $x^+ = \max\{x, 0\}$ and $\log^+ x = \log \max\{x, 1\}$ (see also Davis (1971) and Gundy (1969)). It was shown earlier by Burkholder (1962) that

$$E[X^+ \log^+ X] < \infty \Longleftrightarrow E\left[\sup_n \frac{S_n}{n}\right] < \infty \Longleftrightarrow E\left[\sup_n \frac{X_n}{n}\right] < \infty, \quad (3.1)$$

$$E[|X| \log |X|] < \infty \Longleftrightarrow E\left[\sup_n \frac{|S_n|}{n}\right] < \infty \Longleftrightarrow E\left[\sup_n \frac{|X_n|}{n}\right] < \infty. \quad (3.2)$$

As shown by McCabe and Shepp (1970), (3.1) is also equivalent to

$$\sup_{\nu \in T_+} E\left[\frac{S_\nu}{\nu}\right] < \infty \Longleftrightarrow \sup_{\nu \in T_+} E\left[\frac{X_\nu}{\nu}\right] < \infty, \quad (3.3)$$

while (3.2) is also equivalent to

$$\sup_{\nu \in T_+} E\left[\frac{|S_\nu|}{\nu}\right] < \infty \Longleftrightarrow \sup_{\nu \in T_+} E\left[\frac{|X_\nu|}{\nu}\right] < \infty. \quad (3.4)$$

Closely related to (3.3) is the problem of existence of $\tilde{\nu} \in T_+$ that attains $\sup_{\nu \in T_+} E[X_\nu/\nu]$, analogous to the optimal stopping problem for S_n/n; see Chow and Dvoretzky (1969) and Chow and Lan (1975) for results on the X_n/n problem.

Under the assumption that $E[X^+ \log^+ X] < \infty$, there may not exist $\tilde{\nu} \in T_+$ that attains $\sup_{\nu \in T_+} E[S_\nu/\nu]$. Dvoretzky (1967) conjectured that the moment condition

$$E[|X|^q] < \infty, \quad \text{for some } q > 1, \quad (3.5)$$

would suffice for the existence of such $\tilde{\nu}$. Thompson, Basu and Owen (1971) showed that such $\tilde{\nu}$ exists under an additional assumption on the truncated variance besides (3.5). Davis (1973) subsequently removed this additional assumption and proved Dvoretzky's conjecture indeed holds.

In the remainder of this section we shall assume that $E[X] = 0$ and $E[X^+ \log^+ X] < \infty$. Let $T_{(0,\infty]}$ be the class of all stopping times taking values in $(0, \infty]$. By Theorem 4.5' and Lemma 4.11 of Chow, Robbins and Siegmund (1971), there exists an extended-valued optimal stopping time and

$$\sup_{\nu \in T_+} E\left[\frac{S_\nu}{\nu}\right] = \sup_{\nu \in T_{(0,\infty]}} E\left[\frac{S_\nu}{\nu}\right] < \infty, \quad (3.6)$$

where S_ν/ν is defined as 0 $(= \lim_{n\to\infty} S_n/n)$ on $\{\nu = \infty\}$. Let F denote the distribution function of X and define

$$h(x) = \int_x^\infty \frac{y}{x}\log\frac{y}{x}dF(y) \ \text{ for } \ x > 0, \quad I = \int_0^\infty \frac{dF(x)}{h(x)}, \quad I' = \int_{-\infty}^\infty \frac{dF(x)}{h(|x|)}. \tag{3.7}$$

Concerning when the extended-valued optimal stopping time that attains the supremum in (3.6) terminates with probability 1, Klass (1973) proved the following results:

(a) If $I = \infty$, then every optimal stopping time stops (i.e. is finite-valued) with probability 1.
(b) If $I' < \infty$, then every optimal stopping time has positive probability of never stopping in finite time.

In particular, by (a) and (b), if X is symmetric about 0, then $I = \infty$ is a necessary and sufficient condition for the existence of a finite-valued optimal stopping time. Furthermore, the minimal (extended-valued) optimal stopping time is given by $\inf\{n \geq 1 : S_n \geq b_n\}$, where b_n is determined by

$$\frac{b_n}{n} = \sup_{\nu\in T_{(0,\infty]}} E\left[\frac{b_n + S_\nu}{n+\nu}\right].$$

Klass (1973) also showed that if $E[(X^+)^q] < \infty$ for some $q > 1$, then $P(S_n > b_n \ \text{ i.o.}) = 1$, so that every optimal stopping time stops with probability 1, thereby weakening (3.5) to the more natural assumption $E[(X^+)^q] < \infty$ for Dvoretzky's conjecture.

An important outgrowth of Klass' work leading to (a) and (b) is that for any i.i.d. sequence of mean zero random variables X_n, $P(S_n \geq cE[|S_n|] \ \text{ i.o.}) = 1$ for every $c < 1/4$; see Theorem 14 of Klass (1973). Another byproduct of his investigation of the S_n/n problem is a counter-example to a claim of Feller (1946) for sums of i.i.d. random variables with $E[X] = 0$ and $E[|X|^{1+\delta}] = \infty$ for some $0 < \delta < 1$. Let $\{c_n\}$ be a sequence of positive numbers such that $c_n/n \downarrow$ and $c_n^{2-\epsilon}/n \uparrow \infty$ for some $\epsilon > 0$. After proving that

$$P(|S_n| > c_n \ \text{ i.o.}) = P(|X_n| > c_n \ \text{ i.o.}), \tag{3.8}$$

Feller stated without proof that the absolute values can be removed from both sides of (3.8). Corollary 11 of Klass (1973) gives a density function of X with mean 0 such that $E[|X|^{1+\delta}] = \infty$ for all $\delta > 0$ and

$$P\{S_n > (1-\epsilon)c(n) \ \text{ i.o.}\} = 1, \quad P\{X_n > (1-\epsilon)c(n) \ \text{ i.o.}\} = 0, \tag{3.9}$$

for any $0 < \epsilon < 1$, where $c(x) = \beta^{-1}(x)$ and $\beta(x) = 1/h(x)$. The sequence $c(n)$ plays a fundamental role in Klass' work on the optimal stopping time $\nu^* \in T_{(0,\infty]}$ that attains the supremum in (3.6). In particular, he showed that

$$\begin{aligned} P\{X_n \geq c(n) \text{ i.o.}\} = 1 &\implies P\{\nu^* < \infty\} = 1, \\ P\{|X_n| \geq c(n) \text{ i.o.}\} = 0 &\implies P\{\nu^* = \infty\} > 0. \end{aligned} \tag{3.10}$$

The results (a) and (b) leave the case $I < \infty = I'$ unresolved. Instead of working with I', Klass (1975) introduced the quantity

$$Q = \limsup_{y \to \infty} \frac{E[(-X)\mathbf{1}_{\{-X>y\}}]}{E[X \log(X/y)\mathbf{1}_{\{X>y\}}]}, \tag{3.11}$$

to be used in conjunction with I, and replaced (b) by the following:

(b1) If $I < \infty$ and $Q > 1$, then $P(S_n > b_n \text{ i.o.}) = 1$, implying that every optimal stopping time terminates with probability 1.
(b2) If $I < \infty$ and $Q < 1$, then $P(S_n > b_n \text{ i.o.}) = 0$, implying that every optimal stopping time has positive probability of never stopping in finite time.

In Klass (1974), general conditions are given on the distribution of X and positive nonincreasing constants $c_1 \geq c_2 \geq \cdots$ that guarantee $E[\sup_n c_n S_n] < \infty$ (or $E[\sup_n c_n |S_n|] < \infty$, resp.), which then implies the existence of an extended-valued stopping time τ with

$$E[c_\tau S_\tau] = \sup_{\nu \in T_{(0,\infty]}} E[c_\nu S_\nu] < \infty$$

$$(\text{or } E[c_\tau |S_\tau|] = \sup_{\nu \in T_{(0,\infty]}} E[c_\nu |S_\nu|] < \infty, \text{ resp.}). \tag{3.12}$$

As a consequence, for $1/2 < \alpha < 1$, $E[|X|^{1/\alpha}] < \infty$ is a necessary and sufficient condition for the existence of an extended-valued stopping time τ such that

$$E[|S_\tau|/\tau^\alpha] = \sup_{\nu \in T_{(0,\infty]}} E[|S_\nu|/\nu^\alpha] < \infty. \tag{3.13}$$

Strengthening the condition $E[X^+ \log^+ X] < \infty$ into

$$E[X^+ \exp\{\gamma(\log^+ X)^{1/2}\}] < \infty \text{ for some } \gamma > 0, \tag{3.14}$$

Chow and Lan (1975) proved that (3.14) implies $I = \infty$, which implies by (a) the existence of a finite-valued optimal stopping time. Chow and

Cuzick (1979) subsequently showed that $I = \infty$ under the weaker moment condition

$$E[X^+ \exp\{(\log^+ X/\log^+\log^+ X)^{1/2}\}] < \infty. \tag{3.15}$$

Note that (3.15) is weaker than the condition $E[(X^+)^q] < \infty$ for some $q > 1$ mentioned above in connection with Dvoretzky's conjecture.

4. A Second-Order Approximation and Computational Results for the S_n/n Problem

In this section we return to the case of finite second moment considered in Section 2 and consider computation of the optimal stopping boundary for the S_n/n problem. This is an infinite-horizon optimal stopping problem and numerical computation involves (i) approximation by a finite-horizon problem with a large horizon N and (ii) using backward induction to compute the optimal stopping boundary for the finite-horizon problem, as in Section 5 of Chow and Robbins (1965) for the Bernoulli case. We shall assume, without loss of generality as in Section 2, that $E[X] = 0$ and $E[X^2] = 1$, and use a simpler numerical approximation method based on the fact that the continuous-time optimal stopping problem for $W(t)/t$ has an explicit optimal stopping boundary $r\sqrt{t}$ (with r being the unique root of (2.3)) as well as an explicit value function

$$\begin{aligned} v(x,t) &= [(1-r^2)/\sqrt{t}]\Phi\left(\frac{x}{\sqrt{t}}\right)/\phi\left(\frac{x}{\sqrt{t}}\right), \quad \text{for } x < r\sqrt{t}, \\ &= \frac{x}{t}, \quad \text{for } x \geq r\sqrt{t}, \end{aligned} \tag{4.1}$$

[see Shepp (1969, p. 994)]. By definition,

$$v(x,t) = \sup_{\nu} E[(x + W(\nu))/(t+\nu)] = E[(x+W(\tau))/(t+\tau)],$$

where $\tau = \inf\{s : x + W(s) \geq r\sqrt{t+s}\}$. While Shepp (1969) and Walker (1969) showed that $b_n \sim r\sqrt{n}$ for large n, we introduce here a second-order correction to incorporate discreteness of the stopping boundary and the distribution of X. Specifically, we approximate the optimal stopping boundary b_n in (2.1) by

$$b_n \doteq r\sqrt{n} - \rho, \quad \text{where } \rho = E[S_{\tau_+}^2]/\{2E[S_{\tau_+}]\}, \tag{4.2}$$

in which $\tau_+ = \inf\{n \geq 1 : S_n > 0\}$.

For the symmetric Bernoulli distribution $P(X = 1) = P(X = -1) = 1/2$, $\rho = 1/2$. For the standard normal distribution, $\rho \approx 0.5826$. The use of

ρ to correct the Brownian motion optimal stopping boundary for discrete time was introduced by Chernoff (1965) for normal X, and later also by Chernoff and Petkau (1976) for symmetric Bernoulli X. They arrived at such correction by considering a particular continuous-time optimal stopping problem in which, starting from a negative time t with a horizon at 0, a stopping rule for Brownian motion $\{W(\cdot)\}$ is chosen to maximize $E\{g(W(\nu),\nu) \mid W(t) = y_0\}$ over all stopping rules $t \le \nu \le 0$, where $y_0 < 0$ and

$$g(y,s) = -s\mathbf{1}_{\{s<0\}} + y^2\mathbf{1}_{\{s=0,y\le 0\}}. \tag{4.3}$$

Approximating Brownian motion by a normal or symmetric Bernoulli random walk so that the discrete-time optimal stopping problem maximizes $E[g(y_0 + \sqrt{\delta}S_\nu, t + \delta\nu)]$ over integer-valued stopping rules ν, they showed that the (discrete-time) optimal stopping boundary $a_\delta(t)$ for the approximating random walk is related to the (continuous-time) optimal stopping boundary $a(t)(\equiv 0)$ for Brownian motion by

$$a_\delta(t) = a(t) - \rho\sqrt{\delta} + o(\sqrt{\delta}). \tag{4.4}$$

Subsequently, Hogan (1986) extended the result to general random walks with $E[X] = 0$ and $E[X^2] = 1$. Whereas he (and Chernoff and Petkau before him) established (4.4) only for the special case (4.3), Chernoff (1972) gave some heuristic argument based on Taylor expansions around the optimal stopping boundary to justify the use of (4.4) for more general payoff functions.

Recently, Lai, Yao and AitSahlia (2005) have proved that (4.4) indeed holds for general payoff functions satisfying certain regularity conditions, provided that the backward induction algorithm to compute the (discrete-time) value function and stopping boundary is suitably initialized. For a general payoff function $g(x,t)$, they consider a finite-horizon optimal stopping problem with value function

$$v(x,t) = \sup_{\nu \in T_{[t,\bar{t}]}} E\{g(W(\nu),\nu) \mid W(t) = x\}, \ -\infty < x < \infty, \ \underline{t} \le t < \bar{t}, \tag{4.5}$$

where $\underline{t}$ is the initial time, $\bar{t}$ is the (finite) horizon and $T_{[t,\bar{t}]}$ is the collection of all stopping rules taking values between t and $\bar{t}$.

Suppose that in the continuous-time optimal stopping problem (4.5), the optimal continuation region is of the form $\{(x,t) : x < b(t), \underline{t} \le t < \bar{t}\}$ and one has an approximation $v_\delta(\cdot,t^*)$ to the value function $v(\cdot,t^*)$ and an approximation β_δ to $b(t^*)$ at some t^* ($\underline{t} < t^* < \bar{t}$), such that

$$\beta_\delta = b(t^*) + O(\sqrt{\delta}), \ \ v_\delta(x,t^*) - v(x,t^*) = O(\delta e^{\alpha'(x-b(t^*))^-}) \tag{4.6}$$

uniformly in x for some constant $\alpha' \geq 0$, where $x^- = \max\{-x, 0\}$. (Note that with $b(t^*)$ approximated by β_δ, $v_\delta(x, t^*) = g(x, t^*)$ for $x \geq \beta_\delta$, so $v_\delta(x, t^*) = v(x, t^*) = g(x, t^*)$ for $x \geq \max\{b(t^*), \beta_\delta\}$.) A random walk approximation to the value function $v(\cdot, t)$ and the stopping boundary $b(t)$ for $\underline{t} \leq t \leq t^*$ can be described as follows. Let $t_0 = t^* > t_1 > \cdots > t_{K_\delta} = \underline{t}$ partition the interval $[\underline{t}, t^*] = [t_{K_\delta}, t_0]$ into K_δ subintervals such that

$$t_{i-1} - t_i = \delta \text{ for } 1 \leq i < K_\delta, \quad t_{K_\delta - 1} - t_{K_\delta} \leq \delta.$$

Let $X, X_1, X_2, \ldots$ be i.i.d. random variables such that

$$E[X] = 0 = E[X^3], \quad E[X^2] = 1, \quad Ee^{\alpha''|X|} < \infty \text{ for some } \alpha'' > 0, \tag{4.7}$$

and let $S_n = X_1 + \cdots + X_n$ ($S_0 = 0$). Approximating the Brownian motion $W(t)$ in the optimal stopping problem (4.5) by the random walk S_n, we can use backward induction to compute the value function $v_\delta(x, t_i)$ and stopping boundary $b_\delta(t_i)$ of the corresponding discrete-time optimal stopping problem. Specifically, first initialize by defining $b_\delta(t_i) = \beta_\delta$, for $0 \leq i < |\log \delta|$. Then define recursively

$$v_\delta(x, t_i) = \begin{cases} g(x, t_i) & \text{if } x \geq b_\delta(t_{i-1}) + \sqrt{\delta}|\log \delta|, \\ \max\left\{g(x, t_i), E[v_\delta(x + \sqrt{\delta}X, t_{i-1})]\right\} & \\ & \text{if } |x - b_\delta(t_{i-1})| < \sqrt{\delta}|\log \delta|, \\ E[v_\delta(x + \sqrt{\delta}X, t_{i-1})] & \text{if } x \leq b_\delta(t_{i-1}) - \sqrt{\delta}|\log \delta| \end{cases} \tag{4.8}$$

for $1 \leq i \leq K_\delta$, and define $b_\delta(t_i)$ for $|\log \delta| \leq i \leq K_\delta$ recursively by

$$\begin{aligned} b_\delta(t_i) = \inf\{x : |x - b_\delta(t_{i-1})| < \sqrt{\delta}|\log \delta|, & \\ g(x, t_i) \geq E[v_\delta(x + \sqrt{\delta}X, t_{i-1})]\} & \ (\inf \emptyset = b_\delta(t_{i-1})). \end{aligned} \tag{4.9}$$

Under certain regularity conditions, Lai, Yao and AitSahlia (2005) have shown that

$$b_\delta(t_i) = b(t_i) - \rho\sqrt{\delta} + o(\sqrt{\delta}) \text{ uniformly in } |\log \delta| \leq i \leq K_\delta \tag{4.10}$$

and

$$\sup_{1 \leq i \leq K_\delta,\ -\infty < x < \infty} e^{-\alpha(x - b(t_i))^-} |v_\delta(x, t_i) - v(x, t_i)| = O(\delta) \text{ for some } \alpha \geq 0. \tag{4.11}$$

Returning to the S_n/n problem in which the payoff function is $g(x, t) = x/t$, we now justify the (second-order) approximation (4.2) for standard normal X by making use of (4.10). Unlike the applications considered by Chernoff and Petkau who use random walk approximations to compute the optimal stopping boundary and value function of a continuous-time optimal stopping problem, we approximate, via a family of optimal stopping problems $\{\mathcal{P}_\delta\}$ indexed by $\delta > 0$ (to be defined below), the (discrete-time) S_n/n problem by the corresponding continuous-time optimal stopping problem for which closed-form expressions are available for the value function and optimal stopping boundary. For given $\delta > 0$, $-\infty < x < \infty$ and $t > 0$, the problem $\mathcal{P}_\delta$ is to maximize $E[g(x+\sqrt{\delta}S_\nu, t+\delta\nu)]$ over all stopping rules ν taking values in $\{0,1,2,\dots\}$; $\delta = 1$ corresponds to the original problem. Let $b^{(\delta)}(t)$ denote the optimal stopping boundary, i.e., defining for given $-\infty < x < \infty$ and $t > 0$ the stopping rule $\tau = \inf\{k \geq 0 : x + \sqrt{\delta}S_k \geq b^{(\delta)}(t+\delta k)\}$, we have

$$E[g(x+\sqrt{\delta}S_\tau, t+\delta\tau)] = \sup_\nu E[g(x+\sqrt{\delta}S_\nu, t+\delta\nu)].$$

Note that $b^{(\delta)}(n\delta) = b_n$ for $\delta = 1$. Since g satisfies $\sqrt{\delta}g(\sqrt{\delta}x, \delta t) = g(x,t)$, we have

$$b^{(\delta)}(n\delta) = \sqrt{\delta}b^{(1)}(n) = \sqrt{\delta}b_n, \quad \text{for} \quad n = 1,2,\dots \text{ and } \delta > 0. \tag{4.12}$$

For small $\delta > 0$, the problem $\mathcal{P}_\delta$ is close to the continuous-time optimal stopping problem, so that $b^{(\delta)}(t)$ is expected to be close to $b(t)$. We now assume that X is standard normal, for which the random walk $\{\sqrt{\delta}S_n, n = 1,2,\dots\}$ has the same distribution as $\{W(\delta n), n = 1,2,\dots\}$ and the value function is

$$\begin{aligned} v_\delta(x,t) &= \sup_\nu E[g(x+\sqrt{\delta}S_\nu, t+\delta\nu)] \\ &= \sup_{\eta\in T_\delta} E[g(x+W(\eta), t+\eta)], \end{aligned} \tag{4.13}$$

where T_δ consists of all stopping rules taking values in $\{0,\delta,2\delta,\dots\}$.

Lemma 1. For standard normal X,

$$0 \leq v(x,t) - v_\delta(x,t) \leq rt^{-3/2}\delta.$$

Proof. Consider the optimal stopping rule $\tau := \inf\{s \geq 0 : x + W(s) \geq r(t+s)^{1/2}\}$ for Brownian motion, and the stopping rule $\tau_\delta := \delta(\lfloor \tau/\delta \rfloor + 1)$ which takes values in $\{\delta, 2\delta, \dots\}$. Note that $0 < \Delta := \tau_\delta - \tau \leq \delta$, $v(x,t) = E[(x+W(\tau))/(t+\tau)] = E[r/(t+\tau)^{1/2}]$, and

$$\begin{aligned} v_\delta(x,t) &\geq E\left[\frac{x+W(\tau_\delta)}{t+\tau_\delta}\right] = E\left[\frac{x+W(\tau)+(W(\tau+\Delta)-W(\tau))}{t+\tau+\Delta}\right] \\ &= E\left[\frac{x+W(\tau)}{t+\tau+\Delta}\right] = E\left[\frac{r(t+\tau)^{1/2}}{t+\tau+\Delta}\right]. \end{aligned}$$

Therefore, $v(x,t) - v_\delta(x,t) \leq E[\{r(t+\tau)^{1/2}\Delta\}/\{(t+\tau)(t+\tau+\Delta)\}] \leq r\delta/t^{3/2}$. □

From Lemma 1, it follows for fixed t^* that $0 \leq v(x,t^*) - v_\delta(x,t^*) = O(\delta)$ uniformly in x, so (4.6) is satisfied with $\alpha' = 0$. By (4.10), $b^{(\delta)}(1) = b(1) - \rho\sqrt{\delta} + o(\sqrt{\delta}) = r - \rho\sqrt{\delta} + o(\sqrt{\delta})$. Setting $\delta = 1/n$, we have by (4.12), as $n \to \infty$,

$$b_n = \sqrt{n} b^{(1/n)}(1) = r\sqrt{n} - \rho + o(1), \tag{4.14}$$

justifying (4.2) for standard normal X.

The preceding argument for the S_n/n problem can be extended to more general payoff functions of the form $n^{-\alpha} f(S_n/\sqrt{n})$ for some $\alpha > 0$; see Shepp (1969) and Lerche (1986) for specific examples. Specifically, with $g(x,t) = t^{-\alpha} f(x/\sqrt{t})$, the preceding argument for the S_n/n problem can be applied to show for standard normal X that

$$b_n = \gamma\sqrt{n} - \rho + o(1), \quad \text{as } n \to \infty, \tag{4.15}$$

under the following assumptions on f, which are satisfied by the special case $f(x) = x$ considered by Walker (1969) (cf. (2.4) and (2.5)):

(C1) f is nondecreasing and convex.

(C2) There exist (optimal) stopping boundary points $b_1, b_2, \ldots$ with the property that for any given $-\infty < c < \infty$ and $n = 1, 2, \ldots$,

$$E\left[(n+\tau(n,c))^{-\alpha} f\left(\frac{c + S_{\tau(n,c)}}{\sqrt{n+\tau(n,c)}}\right)\right] = \sup_\nu E\left[(n+\nu)^{-\alpha} f\left(\frac{c+S_\nu}{\sqrt{n+\nu}}\right)\right],$$

where $\tau(n,c) = \inf\{k \geq 0 : c + S_k \geq b_{n+k}\}$.

(C3) The continuous-time optimal stopping boundary takes the form of $b(t) = \gamma\sqrt{t}$ for some $\gamma > 0$, so that the value function is

$$v(x,t) = E\left[(t+\tau)^{-\alpha} f\left(\frac{x + W(\tau)}{\sqrt{t+\tau}}\right)\right]$$

where $\tau = \inf\{s \geq 0 : x + W(s) \geq \gamma\sqrt{t+s}\}$.

4.1. *Numerical results on the adequacy of the second-order approximation*

We examine the adequacy of the approximation (4.2) to the optimal stopping boundary for the S_n/n problem in the symmetric Bernoulli case by comparing it against the benchmark value that is computed as follows. Noting that the Bernoulli S_n/n problem is an infinite-horizon optimal stopping problem, Section 5 of Chow and Robbins (1965) first shows that the value function $V(x,n)$ is the limit of $V_N(x,n)$ as $N \to \infty$, where V_N is the value function of the finite-horizon problem with payoff function $g(x,n) = x^+/n$ for $n \le N$. For general X with mean 0 and satisfying $E[X^+ \log^+ X] < \infty$, we have the following simple bound on the difference between V and V_N:

$$0 \le V(x,n) - V_N(x,n) \le b_N/N \text{ for all } x \text{ and } n \le N. \tag{4.16}$$

In fact, a somewhat stronger result will be proved at the end of this section.

Lemma 2. For $-\infty < x < \infty$ *and* $n \le N$,

$$0 \le V(x,n) - V_N(x,n) \le \frac{b_N}{N} P(x + S_k < b_{n+k}, k = 1, \ldots, N-n).$$

Let $b_{n,N}$ denote the optimal stopping boundary points for the finite-horizon S_n/n problem. When X is Bernoulli, $V_N(x,n)$ can be computed by the following backward induction algorithm:

$$V_N(x,n) = \max\left\{\frac{x^+}{n}, \frac{1}{2}[V_N(x+1,n+1) + V_N(x-1,n+1)]\right\}. \tag{4.17}$$

The optimal stopping boundary points are given by

$$b_{n,N} = \inf\left\{x : \frac{x}{n} > \frac{1}{2}[V_N(x+1,n+1) + V_N(x-1,n+1)]\right\}. \tag{4.18}$$

Clearly both $V_N(x,n)$ and $b_{n,N}$ are nondecreasing in N and bounded, respectively, by $V(x,n)$ and b_n from above.

Since $b_N \sim r\sqrt{N}$ as shown by Shepp (1969) and Walker (1969), it follows from (4.16) that

$$0 \le V(x,n) - V_N(x,n) = O(1/\sqrt{N}) \quad \text{uniformly in } x \text{ and } n \le N.$$

For given n and $\epsilon > 0$, as $N \to \infty$,

$$V_N(b_n - \epsilon, n) - \frac{b_n - \epsilon}{n} \longrightarrow V(b_n - \epsilon, n) - \frac{b_n - \epsilon}{n} > 0,$$

implying $b_n - \epsilon < b_{n,N}$ for large N. It then follows that $b_{n,N} \uparrow b_n$ as $N \to \infty$.

For $N = 5{,}000 \times 2^i, 0 \le i \le 6$, we have used the following method to compute $b_{n,N}$. For each N, by backward induction on $n = N, N-1, \ldots, 1$, we computed $V_N(x,n)$ for x in a grid set $\{k\delta, k = 0, \pm 1, \ldots\}$ for some small $\delta > 0$ (with $1/\delta$ an integer). Defining $d(x,n) = \frac{1}{2}[V_N(x+1,n+1) + V_N(x-1,n+1)] - x/n$, let $x_0 = \inf\{k\delta : d(k\delta, n) < 0\}$, so that $d(x_0,n) < 0$ and $d(x_0 - \delta, n) \ge 0$, implying $x_0 - \delta \le b_{n,N} < x_0$. Using the simple linear interpolation method proposed by Chernoff and Petkau (1984, 1986), we estimated $b_{n,N}$ by the value of x satisfying

$$d(x_0, n)(x - x_0 + \delta) + d(x_0 - \delta, n)(x_0 - x) = 0.$$

This linear interpolation method appears to be very effective for the present study. For example, to estimate $b_{n,N}$ with $n = 10$ and $N = 20,000$, using $\delta = 0.1, 0.05$ and 0.01, the values of x_0 are, respectively, $2.3, 2.25$ and 2.22, while by the linear interpolation method the corresponding estimated values of $b_{n,N}$ are, respectively, 2.21675..., 2.21672... and 2.21662..., agreeing up to the third decimal place. As for $b_{n,N}$ with $n = 10$ and $N = 320,000$, using $\delta = 0.5, 0.2$ and 0.1, the values of x_0 are, respectively, $2.5, 2.4$ and 2.3, while the corresponding "linearly interpolated" values of $b_{n,N}$ are, respectively, 2.22299..., 2.21820... and 2.21816.... The last two numbers agree up to the third decimal place.

Approximating b_n by $b_{n,N}$ has always a negative bias since $b_{n,N} < b_n$. While we have not derived an asymptotic formula for $b_n - b_{n,N}$ for large N, we propose an approximation of the form $b_{n,N} \doteq b_n - c_n/N^{\gamma_n}$, in which b_n, c_n and γ_n are determined by using the values of $b_{n,N}$ for three different N. This corresponds to an extrapolation method; see Chernoff and Petkau (1984). For example, using $\delta = 0.1$ together with the linear interpolation method, the estimated values of $b_{10,80,000}$, $b_{10,160,000}$ and $b_{10,320,000}$ are, respectively, 2.21779..., 2.21802... and 2.21816..., which together yield an extrapolated estimate of 2.21838... for b_{10}. Since this extrapolated estimate of b_n for relatively small n is only slightly greater than $b_{n,N}$ with $N = 320,000$, we estimate b_n simply by $b_{n,N}$ with $N = 320,000$. It should be noted that the error of the estimate (i.e. the difference between b_n and $b_{n,320,000}$) tends to be greater for larger n and that the third decimal place of the estimate may not be completely reliable for moderate n.

Some of the estimated b_n are reported in Table 1 along with the second-order approximation $r\sqrt{n} - 0.5 \approx 0.83992\sqrt{n} - 0.5$. Table 1 shows that the corrected (second-order) approximation $r\sqrt{n} - 0.5$ underestimates b_n by an amount of about 0.05 for $n \geq 50$, implying that the uncorrected (first-order) approximation $r\sqrt{n}$ overestimates b_n by an amount of about 0.45. Therefore the correction term -0.5 is indeed effective for estimating the optimal stopping boundary points.

Table 1. *Optimal stopping boundary and its second–order approximation*

n	1	2	3	4	6	8	10
b_n	0.456	0.775	1.032	1.252	1.624	1.940	2.218
$r\sqrt{n}-0.5$	0.340	0.688	0.955	1.180	1.557	1.876	2.156
difference	0.116	0.087	0.077	0.072	0.067	0.064	0.062
n	12	14	16	18	20	30	40
b_n	2.470	2.702	2.918	3.121	3.313	4.154	4.864
$r\sqrt{n}-0.5$	2.410	2.643	2.860	3.063	3.256	4.100	4.812
difference	0.060	0.059	0.058	0.058	0.057	0.054	0.052
n	50	60	70	80	90	100	
b_n	5.489	6.054	6.574	7.058	7.512	7.942	
$r\sqrt{n}-0.5$	5.439	6.006	6.527	7.012	7.468	7.899	
difference	0.050	0.048	0.047	0.046	0.044	0.043	

Proof of Lemma 2. Clearly $V(x,n) \geq V_N(x,n)$, and $V(x,n) = V_N(x,n) = x/n$ for $x \geq b_n$. For $x < b_n$, let $\tau = \inf\{k \geq 1 : x + S_k \geq b_{n+k}\}$ and $\tau_N = \min\{\tau, N-n\}$. Letting $A = \{\tau > N-n\}(=\{\tau > \tau_N\})$, we have

$$V(x,n) = E\left[\frac{x+S_\tau}{n+\tau}\mathbf{1}_{A^c}\right] + E\left[V(x+S_{N-n},N)\mathbf{1}_A\right],$$

$$E\left[V(x+S_{N-n},N)\mathbf{1}_A\right] < E\left[V(b_N,N)\mathbf{1}_A\right] = E\left[\frac{b_N}{N}\mathbf{1}_A\right],$$

where the inequality follows from the fact that $x + S_{N-n} < b_N$ on A. Also,

$$\begin{aligned} V_N(x,n) &\geq E\left[\frac{(x+S_{\tau_N})^+}{n+\tau_N}\right] \\ &= E\left[\frac{x+S_\tau}{n+\tau}\mathbf{1}_{A^c}\right] + E\left[\frac{(x+S_{N-n})^+}{N}\mathbf{1}_A\right]. \end{aligned}$$

Then $V(x,n) - V_N(x,n) < E[\{b_N/N - (x+S_{N-n})^+/N\}\mathbf{1}_A] \leq (b_N/N)P(A)$. □

4.2. *A more accurate hybrid approximation*

Table 1 shows that the simple second-order approximation (4.2) has a relative error less than 5% in estimating b_n for $n \geq 12$ and less than 1% for $n \geq 50$. The relative error increases to 25% at $n = 1$. To improve the accuracy of the approximation, we propose here a hybrid method that uses backward induction with N of moderate size (say $N = 50$ or 100) and with $g(x,N) = v(x,N)$, the value function of the continuous-time optimal stopping problem for $W(t)/t$; see (4.1). Denoting by $\widetilde{V}_N(x,n)$ the corresponding value function, we have $\widetilde{V}_N(x,N) = v(x,N)$ and

$$\widetilde{V}_N(x,n) = \max\left\{\frac{x}{n}, \frac{1}{2}[\widetilde{V}_N(x+1,n+1) + \widetilde{V}_N(x-1,n+1)]\right\}, \quad (4.19)$$

for $n = N-1,\ldots,1$. Noting that $\widetilde{V}_N(x,N) = v(x,N)$ is convex in x with $\partial v(x,N)/\partial x \le 1/N$, it can be shown that the optimal stopping boundary point $\widetilde{b}_{n,N}$ is the unique root of the equation

$$\frac{x}{n} = \frac{1}{2}[\widetilde{V}_N(x+1,n+1) + \widetilde{V}_N(x-1,n+1)]. \quad (4.20)$$

Since $v(x,N) \ge x^+/N$ for all x, $\widetilde{V}_N(x,n) \ge V_N(x,n)$ for all x and $n \le N$. It seems possible that $\widetilde{V}_N(x,n)$ approximates $V(x,n)$ better than $V_N(x,n)$, since $v(x,N)$ provides a good approximation to $V(x,N)$ when N is of moderate size. Consequently, we expect $\widetilde{b}_{n,N}$ to be a better estimate of b_n than $b_{n,N}$.

Table 2. *Hybrid approximation using* $N = 50$

n	1	2	3	4	5	6	7	8
b_n	0.456	0.775	1.032	1.252	1.447	1.624	1.788	1.940
$\widetilde{b}_{n,50}$	0.456	0.775	1.032	1.252	1.447	1.625	1.788	1.940
n	10	12	14	16	18	20	30	40
b_n	2.218	2.470	2.702	2.918	3.121	3.313	4.154	4.864
$\widetilde{b}_{n,50}$	2.219	2.471	2.704	2.920	3.123	3.316	4.159	4.875

Table 3. *Hybrid approximation using* $N = 100$

n	1	2	3	4	5	6	8
b_n	0.456	0.775	1.032	1.252	1.447	1.624	1.940
$\widetilde{b}_{n,100}$	0.456	0.775	1.032	1.252	1.447	1.624	1.940
n	10	12	14	16	18	20	30
b_n	2.218	2.470	2.702	2.918	3.121	3.313	4.154
$\widetilde{b}_{n,100}$	2.218	2.471	2.703	2.919	3.122	3.314	4.156
n	40	50	60	70	80	90	
b_n	4.864	5.489	6.054	6.574	7.058	7.512	
$\widetilde{b}_{n,100}$	4.866	5.492	6.059	6.581	7.068	7.527	

Tables 2 and 3 compare b_n with $\widetilde{b}_{n,N}$ for $N = 50$ and 100. They show that the hybrid approximation method works remarkably well especially for small n. It is worth noting that whereas the b_n shown in the tables tend to slightly underestimate the true value of the optimal stopping boundary, $\widetilde{b}_{n,N}$ appears to slightly overestimate b_n with the approximation error diminishing as N increases.

5. Conclusion

Since the seminal work of Chow and Robbins (1965), the S_n/n problem has undergone many important developments. The infinite-variance case has led to a variety of beautiful results in the probability theory of sample means when $E[X^2] = \infty$, as we have reviewed in Section 3. Despite the infinite-horizon and time-varying nature of the optimal stopping problem, we have shown in Section 4 that there are surprisingly simple approximate solutions of the S_n/n problem in the finite-variance case via second-order corrections of the explicit solutions of the corresponding problem for Brownian motion. Of particular interest is the hybrid approximation developed in Section 4.2. It uses the value function of the continuous-time optimal stopping problem at $N = 50$ or 100 to initialize the backward induction algorithm, thereby achieving an accuracy comparable to that of the direct finite-horizon approximation to the infinite-horizon problem with $N = 320,000$. The importance of this large horizon reduction becomes more pronounced if we move from the symmetric Bernoulli distribution, with its simple backward induction algorithm (4.17), to more complicated distributions that require numerical integration to evaluate expectations. Brezzi and Lai (2002, p.92) have used a similar idea to compute the optimal stopping boundary of an infinite-horizon optimal stopping problem associated with the Gittins index of Brownian motion.

Acknowledgement

The authors thank Chihchi Hu for discussions and assistance. The first author gratefully acknowledges support from the National Science Foundation under grant DMS-0305749. The second author gratefully acknowledges support from the National Science Council of Taiwan, ROC.

References

1. Basu, A.K. and Chow, Y.S. (1977). On the existence of optimal stopping rules for the reward sequence S_n/n. *Sankhya, Ser. A* **39**, 278–289.
2. Breiman, L. (1964). Stopping-rule problems. *Applied Combinatorial Mathematics* (ed. E.F. Beckenbach), Chapter 10. Wiley, New York.
3. Brezzi, M. and Lai, T.L. (2002). Optimal learning and experimentation in bandit problems. *J. Econ. Dynamics Contr.* **27**, 87–108.
4. Burkholder, D.L. (1962). Successive conditional expectation of an integrable function. *Ann. Math. Statist.* **33**, 887–893.
5. Chernoff, H. (1965). Sequential tests for the mean of a normal distribution IV (discrete case). *Ann. Math. Statist.* **36**, 55–68.
6. Chernoff, H. (1972). *Sequential Analysis and Optimal Design.* CBMS Re-

gional Conference Series in Applied Math. No. 8, Society for Industrial and Applied Mathematics, Philadelphia.

7. Chernoff, H. and Petkau, A.J. (1976). An optimal stopping problem for sums of dichotomous random variables. *Ann. Probab.* **4**, 875–889.
8. Chernoff, H. and Petkau, A.J. (1984). *Numerical methods for Bayes sequential decision problems.* Technical Report ONR 34, MIT Statistics Center.
9. Chernoff, H. and Petkau, A.J. (1986). Numerical solutions for Bayes sequential decision problems. *SIAM J. Scient. Statist. Comput.* **7**, 46–59.
10. Chow, Y.S. and Cuzick, J. (1979). Moment conditions for the existence and nonexistence of optimal stopping rules for S_n/n. *Proc. Amer. Math. Soc.* **75**, 300–307.
11. Chow, Y.S. and Dvoretzky, A. (1969). Stopping rules for X_n/n and related problems. *Israel J. Math.* **3**, 240–248.
12. Chow, Y.S. and Lan, K.K. (1975). Optimal stopping rules for X_n/n and S_n/n.*Statistical Inference and Related Topics,* Vol. 2 (ed. M.L. Puri), 159–177, Academic Press, New York.
13. Chow, Y.S. and Robbins, H. (1961). A martingale system theorem and applications. *Proc. Fourth Berkeley Symp. Math. Statist. Probab.* **1**, 93–104.
14. Chow, Y.S. and Robbins, H. (1963). On optimal stopping rules. *Z. Wahrschein. und Verw. Gebiete* **2**, 33–49.
15. Chow, Y.S. and Robbins, H. (1965). On optimal stopping rules for S_n/n. *Illinois J. Math.* **9**, 444–454.
16. Chow, Y.S., Robbins, H. and Siegmund, D. (1971). *Great Expectations: The Theory of Optimal Stopping,* Houghton Mifflin, Boston.
17. Chow, Y.S. and Teicher, H. (1997). *Probability Theory: Independence, Interchangeability, Martingales,* 3rd ed., Springer-Verlag, New York.
18. Davis, B. (1971). Stopping rules for S_n/n and the class $L \log L$. *Z. Wahrschein. und Verw. Gebiete* **17**, 147–150.
19. Davis, B. (1973). Moments of random walk having infinite variance and the existence of certain optimal stopping rules for S_n/n. *Illinois J. Math.* **17** 75–81.
20. Dvoretzky, A. (1967). Existence and properties of certain optimal stopping rules. *Proc. Fifth Berkeley Symp. Math. Statist. Probab.* **1**, 441–452.
21. Feller, W. (1946). A limit theorem for random variables with infinite moments. *Amer. J. Math.* **68**, 257–262.
22. Gundy, R.F. (1969). On the class $L \log L$ martingales and singular integrals. *Studia Math.* **33**, 109–118.
23. Hogan, M. (1986). Comments on a problem of Chernoff and Petkau. *Ann. Probab.* **14**, 1058–1063.
24. Klass, M.J. (1973) Properties of optimal extended-valued stopping rules for S_n/n. *Ann. Probab.* **1**, 719–757.
25. Klass, M.J. (1974). On stopping rules and the expected supremum of S_n/a_n and $|S_n|/a_n$. *Ann. Probab.* **2**, 889–905.
26. Klass, M.J. (1975). A survey of the S_n/n problem. *Statistical Inference and Related Topics,* Vol. 2 (ed. M.L. Puri), 179–201, Academic Press, New York.
27. Lai, T.L., Yao, Y.-C. and AitSahlia, F. (2005). Corrected random walk ap-

proximations to free boundary problems in optimal stopping. Technical Report, Department of Statistics, Stanford University.
28. Lerche, H.R. (1986). An optimal property of the repeated significance test. *Proc. Nat. Acad. Sci. U.S.A.* **83**, 1546–1548.
29. McCabe, B.J. and Shepp, L.A. (1970). On the supremum of S_n/n. *Ann. Math. Statist.* **41**, 2166–2168.
30. Shepp, L.A. (1969). Explicit solutions to some problems of optimal stopping. *Ann. Math. Statist.* **40**, 993–1010.
31. Siegmund, D., Simons, G. and Feder, P. (1968). Existence of optimal stopping rules for rewards related to S_n/n. *Ann. Math. Statist.* **39**, 1228–1235.
32. Simons, G. and Yao, Y.-C. (1989). Optimally stopping the sample mean of a Wiener process with an unknown drift. *Stoc. Proc. Appl.* **32**, 347–354.
33. Taylor, H.M. (1968). Optimal stopping in a Markov process. *Ann. Math. Statist.* **39**, 1333–1344.
34. Teicher, H. and Wolfowitz, J. (1966). Existence of optimal stopping rules for linear and quadratic rewards. *Z. Wahrschein. und Verw. Gebiete* **5**, 361–368.
35. Thompson, M., Basu, A.K. and Owen, W.L. (1971). On the existence of the optimal stopping rule in the S_n/n problem when the second moment is infinite. *Ann. Math. Statist.* **42**, 1936–1942.
36. Walker, L.H. (1969). Regarding stopping rules for Brownian motion and random walks. *Bull. Amer. Math. Soc.* **75**, 46–50.

BOLD PLAY AND THE OPTIMAL POLICY FOR VARDI'S CASINO

LARRY SHEPP

Department of Statistics, Rutgers University, Piscataway, NJ 08854, U.S.A.

In Dubins and Savage's casino, there are only limited gambles available and the optimal win probability, attained by bold play, is a singular function of the intial fortune, f. Vardi asked the question of how-to-gamble-if-you-must in a casino where any bet is available as long as its expected value is less than a fixed constant $c < 0$ times the amount staked, and one cannot stake more than one has at any moment. Vardi's casino has the optimal probability, $1-(1-f)^{1+c}$, though it is only attainable within ϵ, and the ϵ-optimal betting strategy is far from bold play in some aspects.

Some key words: How-to-gamble-if-you-must; Martingales; Optimal betting strategy.

1. Introduction

A gambler is in a do-or-die situation and must turn his present fortune of size $f \in (0,1)$ into a fortune of size 1 in order to pay off a gambling debt in a casino where only certain sub-fair bets are available to him. In a typical case there is only one table available and this allows a bet of any stake $s \leq f$, gives odds ratio r and has win probability $w < \frac{1}{1+r}$ (which makes it a sub-fair casino). It was shown by Dubins and Savage (1965) that bold play is optimal in the sense that the gambler maximizes the probability of reaching fortune 1 before fortune 0 and thus being able to pay off his debt in a sequence of gambles ending at 0 or 1 by betting at each step, n, the smaller of what he currently has ($s = f_n$) or just enough (namely $s = \frac{1-f_n}{r}$) to reach 1 if the current bet wins. But Dubins & Savage (1965) also showed that the bold play strategy is not the only optimal strategy in general. The maximal probability, $P(f)$, attained by bold play can be obtained by sometimes departing from bold play in a specific way and (Dubins & Savage, 1965) gives $P(f)$ as a rather complicated function of f.

In more realistic casinos the casino allows more than one table or pair

r, w, and then the optimal strategy is usually not known and hard to approximate much less determine without writing a large linear program. Yehuda Vardi raised the question in a conversation of whether bold play would also be optimal in a gambling house where stake $s \leq f$ is allowed at any time so long as the odds are such that the *expected return* is at most sc, where $c \in (-1, 0)$ is negative and given. We show Vardi's question has a neat answer: the supremum over all betting strategies of the probability to reach $f = 1$, in Vardi's casino, although not attained by any strategy, is simply

$$P(f) = 1 - (1 - f)^{1+c}, \quad 0 \leq f \leq 1.$$

The supremum is achieved as the limit of the probability attained by the strategy S_α as $\alpha \downarrow 0$, where S_α (boldly) stakes f when $f \leq \alpha$ and S_α (timidly) stakes $s = \frac{\alpha}{1-\alpha}(1 - f)$ when $\alpha \leq f < 1$. Note that $s \leq f$ as required. Whenever money is bet, all of it is bet (boldly) on the table with the right odds, r, to carry the fortune to unity in case of a win (similar to the Dubins-Savage case). We show the strategy S_α obtains the winning probability

$$P_\alpha(f) = 1 + (1 - (1 + c)\alpha)^n (c - \frac{(1 + c)(1 - f)}{(1 - \alpha)^n})$$

when $f \in I_n \equiv [(1 - (1 - \alpha)^n, 1 - (1 - \alpha)^{n+1}), n \geq 0$. The limit as $\alpha \downarrow 0$ is then seen to be $P(f) = 1 - (1 - f)^{1+c}$.

Vardi's casino is thus nearly optimized by a strategy close to bold play in a sense or at least one with much in common with bold play. Note that in the Dubins-Savage version with a single available table, bold play also sometimes sets money aside; of course no optimal strategy ever overshoots the goal $f = 1$. Vardi's casino differs from the Dubins-Savage casino in that the optimal strategy is only a limiting strategy and so does not exist. Nevertheless, it is in some sense *unique* since one cannot depart too dramatically from it and preserve asymptotic optimality.

2. Bold Play as a General Strategy in Life

Suppose the mafia will kill you if you do not repay your debt, normalized to be of size 1, to them. Your present fortune is $0 \leq f_0$. Suppose you are in a casino where there is only one *subfair* bet with odds r, i.e., you can stake any amount $s \leq f_0$ and reach the fortune

$$f_1 = f_0 + rs \text{ with probability } w < \frac{1}{1 + r},$$

or

$$f_1 = f_0 - s \text{ with probability } 1 - w > \frac{r}{1+r}.$$

It was proved (Dubins & Savage, 1965) that *bold play*, where you stake $s = \min(f_0, \frac{1-f_0}{r})$ at each bet until you either go broke or reach the goal is optimal.

Remark 1. Does it really need proof? The answer is yes. Dubins reported to me that some erroneous proofs of his theorem have been given and he could prove that some of the proofs were erroneous without reading them simply because they claimed uniqueness which is false. The referee of this paper observes that further evidence that bold play is not trivial has been given for casinos where there are betting limits (Heath, Pruitt & Sudderth, 1972; Schweinsberg, 2005), observes that Vardi's casino obeys the technical definition of a casino in Dubins & Savage (1965), and says it "is interesting that, even though Dubins and Savage do not mention the problem of this paper, they do observe that $P(f)$ is a "casino function" on page 78 of their book"

Remark 2. It is metamathematically clear that no proof can avoid using supermartingale arguments because the restriction that the betting strategy not know the future outcomes of the bets must be used by the proof somewhere (or else one can do better) and it is hard to see how this lack of knowledge of the future can be used without somehow using martingale theory.

Remark 3. To see by another argument that the problem is not as easy as it may appear consider the analogous problem where after every bet an interest payment is extracted so that

$$f_1 = \frac{f_0 + rs}{1+a} \text{ with probability } w < \frac{1}{1+r},$$

or

$$f_1 = \frac{f_0 - s}{1+a} \text{ with probability } 1 - w,$$

by dividing by $1 + a > 1$ then it is even more "evident" that bold play is optimal, *but this is in fact false for* $r > 1$*!*, see Chen, Shepp & Zame (2004). In Chen (19778) it is shown that bold play is optimal even when $a > 0$ for $r = 1$ but in Chen, Shepp, Yao & Zhang (2005) it is shown that there are initial fortunes f_0 for which bold play is optimal even with interest payments, but there are other initial fortunes f_0 where one must play non-boldly to obtain the maximal probability to attain level $f = 1$. The optimal strategy is thus usually not bold play but what it is exactly remains open and apparently very difficult. Even whether bold play is optimal or not for some cases of a, r, w also remains open. For more general casinos, where more than one type of bet is available, even when $a = 0$, the problem seems to be difficult. However, we show here that if *very many* bets are available, in one special formulation below, then the problem becomes explicitly solvable and the solution becomes much easier than that of the Dubins and Savage problem. The reason that the problem becomes easier when there is a continuum of bets available seems to be that the problem then becomes less "discrete". In many optimal control problems which can be explicitly solved a large role is played by Ito calculus, see for example, Shepp (2002). Note however that our problem does not involve Ito calculus. Indeed, although the almost-optimal strategy, S_α bets small amounts to reach the goal in one play, the almost-optimal process of fortunes is not at all a diffusion but rather a *jump* process. If we consider the Vardi casino with interest payments $1 + a$, as above, the problem becomes more challenging and will be studied elsewhere.

If the casino allows *any* bet for which the mean return on a bet of size s is at most cs where $c < 0$ is given, is it true that bold play maximizes the probability that the gambler reaches 1 before going broke? Here bold play would be the strategy to bet it all on that table where the odds allow you to just reach your goal. Note that in the alternate case of a "fair" casino, that is if $c = 0$ instead of $c < 0$, the martingale theorem would allow us to conclude that so long as one never bets to win more than $1 - f_n$ where f_n is the fortune at time n, then one ends at either 0 (broke) or 1 (success) at some random time τ and since we have a bounded martingale $Ef_\tau = f_0$. There is no way to get any other payoff unless one makes excessive bets and this will make the probability of success smaller. Finally, in a super-fair casino, that is if $c > 0$, one can ensure that one attains the eventual fortune 1 with probability arbitrarily close to one by making very small bets, as easily follows from the law of large numbers. The referee points out that superfair casinos are handled in great generality in Dubins & Savage (1965), page 72.

3. Proofs of the Assertions

Suppose that for every value of the odds, $r > 0$, there is a table available that pays these odds, but with a win probability, $w = w(r)$, which is such that the mean value of a dollar bet on the r table is exactly c, where

$-1 < c < 0$. That is,

$$w(r) * r - (1 - w(r)) * 1 = c.$$

This makes $w(r) = \frac{1+c}{1+r}$, and $1 - w(r) = \frac{r-c}{1+r}$.

Bold play for this problem might be interpreted as to always bet the full fortune f_0 on that table that will produce the fortune 1 if you win and will make you broke if you lose. This however is not optimal although there is perhaps a way that the true solution might be considered to be bold. The payoff if the gambler uses this strategy is the chance that he wins on the first (and last) bet which is $w(r) = w(\frac{1}{f_0} - 1) = f_0 * (1 + c)$. Other strategies do better than this and achieve a higher probability to reach 1 starting from f_0.

Define $Q(f) = 1 - (1 - f)^{1+c}$. We will show first that $Q(f_0)$ is an upper bound on the probability to reach one under any betting strategy starting from f_0. Then we will give a sequence of betting strategies which in the limit achieves $Q(f_0)$.

Suppose we have some strategy for stakes that is independent of the win outcomes on the various tables at *future* times $n+1, \ldots$. Then the sequence of r.v.'s f_n are the fortunes obtained under the given betting strategy is a stochastic process as is the sequence

$$X_n = Q(f_n), n = 0, 1, \ldots, \tau,$$

where τ is the first time, $n > 0$, that $f_n = 0$ or $f_n = 1$. We need to show that for any f_n sequence, X_n is a supermartingale. Indeed if this is the case then the supermartingale lemma (since X_n is bounded) says that $EX_\tau \le EX_0$. Since $EX_\tau = EQ(f_\tau) = P(f_\tau = 1)$ under this strategy, we see that for any strategy, $P(f_\tau = 1) \le Q(f_0)$. This standard supermartingale argument will then complete the proof that $Q(f_0)$ is an upper bound on the probability to reach one under any strategy.

We still have to check that $E[Q(f_{n+1})|f_n] \le Q(f_n)$, which means that if we have fortune f and decide to stake $s \le f$ on table with odds r, we need to show for every choice of r and $s \le f$, that

$$E[Q(f')|f, r, s] \le Q(f).$$

But for $f + rs < 1$ and $\gamma = 1 + c$, this becomes

$$\begin{aligned} &Q(f + rs)w(r) + Q(f - s)(1 - w(r)) \\ &= (1 - (1 - f - rs)^\gamma)\frac{\gamma}{1+r} + (1 - (1 - f + s)^\gamma)(1 - \frac{\gamma}{1+r}) \\ &\le 1 - (1 - f)^\gamma. \end{aligned}$$

Subtracting 1 from both sides, multiplying by -1, reversing the inequality, replacing $1-f$ by $\frac{s}{x}$, and dividing by $(\frac{s}{x})^\gamma$, this is the same as showing that $g(x) \geq 1$ for $x \in (0, \frac{1}{r})$, where

$$g(x) = (1-rx)^\gamma \frac{\gamma}{1+r} + (1+x)^\gamma)(1 - \frac{\gamma}{1+r}).$$

It's clear that $g(0) = 1$ and it's easy to check that $g'(x) > 0$, so this completes the proof that Q is an upper bound. The proof that the betting strategy S_α achieves Q in the limit as $\alpha \downarrow 0$ will now be given. Let S_α be the betting strategy which stakes f when $f \leq \alpha$ and S_α stakes $\frac{\alpha}{1-\alpha}(1-f)$ when $\alpha \leq f < 1$. Whatever stake s is bet, S_α bets s on the table with odds, $r = \frac{1-f}{s}$, which carries the fortune to unity in case of a win. This win probability is $w = \frac{1+c}{1+r}$ and this gives the recurrence for the value of the probability $P_\alpha(f)$ that one reaches fortune 1 using strategy S_α:

$$P_\alpha(f) = (1+c)f, \text{ for } f \leq \alpha,$$

and

$$P_\alpha(f) = (1+c)\alpha + P_\alpha(\frac{f-\alpha}{1-\alpha})(1-(1+c)\alpha), \quad \text{for } f \geq \alpha.$$

It's easy to see by induction that the solution to this recurrence is:

$$P_\alpha(f) = 1 + (1-(1+c)\alpha)^n (c - \frac{(1+c)(1-f)}{(1-\alpha)^n}),$$

for

$$f \in I_n = [1-(1-\alpha)^n, 1-(1-\alpha)^{n+1}).$$

Now it's easy to see that for fixed f, $P_\alpha(f)$ increases to $Q(f)$ as $\alpha \downarrow 0$. That there is no strategy S which attains $Q(f)$ is clear because if ever a stake $s > 0$ under S is made, the payoff is strictly less than $Q(f)$ because of the supermartingale inequality. But if one never stakes $s > 0$ then $f_n \equiv f_0$ which is not allowed since one must play until one either goes broke or attains fortune 1. In the more general problem where any bet with expectation cs is allowed if the stake is s, one cannot achieve more than $Q(f)$, obtained with only two-valued bets, because the set of distributions of r.v.'s with a given mean is a convex set whose extreme points are the distributions corresponding to two valued r.v.'s and so any random variable with mean cs is a mixture of two valued r.v.'s and the payoff is a convex combination of that obtained for two-valued r.v.'s and so is at most $Q(f)$. This completes the proof of all the claims.

We will study elsewhere the Vardi casino for the more challenging problem when there is an interest payment imposed after each bet as discussed in Remark 3.

References

1. DUBINS, L.E. & SAVAGE, L.J. (1965). *How To Gamble If You Must, Inequalities for Stochastic Processes.* McGraw-Hill, New York.
2. CHEN, R. (1978). Subfair "red-and-black" in the presence of inflation. *Z. Wahr. verw. Gebiete* **42**, 293-301.
3. CHEN, R., SHEPP, L.A. & ZAME, A. (2004). Subfair primitive casino in the presence of inflation. *J. Appl. Prob.* **41**, 587-592.
4. CHEN, R.W., SHEPP, L.A., YAO, Y-C. & ZHANG, C-H. (2005). On optimality of bold play for primitive casinos in the presence of inflation. *J. Appl. Prob.* **42**, 121-137.
5. SHEPP, L.A. (2002). A model for stock price fluctuations based on information. *IEEE Trans. Inf. Th.* **48**, 1372-1378.
6. HEATH, D.C., PRUITT, W.E. & SUDDERTH, W.D. (1972). Subfair red-and-black with a limit. *Proc. Amer. Math. Soc.* **35**, 555-560.
7. SCHWEINSBERG, J. (2005). Improving on bold play when the gambler is restricted. *J. Appl. Prob.* **42**, 321-333.

THE UPPER LIMIT OF A NORMALIZED RANDOM WALK

CUN-HUI ZHANG*

Department of Statistics, Rutgers University
Hill Center, Busch Campus, Piscataway, NJ 08854, U.S.A.
czhang@stat.rutgers.edu

We consider the upper limit of $S_n/n^{1/p}$ where S_n are partial sums of iid random variables. Under the assumption of $E(X^+)^p < \infty$, we provide an integral test which determines the upper limit up to certain universal constant factors depending on p only. The problem is closely related to moment properties of ladder variables. We prove our theorem by considering the lower limit of T_k/k^p where T_k is the k-th epoch of the random walk.

Some key words: Random walk; Strong law of large numbers; Asymmetric random variables; Integral test; Law of the iterated logarithm; Ladder variables; Truncated moment.

1. Introduction

Let X, $X_n, n \geq 1$, be a sequence of iid variables. Set $S_n = \sum_{i=1}^n X_i$. The Kolmogorov and Marcinkiewicz-Zygmund strong laws of large numbers assert that

$$E|X|^p < \infty \Leftrightarrow \frac{S_n}{n^{1/p}} \to 0 \quad \text{a.s.}, \quad 0 < p < 2, \tag{1}$$

provided that $EX = 0$ whenever $E|X| < \infty$. What happens if $E|X|^p = \infty$? If EX is defined, $S_n/n \to EX$ a.s. by (1). The Borel-Cantelli lemma implies $\limsup_n |S_n|/n^{1/p} = \infty$ a.s. for $E|X|^p = \infty$, as in the standard proof of (1). However, what is the value of

$$\limsup_{n\to\infty} \frac{S_n}{n^{1/p}} \qquad \text{a.s.} \tag{2}$$

*Research partially supported by National Science Foundation Grants DMS-0405202, DMS-0504387 and DMS-0604571 when $E|X|^p = \infty$? In this note, we provide a univariate integral test which determines (2) within certain universal constant factors under the assumption $E(X^+)^p < \infty$ for $1 < p < 2$, where $x^\pm = \max(\pm x, 0)$.

We first state our main result. For $y > 0$ define

$$K(y; X) = \inf\Big\{t > 0 : yE\min\big(X^2/t^2, |X|/t\big) \le 1\Big\}. \tag{3}$$

Here and in the sequel, $\inf\emptyset = \infty$ and $\sup\emptyset = 0$. Define

$$\kappa_p^*(X) = \sup\left\{t : \sum_{n=1}^\infty \frac{1}{n}\exp\left[-\Big(\frac{tn^{1/p}}{E|S_n|}\Big)^{p/(p-1)}\right] = \infty\right\}, \tag{4}$$

where S_n are sample sums from X as in (1), and

$$\kappa_p(X) = \sup\left\{t : \sum_{n=1}^\infty \frac{1}{n}\exp\left[-\Big(\frac{tn^{1/p}}{K(n;X)}\Big)^{p/(p-1)}\right] = \infty\right\}. \tag{5}$$

Theorem 1. *Let $1 < p < 2$ and $\kappa_p \in [0, \infty]$ be as in (5). Then, there exist universal constants $0 < c_p < C_p < \infty$ (depending on p only) such that*

$$c_p\,\kappa_p(X^-) \le \limsup_{n\to\infty} \frac{S_n}{n^{1/p}} \le C_p\,\kappa_p(X^-) \tag{6}$$

for all X satisfying $EX = 0$ and $E(X^+)^p < \infty$. Moreover, (6) remains valid if $\kappa_p(X^-)$ is replaced by $\kappa_p(X)$ or $\kappa_p^*(X)$ with the $\kappa_p^*(X)$ in (4).

Remark. Suppose $EX = 0$. Let $1 < p < 2$. If $E|X|^p < \infty$, then $\kappa_p(X^-) = \kappa_p(X) = \kappa_p^*(X) = 0$ by (1) and Theorem 1. However, as shown in Example 1 below, $E(X^+)^p < \infty$ and $\kappa_p(X^-) = 0$ do not imply $E|X|^p < \infty$. It follows from Lemma 1 below that $E(X^+)^p < \infty$ implies

$$2^{-1/p}\kappa_p(X^-) \le \kappa_p(X) \le 2^{1/p}\kappa_p(X^-).$$

It follows from Klass (1980), cf. (20) below, that

$$2^{-1}\kappa_p(X) \le \kappa_p^*(X) \le 2\kappa_p(X).$$

Example 1. *Let $1 < p < 2$ and $L(x)$ be a slowly varying function satisfying $\lim_{x\to\infty} L(cx)/L(x) = 1$ for all $c > 1$. Suppose $EX = 0$, $E(X^+)^p < \infty$ and $P\{X < -t\} = t^{-p}L(\log t)$. Then, $K(n; X^-) = (C_p + o(1))\{nL(\log n)\}^{1/p}$ with $C_p = [p/\{(2-p)(p-1)\}]^{1/p}$ and*

$$\kappa_p(X^-) = C_p \sup\left\{t : \sum_{n=1}^\infty \frac{1}{n}\exp\left[-\Big(\frac{t^p}{L(\log n)}\Big)^{1/(p-1)}\right] = \infty\right\}.$$

Suppose further $L(t) = (C + o(1))(\log t)^{-\beta}$ *as* $t \to \infty$ *for some* $C > 0$. *If* $\beta > p - 1$, *then* $E|X|^p = \infty$ *and* $\kappa_p(X^-) = 0$. *If* $\beta < p - 1$, *then* $\kappa_p(X^-) = \infty$. *If* $\beta = p - 1$, *then* $\kappa_p(X^-) = C_p C^{1/p}$ *and the upper limit (2) is in* $(0, \infty)$ *by Theorem 1.*

Our problem has been considered by Derman and Robbins (1955), Binmore and Katz (1968), Kesten (1970), Pruitt (1981) and more, in addition to references cited below. In Section 2, we describe the results of Erickson (1973), Chow and Zhang (1986) and Zhang (1986), whose integral tests determine (2) up to universal constant factors for $0 < p \leq 1$. We discuss ladder variables in Section 3. We prove Theorem 1 in Section 4 based on the results described in Sections 2 and 3. Section 5 contains some remarks, including discussion about the upper limit of S_n/b_n for more general normalizing constants $b_n \to \infty$, the order of $E|S_n|/n^{1/p}$, and the relationship between Theorem 1 and the universal law of the iterated logarithm of Klass (1976, 1977).

2. The Case of $0 < p \leq 1$

For $p = 1$, Erickson (1973) obtained the following integral test for (2) based on a result of Kesten (1970):

$$E|X| = \infty \Leftrightarrow J_+ + J_- = \infty$$

and in this case

$$J_+ = \infty \;\Leftrightarrow\; \limsup_{n\to\infty} S_n/n = \infty \text{ a.s.} \;\Leftrightarrow\; \limsup_{n\to\infty} S_n/n > -\infty \text{ a.s.}$$

$$J_- = \infty \;\Leftrightarrow\; \liminf_{n\to\infty} S_n/n = -\infty \text{ a.s.} \;\Leftrightarrow\; \liminf_{n\to\infty} S_n/n < \infty \text{ a.s.}$$

where $J_\pm$ are integrals defined as

$$J_\pm = \int_0^\infty \min\left\{x, \frac{x}{E(X^\mp \wedge x)}\right\} dP\{\pm X \leq x\}. \tag{7}$$

Unlike the moment condition $E|X|^p < \infty$ for (1), (7) describes the interplay between the positive and negative parts of the variables X_i and determines if one part is dominant in the fluctuation of S_n. We will see more integrals of this form below.

Erickson's (1973) result was extended by Chow and Zhang (1986) to $0 < p < 1$ as follows:

$$E|X|^p = \infty \Leftrightarrow J_+(p) + J_-(p) = \infty$$

and in this case

$$J_+(p) = \infty \;\Leftrightarrow\; \limsup_{n\to\infty} S_n/n^{1/p} = \infty \text{ a.s.}$$

$$J_-(p) = \infty \;\Leftrightarrow\; \liminf_{n\to\infty} S_n/n^{1/p} = -\infty \text{ a.s.}$$

where $J_\pm(p)$ are defined as

$$J_\pm(p) = \int_0^\infty \min\left\{x^p, \frac{x}{E(X^\mp \wedge x)}\right\} dP\{\pm X \le x\}. \qquad (8)$$

However, while (7) completely determines (2) for $p = 1$, (8) only determines whether the upper limit (2) is infinity for $0 < p < 1$, since (2) could take a finite value when $J_+(p) < J_-(p) = \infty$.

Suppose $0 < p < 1$ and $J_+(p) < J_-(p) = \infty$. The upper limit (2) can be determined up to universal constant factors as follows. Chow and Zhang (1986) proved that, as far as (2) is concerned, the random walk is dominated by its negative parts in the sense that

$$\begin{aligned}\limsup_{n\to\infty} \frac{S_n}{n^{1/p}} &= \limsup_{n\to\infty} \frac{-1}{n^{1/p}} \sum_{i=1}^n X_i^- \\ &= -\liminf_{n\to\infty} \frac{1}{n^{1/p}} \sum_{i=1}^n X_i^- . \qquad (9)\end{aligned}$$

Thus, determination of the upper limit (2) for $\sum_{i=1}^n X_i/n^{1/p}$ is equivalent to that of the lower limit for $\sum_{i=1}^n X_i^-/n^{1/p}$. For nonnegative variables X, define

$$\theta_q(X) = \inf\left\{\lambda : \int_1^\infty \frac{1}{x} \exp\left[-\left\{\frac{E(X\wedge x)}{x^{1-q}\lambda^q}\right\}^{1/(1-q)}\right] dx = \infty\right\}. \qquad (10)$$

For $q < 1$ and $P(X \ge 0) = 1$, Zhang (1986) proved

$$\begin{aligned}2^{-1/q}\max(1.17, 2^{2-1/q})\theta_q(X) &\le \liminf_{n\to\infty} S_n/n^{1/q} \\ &\le 2.7183\,\theta_q(X)2^{1/q} . \qquad (11)\end{aligned}$$

It becomes clear now that the combination of (9) and (11) results in the following statement: for $0 < p < 1$ and $J_+(p) < J_-(p) = \infty$,

$$\begin{aligned}-2.7183\,\theta_p(X^-)2^{1/p} &\le \limsup_{n\to\infty} S_n/n^{1/p} \\ &\le -2^{-1/p}\max(1.17, 2^{2-1/p})\theta_p(X^-) . \qquad (12)\end{aligned}$$

Example 2. *Let* $0 < p < 1$. *Suppose* $J_+(p) < J_-(p) = \infty$ *and* $P\{X < -t\} = (C + o(1))t^{-p}\big(\log(\log t)\big)^{1-p}$. *Then,* $\theta_p(X^-) = \{C/(1-p)\}^{1/p}$ *and (2) is in* $(-\infty, 0)$ *by (12).*

The truncated moments $E(X^{\pm} \wedge x)$ appear prominently in all three integrals (7), (8) and (10) as $\sup\{x : nE(X^{\pm} \wedge x) \geq x\}$ represent typical orders of magnitude for $\sum_{i=1}^{n} X_i^{\pm}$. This provides the probabilistic content of these integrals and is a useful fact for the investigation of the fluctuation of random walks.

3. Moment Properties of Ladder Variables

Consider $EX = 0$ in this section. The fluctuation of the random walk S_n is closely related to moment properties of the ladder variables

$$S_{\tau}, \quad S_{\tau_-}, \quad \tau = \inf\{n : S_n > 0\}, \quad \tau_- = \inf\{n : S_n \leq 0\}. \tag{13}$$

Spitzer (1960) proved $E|X|^{p+1} < \infty \Rightarrow ES_{\tau}^{p} < \infty$ for $p = 1$ and $p = 2$ with explicit formulas for ES_{τ}^{p}. Lai (1976) extended Spitzer's result to all positive integers p via a Tauberian argument. Chow and Lai (1979) weakened the moment condition by proving $E(X^+)^{p+1} \Rightarrow ES_{\tau}^{p} < \infty$ for all $p > 0$. The problem of the finiteness of ES_{τ}^{p} was completely solved in Chow (1986) and further investigated in Chow (1997). Chow (1986) proved

$$ES_{\tau}^{p} < \infty \Leftrightarrow \int_0^{\infty} \frac{x^{p+1}}{EX^-(X^- \wedge x)} dP\{X \leq x\} < \infty.$$

The integration above again involves the interplay between the positive and negative parts of X as in (7).

The direct connection between ladder variables and (2) can be easily seen in the simpler case of integer valued X with $P(X \leq 1) = 1$. Define

$$\tau_k = T_k - T_{k-1}, \quad Y_k = S_{T_k} - S_{T_{k-1}}, \quad k = 1, 2, \ldots, \tag{14}$$

with $T_0 = S_0 = 0$ and $T_k = \inf\{n > T_{k-1} : S_n > S_{T_{k-1}}\}$, so that (τ_k, Y_k) are iid copies of (τ, S_τ). Since $S_n/n^{1/p} \leq S_{T_k}/T_k^{1/p} = k/T_k^{1/p}$ for $T_k \leq n < T_{k+1}$ in the simpler case,

$$\begin{aligned}\limsup_{n\to\infty} \frac{S_n}{n^{1/p}} &= \limsup_{k\to\infty} \frac{\sum_{j=1}^{k} Y_j}{(\sum_{j=1}^{k} \tau_k)^{1/p}} \\ &= \left(\liminf_{n\to\infty} \frac{1}{n^{1/q}} \sum_{i=1}^{n} \tau_i \right)^{-1/p},\end{aligned} \tag{15}$$

where $q = 1/p$. Thus, determination of the upper limit (2) for $\sum_{i=1}^{n} X_i/n^{1/p}$ and $1 < p < 2$ can be done up to universal constant factors by calculating $\theta_q(\tau)$ with $q = 1/p < 1$ as in (11). This involves finding upper and lower bounds for the truncated mean $E(\tau \wedge x)$ of the first epoch τ in (13).

An argument similar to (15) was used in Klass and Zhang (1994) to investigate the fluctuation of the partial sum maxima $S_n^* = \max_{k\le 0} S_k$. Based on an adaptation of Erickson's (1973) method to ladder variables, they proved the following result: for $EX = 0$

$$\begin{aligned} J^*(p) = \infty &\Leftrightarrow \liminf_n S_n^*/n^{1/p} = 0 \\ &\Leftrightarrow \liminf_n S_n^*/n^{1/p} < \infty \\ &\Leftrightarrow \limsup_n \sum_{k=1}^n \tau_k^{1/p} \Big/ \sum_{k=1}^n Y_k = \infty, \end{aligned}$$

where (τ_k, Y_k) are as in (14) and

$$J^*(p) = \int_0^\infty \frac{x}{E(S_\tau \wedge x)} dP\{\tau^{1/p} \le x\}. \tag{16}$$

This is again an integral of the form (7), but it involves τ and S_τ instead of $X^\pm$. Duality inequalities

$$n \le E\Big(\tau \wedge n\Big) E\Big(\tau_- \wedge n\Big) \le 2n, \tag{17}$$

used in Klass and Zhang (1994) to derive more explicit integral tests for (16), will also be used in Section 4 to proof Theorem 1.

4. Proof of Theorem 1

Suppose $EX = 0$ and $E(X^+)^p < \infty$ with $1 < p < 2$. Assume further through out this section (without loss of generality through proper normalization of X if necessary) that $EX^+ = P\{X > 0\}$ and $X = \psi(U)$ for certain uniform variable U and increasing function ψ. For constants or functions f and g, we write $f \asymp g$ if $cf \le g \le Cf$ for constants $0 < c < C < \infty$ possibly dependent on p.

Let $X' \le 1$ be an integer-valued random variable satisfying

$$EX' = 0, \quad X > 0 \Leftrightarrow X' = 1, \quad X \le 0 \Rightarrow |X - X'| \le 1.$$

Such X' certainly exists since $\lfloor X \rfloor \le X \le \lfloor X \rfloor + 1$, where $\lfloor a \rfloor$ is the integer part of a. Let (X_i, X_i') be iid copies of (X, X') and $S_n' = \sum_{i=1}^n X_i'$. Since $E(X^+)^p < \infty$, $E|X - X'|^p < \infty$, so that $(S_n - S_n')/n^{1/p} \to 0$ a.s. Thus, the upper limit (2) for X is the same as that for X'. It then follows from (15) and (11) that

$$\limsup_{n\to\infty} \frac{S_n}{n^{1/p}} \asymp \left(\liminf_{n\to\infty} \frac{1}{n^{1/q}} \sum_{i=1}^n \tau_i' \right)^{-1/p} \asymp \{\theta_q(\tau')\}^{-1/p} \quad \text{a.s.,} \tag{18}$$

where $q = 1/p$ and τ_i' are iid copies of $\tau' = \inf\{n \geq 1 : S_n' > 0\}$.

The next step is to calculate $E(\tau' \wedge n)$ as it is used to determine $\theta_q(\tau')$ in (10). Since X' is an integer valued variable with $P(X' \leq 1) = 1$, it follows from the ballot problem (Chow and Teicher, 1988, page 242) that

$$P\{\tau_-' > n | S_n'\} = (S_n')^+/n, \quad \tau_-' = \inf\{n \geq 1 : S_n' \leq 0\}.$$

This implies $E(\tau_-' \wedge n) \asymp E|S_n'|$, so that by (17)

$$E(\tau' \wedge n) \asymp \frac{n}{E|S_n'|} .$$

Thus, by (10), we find

$$\begin{aligned}\theta_q(\tau') &= \inf\left\{\lambda : \int_1^\infty \frac{1}{x} \exp\left[-\left\{\frac{E(\tau' \wedge x)}{x^{1-q}\lambda^q}\right\}^{1/(1-q)}\right] dx = \infty\right\} \\ &\asymp \inf\left\{\lambda : \sum_{n=1}^\infty \frac{1}{n} \exp\left[-\left(\frac{\lambda^{-q} n^q}{E|S_n'|}\right)^{1/(1-q)}\right] = \infty\right\} \\ &\asymp \inf\left\{\lambda : \sum_{n=1}^\infty \frac{1}{n} \exp\left[-\left(\frac{\lambda^{-1/p} n^{1/p}}{E|S_n'|}\right)^{p/(p-1)}\right] = \infty\right\} .\end{aligned}$$

Or, equivalently

$$\begin{aligned}&\{\theta_q(\tau')\}^{-1/p} \\ &\asymp \sup\left\{t : \sum_{n=1}^\infty \frac{1}{n} \exp\left[-\left(\frac{tn^{1/p}}{E|S_n'|}\right)^{p/(p-1)}\right] = \infty\right\} \\ &= \kappa_p^*(X') .\end{aligned} \tag{19}$$

This makes sense, since the integral test takes a critical value in $(0, \infty)$ when $E|S_n'|/n^{1/p}$ is of an iterated logarithmic order.

It remains to show that we may replace $E|S_n'|$ in (19) by $K(n; X^-)$ or $K(n; X)$ in (3), and eventually by $E|S_n|$ in (4). For random variables X with $EX = 0$, Klass (1980) proved that for sample sums S_n from X

$$K(n; X)/2 \leq E|S_n| \leq 2K(n; X). \tag{20}$$

We shall use these inequalities and a symmetrization argument. Let $\{\varepsilon, \varepsilon_i, i \geq 1\}$ be iid variables independent of $\{X, X_i, i \geq 1\}$ with $P(\varepsilon_i = \pm 1) = 1/2$. By (20) and (3)

$$E|S_n'| \asymp K(n; X') = K(n; \varepsilon X') \asymp E\left|\sum_{i=1}^n \varepsilon_i X_i'\right|.$$

Since $|X' - X^-| \le 1$, we have again by (20) and (3) that

$$\begin{aligned} E\Big|\sum_{i=1}^n \varepsilon_i X_i'\Big| &\asymp E\Big|\sum_{i=1}^n \varepsilon_i X_i^-\Big| + \eta_n\sqrt{n} \\ &\asymp K(n;\varepsilon X^-) + \eta_n\sqrt{n} \\ &= K(n;X^-) + \eta_n\sqrt{n} \end{aligned}$$

with certain $|\eta_n| \le 1$. Thus, in view of (18) and (19),

$$\begin{aligned} &\limsup_{n\to\infty} S_n/n^{1/p} \qquad (21) \\ &\asymp \sup\Big\{t : \sum_{n=1}^\infty \frac{1}{n}\exp\Big[-\Big(\frac{tn^{1/p}}{K(n;X^-)+\eta_n\sqrt{n}}\Big)^{p/(p-1)}\Big] = \infty\Big\}. \end{aligned}$$

It follows easily from (3) that $K(y;X)/\sqrt{y}$ is increasing in y (Klass, 1980). Thus, the sum on the right-hand side of (21) is infinity in a neighborhood of $t = t_0$ if and only if

$$\sum_{n=1}^\infty \frac{1}{n}\exp\Big[-\Big(\frac{tn^{1/p}}{K(n;X^-)}\Big)^{p/(p-1)}\Big] = \infty$$

a neighborhood of $t = t_0$, since $K(n;X^-)/\sqrt{n} \to \infty$ necessarily in both cases. Therefore, (6) follows from (21).

If we consider iid copies of $X'' = \varepsilon X^+$, (1) and (6) imply $\kappa_p(X^+) = \kappa_p(\varepsilon X^+) = 0$. Thus, by Lemma 1 (ii) below,

$$\kappa_p(X) \le 2^{1/p}\kappa_p(X^-), \quad \kappa_p(X^-) \le 2^{1/p}\kappa_p(X),$$

so that $\kappa_p(X^-)$ can be replaced by $\kappa_p(X)$ in (6). It follows from (20), (5) and (4) that $2^{-1}\kappa_p(X) \le \kappa_p^*(X) \le 2\kappa_p(X)$, so that $\kappa_p(X^-)$ can be also replaced by $\kappa_p^*(X)$ in (6). □

Lemma 1. (i) Let $K(y;X)$ be as in (3). Then,

$$K(y;Z_1+Z_2) \le K(2y;Z_1) + K(2y;Z_2) \qquad (22)$$

(ii) Let $\kappa_p(X)$ be defined as in (5). Then,

$$\kappa_p(Z_1+Z_2) \le 2^{1/p}\Big\{\kappa(Z_1)+\kappa(Z_2)\Big\}.$$

Proof of Lemma 1. (i) Let $g(x) = \min(x^2,|x|)$. For $w_j \ge 0$ with $w_1 + w_2 = 1$,

$$g(w_1x_1 + w_2x_2) \le g\big(\max(|x_1|,|x_2|)\big) \le g(x_1) + g(x_2).$$

Let $t_j = K(2y; Z_j)$ and $w_j = t_j/(t_1+t_2)$. It follows from (3) that $Eg(Z_j/t_j) = 1/(2y)$, so that

$$Eg((Z_1+Z_2)/(t_1+t_2)) \le \sum_{j=1}^{2} Eg(Z_j/t_j) = 2/(2y) = 1/y.$$

Thus, $K(y; Z_1+Z_2) \le t_1+t_2$.

(ii) Let $h(x) = \exp\big[-x^{-p/(p-1)}\big]$, $a_n = K(n; Z_1)/n^{1/p}$ and $b_n = K(n; Z_2)/n^{1/p}$. Since either $a_n \le Mb_n$ or $b_n < a_n/M$ for positive a_n and b_n, due to the monotonicity of $h(x)$ ($\uparrow$ in x),

$$\begin{aligned}&\sum_{n=1}^{\infty} n^{-1} h((a_n+b_n)/t)\\ &\le \sum_{n=1}^{\infty} n^{-1} h((M+1)b_n/t) + \sum_{n=1}^{\infty} n^{-1} h(a_n(1+1/M)/t).\end{aligned} \tag{23}$$

for all $t>0$. Set $t = \kappa(Z_1)+\kappa(Z_2)+\epsilon_0$ with $\epsilon_0>0$ and $M = \kappa(Z_1)/\kappa(Z_2)$. We find

$$(1+1/M)\kappa_p(Z_1) = (M+1)\kappa(Z_2) = \kappa(Z_1)+\kappa(Z_2) < t.$$

Since $t/(1+1/M) > \kappa_p(Z_1)$ and $t/(M+1) > \kappa_p(Z_2)$, (22) and (23) imply

$$\begin{aligned}&\sum_{n=1}^{\infty} \frac{1}{2n} \exp\Big[-\Big(\frac{2^{1/p} t n^{1/p}}{K(n; Z_1+Z_2)}\Big)^{p/(p-1)}\Big]\\ &\le \sum_{n=1}^{\infty} \frac{1}{n} \exp\Big[-\Big(\frac{t n^{1/p}}{K(n; Z_1)+K(n; Z_2)}\Big)^{p/(p-1)}\Big]\\ &\le \sum_{n=1}^{\infty} n^{-1} h((M+1)b_n/t) + \sum_{n=1}^{\infty} n^{-1} h(a_n(1+1/M)/t)\\ &= \sum_{n=1}^{\infty} \frac{1}{n} \exp\Big[-\Big(\frac{\{t/(M+1)\} n^{1/p}}{K(n; Z_2)}\Big)^{p/(p-1)}\Big]\\ &\quad + \sum_{n=1}^{\infty} \frac{1}{n} \exp\Big[-\Big(\frac{\{t/(1+1/M)\} n^{1/p}}{K(n; Z_1)}\Big)^{p/(p-1)}\Big]\\ &< \infty,\end{aligned}$$

in view of (5). It then follows from (5) that

$$\kappa_p(Z_1+Z_2) \le 2^{1/p} t = 2^{1/p}\{\kappa(Z_1)+\kappa(Z_2)+\epsilon_0\}.$$

Since $\epsilon_0>0$ is arbitrary, this completes the proof of the lemma. □

5. Remarks

We discuss possible extensions, the order of $E|S_n|/n^{1/p}$, and the relationship between Theorem 1 and the universal law of the iterated logarithm.

Extensions of Theorem 1 can be obtained to cover

$$\limsup_{n\to\infty} \frac{S_n}{b_n} \tag{24}$$

for more general normalizing sequences $0 < b_n \to \infty$ satisfying certain regularity conditions, since the idea of our proofs does not depend on $b_n = n^{1/p}$ and Zhang (1986) does cover more general normalizing constants. It seems that

$$\liminf_{n\to\infty} \frac{b_{2n}}{b_n} > \sqrt{2}, \qquad \limsup_{n\to\infty} \frac{b_{2n}}{b_n} < 2,$$

provide a possible set of such regularity conditions on b_n.

Consider the case $E|X|^p < \infty$ with $1 < p < 2$ and $EX = 0$. Since $\min(x^2/n^{2/p}, |x|/n^{1/p}) \le x^p/n$, by the dominated convergence theorem

$$n\, E \min\left(\frac{|X|^2}{n^{2/p}}, \frac{|X|}{n^{1/p}}\right) = o(1),$$

so that $E|S_n| \asymp K(n; X) = o(n^{1/p})$ by (20) and (3). However, the behavior of the $o(1)$ is not clear from this argument. Theorem 1 implies $\kappa_p^*(X) = 0$ with (4), or equivalently

$$\sum_{k=1}^{\infty} \exp\left[-\left(\frac{t2^{k/p}}{E|S_{2^k}|}\right)^{p/(p-1)}\right] < \infty, \quad \forall t > 0,$$

since $E|S_n|$ is increasing in n and $E|S_{2n}| \le 2E|S_n|$. Chow (1988) proved

$$\sum_{n=1}^{\infty} \frac{E|S_{2^k}|}{2^{k/p}} < \infty \Leftrightarrow \int_0^{\infty} \left\{P(|X| > t)\right\}^{1/p} dt < \infty.$$

Klass (1976, 1977) proved the following universal law of the iterated logarithm: for $EX = 0$

$$1 \le \limsup_{n\to\infty} \frac{S_n}{a_n} \le 1.5, \quad a_n = (\log(\log n))K(n/\log(\log n); X), \tag{25}$$

provided $P\{X_n > a_n, i.o.\} = 0$, where $K(y; X)$ is as in (3). While (25) seeks a *sharp* normalizing sequence a_n for a given variable X, our approach tries to find the upper limit (2) or (24) given the normalizing constants. It seems that Theorem 1 is almost a consequence of (25), but we are unable to derive it from (25) since it is not clear if the lower and upper limits of $a_n/n^{1/p}$ could be (c, ∞) or $(0, c)$ for some $0 < c < \infty$.

If $a_n \asymp n^{1/p}$ in (25), then $K(n;X) \asymp n^{1/p}(\log(\log n))^{1/p-1}$ and $\kappa_p(X)$ is positive and finite as in Example 1. In this case, Theorem 1 and (25) are equivalent up to the constant factors, but (25) provides sharper bounds.

Theorem 1 assumes the condition $E(X^+)^p < \infty$. If $EX = 0$ and $E(X^+)^p = \infty$, $1 < p < 2$, is it possible to have $\limsup_n S_n/n^{1/p} < \infty$ a.s.? We do not know the answer to the question.

References

1. Binmore, K.G. and Katz, M. (1968). A note on the strong law of large numbers. *Bull. Amer. Math. Soc.* **74** 941-943.
2. Chow, Y. S. (1986). On moments of ladder height variables. *Adv. Appl. Math.* **7** 46-54.
3. Chow, Y. S. (1988). On the rate of moment convergence of sample sums and extremes. *Bull. Inst. Math. Acad. Sinica* **16** 219-243.
4. Chow, Y. S. (1997). On Spitzer's formula for the moment of ladder variables. *Statistica Sinica* **7** 149-156.
5. Chow, Y. S. and Lai, T. L. (1979). Moments of ladder variables for driftless random walks. *Z. Wahrscheinlichkeitstheorie verw. Gebiete* **48** 253-257.
6. Chow, Y. S. and Teicher, H. (1988). *Probability Theory.* 2nd edition. Springer-Verlag, New York.
7. Chow, Y. S. and Zhang, C.-H. (1986). A note on Feller's strong law of large numbers. *Ann. Probab.* **14** 1088-1094.
8. Derman, C. and Robbins, H. (1955). The strong law of large numbers when the first moment does not exist. *Proc. Nat. Acad. Sci. U.S.A.* **41** 586-587.
9. Erickson, K. B. (1973). The SLLN when the mean is undefined. *Trans. Amer. Math. Soc.* **185** 371-381.
10. Kesten, H. (1970). The limit points of a normalized random walk. *Ann. Math. Statist.* **41** 1173-1205.
11. Klass, M. J. (1976). Toward a universal law of the iterated logarithm. Part I. *Z. Wahrscheinlichkeitstheorie verw. Gebiete* **36** 165-178.
12. Klass, M. J. (1977). Toward a universal law of the iterated logarithm. Part II. *Z. Wahrscheinlichkeitstheorie verw. Gebiete* **39** 151-165.
13. Klass, M. J. (1980). Precision Bounds for the Relative Error in the Approximation of $E|Sn|$ and Extensions. *Ann. Probab.* **8** 350-367.
14. Klass, M. J. and Zhang, C.-H. (1994). On the almost sure minimal growth rate of partial sum maxima. *Ann. Probab.* **22** 1857-1878.
15. Lai, T. L. (1976). Asymptotic moments of random walks with applications to ladder variables and renewal theory. *Ann. Probab.* **4** 51-66.
16. Pruitt, W. E. (1981). General one-sided laws od iterated logarithm. *Ann.Probab.* **9** 1-48.
17. Spitzer, F. (1960). A Tauberian theorem and its probability interpretation. *Trans. Amer. Math. Soc.* **94** 150-169.
18. Zhang, C.-H. (1986). The lower limit of a normalized random walk. *Ann. Probab.* **14** 560-581.

Semiparametric Statistics

ANALYSIS OF A SEQUENCE OF DEPENDENT 2 × 2 TABLES

S. G. KOU

Department of IEOR, 312 Mudd Building, Columbia University
500 West 120th Street, New York, NY 10027, U.S.A.

ZHILIANG YING

Department of Statistics, MC 4690, Columbia University
1255 Amsterdam Avenue, 10th Floor, New York, NY 10027, U.S.A.

A sequence of dependent 2 × 2 contingency tables often arises in epidemiologic cohort studies, controlled clinical trials, and other follow-up studies. Due to dependence, it is unclear, however, whether and how the conditional approach for a single 2 × 2 table can be extended to analyze a sequence of dependent 2 × 2 tables. We show that distributional properties can be derived by considering a "tangent" sequence of independent 2 × 2 tables, with each 2 × 2 table being represented by a sum of independent, yet not identically distributed, Bernoulli trials, whose success probabilities can easily be computed (Kou and Ying, 1996). The method has four applications: (1) We provide a characterization of the validity of a weighted log-rank test. (2) The method leads to a simple algorithm to compute the maximum partial likelihood estimator of the common odds ratio, as well as its variance estimator. The efficiency over the traditional Mantel-Haenszel estimator is also demonstrated. (3) We show how to use the method to provide estimation in Breslow-Zelen model for regression with nonhomogeneous odd ratios. (4) The method is applied to analysis of a proportional hazards model with tied observations. The computation is straightforward by using a link with the roots of Jacobi polynomials.

1. Introduction

Suppose two coins with success probabilities p_1 and p_2 are tossed N_1 and N_2 times. Let M_1 and M_2 be the numbers of heads and tails, respectively, in the $N = N_1 + N_2$ tosses, and X the number of heads from the first coin. It is known that the conditional distribution of X given the M_i and N_i is noncentral hypergeometric

$$P(X = x) = \frac{\binom{N_1}{x}\binom{N_2}{M_1-x}\theta^x}{\sum_{u=L}^{S}\binom{N_1}{u}\binom{N_2}{M_1-u}\theta^u}, \quad L \le x \le S, \tag{1.1}$$

where $\theta = \{p_1/(1-p_1)\}/\{p_2/(1-p_2)\}$ is the odds ratio parameter, $L = \max(0, M_1 - N_2)$ and $S = \min(N_1, M_1)$; see Breslow and Day (1980, p.125). In particular, if $p_1 = p_2$, then $\theta = 1$ and (1.1) reduces to the (central) hypergeometric distribution

$$P(X = x) = \binom{N_1}{x}\binom{N_2}{M_1 - x} \bigg/ \binom{N}{M_1}, \quad L \le x \le S, \tag{1.2}$$

giving rise to the celebrated Fisher's exact test for the hypothesis $p_1 = p_2$.

The noncentral hypergeometric family (1.1) is frequently used in epidemiological studies to investigate relationship between occurrence of a disease and exposure to a possible risk factor. Such studies may be summarized into a 2×2 table (see below), in which N_1 and N_2 are taken as the numbers of persons in the exposed and the unexposed groups, respectively, and M_1 and M_2 the numbers of diseased and disease-free individuals, respectively (Breslow and Day, 1980, p. 124).

	Diseased	Disease-free	
Exposed	X	$N_1 - X$	N_1
Unexposed	$M_1 - X$	$X + N_2 - M_1$	N_2
	M_1	M_2	N

The hypothesis that the disease rates for the two groups, the exposed and the unexposed, are the same is therefore tantamount to $\theta = 1$, and Fisher's exact test applies. The same table arises in controlled clinical experiments as well, where the exposure factor becomes the treatment/control indicator.

Of pivotal concern in the current paper is a sequence of K dependent 2×2 tables, where K may be a random variable or stopping time, and the dependence structure among the tables may not be fully known (e.g. due to possible censorship in the data). One motivation comes from epidemiological cohort studies, in which individuals are identified along with their exposure history, and are followed forward in time to ascertain the occurrences of the diseases of interest so that the exposure information can be related to subsequent disease experience (Breslow and Day, 1987). Such studies are useful to establish causality between exposure to a possible risk factor and occurrence of a disease. Stratification in time is often needed, giving rise to a sequence of dependent 2×2 tables. Such dependent sequences also appear in controlled clinical trials, where patients receiving treatment/control are followed in time until occurrence of certain clinical endpoints. See Miller (1981, Chapter 4) for an excellent introduction on how such sequences of dependent 2×2 tables arise from medical follow-up studies.

Due to the nature of follow-up studies, the 2×2 tables thus constructed are typically dependent in such a way that the conditional distribution of the kth table given its margin is noncentral hypergeometric of form (1.1). This property allows us to make use of the decoupling method (Kwapień and Woyczyński, 1992; de la Peña and Giné, 1999) to connect the sequence of such *dependent* 2×2 tables to a "tangent" sequence of *independent* tables. In conjunction with a representation that a noncentral hypergeometric random variable may be expressed as a sum of independent Bernoulli random variables (Kou and Ying, 1996), the decoupling method enables us to establish validity of normal approximations to weighted sums of possibly dependent 2×2 tables.

Our method has implications in several aspects. First, it allows us to rigorously investigate the weighted log-rank statistic to test the hypothesis that the odds ratio are equal to one, i.e. no treatment effect. In the existing literature, the asymptotic theory for the log-rank statistic applies only to the situation of a large collection of small tables or a small number of large tables. We establish here the usual asymptotic properties under a minimal condition that the total conditional variance goes to infinity, no matter how many large or small tables are. See Section 3.

Second, our approach provides a simple and efficient way to compute the maximum (partial) likelihood estimator of the common odds ratio. A widely used estimator for the common odds ratio for a sequence of independent 2×2 tables is due to Mantel and Haenszel (1959). However, complication could arise in estimating its variance when the tables are a mixture of large and small frequencies (Robins, Breslow and Greenland, 1986; Phillips and Holland, 1987). For a review of the Mantel-Haenszel method, see Breslow (1996). Alternatively, one may use the likelihood approach to estimate the odds ratio. Assuming the tables to be independent, the conditional likelihood given all the margins is a product of hypergeometric probability mass functions of form (1.1). When all tables are independent, it is possible, though computationally demanding, to implement certain exact inference procedures (Cytel Inc., 2003), or use saddle point approximation (Strawderman and Wells, 1998). However, this approach obviously fails when the independent assumption is not valid. Our method leads to a fast way to compute the maximum partial likelihood estimator as well its asymptotic variance via a connection with the roots of Jacobi polynomials. This simplifies the computation effort significantly. The calculation is especially useful in survival analysis where the dependence among tables arises naturally. See Section 4.

Third, when the odds ratios are not equal, but dependent on covariates, Breslow (1976) and Zelen (1971) proposed a regression modeling approach. Our method provides an easy way to compute the estimators for such a model, along with their asymptotic confidence intervals. In addition, we are able to justify the asymptotic results even for dependent tables. See Section 5.

Finally, it is common for survival data to have ties, which may be due to the discreteness of the underlying survival distribution, or to grouping of data. In his fundamental paper on the proportional hazards model, Cox (1972) defined an extension to cover possible discontinuity of the survival distribution. The exact partial likelihood in this case appears to be complicated, and various approximations have been proposed (Breslow, 1976, Efron, 1977). The approach proposed in this paper for dependent 2×2 tables is also an effective tool to handle the extension. In particular, it can be used to compute the maximum partial likelihood estimator exactly, not via approximation, and to derive its asymptotic properties; see Section 6.

The paper is organized as follows. Basic decomposition for single table, and tangle sequence is established in the next section. Weighted log-rank test is analyzed in Section 3. Section 4 deals with the problem of maximum partial likelihood estimation, whose asymptotic properties are studied. A useful and slightly more general model proposed by Zelen (1971) and Breslow (1976) is studied in Section 5. Applications to survival data are given in Section 6. Several examples are presented in Section 7 to illustrate the methods. Some technical proofs are given in the Appendix.

2. Preliminary Results

After introducing basic notation, we shall present two results necessary for our discussion. The first is a decomposition of the non-central hypergeometric random variable into a sum of independent, yet not identically distributed, Bernoulli random variables. The second is a "decoupling" result which reduces the study of dependent 2×2 tables to a sequence of "tangent" yet independent sequences.

2.1. *Notation*

Throughout the rest, $\{X^{(k)}, N_i^{(k)}, M_i^{(k)},\ i = 1,2,\ k \geq 1\}$ will be used to describe a sequence of 2×2 tables in which $N_i^{(k)}$ and $M_i^{(k)}$, $i = 1, 2$, are the margins of the kth table and $X^{(k)}$ its upper left corner. Set $N^{(k)} = N_1^{(k)} + N_2^{(k)}$. Note that $N_1^{(k)} + N_2^{(k)} = M_1^{(k)} + M_2^{(k)}$. Associated with the sequence is a σ-filtration $\{\mathcal{F}_k, k \geq 0\}$ such that $X^{(k)}$ is measurable with respect to $\mathcal{F}_k$ and $N_i^{(k)}$ and $M_i^{(k)}$ are predictable, i.e., measurable with respect to $\mathcal{F}_{k-1}$). The observed data will be K such tables, $\{X^{(k)}, N_i^{(k)}, M_i^{(k)}, i = 1, 2, 1 \leq k \leq K\}$. It is assumed that K is a stopping time with respect to $\mathcal{F}_k$ and the conditional distribution of $X^{(k)}$ given $\mathcal{F}_{k-1}$ is noncentral hypergeometric as defined by (1.1), i.e.,

$$P(X^{(k)} = x | \mathcal{F}_{k-1}) = \frac{\binom{N_1^{(k)}}{x}\binom{N_2^{(k)}}{M_1^{(k)}-x}\theta_k^x}{\sum_{u=L^{(k)}}^{S^{(k)}} \binom{N_1^{(k)}}{u}\binom{N_2^{(k)}}{M_1^{(k)}-u}\theta_k^u} \tag{2.1}$$

for $L^{(k)} \leq x \leq S^{(k)}$, where $L^{(k)} = \max(0, M_1^{(k)} - N_2^{(k)})$, $S^{(k)} = \min(N_1^{(k)}, M_1^{(k)})$ and $\theta_k \in (0, \infty)$ is the odds ratio parameter of the kth table. This assumption is satisfied by tables arising from analysis of survival data (Miller, 1981, Chapter 4) as well as medical follow-up studies.

2.2. *A decomposition for a single 2 × 2 table*

One of the two key elements to derive the main result is the following lemma, which can be found in Kou and Ying (1996).

Lemma 2.1. *Consider a single noncentral hypergeometric random variable X specified by (1.1). Let $\eta_1, \eta_2, \ldots$ be independent uniform* $(0, 1)$ *random variables. Then*

$$X \stackrel{d}{=} \sum_{i=1}^{S} I(\eta_i \leq (1 + \theta^{-1}\lambda_i)^{-1}), \tag{2.2}$$

where "$\stackrel{d}{=}$" denotes equality in distribution, $I(\cdot)$ the indicator function and $-\lambda_1, \ldots, -\lambda_S$ the roots of polynomial

$$\phi(z) = \sum_{u=L}^{S} \binom{N_1}{u}\binom{N_2}{M_1 - u} z^u. \tag{2.3}$$

Polynomial ϕ is, up to a scale constant, the probability generating function of (central) hypergeometric distribution (1.2). A crucial aspect of this lemma is that all roots of ϕ are real, which can be obtained by connecting (2.3) to the classical Jacobi polynomials (Szegö, 1959). Lemma 2.1 is useful not only because it decomposes any noncentral hypergeometric distribution into a sum of independent Bernoulli random variables, but also that all the λ_i are *independent* of the odds ratio θ, as (2.3) does not involve θ. In other words, they are the same for both hypergeometric and noncentral hypergeometric distributions. This fact greatly reduces computational burden in dealing with the noncentral hypergeometric distribution.

Specifically, for the kth table, the roots $\lambda_i^{(k)}$ are simple functions of roots of the Jacobi polynomials, which are widely used in mathematics and engineering literature (Kou and Ying, 1996). The roots can be determined easily by using software packages such as *Mathematica* (Wolfram, 1991). Alternatively, one can utilize the associated matrix of which ϕ is the characteristic polynomial. Writing the probability generating function (2.3) as $\phi(z) = \sum_{i=0}^m a_i z^i$ ($a_m \neq 0$), its roots are exactly the eigenvalues of its associate matrix

$$Q = \begin{pmatrix} -\frac{a_{m-1}}{a_m} & -\frac{a_{m-2}}{a_m} & \cdots & -\frac{a_1}{a_m} & -\frac{a_0}{a_m} \\ 1 & 0 & \cdots & 0 & 0 \\ 0 & 1 & \cdots & 0 & 0 \\ \vdots & & \ddots & & \vdots \\ 0 & 0 & \cdots & 1 & 0 \end{pmatrix}$$

See Johnson and Riess (1982, Section 4.4.2). Again, the eigenvalues can be determined easily by the existing software packages. For example, a relevant command in *Splus* (MathSoft, 1995) is `eigen(Q)$values` for obtaining eigenvalues of Q.

From (2.2), it follows that the mean and variance of X can be expressed in terms of λ_i and θ:

$$\mathsf{E}_\theta X = \sum_{i=1}^{S} \frac{1}{1+\theta^{-1}\lambda_i}, \quad \mathsf{Var}_\theta(X) = \sum_{i=1}^{S} \frac{\theta^{-1}\lambda_i}{(1+\theta^{-1}\lambda_i)^2}, \tag{2.4}$$

where the subscript θ in E_θ and Var_θ indicates that the expectation and variance are taken with θ being the true parameter.

2.3. *Decoupling method*

The second key element in our analysis is the "decoupling" method, a recent development in probability theory. A systematic exposure of the method can be found in Kwapień and Woyczyński (1992) and de la Peña and Giné (1999). To apply the method, we need to introduce two definitions, which are followed by a lemma.

Definition 2.1. Two sequences of random variables, $\{Y_k, k \geq 1\}$ and $\{Z_k, k \geq 1\}$, are said to be tangent with respect to $\{\mathcal{H}_k\}$ if, for each k,

Y_k and Z_k are $\mathcal{H}_k$ measurable and have the same conditional distribution given $\mathcal{H}_{k-1}$, i.e.,

$$Y_k|\mathcal{H}_{k-1} \stackrel{d}{=} Z_k|\mathcal{H}_{k-1}.$$

Definition 2.2. A sequence of random variables $\{Y_k\}$ adapted to $\{\mathcal{H}_k\}$ (i.e. Y_k is measurable with respect to $\mathcal{H}_k$) is said to satisfy condition (CI) with respect to $\mathcal{T}$, if there exists a σ-field $\mathcal{T} \subset \bigvee_{k\geq 1} \mathcal{H}_k$, where $\bigvee_{k\geq 1} \mathcal{H}_k$ denotes the minimum σ-field spanned by $\{\mathcal{H}_l, l \geq 1\}$, such that $Y_k|\mathcal{H}_{k-1} \stackrel{d}{=} Y_k|\mathcal{T}$ for all $k \geq 1$ and that Y_k are conditionally mutually independent given $\mathcal{T}$.

Lemma 2.2. *(See Theorem 5.8.3 of Kwapień and Woyczyński (1992) and also de la Peña and Giné (1999)). For each $n \geq 1$, let $\{Y_{nk}, k \geq 1\}$ be a tangent sequence of $\{Z_{nk}, k \geq 1\}$ with respect to σ-filtration $\{\mathcal{H}_{nk}, k \geq 1\}$. Suppose, for each $n \geq 1$, the sequence $\{Y_{nk}\}$ also satisfies condition (CI) with respect to $\{\mathcal{T}_n\}$. If the conditional distribution of $\sum_k Y_{nk}$ given $\mathcal{T}_n$ converges to a non-random probability distribution μ, then $\sum_k Z_{nk}$ also converges in distribution to μ.*

3. Weighted Log-Rank Test Statistic

Consider now a normalized sum of weighted 2×2 tables

$$U = \sum_{k=1}^{K} w_k \left(X^{(k)} - \mathsf{E}_{\theta_k}^{(k-1)}\left(X^{(k)}\right)\right) \Big/ \left(\sum_{k=1}^{K} w_k^2 \mathsf{Var}_{\theta_k}^{(k-1)}(X^{(k)})\right)^{1/2}, \tag{3.1}$$

where w_k are $\mathcal{F}_{k-1}$-measurable weights, and $\mathsf{E}_{\theta_k}^{(k-1)}$, $\mathsf{Var}_{\theta_k}^{(k-1)}$ denote conditional expectation and variance given $\mathcal{F}_{k-1}$ with θ_k being the odds ratio of the kth table. An important special case is $\theta_k \equiv 1$ for all k. In such a case, $\mathsf{E}^{(k-1)}(X^{(k)}) = M_1^{(k)} N_1^{(k)}/N^{(k)}$, $\mathsf{Var}^{(k-1)}(X^{(k)}) = M_1^{(k)} M_2^{(k)} N_1^{(k)} N_2^{(k)}/\{(N^{(k)}-1)(N^{(k)})^2\}$ and U is known as the normalized weighed log-rank test statistic for testing null hypothesis $\theta_k = 1$ (Tarone and Ware, 1977). Because the exact distribution for such a statistic is practically impossible to derive, and furthermore the dependency structure among tables may be complicated, the asymptotic approximation becomes essential for inference.

To apply the decoupling method, we need to construct a tangent sequence of $\{X^{(k)}\}$ that also satisfies condition (CI). Let $\{\eta_j, j \geq 1\}$ be a sequence of independent random variables with the uniform distribution on $(0, 1)$. They are chosen, in particular, to be independent of filtration $\{\mathcal{F}_k, k \geq 0\}$, thus also of the original sequence of 2×2 tables. For each

k and each $i = 1, \ldots, S^{(k)}$, let $\eta_{k,i} = I(k \le K)\eta_{S^{(1)}+\cdots+S^{(k-1)}+i}$, and let $\mathcal{G}_k$ denote the σ-field generated by $\mathcal{F}_k$ and $\{\eta_{l,j}, j = 1, \ldots, S^{(l)}, l = 1, \ldots, k\}$. Then by Lemma 2.1, there exist, for each $1 \le k \le K$, nonnegative numbers $\lambda_1^{(k)}, \ldots, \lambda_{S^{(k)}}^{(k)}$, which depend *only* on $N_1^{(k)}, N_2^{(k)}, M_1^{(k)}, M_2^{(k)}$, such that conditional on these margins,

$$\sum_{i=1}^{S^{(k)}} I(\eta_{k,i} \le (1 + \theta_k^{-1}\lambda_i^{(k)})^{-1})$$

has the same hypergeometric distribution as that of $X^{(k)}$. Define

$$X_k^* = w_k[X^{(k)} - \mathsf{E}_{\theta_k}^{(k-1)}(X^{(k)})]I(K \ge k),$$

$$W_k^* = w_k \sum_{i=1}^{S^{(k)}} \left[I(\eta_{k,i} \le (1 + \theta_k^{-1}\lambda_i^{(k)})^{-1}) - (1 + \theta_k^{-1}\lambda_i^{(k)})^{-1} \right] I(K \ge k).$$

Then, recalling the expression of the expectation in (2.4) and $I(K \ge k) \in \mathcal{F}_{k-1}$, we see that W_k^* and X_k^* have the same conditional distribution given $\mathcal{G}_{k-1}$. We therefore have the following conclusion.

Lemma 3.1. *The sequences of random variables* $\{X_k^*\}$ *and* $\{W_k^*\}$ *are tangent with respect to* $\{\mathcal{G}_k\}$.

Next let $\mathcal{T} = \bigvee_{k \ge 1} \mathcal{F}_k$. Then $\mathcal{T} \subset \bigvee_{k \ge 1} \mathcal{G}_k$. Clearly by examining the probability generating functions, we know that $W_k^* | \mathcal{G}_{k-1} \stackrel{d}{=} W_k^* | \mathcal{T}$. Moreover, conditional on $\mathcal{T}$, the $\lambda_i^{(k)}$ are fixed and the $\eta_{k,j}$ are independent, whence W_k^* are independent. So we have the following lemma.

Lemma 3.2. *The sequence* $\{W_k^*\}$ *satisfies condition (CI) with, using notation in Definitions 2.1 and 2.2,* $\mathcal{H}_k = \mathcal{G}_k$ *and* $\mathcal{T} = \bigvee_{k \ge 1} \mathcal{F}_k$.

In view of the preceding construction, we see that the weighted sum of the X_k can be approximated by a sum of weighted independent Bernoulli random variables, whose asymptotic distribution is easy to characterize and which is also quite simple to simulate. The asymptotic results to be presented below is interpreted in the sense of double arrays: there is an additional index n going to ∞ and all the quantities $(X^{(k)}, M_i^{(k)}, N_i^{(k)}, i = 1, 2, w_k, \mathcal{F}_k), k = 1, \ldots, K$, depend on n.

Theorem 3.1. *Suppose that there exists a nonrandom sequence* $q_n \to \infty$ *such that*

$$\frac{\sum_{k=1}^K w_k^2 \mathsf{Var}_{\theta_k}^{(k-1)}(X^{(k)})}{q_n} \xrightarrow{p} 1 \tag{3.2}$$

and that $\max_{1\le k\le K} w_k^2/q_n \to 0$ *in probability. Then* U *defined by (3.1) converges in distribution to* $N(0,1)$.

Proof. Let $\tilde{X}_k^* = X_k^*/\sqrt{q_n}$ and $\tilde{W}_k^* = W_k^*/\sqrt{q_n}$. By Lemma 3.1, $\{\tilde{X}_k^*\}$ and $\{\tilde{W}_k^*\}$ are tangent with respect to $\{\mathcal{G}_k\}$; by Lemma 3.2, $\{\tilde{W}_k^*\}$ satisfies condition (CI) with $\mathcal{T} = \bigvee_{k\ge1}\mathcal{F}_k$. Thus by Lemma 2.2, $\sum_{k\ge1}\tilde{X}_k^*$ converges to $N(0,1)$, provided we can show that $\sum_{k\ge1}\tilde{W}_k^*|\mathcal{T}$ converges to standard normal distribution. The latter is obvious because, conditional on $\mathcal{T}$, $\sum_{k\ge1} W_k^* = \sum_{j=1}^{S^{(k)}} w_j[I(\eta_{k,j}\le(1+\theta^{-1}\lambda_j^{(k)})^{-1}) - (1+\theta^{-1}\lambda_j^{(k)})^{-1}]$ is a sum of independent random variables and, by (3.2), $\mathsf{Var}(\sum_{k\ge1}W_k^*|\mathcal{T})/q_n \xrightarrow{p} 1$ which, in conjunction with the assumption $\max_j w_j^2/q_n \to 0$, implies the Lindeberg condition. In view of of (3.2), U also converges to $N(0,1)$, via Slutsky's Theorem. □

It seems that the variance stability by q_n in the theorem is indispensable even in some simplest cases. To give an example, consider a sequence of independent 2×2 tables with common odds ratio $\theta=1$, and the margins $M_1^{(k)} = M_2^{(k)} = N_1^{(k)} = N_2^{(k)} = 1$. Let K be the first time that $\sum_{k=1}^K(X^{(k)} - 1/2) \ge n$. Then $\sum_{k=1}^K \mathsf{Var}_\theta^{(k-1)}(X^{(k)}) = K/4 \to \infty$ as $n\to\infty$. However $U = (\sum_{k=1}^K(X^{(k)}-1/2))(K/4)^{-1/2} \ge 0$ and therefore does not converge to $N(0,1)$.

It may be revealing to compare Theorem 3.1 with some known results in special cases. If we consider a special case that K is nonrandom and does not increase to ∞, and that the K tables are independent, then we can apply Lemma 2.1 K times to express U (in distribution) as a weighted sum of independent Bernoulli random variables. Thus U converges in distribution to $N(0,1)$ via the Lindeberg central limit theorem. In another extreme case in which $M_1^{(k)} \equiv 1$, implying $X^{(k)}$ is either 0 or 1, the Lindeberg-type condition for U becomes trivial. It follows from a martingale central limit theorem as given in Pollard (1984, p.171) that U converges in distribution to $N(0,1)$ provided that $K\to\infty$ in probability and an additional variance stability condition holds. In survival analysis, the last situation arises when one constructs a two-sample log-rank test under the assumption that the underlying survival distribution is continuous (therefore no ties). See Miller (1981) and Fleming and Harrington (1991).

In general, for some k, $X^{(k)}$ may have large conditional variances, while for others $X^{(k)}$ may be bounded (thus having relatively small variances). For cases of the first kind, each normalized $X^{(k)}$ should already be closed to normal; and for the latter ones, the martingale central limit theorem suggests intuitively that their normalized sum should also be close to normal. Therefore it appears that U should always be approximately normal, no matter how the tables are arranged. Yet neither the representation by independent Bernoulli random variables nor the martingale central limit theorem can be applied directly. In these respects, Theorem 3.1 presented here becomes an effective tool, complementing the available methods.

4. Maximum Partial Likelihood Estimator for the Common Odds Ratio

In many applications, especially medical follow-up studies, it is common to assume homogeneity of the odds ratio parameters, i.e., $\theta_k = \theta$ (Mantel and Haenszel, 1959; Breslow, 1996). Therefore, an important statistical problem is to estimate the common odds ratio θ and that will be the concern of this section.

4.1. *The estimator*

The conditional probability mass function of the kth table given its margins and the $k-1$ preceding tables can be written as

$$g_k(M_1^{(k)}, M_2^{(k)}, N_1^{(k)}, N_2^{(k)}) \frac{\binom{N_1^{(k)}}{X^{(k)}}\binom{N_2^{(k)}}{M_1^{(k)}-X^{(k)}}\theta^{X^{(k)}}}{\sum_{u=L^{(k)}}^{S^{(k)}}\binom{N_1^{(k)}}{u}\binom{N_2^{(k)}}{M_1^{(k)}-u}\theta^u},$$

where g_k is the conditional probability mass of $(M_1^{(k)}, M_2^{(k)}, N_1^{(k)}, N_2^{(k)})$ given the $k-1$ preceding tables. Following Cox (1975), we ignore the g's and obtain the following partial likelihood

$$\prod_{k=1}^{K} \frac{\binom{N_1^{(k)}}{X^{(k)}}\binom{N_2^{(k)}}{M_1^{(k)}-X^{(k)}}\theta^{X^{(k)}}}{\sum_{u=(k)}^{S^{(k)}}\binom{N_1^{(k)}}{u}\binom{N_2^{(k)}}{M_1^{(k)}-u}\theta^u}. \tag{4.1}$$

Notice that (4.1) becomes the full likelihood when the tables are independent and their margins are fixed. By taking the logarithm of the partial likelihood and then setting its derivative with respect to θ equal to 0, we see that the maximum partial likelihood estimator (MPLE), denoted by $\hat{\theta}$, satisfies the following equation

$$\sum_{k=1}^{K}\left[X^{(k)} - E_\theta^{(k-1)}X^{(k)}\right] = 0,$$

which, through (2.4), is the same as

$$\sum_{k=1}^{K} X^{(k)} - \sum_{k=1}^{K} \sum_{i=1}^{S^{(k)}} \frac{1}{1+\theta^{-1}\lambda_i^{(k)}} = 0. \tag{4.2}$$

The simple form of (4.2) allows the following characterization of existence and uniqueness of the MPLE.

Proposition 4.1. *A necessary and sufficient condition that guarantees the existence and uniqueness of the MPLE $\hat{\theta}$ is*

$$\sum_{k=1}^{K} L^{(k)} < \sum_{k=1}^{K} X^{(k)} < \sum_{k=1}^{K} S^{(k)}. \tag{4.3}$$

Since $L^{(k)} \le X^{(k)} \le S^{(k)}$ for all k, it follows that (4.3) is equivalent to that there exist k and k' such that $L^{(k)} < X^{(k)}$ and $X^{(k')} < S^{(k')}$.

Proof. We first show that (4.3) is sufficient. It is obvious that, on $(0, \infty)$, $(1 + \theta^{-1}\lambda_i^{(k)})^{-1}$ is strictly increasing in θ, if $\lambda_i^{(k)} > 0$. Therefore, $\sum_{k=1}^{K}\sum_{i=1}^{S^{(k)}}\{1+\theta^{-1}\lambda_i^{(k)}\}^{-1}$ goes to $\sum_{k=1}^{K} L^{(k)}$ as $\theta \to 0$, and goes to $\sum_{k=1}^{K} S^{(k)}$ as $\theta \to \infty$, because for each k there are exactly $L^{(k)}$ zero among $\{\lambda_i^{(k)},\ 1 \le i \le S^{(k)}\}$. We have then that, for any $X \in (\sum_{k=1}^{K} L^{(k)}, \sum_{k=1}^{K} S^{(k)})$, there exists a unique $\hat{\theta}$ such that (4.2) holds. The same reason also leads to the conclusion that no such $\hat{\theta}$ exists for $X = \sum_{k=1}^{K} L^{(k)}$ or $X = \sum_{k=1}^{K} S^{(k)}$, whence the necessity is obtained as well. □

It is not difficult to see that the partial likelihood estimating function is monotone decreasing and convex. Thus solving the MPLE $\hat{\theta}$ in (4.2) is rather straightforward numerically (for example, by the standard Newton-Raphson method), once all $\lambda_i^{(k)}$ are calculated. The large sample properties of $\hat{\theta}$ are summarized by the following theorem. Its proof is given in Appendix.

Theorem 4.1. *Suppose that (3.2) is satisfied with $w_k \equiv 1$. Then $\hat{\theta}$ is consistent and asymptotically normal. More precisely, we have $\hat{\theta} \to \theta$ in probability and*

$$\theta^{-1}\left(\sum_{k=1}^{K} \mathsf{Var}_\theta^{(k-1)}(X^{(k)})\right)^{1/2} (\hat{\theta} - \theta) \to_{\mathcal{L}} N(0,1), \tag{4.4}$$

$$\left(\sum_{k=1}^{K} \mathsf{Var}_\theta^{(k-1)}(X^{(k)})\right)^{1/2} (\log\hat{\theta} - \log\theta) \to_{\mathcal{L}} N(0,1). \tag{4.5}$$

Furthermore, the normalizing factor, $\left(\sum_{k=1}^{K} \mathsf{Var}_{\theta}^{(k-1)}(X^{(k)})\right)$*, may be replaced by its empirical counterpart* $\widehat{\mathsf{Var}}$*, where*

$$\widehat{\mathsf{Var}} = \sum_{k=1}^{K} \mathsf{Var}_{\hat{\theta}}^{(k-1)}(X^{(k)}) = \sum_{k=1}^{K}\sum_{i=1}^{S^{(k)}} \frac{\hat{\theta}^{-1}\lambda_i^{(k)}}{(1+\hat{\theta}^{-1}\lambda_i^{(k)})^2}, \tag{4.6}$$

and (4.4) and (4.5) still hold with the replacement.

Because of skewness (θ is positive), it is often more desirable to construct asymptotically valid confidence intervals using (4.5) rather than (4.4). In particular, an asymptotic $100(1-\alpha)\%$ confidence interval for $\log\theta$ is

$$\log\hat{\theta} \pm z_{\alpha/2}(\widehat{\mathsf{Var}})^{-1/2}, \tag{4.7}$$

which can be exponentiated to get the corresponding interval for θ

$$\hat{\theta}\exp\{\pm z_{\alpha/2}(\widehat{\mathsf{Var}})^{-1/2}\}.$$

Using (4.2) and (4.5) we can also get a simple and asymptotically valid resampling method to approximate the distribution of $\hat{\theta}$. Specifically, we first calculate $\lambda_i^{(k)}$ and $\hat{\theta}$. We generate uniform(0,1) random variables $\eta_{k,i}$. With $\lambda_i^{(k)}$, $\hat{\theta}$ and $\eta_{k,i}$, we get simulated data via

$$\tilde{X}^{(k)} = \sum_{i=1}^{S^{(k)}} I(\eta_{k,i} \le (1+\hat{\theta}_k^{-1}\lambda_i^{(k)})^{-1}).$$

We then compute $\tilde{\theta}$ from the simulated data $\tilde{X}^{(k)}$ by solving

$$\sum_{k=1}^{K}\left[\tilde{X}^{(k)} - \sum_{i=1}^{S^{(k)}} \frac{1}{1+\tilde{\theta}^{-1}\lambda_i^{(k)}}\right] = 0.$$

It is not difficult to justify that the conditional distribution of $\tilde{\theta}-\hat{\theta}$ does approximate the unconditional distribution of $\hat{\theta}-\theta$. So we can repeatedly generate $\tilde{\theta}$ to obtain empirically the (conditional) distribution of $\tilde{\theta}-\hat{\theta}$, which can then be used for inference of θ.

4.2. *Comparison with other estimators*

Various other methods have been proposed for the inference of the common odds ratio parameter. Woolf (1955) suggested to estimate logarithm of the common odds ratio by a weighted sum of logarithms of the empirical odds ratios from different tables. Validity of his method requires that sizes of all tables be reasonably large. Inconsistency could arise if it is applied to a large number of sparse tables (Breslow, 1981).

The most widely used method is, perhaps, due to Mantel and Haenszel (1959). Let $X_{ij}^{(k)}$, $1 \leq i, j \leq 2$ be the four cell counts in the kth tables, i.e., $X_{11}^{(k)} = X^{(k)}$, $X_{12}^{(k)} = N_1^{(k)} - X^{(k)}$, etc. The Mantel-Haenszel (M-H) estimator is defined by

$$\hat{\theta}_{MH} = \frac{\sum_{k=1}^{K} X_{11}^{(k)} X_{22}^{(k)} / N^{(k)}}{\sum_{k=1}^{K} X_{12}^{(k)} X_{21}^{(k)} / N^{(k)}}. \tag{4.8}$$

It is known that the estimator is consistent in the case of either a small number of large tables or a large number of sparse tables and is robust against indivisual zero cell entries. The estimator is also obviously easy to compute.

Comparing with the M-H estimator, the computation of the MPLE is more involved. However, as we discussed before, with the modern computing capability, finding MPLE $\hat{\theta}$ has become manageable as the roots λ's can be found easily.

There are two main advantages of the MPLE. First, as Theorem 3.1 shows, the MPLE is consistent for both a finite number of large tables and a large number of sparse tables. In addition, it is also valid for a combination of large and small tables. In contrast, it has been no theoretic justification for the M-H estimator in the case of mixture of large and small tables. Secondly, the variance estimation for $\hat{\theta}_{MH}$ could be quite complicated (Robins et al., 1986; Phillips and Holland, 1987), whereas the variance estimator for the MPLE is given in (4.7).

It is worth mentioning that the MPLE is also robust against individual zero cell entries. In fact, the following theorem shows that the MPLE exists if and only if the M-H exists.

Theorem 4.2. *A necessary and sufficient condition for the existence and uniqueness of MPLE $\hat{\theta}$ is both $\sum_k X_{11}^{(k)} X_{22}^{(k)}$ and $\sum_k X_{12}^{(k)} X_{21}^{(k)}$ are not zero. Therefore, MPLE is well defined if and only if $\hat{\theta}_{MH}$ is well defined.*

Proof. By Proposition 3.1 and Remark 3.1, it suffices to show that, for any k and k', $L^{(k)} < X^{(k)}$ and $X^{(k')}$ if and only if $X_{11}^{(k)} X_{22}^{(k)} > 0$ and $X_{12}^{(k')} X_{21}^{(k')} > 0$. But it is straightforward to see from the 2×2 table that $X_{11}^{(k)} X_{22}^{(k)} > 0$ is equivalent to $L^{(k)} < X^{(k)}$ and $X_{12}^{(k')} X_{21}^{(k')} > 0$ is equivalent to $X^{(k')} < S^{(k')}$. So the theorem follows. □

5. Analysis of the Breslow-Zelen Model

The homogeneity assumption about odds ratio parameters θ_k of the K tables may be violated. To accommodate the possible nonhomogeneity, a useful regression model was proposed and studies by Zelen (1971) and Breslow (1976). Their model assumes relationship

$$\log(\theta_k) = \alpha + \beta' \mathbf{z_k}, \qquad 1 \le k \le K,$$

where $\mathbf{z_k}$ is a vector of covariates associated with the kth table, and α and β the intercept and the regression parameters. Typically, the $\mathbf{z_k}$ include the stratification variable used to obtain the sequence of tables. See Breslow and Cologne(1986). Note that without including the $\mathbf{z_k}$, the model reduces to the setup of the preceding section with $\theta = e^\alpha$.

Our method yields a direct way for analyzing and computing the estimators for the Breslow-Zelen model. More precisely, the partial likelihood function may be obtained analogously to (4.1), with its θ replaced by $\exp(\alpha + \beta' z_k)$. Differentiating with respect to α and β, we arrive at the following estimating equation, which is analogous to (4.2),

$$\sum_{k=1}^{K} \left(X^{(k)} - \sum_{i=1}^{S^{(k)}} \frac{1}{1 + \lambda_i^{(k)} / \exp(\alpha + \beta' \mathbf{z_k})} \right) \begin{pmatrix} 1 \\ \mathbf{z_k} \end{pmatrix} = 0. \qquad (5.1)$$

The $\lambda_i^{(k)}$ are the same as those in (4.2) and thus depend neither on the parameter value nor on the covariates.

Because the derivative matrix of left-hand side of (5.1) is negative definite, the log-likelihood function is concave so that numerically it is straightforward to compute MPLE $(\hat{\alpha}, \hat{\beta})$ of (α, β). The asymptotic variance-covariance matrix can be estimated quiet easily by $\mathbf{I}^{-1}(\hat{\alpha}, \hat{\beta})$, where

$$\mathbf{I}(\alpha, \beta) = \sum_{k=1}^{K} \begin{pmatrix} 1 \\ \mathbf{z_k} \end{pmatrix} \begin{pmatrix} 1 \\ \mathbf{z_k} \end{pmatrix}' \sum_{i=1}^{S^{(k)}} \frac{\lambda_i^{(k)} / \exp(\alpha + \beta' \mathbf{z_k})}{[1 + \lambda_i^{(k)} / \exp(\alpha + \beta' \mathbf{z_k})]^2}.$$

Suppose that there exists a nonrandom sequence q_n such that $q_n^{-1} \mathbf{I}(\alpha, \beta)$ converges to a nonrandom positive definite matrix. Then it is not difficult to show that $(\hat{\alpha}, \hat{\beta})$ is consistent and asymptotically normal under suitable normalization. Hence, inference procedures such as testing and interval estimation based on $(\hat{\alpha}, \hat{\beta})$ and its covariance estimator $\mathbf{I}^{-1}(\hat{\alpha}, \hat{\beta})$ are asymptotically correct.

Because $\beta = 0$ corresponds to the homogeneity of odds ratios θ_k, the Breslow-Zelen model provides a natural tool to test $\theta_1 = \cdots = \theta_K$. In particular, letting $(\mathbf{I}^{-1}(\alpha, \beta))_{22}$ denote the lower right corner of $\mathbf{I}^{-1}(\alpha, \beta)$,

$\hat{\beta}'(\mathbf{I}^{-1}(\hat{\alpha},\hat{\beta}))_{22}^{-1}\hat{\beta}$ follows a χ^2 distribution under the homogeneity assumption and gives a Wald-type test. On the other hand, a score test can be obtained by replacing α by $\hat{\alpha}$ and β by 0 in (5.1) with a suitable normalization.

6. Applications to Survival Analysis

In this section we apply the results developed earlier to hypothesis testing and parameter estimation problems in survival analysis, where survival distributions may be discontinuous. We first consider an extension of the proportional hazards model that also covers discontinuity and discuss a normal approximation to the log-rank test statistic. Then we apply the results to a log-rank-type test in survival analysis, which is connected to a group sequential design.

To fix notation, let $T_1, T_2, ..., T_{n_1}$ denote i.i.d. survival times from the first population with a possibly discontinuous distribution function F_1 and $T_{n_1+1}, ..., T_{n_1+n_2}$ be i.i.d. survival times from the second population with distribution function F_2. Let Λ_i, $i = 1, 2$ be the corresponding cumulative hazard functions. The usual right censorship is incorporated as we only observe $\tilde{T}_i = T_i \wedge C_i$, and $\Delta_i = I(T_i \le C_i)$, $i = 1, ..., n_1 + n_2$, where the C_i are the censoring times, assumed as usual to be independent of the T_i.

In his fundamental paper on the proportional hazards model, Cox (1972) also introduced an extension to accommodate discontinuity by assuming that the odds ratio of the hazard functions to be proportional

$$\frac{\mathrm{d}\Lambda_1(t)}{1 - \mathrm{d}\Lambda_1(t)} = \theta \frac{\mathrm{d}\Lambda_2(t)}{1 - \mathrm{d}\Lambda_2(t)}. \tag{6.1}$$

Note that when the baseline cumulative hazard function Λ_2 is continuous, (6.1) effectively reduces to $\mathrm{d}\Lambda_1(t) = \theta \mathrm{d}\Lambda_2(t)$, which is the usual formulation of the proportional hazards model.

Suppose that there are K distinct time points, to be denoted by $\tau_1 < \cdots < \tau_k$, at which one or more failures have occurred. This means that for each τ_k, we can find at least one i with $\Delta_i = 1$ and $\tilde{T}_i = \tau_k$. Let

$$\begin{aligned}
X^{(k)} &= \#\{i \le n_1 : \tilde{T}_i = \tau_k \text{ and } \Delta_i = 1\}, \\
N^{(k)} &= \#\{i \le n_1 + n_2 : \tilde{T}_i \ge \tau_k\}, \\
N_1^{(k)} &= \#\{i \le n_1 : \tilde{T}_i \ge \tau_k\}, \quad N_2^{(k)} = N^{(k)} - N_1^{(k)}, \\
M_1^{(k)} &= \#\{i \le n_1 + n_2 : \tilde{T}_i = \tau_k, \ \Delta_i = 1\}, \quad M_2^{(k)} = N^{(k)} - M_1^{(k)}.
\end{aligned} \tag{6.2}$$

Since the underlying distributions may be discontinuous, $X^{(k)}$ could take any nonnegative integer value. It is not difficult to show that the quantities so defined constitute a sequence of 2×2 tables as described at the beginning of Section 2, satisfying (2.1) with common odds ratio θ. Therefore, the partial likelihood estimator of θ can be obtained by maximizing (4.1) or solving (4.2). The asymptotic properties will remain valid if the variance stability condition in Theorems 3.1 and 4.1 are satisfied. The following result verifies the condition; the proof will be given in Appendix.

Theorem 6.1. *Suppose* $\min\{n_1, n_2\}/n$ *converges to a positive number. Then, under the extended proportional hazards model assumption (6.1), the variance stability condition is satisfied. More precisely*

$$\frac{\sum_{k=1}^{K} \mathsf{Var}_\theta^{(k-1)}(X^{(k)})}{n} \xrightarrow{p} \alpha > 0,$$

for some positive constant α. *Here* $n = n_1 + n_2$ *is the total sample size.*

Corollary 6.1. *Suppose* $\min\{n_1, n_2\}/n$ *converges to a positive number. Then the maximum partial likelihood estimator* $\hat{\theta}$, *which maximizes (4.1) or solves (4.2) is consistent and asymptotically normal.*

Recall that (4.2) is the same as $\sum_{k=1}^{K}(X^{(k)} - \mathsf{E}_\theta^{(k-1)} X^{(k)}) = 0$. But $\mathsf{E}_\theta X^{(k)} = \sum_i 1/(1+\theta^{-1}\lambda_i^{(k)})$. So an essential step to compute $\hat{\theta}$ is to find $\lambda_i^{(k)}$ first. However, if $M_1^{(k)} = 1$, i.e. only one failure occurs at τ_k, then $\mathsf{E}_\theta^{(k-1)} X^{(k)} = 1/(1+\theta^{-1}N_2^{(k)}/N_1^{(k)})$. In particular, if the underlying distributions are continuous, then $M_1^{(k)} = 1$ for all k and (4.2) becomes

$$\sum_{k=1}^{K}\left(X^{(k)} - \frac{1}{1+\theta^{-1}N_2^{(k)}/N_1^{(k)}}\right) = 0,$$

which is exactly Cox's partial likelihood estimating equation for the two sample problem and whose asymptotic properties can be established via the elegant martingale theory. Without the continuity assumption, however, the martingale and stochastic integration approaches do not appear to be applicable, yet our Corollary 6.1 still applies.

At the end of Section 4.1, we proposed a resampling scheme to approximate distribution of $\hat{\theta}$. The proposal is certainly applicable here. Note that such a resampling scheme is very different from bootstrapping the survival times as one would ordinarily do. It will be interesting to compare the two resampling schemes.

Next, we consider the log-rank test statistic under the current setting. Letting $\theta = 1$, (4.2) may be used for testing null hypothesis $F_1 = F_2$. Under the null hypothesis, $X^{(k)}$ follows the central hypergeometric distribution and thus $\mathsf{E}(X^{(k)}|M_i^{(k)}, N_i^{(k)},\ i = 1, 2) = M_1^{(k)}N_1^{(k)}/N^{(k)}$ and

$\mathsf{Var}(X^{(k)}|M_i^{(k)}, N_i^{(k)}) = \sum_{k=1}^K M_1^{(k)} M_2^{(k)} N_1^{(k)} N_2^{(k)} (N^{(k)} - 1)^{-1} (N^{(k)})^{-1}$.
Hence, a natural test statistic is

$$U = \frac{\sum_{k=1}^K \left(X^{(k)} - M_1^{(k)} N_1^{(k)} / N^{(k)}\right)}{\left[\sum_{k=1}^K M_1^{(k)} M_2^{(k)} N_1^{(k)} N_2^{(k)} \left(N^{(k)} - 1\right)^{-1} \left(N^{(k)}\right)^{-2}\right]^{1/2}}. \tag{6.3}$$

From Theorems 3.1 and 6.1, we get the following corollary.

Corollary 6.2. *Suppose that n_1/n converges to a positive number and that the variance stability condition holds. Then U has asymptotically $N(0,1)$ distribution under the null hypothesis.*

Instead of forming a 2×2 table at each failure time point τ_k, one may consider other ways of grouping data. In group sequential design, it is often desirable to conduct the interim analysis at time points so that information accumulated between two consecutive analyses stays constant (Pocock, 1977). Since the information under the proportional hazard are approximately proportional to the number of failures, such a goal may be achieved by setting interim analysis so that the number of failures between any two consecutive interim analyses is equal or close to a prefixed number. It can be shown that under such a design, U, the test statistic defined by (6.3), converges to the standard normal under the null hypothesis.

7. Examples

Example 1. To illustrate the preceding inference methods for the common odds ratio, we shall first consider in Table 1 the following data set, taken from Mantel (1963), about a comparison of the effectiveness of 1.5-hour-delayed versus immediately injected Penicillin to protect rabbits against lethal injection with β-hemolytic streptococci.

Note that the individual likelihood for the first and last tables are identical to one, whence do not contribute to (4.1). The roots of (2.3) for the other three tables are listed in Table 2. Therefore, solving (4.2) by Newton-Raphson method yields our estimator of the common odds ratio as well as its asymptotic confidence interval.

It can be seen from Table 3 that there is a noticeable difference between the MPLE and the M-H estimate, in terms of both the estimators and the confidence intervals. In this special case, our maximum partial likelihood estimator is indeed the conditional maximum likelihood estimator (CMLE), since all the tables are independent. It is suggested in Breslow

Penicillin Level	Delay	Response	
		Cured	Died
1/8	None	0	6
	1.5 h	0	5
1/4	None	3	3
	1.5 h	0	6
1/2	None	6	0
	1.5 h	2	4
1	None	5	1
	1.5 h	6	0
4	None	2	0
	1.5 h	5	0

Penicillin Level	Lambda's
1/4	3.186, 1, 0.314
1/2	5.552, 1.669, 0.599, 0.180, 0, 0
1	1, 0, 0, 0, 0, 0

Method	Estimator of θ	95% C. I. for θ
MPLE	10.36	(1.13, 94.77)
M-H method	7	(1.03, 47.73)

(1981), Hauck (1988), and Santner and Duffy (1989, section 5.5) that for independent tables, CMLE is better than the M-H estimator in terms of asymptotic variance and efficiency. For this special example, to see which method is better numerically, we shall perform a simulation, of 10,000 runs, for estimation of $\log\theta$, with the same margins as specified in the example; and the result is shown in Table 4.

Because of the very small sample size, in simulating the tables we encountered cases violating the equation (4.3), in which both our estimator and the M-H estimator will be undefined, as they will give either $-\infty$ or ∞. We deleted all these cases (in particular, the percentage of number of vio-

Ture Value	Method	Sample Mean	Sample Variance	MSE	Coverage Prob.
$\log\theta = 0$	MPLE Method	-0.007	0.729	0.729	0.965
$(\theta = 1)$	M-H method	-0.008	0.789	0.789	0.965
$\log\theta = 0.5$	MPLE Method	0.529	0.739	0.740	0.966
$(\theta = 1.649)$	M-H method	0.548	0.800	0.802	0.966
$\log\theta = -0.5$	MPLE Method	-0.522	0.723	0.723	0.970
$(\theta = 0.607)$	M-H method	-0.541	0.783	0.785	0.970
$\log\theta = 1.0$	MPLE Method	1.023	0.693	0.694	0.954
$(\theta = 2.718)$	M-H method	1.062	0.760	0.764	0.954
$\log\theta = -1.0$	MPLE Method	-1.012	0.689	0.689	0.950
$(\theta = 0.368)$	M-H method	-1.051	0.758	0.761	0.950
$\log\theta = 1.5$	MPLE Method	1.426	0.587	0.593	0.988
$(\theta = 4.82)$	M-H method	1.479	0.649	0.650	0.985
$\log\theta = -1.5$	MPLE Method	-1.445	0.581	0.583	0.990
$(\theta = 0.223)$	M-H method	-1.499	0.645	0.645	0.986
$\log\theta = 2.0$	MPLE Method	1.725	0.455	0.531	0.979
$(\theta = 7.389)$	M-H method	1.792	0.514	0.557	0.978
$\log\theta = -2.0$	MPLE Method	-1.730	0.456	0.529	0.976
$(\theta = 0.135)$	M-H method	-1.799	0.518	0.558	0.976
$\log\theta = 2.5$	MPLE Method	1.950	0.325	0.627	0.939
$(\theta = 12.182)$	M-H method	2.025	0.378	0.604	0.939
$\log\theta = -2.5$	MPLE Method	-1.942	0.329	0.641	0.932
$(\theta = 0.082)$	M-H method	-2.020	0.385	0.616	0.932

lation for the simulations listed in Table 4 are 0.19%, 0.81%, 0.77%, 3.29%, 3.20%, 10.33%, 9.90%, 21.96%, 21.31%, 37.39%, 37.37%, corresponding to $\log(\theta) = 0$, 0.5, -0.5, 1.0, ..., -2.5, respectively).

Drug 6-MP	6+, 6, 6, 6, 7, 9+, 10+, 10, 11+, 13, 16, 17+, 19+, 20+, 22, 23, 25+, 32+, 32+, 34+, 35+
Control	1, 1, 2, 2, 3, 4, 4, 5, 5, 8, 8, 8, 8, 11, 11, 12, 12, 15, 17, 22, 23

It is evident from Table 4 that an advantage of our approach is that it has a smaller variance than that of the M-H estimator, uniformly for the cases shown in Table 4. Indeed, the former seems to have smaller mean squared error if the odds ratio is not too big or too small ($|\log(\theta)| < 2.5$); otherwise both estimators are substantially biased (it might well be these cases are too extreme in view of the small sample size, as in such cases about 37% of the simulations lead to undefined estimators). Because of symmetry, the results are similar for positive and negative $\log(\theta)$. It also appears that both confidence intervals have roughly same coverage probabilities.

Example 2. The data set of Table 5 is taken from Freireich et al. (1963), containing times (weeks) of remission of leukemia patients from two groups, drug 6-MP group and control group. The plus signs denote censored values. It has been used by Gehan (1965), Cox (1972) among others.

Note that there are many ties in this data set. To analyze the data, we construct a 2×2 table at each failure time, resulting in 17 tables as shown in Table 6.

Because of the censorship, the tables so constructed are *dependent*, and, furthermore, the dependent structures are not specified; the only information we have is that for each table conditioning on the margins (N_1, N_2, M_1, M_2), the first cell (X) is hypergeometrically distributed. Let $\Lambda_1(t)$ and $\Lambda_0(t)$ be cumulative hazard rates for the control and 6-MP groups, respectively. Assume that the hazard rates are proportional for the two groups

$$\frac{d\Lambda_1(t)}{1 - d\Lambda_1(t)} = e^{\beta} \frac{d\Lambda_0(t)}{1 - d\Lambda_0(t)},$$

where $e^{\beta} = \theta$ is the common odds ratio for the 17 tables.

We can use the maximum partial likelihood estimator via (4.2) and (4.7). The results are reported in Table 7. Note that without further assumptions about censorship, it is impossible to provide an exact confidence interval, as the dependent structures are unknown.

Interestingly, our approach for point and interval estimation corresponds to the "exact partial likelihood" for censored data in the presence of ties,

Failure time		Parameters of tables				
Control	6-MP	X	N_1	N_2	M_1	M_2
1, 1		2	21	21	2	40
2, 2		2	19	21	2	38
3		1	17	21	1	37
4, 4		2	16	21	2	35
5, 5		2	14	21	2	33
	6, 6, 6	0	12	21	3	30
	7	0	12	17	1	28
8, 8, 8, 8		4	12	16	4	24
	10	0	8	15	1	22
11, 11		2	8	13	2	19
12, 12		2	6	12	2	16
	13	0	4	12	1	15
15		1	4	11	1	14
	16	0	3	11	1	13
17		1	3	10	1	12
22	22	1	2	7	2	7
23	23	1	1	6	2	5

as suggested in Cox (1972, section 6), a challenging problem in terms of computation. In this connection, we provide a way to perform this exact partial likelihood inference using the Jacobi polynomial. To make a comparison, we also list in Table 7 some other estimators and confidence intervals obtained by the methods suggested by Breslow (1976) and Efron (1977) to approximate the "exact partial likelihood".

Method	Estimator of θ	95% C. I. for θ
Exact Partial Likelihood Inference	5.09	(2.18, 11.91)
Efron Approximation	4.82	(2.15, 10.80)
Breslow Approximation	2.13	(1.42, 3.18)

Table 7 shows that there is a general agreement between our method and Efron's method, and there is also a clear difference between Breslow's method and the others. In particular, Breslow's estimator is not contained in any of the confidence intervals provided by our method and Efron's method; nor do estimators given by our method and Efron method lie within Breslow's confidence interval.

We would also like to point out that, ignoring the dependence between these two by two tables, and formally using the M-H estimator of the common odds ratio, and its variance estimator given in Robins et al. (1986), gives point estimator 5.22 and a 95% confidence interval (2.19, 12.43), which are similar to the ones obtained by the partial likelihood inference and Efron method. However, it remains to be seen whether the Mantel-Haenszel method still gives an asymptotically valid inference, in presence of unknown dependent structures between the tables.

Example 3. To show that our method can be used to handle computation involving large tables, we take the following example from Tuyns et al. (1977) and Breslow and Day (1980). Cases were 200 males with esophageal cancer in a French hospital between January 1972 and April 1974; controls were randomly selected 775 adult males. The table below refer exclusively the role of alcohol for esophageal cancer.

Age		Daily alcohol consumption	
(years)		80+ g	0-79 g
25-34	Case	1	0
	Control	9	106
35-44	Case	4	5
	Control	26	164
45-54	Case	25	21
	Control	29	138
55-64	Case	42	34
	Control	27	139
65-74	Case	19	36
	Control	18	88
75+	Case	5	8
	Control	0	31
	Total sample size: 975		

As indicated in Breslow and Day (1980, p. 146), standard tests show no evidence of heterogeneity of the odds ratio. Therefore we may assume that $\theta_1 = \cdots = \theta_6$. The common odds ratio estimator given by (4.2) is 5.25, along with 95% confidence interval by (4.7) being (3.63, 7.60), compared with M-H estimator 5.16, and the Robins, Breslow, and Greenland (1986) confidence interval (3.56, 7.47). Because of the relatively large sample size, the difference between two methods is not big. The main point here is that it only took less than one second on a Pentium 1.6Mhz CPU to get the estimator and confidence interval, even for this case with relatively large sample size.

Appendix: Proofs

Proof of Theorem 4.1. To prove consistency of $\hat{\theta}$, it suffices to show, in view of (4.2), that for any $0 < \epsilon < \theta$, there exists a constant $M(\epsilon) > 0$ such that

$$P\left(\inf_{\tilde{\theta}:|\tilde{\theta}-\theta|\geq\epsilon} \frac{|\sum_{k=1}^K X^{(k)} - \sum_{i=1}^S \sum_{k=1}^K (1+\tilde{\theta}^{-1}\lambda_i^{(k)})^{-1}|}{\sum_{k=1}^K \mathsf{Var}_\theta^{(k-1)}(X^{(k)})} > M(\epsilon) \right) \longrightarrow 1, \tag{A.1}$$

as (3.2) is assumed. In the mean time, by Theorem 3.1 and again (3.2),

$$\frac{\sum_{k=1}^K X^{(k)} - \sum_{k=1}^K \mathsf{E}_\theta^{(k-1)}(X^{(k)})}{\sum_{k=1}^K \mathsf{Var}_\theta^{(k-1)}(X^{(k)})} \xrightarrow{p} 0.$$

Therefore, (A.1) holds if we can show that

$$\liminf_{n\to\infty} \inf_{\tilde{\theta}:|\tilde{\theta}-\theta|\geq\epsilon} \frac{|\sum_{k=1}^K \sum_{i=1}^S (1+\theta^{-1}\lambda_i^{(k)})^{-1} - \sum_{k=1}^K \sum_{i=1}^S (1+\tilde{\theta}^{-1}\lambda_i)^{-1}|}{\sum_{k=1}^K \mathsf{Var}_\theta^{(k-1)}(X^{(k)})} \geq \frac{M(\epsilon)}{2}. \tag{A.2}$$

Notice that, in the view of the monotonicity of $(1+\theta^{-1}\lambda)^{-1}$, the "inf" in (A.2) can only be achieved at $\tilde{\theta} = \theta \pm \epsilon$, and also by the mean-value theorem

$$\begin{aligned} &\sum_{k=1}^K \sum_{i=1}^S (1+\theta^{-1}\lambda_i^{(k)})^{-1} - \sum_{k=1}^K \sum_{i=1}^S (1+\tilde{\theta}^{-1}\lambda_i^{(k)})^{-1} \\ &= \sum_{k=1}^K \sum_{i=1}^S \frac{\tilde{\theta}_*^{-2}\lambda_i^{(k)}}{(1+\tilde{\theta}_*^{-1}\lambda_i^{(k)})^2}(\theta - \tilde{\theta}) \end{aligned}$$

for some $\tilde{\theta}_*$ between θ and $\tilde{\theta}$. Thus we get

$$\text{left side of (A.2)} \geq \liminf_{n\to\infty} \frac{\sum_{k=1}^K \sum_{i=1}^S (\theta+\epsilon)^{-2}\lambda_i^{(k)}/(1+(\theta-\epsilon)^{-1}\lambda_i^{(k)})^2}{\sum_{k=1}^K \sum_{i=1}^S \theta^{-1}\lambda_i^{(k)}/(1+\theta^{-1}\lambda_i^{(k)})^2}\epsilon$$
$$\geq \frac{(\theta-\epsilon)(\theta+\epsilon)^{-2}\min(\theta-\epsilon, \frac{1}{\theta-\epsilon})}{\max(\theta, \frac{1}{\theta})}\epsilon,$$

where the second inequality follows from the following elementary inequality

$$\min(a^2/b^2, b^2/a^2)\frac{b^{-2}\lambda_i}{(1+b^{-1}\lambda_i)^2} \leq \frac{a^{-2}\lambda_i}{(1+a^{-1}\lambda_i)^2}$$
$$\leq \max(a^2/b^2, b^2/a^2)\frac{b^{-2}\lambda_i}{(1+b^{-1}\lambda_i)^2},$$

for any $0 < a < \infty$ and any $0 < b < \infty$. Hence (A.2) and (A.1) hold.

To show the asymptotic normality, we observe again by the mean-value theorem

$$X^{(k)} - \sum_{k=1}^K \mathsf{E}_\theta^{(k-1)}(X^{(k)}) = \sum_{k=1}^K\sum_{i=1}^S \frac{1}{1+\hat{\theta}^{-1}\lambda_i^{(k)}} - \sum_{k=1}^K\sum_{i=1}^S \frac{1}{1+\theta^{-1}\lambda_i^{(k)}}$$
$$= \sum_{k=1}^K\sum_{i=1}^S \frac{\theta_*^{-2}\lambda_i^{(k)}}{(1+\theta_*^{-1}\lambda_i^{(k)})^2}(\hat{\theta}-\theta)$$

for some θ_* between θ and $\hat{\theta}$. This, along with the fact

$$\min(\hat{\theta}^2/\theta^2, \theta^2/\hat{\theta}^2)\frac{\theta^{-2}\lambda_i^{(k)}}{(1+\theta^{-1}\lambda_i^{(k)})^2} \leq \frac{\theta_*^{-2}\lambda_i^{(k)}}{(1+\ \theta_*^{-1}\lambda_i^{(k)})^2}$$
$$\leq \max(\hat{\theta}^2/\theta^2, \theta^2/\hat{\theta}^2)\frac{\theta^{-2}\lambda_i^{(k)}}{(1+\theta^{-1}\lambda_i^{(k)})^2}, \tag{A.3}$$

from the inequality (A.3), and the fact $\hat{\theta}/\theta \to 1$ in probability, yields

$$\theta^{-1}\left(\sum_{k=1}^K \mathsf{Var}_\theta^{(k-1)}(X^{(k)})\right)^{1/2}\cdot(\hat{\theta}-\theta)$$
$$= \frac{\sum_{k=1}^K\sum_{i=1}^S \theta^{-2}\lambda_i^{(k)}/(1+\theta^{-1}\lambda_i^{(k)})^2}{\sum_{k=1}^K\sum_{i=1}^S \theta_*^{-2}\lambda_i^{(k)}/(1+\theta_*^{-1}\lambda_i^{(k)})^2}\cdot\frac{X^{(k)} - \sum_{k=1}^K \mathsf{E}_\theta^{(k-1)}(X^{(k)})}{\sqrt{\sum_{k=1}^K \mathsf{Var}_\theta^{(k-1)}(X^{(k)})}},$$

which converges to $N(0,1)$, via Theorem 3.1.

Finally, we can easily get from (A.3),

$$\min(\hat{\theta}^2/\theta^2, \theta^2/\hat{\theta}^2) \leq \frac{\widehat{\mathsf{Var}}}{\sum_{k=1}^K \mathsf{Var}_\theta^{(k-1)}(X^{(k)})} \leq \max(\hat{\theta}^2/\theta^2, \theta^2/\hat{\theta}^2).$$

From this and the fact that $\hat{\theta} \to \theta$ in probability we conclude that (4.4) holds with the normalizing factor replaced by (4.6). □

Proof of Theorem 6.1. To prove the theorem, it suffices to verify the

variance stability condition (3.2) in view of Theorem 4.1. Let $D = \{t \in [0, \infty) : F_1(t+) - F_1(t-) = 0\}$. Because of (6.1), $F_2(t)$ is also continuous on the set D. For any $\epsilon > 0$, let $D_\epsilon = \{t \in [0, \infty) : F_1(t+) - F_1(t-) \geq \epsilon\}$ and $D_\epsilon^c = [0, \infty)\backslash(D \cup D_\epsilon)$.

Clearly, for those τ_k in D, $M_1^{(k)} = 1$, as there is no tied observation. When $M_1^{(k)} = 1$, the noncentral hypergeometric distribution becomes Bernoulli distribution with success probability $\theta N_1^{(k)}/(\theta N_1^{(k)} + N_2^{(k)})$. Therefore, for the conditional variance of $X^{(k)}$, we get

$$\begin{aligned}
\frac{1}{n} \sum_{k:\, \tau_k \in D} \mathsf{Var}_\theta^{(k-1)}(X^{(k)}) &= \frac{1}{n} \sum_{k:\, \tau_k \in D} \frac{\theta N_1^{(k)} N_2^{(k)}}{\left(\theta N_1^{(k)} + N_2^{(k)}\right)^2} \\
&= \frac{1}{n} \int_D \frac{\theta N_1(t) N_2(t)}{(\theta N_1(t) + N_2(t))^2} \mathsf{d}J(t),
\end{aligned}$$

which converges to a nonrandom constant by the law of large numbers, where $J(t) = \#\{i : \tilde{T}_i \leq t, \Delta_i = 1\}$.

We next deal with set D_ϵ. By definition, the number of points in D_ϵ, denoted by $r = r(\epsilon)$, is finite. Denote all the points in D_ϵ to be t_1^ϵ, t_2^ϵ, ... , t_r^ϵ. Then, as $n \to \infty$, the conditional variances of all tables constructed on the times t_1^ϵ, ... , t_r^ϵ will all go to ∞. Therefore, from Kou and Ying (1996), the conditional variances can be approximated by

$$\begin{aligned}
&\frac{1}{n} \sum_{k:\, \tau_k \in D_\epsilon} \mathsf{Var}_\theta^{(k-1)}(X^{(k)}) \\
&= \frac{1}{n} \sum_{j=1}^{r} \left(\frac{1}{X_\epsilon^{(j)}} + \frac{1}{M_1\left(t_j^\epsilon\right) - X_\epsilon^{(j)}} + \frac{1}{N_1\left(t_j^\epsilon\right) - X_\epsilon^{(j)}} \right. \\
&\qquad \left. + \frac{1}{N_2\left(t_j^\epsilon\right) - M_1\left(t_j^\epsilon\right) - N_1\left(t_j^\epsilon\right) + X_\epsilon^{(j)}} \right)^{-1} + o_p(1),
\end{aligned}$$

where $X_\epsilon^{(j)} = \#\{i \leq n_1 : \tilde{T} = t_j^\epsilon \text{ and } \Delta_i = 1\}$. This, again by the law of large numbers, converges to a nonrandom constant.

Finally, from (A.3) we can bound the conditional variance of $X^{(k)}$ at θ

with that at $\theta = 1$ to get

$$\frac{1}{n}\sum_{k:\,\tau_k\in D_\epsilon^c} \mathsf{Var}_\theta^{(k-1)}(X^{(k)}) \le \max(\theta, \frac{1}{\theta})\frac{1}{n}\sum_{k:\,\tau_k\in D_\epsilon^c} \frac{N_1^{(k)}N_2^{(k)}M_1^{(k)}M_2^{(k)}}{\left(N^{(k)}\right)^2\left(N^{(k)}-1\right)}$$
$$\le \max(\theta, \frac{1}{\theta})\frac{1}{n}\sum_{k:\,\tau_k\in D_\epsilon^c} M_1^{(k)}$$
$$\le \max(\theta, \frac{1}{\theta})\int_{D_\epsilon^c}(\mathsf{d}F_1(t)+\mathsf{d}F_2(t)) + o_p(1),$$

as $n \to \infty$, where the last inequality follows from the law of large numbers. The above quantity can be made arbitrarily small by letting $\epsilon \to 0$.

Putting these pieces together, we can conclude that $n^{-1}\sum_k \mathsf{Var}_\theta^{(k-1)}(X^{(k)})$ converges to a constant which is obviously positive. Therefore, the variance stability condition is verified. □

Acknowledgements

This research was supported in part by grants from the National Science Foundation and the National Institutes of Health.

References

1. Andersen, P. K., Borgan, Ø., Gill, R. D. and Keiding, N. (1993), *Statistical Models Based on Counting Processes*, New York: Springer-Verlag.
2. Breslow (1976), Regression analysis of the log odds ratio: a method for retrospective studies, *Biometrics*, **32**, 409-416.
3. Breslow (1981), Odds ratio estimators when the data are sparse, *Biometrika*, **68**, 73-84.
4. Breslow (1996), Statistics in epidemiology: the case-control studies, *J. Amer. Statist. Assoc.*, **91**, 14-28.
5. Breslow, N. E. and Cologne, J. (1986), Methods of estimation in log odds ratio regression models, *Biometrics*, **42**, 949-954.
6. Breslow, N. E. and Day, N. E. (1980), *Statistical Methods in Cancer Research, Vol. I, The Design and Analysis of Case-Control Studies*, Lyon: IARC.
7. Breslow, N. E. and Day, N. E. (1987), *Statistical Methods in Cancer Research, Vol. II, The Design and Analysis of Cohort Studies*, Lyon: IARC.
8. Chow, Y. S., and Teicher, H. (1988), *Probability Theory*, New York: Springer-Verlag.
9. Cox, D.R. (1972), Regression models and life tables (with discussion), *J. R. Statist. Soc. B*, **74**, 187-220.
10. Cox, D.R. (1975), Partial likelihood, *Biometrika*, **62**, 269-276.
11. Cox, D.R. and Snell, E.J. (1989), *Analysis of Binary Data*, London: Chapman and Hall.
12. Cytel Inc. (2003), *StatXact: A Statistical Package for Exact Nonparametric inference, Version 6*, Cambridge, MA.

13. de la Peña, V.H. and Giné, E. (1999), *Decoupling: From Dependence to Independence*, New York: Springer-Verlag.
14. Efron (1977), The efficiency of Cox's likelihood function for censored data, *J. of Amer. Stat. Assoc.*, **72**, 557-65.
15. Freireich, E. J. *et al.* (1963), The effect of 6-mercaptopurine on the duration of steroid-induced remissions in acute leukemia. *Blood*, **21**, 699-0716.
16. Fleming, T. R. and Harrington, D. P. (1991), *Counting Processes and Survival Analysis*, New York: Wiley.
17. Gart, J. (1970), Point and interval estimation of the common odds ration in the combination of 2 × 2 tables, *Biometrika*, **57**, 471-475.
18. Gehan, E. A. (1965), A generalized Wilcoxon test for comparing arbitrarily single-censored samples, *Biometrika*, **52**, 203-23.
19. Hauck, W. W. (1989), Odds ratio inference from stratified samples, *Comm. in Statis.-Theory and Method*, **18**, 767-800.
20. Johnson, L. D. and Riess, R. D. (1982), *Numerical Analysis*, New York: Addison-Wesley.
21. Kou, S. G. and Ying, Z. (1996), Asymptotics for a 2 × 2 table with fixed margins, *Statistica Sinica*, **6**, 809-829.
22. Kwapień, S. and Woyczyński (1992), *Random Series and Stochastic Integrals: Single and Multiple*, Boston: Birkhäuser.
23. Liang, K. Y. and Self, S. (1985), Tests for homogeneity of odds ratio when the data are sparse. *Biometrika*, **72**, 353-358.
24. Loève, M. (1977), *Probability Theory Vol. I*, New York: Springer-Verlag.
25. Mantel, N. (1963), Chi-square tests with one degree of freedom: extensions of the Mantel-Haenszel procedure. *J. of Amer. Stat. Assoc.* **58**, 690-700.
26. Mantel, N. and Haenszel, W. (1959), Statistical aspects of the analysis of data from retrospective studies of disease, *J. Nat. Cancer Inst.* **22**, 719–748.
27. MathSoft (1995), *S-plus Manuals*, MathSoft: Seattle
28. McCullagh, P. (1984), On the elimination of nuisance parameters in the proportional odds model, *J. of the Royal Stat. Society, Series B*, **46**, 250-256.
29. McCullagh, P. and Nelder, J. A. (1989), *Generalized Linear Models*, London: Chapman and Hall.
30. Mehta, C. R., Patel, N. R., and Gray, R. (1985), Computing an exact confidence interval for the common odds ratio in several 2 × 2 contingency tables, *J. of Amer. Stat. Assoc.*, **80**, 969-973.
31. Miller, R. G. (1981), *Survival Analysis*, New York: Wiley.
32. Pocock, S. J. (1977), Group sequential methods in the design and analysis of clinical trials, *Biometrika*, **64**, 175-188.
33. Pollard, D. (1984), *Convergence of Stochastic Processes*, New York: Springer-Verlag.
34. Robins, J., Breslow, N. and Greenland, S. (1986), A Mantel-Haenszel variance consistent under both large strata and sparse data limiting models, *Biometrics*, **42**, 311-323.
35. Santner, T. and Duffy. D. (1989), *The statistical Analysis of Discrete Data*, New York: Springer-Verlag.
36. Strawderman, R. and Wells, M. (1998). Approximately exact inference for

the common odds ratio in several 2 X 2 tables, with discussion. *J. of Amer. Stat. Assoc.*, **93**, 1294 - 1320.

37. Szegö, G. (1959), *Orthogonal Polynomials*, Amer. Math. Soc. Colloq. Pub., Vol. **23**, Providence: Amer. Math. Soc.
38. Tarone, R. E. (1985), On heterogeneity tests based on efficient scores, *Biometrika*, **72**, 91-95.
39. Tarone, R. E. and Ware, J. (1977), On distribution-free tests for equality of survival distributions, *Biometrika*, **64**, 156-160.
40. Wolfram, S. (1991), *Mathematica: A System for Doing Mathematics by Computer*, New York: Andison-Wesley.
41. Woolf, B. (1955), On estimating the relationship between blood group and disease, *Ann. Human Genet.*, **19**, 251-253.
42. Zelen, M. (1971), The analysis of several 2×2 contingency tables. *Biometrika*, **58**, 129-37.

A NEW TEST OF SYMMETRY ABOUT AN UNKNOWN MEDIAN

WEIWEN MIAO

Macalester College, Department of Mathematics and Computer Science
St.Paul, MN,55105, U.S.A.

YULIA R. GEL

Department of Statistics and Actuarial Science, University of Waterloo
Waterloo, ON, Canada N2L 3G1

JOSEPH L.GASTWIRTH[3]

Department of Statistics, George Washington University
Washington, DC, 20052, U.S.A.

Many robust estimators of location, e.g. trimmed means, implicitly assume that the data come from a symmetric distribution. Consequently, it is important to check this assumption with an appropriate statistical test that does not assume a known value of the median or location parameter. This article replaces the mean and standard deviation in the classic Hotelling-Solomons measure of asymmetry by corresponding robust estimators; the median and mean deviation from the median. The asymptotic distribution theory of the test statistic is developed and the new procedure is compared to tests recently proposed by Cabilio and Masaro (1996) and Mira (1999). Using their approach to approximating the variance of this class of statistics, it is shown that the new test has greater power than the existing tests to detect the asymmetry of skewed contaminated normal data as well as a majority of skewed distributions belonging to the lambda family. The increased power of the new test suggests that the use of robust estimators in goodness of fit type tests deserves further study.

Some key words: Contaminated data; Large sample theory; Mean deviation from the median; Robust estimators; Skewness; Testing symmetry.

1. Introduction

Let $X_1, \ldots, X_n$ be an independent and identically distributed (i.i.d.) sample from an absolutely continuous distribution F with unknown mean μ, median ν and standard deviation σ. Let $f(x)$, $x \in R$ be the corresponding density function. We are interested in testing whether the distribution F is symmetric about an unknown median ν, i.e.

$$\begin{aligned} H_0 &: f(\nu - x) = f(\nu + x) \\ H_a &: f(\nu - x) \neq f(\nu + x). \end{aligned} \tag{1}$$

Our interest in this problem arose from our examination of two methods for estimating the profit customers made from initial public offerings in a securities law case, where the authors served as consultants to the defendant. One approach used the difference between the offering price of the stock and the price a share sold for at its first trade on the day it was available to the public on the stock exchange. The second method used the difference between the price of a share on the last trade of the first day it was publicly traded and the offering price. Since the stock market had a strong upward trend during the relevant time period, one should adjust for the average daily increase before examining whether the difference between the two methods was symmetric, i.e. equally likely to yield an over or under-estimate of the potential profit. Only at the trial did the regulatory agency provide additional data on actual profits made by some customers. As both of the methods of estimating profit were highly correlated with this actual profit data, the issue of evaluating the relative merits of the two methods dropped out of the case.

Einmahl and McKeague (2003) review the early literature on testing symmetry about a *known* median ν and propose a new test. Other tests that have been recently proposed are discussed in Modarres and Gastwirth (1998) and Thas, Rayner and Best (2005). As noted by Lehmann and Romano (2005), the problem becomes more difficult when the median is unknown. While several tests of symmetry when ν is *unknown* have been proposed by Gastwirth (1971); Antille et al. (1982); Bhattacharya et al. (1982) and Boos (1982), the problem remains an active area of investigation and new tests have been proposed recently (Cabilio and Masaro, 1996; Mira, 1999). In particular, Cabilio and Masaro (1996) consider the test statistic C defined as

$$C = \frac{\bar{X} - M}{s},$$

where $\bar{X}$, M and s are the sample mean, median and standard deviation respectively. It has been shown that $\sqrt{n}C$ is asymptotically normally distributed. However, the asymptotic variance of $\sqrt{n}C$ depends on the underlying distribution F, which generally is unknown. For practical purposes

Cabilio and Masaro (1996) suggest the use of the asymptotic variance of 0.5708 that is derived under the assumption that F has a standard normal distribution. They claim that the effect of any misspecification in the asymptotic variance is minor. Hence, the implied rejection region of the test statistic C is

$$\text{reject } \mathrm{H_0} \text{ if } |\mathrm{C}| \geq \frac{\mathrm{z}_{\alpha/2}\sqrt{0.5708}}{\sqrt{\mathrm{n}}}, \tag{2}$$

where $z_{\alpha/2}$ is the upper $\alpha/2$ percentile of a standard normal distribution.

Mira (1999) considers a related test of symmetry based on the test statistic γ_1, the numerator of the statistic C, or the Bonferroni measure of skewness, i.e.

$$\gamma_1 = 2(\bar{X} - M).$$

The asymptotic distribution of $\sqrt{n}\gamma_1$ is also normal and its variance also depends on the underlying distribution F. Mira (1999) also suggests the use of the asymptotic variance of $\sqrt{n}\gamma_1$ obtained when F follows a standard normal distribution, and indicates that the effect of the incorrect choice of the asymptotic variance is negligible in most practical cases.

In this paper we introduce a new test of symmetry which can be considered as a robustified version of C, i.e.

$$T = \frac{\bar{X} - M}{J},$$

where

$$J = \sqrt{\frac{\pi}{2}} \cdot \frac{1}{n} \sum_{i=1}^{n} |X_i - M|$$

is a robust estimate of standard deviation (Kendall, Stuart and Ord, 1992).

The paper is organized as follows: in the section 2 we derive the asymptotic distribution of the new test statistic T; section 3 is devoted to the size and power of the new symmetry test. This section includes a small simulation study comparing the power of T with the related tests C of Cabilio and Masaro (1996) and γ_1 of Mira (1999). The paper concludes with a discussion and an appendix providing the proofs.

2. Asymptotic Distribution of the Statistics T

The derivation of the asymptotic distribution of the statistic T is based on observing that T is a function of three L-statistics: $\bar{X}$, M and J. The asymptotic properties of L-statistics have been widely studied in the statistical literature (see, for example, Mason, 1992, and references therein). It can be shown that under relatively mild conditions (Chernoff, Gastwirth and Johns Jr., 1967, hereafter cited as CGJ), the joint distribution of those three L-statistics is asymptotically normal. Combing this result and the δ-method of Rao (1965), one obtains the asymptotic distribution of T.

Define

$$D = \frac{1}{n}\sum_{i=1}^{n} |X_i - M|.$$

Then $\tau \equiv E(D) = \mu - 2\gamma$ (Gastwirth, 1982). First, we derive the asymptotic covariances between $\bar{X}$, M and J.

Lemma 2.1. *Let μ, ν and σ be respectively mean, median and standard deviation of an absolutely continuous distribution F. Let*

$$\gamma \equiv \int_{-\infty}^{\nu} x f(x)dx, \qquad \eta \equiv \int_{-\infty}^{\nu} x^2 f(x)dx.$$

Then the asymptotic covariances between $\overline{X}$, M and J are

$$\mathrm{Cov}(\overline{\mathrm{X}}, \mathrm{M}) = \frac{\tau}{2\mathrm{nf}(\nu)}, \qquad \mathrm{Cov}(\overline{\mathrm{X}}, \mathrm{J}) = \sqrt{\frac{\pi}{2}}\frac{1}{\mathrm{n}}\left[\sigma^2 - 2\eta + 2\mu\gamma - \nu\tau\right],$$

$$\mathrm{Cov}(\mathrm{M}, \mathrm{J}) = \sqrt{\frac{\pi}{2}}\frac{\mu - \nu}{2\mathrm{nf}(\nu)}. \tag{3}$$

Proof of Lemma 2.1 is given in the Appendix. The following Lemma presents the joint asymptotic distribution of the statistics $\bar{X}$, M and J:

Lemma 2.2. *Let the underlying distribution F satisfy the conditions listed below*

A. F^{-1} is continuous on the open unit interval $(0, 1)$ and satisfies a first order Lipschitz condition on every interval bounded away from 0 and 1; $(F^{-1})'$ exists and is continuous except on a set of Jordan content 0.

B. $f(\nu) \neq 0$ and the Riemann integral $\int_0^1 (F^{-1})'(u)[u(1-u)]^{\frac{1}{2}}du$ converges absolutely.

C. For $i = 1, 2$ there exist $\delta_i \in (0, 1)$ such that $\forall k_i > 0$ $\exists M_i < \infty$ such that (C1) if $0 < u_1, u_2 < \delta_1$ and $k_1^{-1} < u_1/u_2 < k_1$, then

$$M_1^{-1} < \frac{(F^{-1})'(u_1)}{(F^{-1})'(u_2)} < M_1,$$

(C2) if $1-\delta_2 < u_1, u_2 < 1$ *and* $k_2^{-1} < (1-u_1)/(1-u_2) < k_2$, *then*

$$M_2^{-1} < \frac{(F^{-1})'(u_1)}{(F^{-1})'(u_2)} < M_2.$$

Let $d = \sqrt{\sigma^2 + (\mu-\nu)^2 - \tau^2}$. *Then the statistics*

$$\frac{\overline{X}-\mu}{\sigma}, \qquad \frac{M-\nu}{1/(2f(\nu))} \quad \text{and} \quad \frac{J-\sqrt{\frac{\pi}{2}}\tau}{\sqrt{\frac{\pi}{2}}d} \tag{4}$$

are asymptotically jointly normally distributed with mean 0 and variance covariance matrix given by

$$\frac{1}{n}\begin{pmatrix} 1 & \frac{\tau}{\sigma} & \frac{(\sigma^2-2\eta+2\mu\gamma-\nu\tau)}{\sigma d} \\ \frac{\tau}{\sigma} & 1 & \frac{(\mu-\nu)}{d} \\ \frac{(\sigma^2-2\eta+2\mu\gamma-\nu\tau)}{\sigma d} & \frac{(\mu-\nu)}{d} & 1 \end{pmatrix} \tag{5}$$

The proof of Lemma 2.2 is given in the Appendix. Combining the results of Lemmas 2.1 and 2.2, one obtains the asymptotic distribution of the statistic T.

Theorem 2.1. *Let the distribution* F *satisfy the conditions (A)–(C) listed in Lemma 2.2. Let* $\lambda = \mu - \nu$. *Then*

$$\sqrt{n}(T - \frac{\lambda}{\sqrt{\frac{\pi}{2}}\tau}) \Rightarrow N(0, \sigma_T^2), \tag{6}$$

with

$$\sigma_T^2 = \tfrac{2}{\pi\tau^2}\Bigg(\sigma^2 + \tfrac{1}{4f^2(\nu)} - \tfrac{\tau}{f(\nu)} + \tfrac{\lambda^2(\sigma^2+\lambda^2)}{\tau^2} - \lambda^2$$
$$-\tfrac{2\lambda}{\tau}[\sigma^2 - 2\eta + 2\mu\gamma - \nu\tau] + \tfrac{\lambda^2}{\tau f(\nu)}\Bigg). \tag{7}$$

Proof. The statistic T is a function of three L-statistics: $\overline{X}, M$ and J, which converge in probability, respectively, to μ, ν and $\sqrt{\frac{\pi}{2}}\tau$. The Taylor expansion of T in the neighborhood of $(\mu, \nu, \sqrt{\frac{\pi}{2}}\tau)$ is given by

$$T = \frac{\mu-\nu}{\sqrt{\frac{\pi}{2}}\tau} + \frac{\bar{X}-\mu}{\sqrt{\frac{\pi}{2}}\tau} - \frac{M-\nu}{\sqrt{\frac{\pi}{2}}\tau} - \frac{\mu-\nu}{\frac{\pi}{2}\tau^2}(J - \sqrt{\frac{\pi}{2}}\tau) + o_p(n^{-\frac{1}{2}}). \tag{8}$$

By Lemma 2.2, these statistics in (4) have an asymptotic joint normal distribution with mean 0 and variance covariance matrix (5). Consequently, applying the δ-method yields that

$$\sqrt{n}(T - \frac{\mu - \nu}{\sqrt{\frac{\pi}{2}}\tau}) \xrightarrow{d} N(0, \sigma_T^2), \tag{9}$$

where σ_T^2 is given in (7).

Corollary 2.1. *In the special case when F is symmetric, i.e. $f(\nu - x) = f(\nu + x)$. Then $\mu - \nu = \lambda = 0$; hence, the asymptotic distribution of the statistic T is normal with mean 0 and variance given by*

$$\sigma_T^2 = \frac{2}{\pi\tau^2}\left[\sigma^2 + \frac{1}{4f^2(\nu)} - \frac{\tau}{f(\nu)}\right]. \tag{10}$$

Comments: (1) At first glance, the asymptotic variance of T in Corollary 1 appears to depend on mean μ and variance σ^2 of the underlying density even though T is location-scale invariant. Since the values of $f(\nu)$ and τ are also functions of of these parameters, the values of μ and σ actually do not affect the value of the final formula. In practice, however, one may need to studentize the statistic by estimating the variance of T. Formula (10) is more convenient for this purpose. In particular, one can substitute the empiric estimates of τ calculated from the data along with a kernel estimate of $f(\nu)$ to obtain an asymptotically distribution-free test. (2) As pointed out by a referee, the large sample theory can also be derived using the influence function of the difference between $\bar{X}$ and M as in Cabilio and Masaro (1996).

Table 2.1 lists the asymptotic variances of T for samples from some selected symmetric distributions.

Distribution	Variance	Distribution	Variance
Normal	0.5708	t_3	0.9689
Logistic	0.5779	Uniform	0.8488
Double Exponential	0.6366	Beta(2,2)	0.6539
t_5	0.6373	$0.9N(0,1) + 0.1N(0,3^2)$	0.7165

3. Size, Power and Comparison with Other Existing Tests of Symmetry

The results stated in Theorem 2.1 enable one to use T as a test of hypothesis of symmetry stated in (1). Under H_0 the statistic T is asymptotically normal with mean 0 and variance (10). Hence, the rejection region for (1) becomes

$$\text{reject } \mathrm{H_0} \text{ if } |\mathrm{T}| \geq \frac{\mathrm{z}_{\alpha/2}\sigma_\mathrm{T}}{\sqrt{\mathrm{n}}}.$$

Notice that the asymptotic variance σ_T^2 of the statistic T depends on the underlying density function f, which usually is not known. Therefore, we adopt the approach proposed by Cabilio and Masaro (1996) and Mira (1999) and approximate the unknown σ_T^2 by 0.5708, the asymptotic variance of T when the data come from a normal density. Hence, our rejection region may be rewritten in the form:

$$\text{reject } \mathrm{H_0} \text{ if } |\mathrm{T}| \geq \frac{\mathrm{z}_{\alpha/2}\sqrt{0.5708}}{\sqrt{\mathrm{n}}}. \tag{11}$$

To compare the size of the test T using the actual asymptotic variance σ_T^2 to the approximation described above, we perform a Monte Carlo simulation study on data from several symmetric distributions. Table 2 presents the results. Notice that there is no appreciable deviation from the nominal level of 0.05 for many distributions, e.g. the t_5. However, for distributions with heavier and shorter tails, e.g. the t_3 and uniform, the size of the T test based on the normal approximation tends to be higher than the nominal 5% level. This is due to the fact that the true asymptotic variance σ_T^2 for those distributions are larger than 0.5708, the normal asymptotic variance used in (11). (see Table 2.1). The further the distance between the true σ_T^2 and 0.5708, the further away the actual size of the test to its nominal value.

If one desires to ensure that the size of the test T is $\alpha \pm \epsilon, 0 < \epsilon \ll 1$, then Corollary 1 implies that asymptotically,

$$\begin{aligned}\alpha - \epsilon &\leq P\Big(|T| \geq z_{\alpha/2}\sqrt{0.5708}/\sqrt{n}\Big) \approx P\Big(N(0,1) \geq \frac{z_{\alpha/2}\sqrt{0.5708}}{\sigma_T}\Big)\\ &\leq \alpha + \epsilon.\end{aligned}$$

Hence,

$$z_{\alpha/2}^2 * 0.5708/z_{(\alpha-\epsilon)/2}^2 \leq \sigma_T^2 \leq z_{\alpha/2}^2 * 0.5708/z_{(\alpha+\epsilon)/2}^2.$$

For example, if one wants to guarantee that the actual size of a 0.05 level test is between 0.04 to 0.06, then the asymptotic variance of σ_T^2 should between 0.5199 and 0.6199.

Distribution	n=30		n=50		n=100	
	Normal Appr.	True σ_T^2	Normal Appr.	True σ_T^2	Normal Appr.	True σ_T^2
Normal[a]	-	0.0411	-	0.0415	-	0.0467
Logistic	0.0413	0.0386	0.0462	0.0415	0.0475	0.0452
DE[b]	0.0560	0.0455	0.0581	0.0464	0.0624	0.0465
t_5	0.0496	0.0392	0.0571	0.0437	0.0587	0.0491
t_3	0.0841	0.0219	0.0943	0.0303	0.1089	0.0309
Uniform[c]	0.0805	0.0280	0.0907	0.0373	0.1010	0.0461
Beta(2,2)	0.0484	0.0363	0.0550	0.0418	0.0608	0.0431
$CN(0.1,3)^d$	0.0646	0.0356	0.0704	0.0366	0.0757	0.0463

[a]Case 2 of the Generalized Lambda Distribution (GLD) with $\lambda_1 = 0$, $\lambda_2 = 0.197454$, $\lambda_3 = \lambda_4 = 0.134915$.

[b]Double exponential

[c]Case 2 of the Generalized Lambda Distribution (GLD) with $\lambda_1 = 0$, $\lambda_2 = 2$, $\lambda_3 = \lambda_4 = 1$.

[d]$CN(0.1, 3)$ is the 10% Contaminated Normal,i.e. $0.9N(0,1) + 0.1N(0,3^2)$.

Following Mira (1999) and Cabilio and Masaro (1995), we conduct a simulation study on 8 members of the generalized lambda distribution (Ramberg and Schmeiser, 1974) (GLD) to evaluate the power of the T test. For comparative purposes, we have also simulated the power of the C test of Cabilio and Masaro (1996) and γ_1 test of Mira (1999). From Table 3, we observe that for small sample sizes ($n \leq 50$) our test, T, is significantly more powerful than other two tests in the case 12 of unimodal distribution and is noticeably more powerful in the cases 13-14 of exponential distribution, while showing a slight improvement over other two tests in the case 8 which also belongs to the family of exponential distributions. The test T performs worse for all sample sizes on data from the unimodal distribution case 10 and shows relatively the same power for the case 7 with asymmetry contaminations in the tails and for the unimodal distribution case 9. Notice that all three tests lack power in the unimodal case 11 although the test T is slightly superior here. For smaller sample sizes ($n \leq 50$) the new test T appears to be preferable to other two tests.

4. Discussion

This paper proposes a new test of symmetry that replaces the sample standard deviation in a classical measure of skewness (Hotelling and Solomons,

Distribution	$n = 30$ γ_1, T, C	$n = 50$ γ_1, T, C	$n = 100$ γ_1, T, C
Case 7: $\lambda_1 = 0$ $\lambda_2 = 1$ $\lambda_3 = 1.4$ $\lambda_4 = 0.25$	0.1793 0.2403 0.2348	0.2615 0.3724 0.3448	0.4741 0.5885 0.5596
Case 8: $\lambda_1 = 0$ $\lambda_2 = 1$ $\lambda_3 = 0.00007$ $\lambda_4 = 0.1$	0.4942 0.5259 0.6002	0.7339 0.7815 0.8198	0.9673 0.9795 0.9765
Case 9: $\lambda_1 = 3.5865$ $\lambda_2 = 0.0431$ $\lambda_3 = 0.0252$ $\lambda_4 = 0.0940$	0.1517 0.1483 0.1903	0.2682 0.2896 0.3264	0.5503 0.5476 0.5926
Case 10: $\lambda_1 = 0$ $\lambda_2 = -1$ $\lambda_3 = -0.0075$ $\lambda_4 = -0.03$	0.2362 0.2426 0.2374	0.4278 0.4450 0.3467	0.7850 0.7895 0.5609
Case 11: $\lambda_1 = -0.1167$ $\lambda_2 = -0.3517$ $\lambda_3 = -0.13$ $\lambda_4 = -0.16$	0.0362 0.0391 0.0706	0.0528 0.0484 0.0961	0.0911 0.0778 0.1361
Case 12: $\lambda_1 = 0$ $\lambda_2 = -1$ $\lambda_3 = -0.1$ $\lambda_4 = -0.18$	0.0922 0.1024 0.1851	0.1770 0.1843 0.3007	0.4050 0.4072 0.5317
Case 13: $\lambda_1 = 0$ $\lambda_2 = -1$ $\lambda_3 = -0.001$ $\lambda_4 = -0.13$	0.6270 0.6924 0.8094	0.8858 0.9146 0.9550	0.9958 0.9986 0.9993
Case 14: $\lambda_1 = 0$ $\lambda_2 = -1$ $\lambda_3 = -0.0001$ $\lambda_4 = -0.17$	0.6522 0.7144 0.8409	0.9029 0.9399 0.9722	0.9974 0.9988 0.9993

1932) appearing in the denominator of the recent test of Cabilio and Masaro (1996) by a robust estimator of the standard deviation. The idea continues a graphical approach for checking whether data follows a normal distribution, developed by Gel et al. (2005), which standardizes the data by robust estimators of both the mean and variance.

As is true for the previously proposed tests, the asymptotic distribution of the new test statistic depends on the density of the distribution underlying the data. Indeed, formula (10) for the asymptotic variance shows that the value of the unknown density at the population median is a major factor. Future research is needed to determine whether kernel estimation of the density or an alternative approach will yield a reliable estimate of the variance of our statistic, T, in the moderate sample sizes that are often seen in practice. In order to compare our test with those of Cabilo and Masaro (1996) and Mira (1999) we have adopted the approximation they used. Namely, one uses the asymptotic variance of the statistic when the data follows a normal distribution. We show that a nominal .05 level is approximately preserved for a variety of distributions, e.g., logistic, double-exponential and t_5 that are not far from the normal. For symmetric distributions that differ noticeably from the normal, e.g. t_3 or uniform, this approximation does not work, implying that the current version of our statistic should not be used.

Comparing the new test, T, to C and γ_1 for distributions that are not far from a normal distribution and for which the approximation of the null distribution of all three tests is reliable, we have found that T has higher power against a skewed contaminated normal distribution than C and γ_1. For some of the same members of the Lambda family previously considered, the new test fares well in comparison to C and γ_1 except for case 10 of Table 3. While further research is needed to determine the range of applicability of the current approximation to the variance of the statistic, the results presented here indicate that the general approach of using robust estimators in goodness of fit tests is promising and deserves further investigation.

Acknowledgements

The research of Professor Gastwirth was supported in part by grant SES-0317956 from the National Science Foundation and the research of Professor Gel was supported by a grant from the National Science and Engineering Research Council of Canada.

Appendix

Proof of Lemma 2.1. First, we derive $\text{Cov}(\overline{\text{X}}, \text{M})$. Define an indicator function

$$I_i = \begin{cases} 1 \text{ if } X_i < \nu \\ 0 \text{ otherwise} \end{cases}. \text{ Clearly, } E(I_i) = 1/2 \text{ and } Var(I_i) = 1/4.$$

Without loss of generality, we assume that $\nu \leq M$ and consider the open interval (ν, M). This interval contains $\frac{n}{2} - \sum I_i$ observations and $f(\nu)(M - \nu) \approx P(X \in (\nu, M))$. Hence, $nf(\nu)(M-\nu) \sim \frac{n}{2} - \sum I_i = -\sum_{i=1}^{n}(I_i - \frac{1}{2})$. This implies that

$$M - \nu \sim -\frac{1}{nf(\nu)}\sum_{i=1}^{n}(I_i - \frac{1}{2}). \tag{12}$$

Note that since $X_1, X_2, \ldots, X_n$ are i.i.d., for every $k, 1 \le k \le n$, $E\big[(X_k - \mu)(I_k - \frac{1}{2})\big] = E(X_k I_k) - \mu E(I_k) - \frac{1}{2}E(X_k) + \frac{1}{2}\mu = \gamma - \frac{1}{2}\mu$ and $E(X_k I_k) = \int_{-\infty}^{\nu} x dF(x) \equiv \gamma$. Hence, asymptotically,

$$\begin{aligned} &\mathrm{Cov}(\overline{\mathrm{X}}, \mathrm{M}) = \mathrm{Cov}(\overline{\mathrm{X}} - \mu, \mathrm{M} - \nu) \\ &= \mathrm{Cov}\Big(\frac{1}{\mathrm{n}}\sum_{\mathrm{i=1}}^{\mathrm{n}}(\mathrm{X_i} - \mu), -\frac{1}{\mathrm{nf}(\nu)}\sum_{\mathrm{i=1}}^{\mathrm{n}}(\mathrm{I_i} - \frac{1}{2})\Big) \\ &= -\frac{n}{n^2 f(\nu)}E\big[(X_k - \mu)(I_k - \frac{1}{2})\big] \\ &= \frac{1}{2nf(\nu)}(\mu - 2\gamma) = \frac{\tau}{2nf(\nu)}. \end{aligned} \tag{13}$$

Next we obtain $\mathrm{Cov}(\overline{\mathrm{X}}, \mathrm{J})$. From Gastwirth (1982) we have

$$D = \bar{X} - \frac{2}{n}\sum_{i=1}^{n} X_i I_i + \frac{\nu\varepsilon}{\sqrt{n}},$$

where $\varepsilon = [\sum_{i=1}^{n}(I_i - \frac{1}{2})]/\frac{1}{2}\sqrt{n}$. Hence,

$$D = (\mu - 2\gamma) + \frac{1}{n}\sum_{i=1}^{n}(X_i - \mu) - \frac{2}{n}\sum_{i=1}^{n}(X_i I_i - \gamma) + \frac{2\nu}{n}\sum_{i=1}^{n}(I_i - \frac{1}{2}),$$

or

$$D - \tau \sim \frac{1}{n}\sum_{i=1}^{n}(X_i - \mu) - \frac{2}{n}\sum_{i=1}^{n}(X_i I_i - \gamma) + \frac{2\nu}{n}\sum_{i=1}^{n}(I_i - \frac{1}{2}) \tag{14}$$

Therefore,

$$\begin{aligned} &\mathrm{Cov}(\overline{\mathrm{X}}, \mathrm{D}) = \mathrm{E}\big[(\mathrm{D} - \tau)(\overline{\mathrm{X}} - \mu)\big] \\ &= E\big[\frac{1}{n}\sum_{i=1}^{n}(X_i - \mu)\cdot\frac{1}{n}\sum_{i=1}^{n}(X_i - \mu)\big] \\ &-E\big[\frac{1}{n}\sum_{i=1}^{n}(X_i - \mu)\cdot\frac{2}{n}\sum_{i=1}^{n}(X_i I_i - \gamma)\big] \\ &+E\big[\frac{1}{n}\sum_{i=1}^{n}(X_i - \mu)\cdot\frac{2\nu}{n}\sum_{i=1}^{n}(I_i - \frac{1}{2})\big] \end{aligned} \tag{15}$$

Since $X_1, X_2, \ldots, X_n$ are i.i.d., $E[(X_i - \mu)(X_j - \mu)] = 0$. Let $\eta \equiv \int_{-\infty}^{\nu} x^2 dF(x)$. Then

$$E[\frac{1}{n}\sum_{i=1}^{n}(X_i - \mu) \cdot \frac{1}{n}\sum_{i=1}^{n}(X_i - \mu)] = \frac{1}{n^2}E[n(X_1 - \mu)^2] = \frac{\sigma^2}{n},$$

$$E[\frac{1}{n}\sum_{i=1}^{n}(X_i - \mu) \cdot \frac{2}{n}\sum_{i=1}^{n}(X_i I_i - \gamma)] = \frac{2}{n}E[(X_1 I_1 - \gamma)(X_1 - \mu)]$$

$$= \frac{2}{n}[\eta - \mu\gamma],$$

$$E[\frac{1}{n}\sum_{i=1}^{n}(X_i - \mu) \cdot \frac{2\nu}{n}\sum_{i=1}^{n}(I_i - \frac{1}{2})]$$

$$= \frac{2\nu}{n}E[(I_1 - \frac{1}{2})(X_1 - \mu)] = \frac{2\nu}{n}(\gamma - \frac{1}{2}\mu) = -\frac{\nu\tau}{n}. \qquad (16)$$

Consequently,

$$\mathrm{Cov}(\overline{\mathrm{X}}, \mathrm{D}) = \frac{1}{\mathrm{n}}[\sigma^2 - 2\eta + 2\mu\gamma - \nu\tau]$$

and

$$\mathrm{Cov}(\overline{\mathrm{X}}, \mathrm{J}) = \sqrt{\frac{\pi}{2}}\frac{1}{\mathrm{n}}[\sigma^2 - 2\eta + 2\mu\gamma - \nu\tau].$$

We conclude the proof of Lemma 2.1 by obtaining Cov(M, J). Since X_i and X_j are independent for $i \neq j$,

$$\mathrm{Cov}(\frac{1}{\mathrm{n}}\sum_{\mathrm{i}=1}^{\mathrm{n}}(\mathrm{X_i} - \mu), -\frac{1}{\mathrm{nf}(\nu)}\sum_{\mathrm{i}=1}^{\mathrm{n}}(\mathrm{I_i} - \frac{1}{2})) = -\frac{1}{\mathrm{nf}(\nu)}\mathrm{E}[(\mathrm{X_1} - \mu)(\mathrm{I_1} - \frac{1}{2})]$$

Using representations (12) and (14), we have

$$\begin{aligned}\mathrm{Cov}(\mathrm{M}, \mathrm{D}) &= -\frac{1}{nf(\nu)}E[(X_1 - \mu)(I_1 - \frac{1}{2})] \\ &+ \frac{2}{nf(\nu)}E[(X_1 I_1 - \gamma)(I_1 - \frac{1}{2})] - \frac{2\nu}{nf(\nu)}E(I_1 - \frac{1}{2})^2 \\ &= \frac{\mu - \nu}{2nf(\nu)},\end{aligned} \qquad (17)$$

where $E[(X_1 I_1 - \gamma)(I_1 - \frac{1}{2})] = \frac{1}{2}\gamma$ and $E(I_1 - \frac{1}{2})^2 = \frac{1}{4}$. Since $J = \sqrt{\pi/2}D$, that implies

$$\mathrm{Cov}(\mathrm{M}, \mathrm{J}) = \sqrt{\frac{\pi}{2}}\frac{\mu - \nu}{2\mathrm{nf}(\nu)}.$$

Note that for symmetric distributions, $\mu = \nu$, which implies $\mathrm{Cov}(\mathrm{M}, \mathrm{J}) = 0$.

Proof of Lemma 2.2. First note that

$$\mathrm{Var}\big(|\mathrm{X_i} - \nu|\big) = \mathrm{E}(\mathrm{X_i} - \nu)^2 - \tau^2 = \sigma^2 + (\mu - \nu)^2 - \tau^2.$$

Hence,

$$\mathrm{Var(J)} = \frac{\pi}{2}\frac{1}{\mathrm{n}}[\sigma^2 + (\mu - \nu)^2 - \tau^2].$$

Let $X_{(i)}$ be the *order* statistics of the sample. Let $[(n+1)/2]$ be the largest integer such that $[(n+1)/2] \leq (n+1)/2$. Then

$$D = \frac{1}{n}\sum |X_i - M| = \frac{1}{n}\Bigg[\sum_{i\geq[(n+1)/2]} (X_{(i)} - M) - \sum_{i<[(n+1)/2]} (X_{(i)} - M)\Bigg]$$

$$= \frac{1}{n}\Bigg[- \sum_{i<[(n+1)/2]} X_{(i)} + \sum_{i\geq[(n+1)/2]} X_{(i)}\Bigg].$$

Recall that $J = \sqrt{\frac{\pi}{2}}D$. Hence, the three L-statistics J, $\overline{X}$ and M can be represented as

$$\frac{1}{n}\sum_{i=1}^{n} c(\frac{i}{n+1})X_{(i)} + \sum_{i=1}^{n} aX_{(\frac{n+1}{2})}$$

with the appropriate weight-generating function $c(u)$ and a constant a. The functions c and constants a corresponding to statistics J, $\overline{X}$ and M are:
(i) For J, $c(u) = -\sqrt{\frac{\pi}{2}}$, 0 and $\sqrt{\frac{\pi}{2}}$ for respectively $u < 0.5$, $u = 0.5$ and $u > 0.5$ and $a = 0$;
(ii) for $\overline{X}$, $c(u) = 1$ and $a = 0$;
(iii) for M, $c(u) = 0$ and $a = 1$.

Notice that the conditions (A)–(C) listed in Lemma 2.2 are equivalent to the conditions listed in Theorem 3, Corollary 3 and Remark 9 in CGJ. When those conditions are satisfied, the results of CGJ imply that asymptotically, the joint distribution of the $\overline{X}, M$ and J is normal with mean $(\mu, \nu, \sqrt{\frac{\pi}{2}}\tau)$ and variance covariance matrix (5). This concludes the proof of Lemma 2.2.

Verification of conditions (A)–(C). We consider here the case when the data come from a standard normal distribution as derivation of (A)–(C) is similar for other absolutely continuous distributions.

For $N(0,1)$ $F^{-1}(u) = \Phi^{-1}(u)$. Denote $\phi(t) = \Phi'(t)$. Clearly the condition (A) is satisfied. Notice that

$$\int_0^1 (\Phi^{-1})'(u)[u(1-u)]^{\frac{1}{2}}du = \int_0^1 \frac{[u(1-u)]^{1/2}}{\phi(\Phi^{-1}(u))}du$$

$$= \int_{-\infty}^{\infty} \big[\Phi(t)(1-\Phi(t))\big]^{1/2}dt. \tag{18}$$

Since at ∞, $1-\Phi(t)\sim\frac{1}{t}\phi(t)$, the integral (18) converges at ∞. By symmetry, it converges at $-\infty$, condition (B) is satisfied.

Let $\delta_1=1/4$. Then for any $k>0$, $0\le u_1,u_2\le 1/4$, $k^{-1}\le u_1/u_2\le k$, we have

$$\frac{\phi(\Phi^{-1}(k^{-1}u_1))}{\phi(\Phi^{-1}(u_1))}<\frac{(F^{-1})'(u_1)}{(F^{-1})'(u_2)}=\frac{\phi(\Phi^{-1}(u_2))}{\phi(\Phi^{-1}(u_1))}<\frac{\phi(\Phi^{-1}(ku_1))}{\phi(\Phi^{-1}(u_1))}$$

Consider the function $g(u)\equiv\phi(\Phi^{-1}(ku))/\phi(\Phi^{-1}(u))$, for $0\le u\le 1/4$ for some fixed constant k such that $ku\le 1/4$. We now show that

$$\lim_{u\to 0}g(u)=\frac{\phi(\Phi^{-1}(ku))}{\phi(\Phi^{-1}(u))}=k.$$

Let $t_1=\Phi^{-1}(u)$ and $t_2=\Phi^{-1}(ku)$. Then $t_i<0, i=1,2$ and $t_1\to-\infty\iff t_2\to-\infty$. The L'Hospital Rule implies that

$$\lim_{u\to 0}g(u)=\lim_{u\to 0}\frac{\phi(\Phi^{-1}(ku))}{\phi(\Phi^{-1}(u))}=k\cdot\lim_{u\to 0}\frac{\Phi^{-1}(ku)}{\Phi^{-1}(u)}=k\cdot\lim_{t_1\to-\infty}\frac{t_2}{t_1}.$$

We consider three cases:

(i) $k=1$, then $t_1=t_2$ and $\lim_{\mu\to 0}g(u)=1=k$;

(ii) $k<1$. In this case, $ku<u$, $t_2=\Phi^{-1}(ku)<\Phi^{-1}(u)=t_1<0$ and $\frac{t_2}{t_1}=\frac{\Phi^{-1}(ku)}{\Phi^{-1}(u)}>1$. On the other hand, since $\phi(t)$ is increasing for $t<0$, we have

$$(t_1-t_2)\phi(t_2)<\int_{t_2}^{t_1}\phi(x)dx=\Phi(t_1)-\Phi(t_2)=u-ku=\frac{1-k}{k}\cdot ku=\frac{1-k}{k}\Phi(t_2).$$

This implies that

$$t_1-t_2<\frac{1-k}{k}\cdot\frac{\Phi(t_2)}{\phi(t_2)}\Rightarrow\frac{t_1}{t_2}>1+\frac{1-k}{k}\cdot\frac{\Phi(t_2)}{\frac{\phi(t_2)}{t_2}}\cdot\frac{1}{t_2^2}$$

$$\Rightarrow\frac{t_2}{t_1}<\left\{1+\frac{1-k}{k}\cdot\frac{\Phi(t_2)}{\frac{\phi(t_2)}{t_2}}\cdot\frac{1}{t_2^2}\right\}^{-1}$$

Note that

$$\lim_{t\to-\infty}\frac{\Phi(t)}{\frac{\phi(t)}{t}}=\lim_{t\to-\infty}\frac{\phi(t)}{\frac{-t^2\phi(t)-\phi(t)}{t^2}}=\lim_{t\to\infty}\frac{-t^2}{t^2+1}=-1.$$

Hence, for fixed k,

$$\frac{t_2}{t_1}<\left\{1+\frac{1-k}{k}\cdot\frac{\Phi(t_2)}{\frac{\phi(t_2)}{t_2}}\cdot\frac{1}{t_2^2}\right\}^{-1}\to 1,\quad t_2\to-\infty$$

Together with the fact that $t_2/t_1>1$, we obtain

$$\lim_{\mu\to 0}g(u)=k\cdot\lim_{t_1\to-\infty}\frac{t_2}{t_1}=k.$$

(iii) The proof for $k < 1$ is similar and hence is omitted.
Thus the function $g(u) \equiv \phi(\Phi^{-1}(ku))/\phi(\Phi^{-1}(u))$ is continuous on the closed interval $[0, 1/4]$ and hence reaches its maximum and minimum on the interval.

When $3/4 \leq u_i \leq 1$, by the symmetry properties of the density and distribution functions of the N(0,1), we have

$$\phi(\Phi^{-1}(u)) = \phi(-\Phi^{-1}(u)) = \phi(\Phi^{-1}(1-u))$$

with $1 - u \in [0, 1/4]$. Thus $(F^{-1})'(u_1)/(F^{-1})'(u_2)$ will also be bounded above and below. This verifies that the condition (C) holds.

References

1. ANTILLE A., KERSTING G. and ZUCCINI W. (1982) Testing symmetry. *Journal of American Statistical Association* **77**, 639–651.
2. BHATTACHARYA P. K., GASTWIRTH, G. L. and WRIGHT A. L. (1982) Two modified Wilcoxon tests for symmetry about an unknown location parameters. *Biometrika* **69**, 377–382.
3. BOOS D. (1982) A new test for asymmetry with Hodges-Lehmann estimator. *Journal of American Statistical Association* **77**, 647–651.
4. CABILIO, P. and MASARO, J. (1996). A simple test of symmetry about an unknown median. *The Canadian Journal of Statistics* **24**, 349–361.
5. CHERNOFF, H., GASTWIRTH, J.L., and M.V. JOHNS Jr. (1967). Asymptotic distribution of linear combinations of functions of order statistics with applications to estimation. *The Annals of Mathematical Statistics* **38**, 352–372.
6. EINMAHL H. H. L. and MCKEAGUE I. W. (2003) Empirical likelihood based hypothesis testing. *Bernoulli*, **9**, 267–290.
7. GASTWIRTH, J.L. (1971). On the sign test of symmetry. *Journal of American Statistical Association* **66**, 821–828.
8. GASTWIRTH, J.L. (1982). Statistical properties of a measure of tax assessment uniformity. *Journal of Statistical Planning and Inference* **6**, 1–12.
9. GEL Y. R., MIAO W. and GASTWIRTH J. L. (2005) The importance of checking the assumptions underlying statistical analysis: Graphical methods for assessing normality. *To appear in Jurimetrics.*
10. HOTELLING, H. and SOLOMONS, L. M. (1932). The limits of a measure of skewness. *Annals of Mathematical Statistics* **3**, 141–142.
11. KENDALL, M. G., STUART, A. and ORD K. (1992). *Kendall's Advanced Theory of Statistics, Vol I: Distribution Theorem. 5th edition.* Oxford Univ Press.
12. LEHMANN, E.L. and ROMANO, J.P. (2005). *Testing Statistical Hypotheses.* Springer.
13. MASON, D. (1992). Necessary and sufficient conditions for asymptotic normality of L-statistics. *Annals of Probability* **20**, 1779–1804.
14. MIRA, A. (1999). Distribution-free test for symmetry based on Bonferroni's measure. *Journal of Applied Statistics* **26**, 959–972.
15. MODARRES R. and GASTWIRTH J. L. A hybrid test for the hypothesis of symmetry. *Journal of Applied Statistics* **25**, 777–783.
16. RAMBERG, J.S. and SCHMEISER, B.W. (1974). An approximate method for generating asymmetric random variables. *Communications of the ACM* **17**, 78–82.

17. RAO, C.R. (1965). *Linear statistical inference and applications.* John Wiley, New York.
18. THAS, O., RAYNER, J.C.W. and BEST, D.J. (2005). Tests for Symmetry Based on the One-sample Wilcoxon Signed Rank Statistic. *Communications in Statistics: Simulation and Computation* **34** (to appear).

THE COX PROPORTIONAL HAZARDS MODEL WITH A PARTIALLY KNOWN BASELINE

MAI ZHOU

Department of Statistics, University of Kentucky
Lexington, KY 40506-0027, U.S.A.

The Cox proportional hazards regression model has been widely used in the analysis of survival/duration data. It is semiparametric because the model includes a baseline hazard function that is completely unspecified. We study here the statistical inference of the Cox model where some information about the baseline hazard function is available, but it still remains as an infinite dimensional nuisance parameter. We incorporate the information about the baseline hazard into the inference for regression coefficient by using the empirical likelihood method (Owen 2001) and obtained the modified test/estimator and their asymptotic distributions. The modified estimator is shown to be better than the regular Cox partial likelihood estimator in theory and in several simulations.

Some key words: Empirical likelihood; Information matrix; Log-rank test; Wilks theorem.

1. Introduction and Background

One of the most widely used regression models in survival analysis is the Cox proportional hazards model suggested by Cox (1972, 1975). Let $X_1, \ldots, X_n$ and $C_1, \ldots, C_n$ be nonnegative independent random variables. Think of C_i as the censoring time associated with the survival time X_i. Due to censoring, we can only observe (T_1, δ_1), ..., (T_n, δ_n) where

$$T_i = \min(X_i, C_i) \quad \text{and} \quad \delta_i = I_{[X_i \le C_i]}. \tag{1}$$

Also available are $z_1, \ldots, z_n$, which are covariates associated with the survival times $X_1, \ldots, X_n$ and we assume z_i do not change with time in this paper.

According to the Cox's proportional hazards model, the cumulative hazard function, $\Lambda_i(t)$, of the ith survival time is related to the covariate z_i.

That relation is given by

$$\Lambda_{X_i}(t) = \Lambda_i(t) = \Lambda(t|z_i) = \Lambda_0(t)\exp(\beta_0 z_i) \ , \tag{2}$$

where β_0 is an unknown regression coefficient and $\Lambda_0(t)$ is the so called baseline cumulative hazard function. Another way to think of $\Lambda_0(t)$ is that it is the cumulative hazard function for an individual with zero covariate, $z = 0$.

The semiparametric Cox proportional hazards model assumes that the baseline cumulative hazard function $\Lambda_0(t)$ is completely unknown and arbitrary.

In this paper we study the statistical inference in the Cox model where we have some information about the baseline hazard. For example, we may know that the baseline hazard function has median equal to 40; or that the cumulative hazard is linear within the time interval $(25, 50)$. For the stratified Cox model, we may know that the two baseline cumulative hazards cross at $t = 50$, etc. In practice, when comparing a placebo against a new treatment in a two sample test, we often have additional/prior knowledge about the survival/hazard experience for the placebo group. Similarly, if a desease is well studied before then there often are some information about the baseline hazard available from prior studies. By using these information, we strike a compromise between the complete nonparametric (Cox model) and parametric models.

Empirical likelihood method is used in this paper to give inference about β_0 in the presence of this additional information. We show that the maximum empirical likelihood estimator has asymptotically a normal distribution and the (profile) empirical likelihood ratio also follows a Wilks type theorem under null hypothesis.

It is worth pointing out that in the regular Cox model, the partial likelihood estimator of β is "free" of the baseline, yet the information on the baseline does help improve the estimation of β. Our modified estimator of β is shown to be more accurate and the test have better power compared to the regular Cox partial likelihood estimator/test.

What we propose to do in this paper with the additional information is to shrink the space of the nuisance parameter, and show that this leads to improved estimation/testing for the regression parameter via empirical likelihood. Furthermore, if we keep shrinking the nuisance parameter space by adding more information about the baseline hazard, we would eventually get to the case where there is no nuisance parameter – the parametric proportional hazard model. Therefore the models and methods we study here bridges the gap between an empirical likelihood with a completely unspecified infinite dimensional nuisance parameter and a parametric likelihood with no nuisance parameter, or finite dimensional nuisance parameters. In fact we show the Fisher information of one model approaches that of the other model with increasing knowledge about the nuisance parameter (Theorem 6).

Similar idea also work for many other semi-parametric models with infinite dimensional nuisance parameters. If there are additional information available for the nuisance parameter then by incorporating them into the empirical likelihood (by shrinking the parameter space) we can improve the estimation of the finite dimensional parameter of interest. See Chen (2005), chapters four and five for details of this approach in other models.

We end this section by presenting a few known results about the regular Cox partial likelihood estimator of the regression coefficient β_0, which can be found in Andersen and Gill (1982), and Pan (1997). For simplicity we gave detailed formula only for the case $\dim(z_i) = 1$. When $\dim(z_i) = k$, parallel results to those obtained here can be obtained similarly.

Let $\Re_i = \{j : T_j \geq T_i\}$, the risk set at time T_i. Define

$$\ell(\beta) = \sum_{i=1}^{n} \delta_i z_i - \sum_{i=1}^{n} \delta_i \frac{\sum\limits_{j\in\Re_i} z_j \exp(\beta z_j)}{\sum\limits_{j\in\Re_i} \exp(\beta z_j)}, \tag{3}$$

and

$$I(\beta_0) = \sum_{i=1}^{n} \delta_i \left(\frac{\sum\limits_{j\in\Re_i} z_j^2 \exp(\beta_0 z_j)}{\sum\limits_{j\in\Re_i} \exp(\beta_0 z_j)} - \left[\frac{\sum\limits_{j\in\Re_i} z_j \exp(\beta_0 z_j)}{\sum\limits_{j\in\Re_i} \exp(\beta_0 z_j)} \right]^2 \right) = -\ell'(\beta_0). \tag{4}$$

If $\hat{\beta}_c$ is the solution of (3), i.e. $\ell(\hat{\beta}_c) = 0$, then $\hat{\beta}_c$ is called the Cox partial likelihood estimator of the regression coefficient β_0.

Theorem 1 (Andersen and Gill 1982) *Under some mild regularity conditions we have the following results:*
(1). If $\hat{\beta}_c$ is the solution of (3), then, as $n \to \infty$,

$$\sqrt{n}(\hat{\beta}_c - \beta_0) \xrightarrow{\mathcal{D}} N(0, \Sigma^{-1}), \tag{5}$$

where $\Sigma = Plim_{n\to\infty} \frac{1}{n} I(\beta_0)$.
(2). If $\beta_n^\star \xrightarrow{P} \beta_0$, then $\frac{1}{n} I(\beta_n^\star) \xrightarrow{P} \Sigma$.

Before we present the next theorem we need some definitions of the empirical likelihood. The contribution to the asymptotic (Poisson) empirical likelihood function from (T_i, δ_i) is

$$(\Delta\Lambda_i(T_i))^{\delta_i} \exp\{-\Lambda_i(T_i)\}.$$

Under Cox's proportional hazards model,

$$\Delta\Lambda_i(T_i) = \Delta\Lambda_0(T_i)\exp(\beta z_i), \quad \text{and} \quad \Lambda_i(T_i) = \Lambda_0(T_i)\exp(\beta z_i).$$

Also, we write $\Lambda_0(T_i) = \sum_{j:T_j \le T_i} \Delta\Lambda_0(T_j)$. Thus the empirical likelihood function under the Cox's model is

$$\mathcal{A}L^c(\beta, \Lambda_0) = \prod_{i=1}^{n} (\Delta\Lambda_0(T_i)e^{\beta z_i})^{\delta_i} \exp\{-e^{\beta z_i} \sum_{j:T_j \le T_i} \Delta\Lambda_0(T_j)\}, \tag{6}$$

where we shall require $\Lambda_0 \ll \hat{\Lambda}_{NA}$, the Nelson-Aalen estimator based on the data $(T_i, \delta_i), i = 1, 2, \cdots, n$. This requirement is similar to that of $F \ll \hat{F}_n$ for CDFs imposed by Owen, see Owen (1988) for some discussions.

Theorem 2 (Pan 1997) *Under the same conditions as in Theorem 1, we have the following empirical likelihood ratio result:*

$$-2\log \frac{\max_{\{\Lambda_0:\ \Lambda_0 \ll \hat{\Lambda}_{NA}\}} \mathcal{A}L^c(\beta_0, \Lambda_0)}{\max_{\{\beta,\ \Lambda_0:\ \Lambda_0 \ll \hat{\Lambda}_{NA}\}} \mathcal{A}L^c(\beta, \Lambda_0)} = I(\xi)(\beta_0 - \hat{\beta}_c)^2 \xrightarrow{\mathcal{D}} \chi^2_{(1)} ,$$

where ξ is between β_0 and $\hat{\beta}_c$.

2. Inference of β_0 with Information on the Baseline

The simplest form of the additional information on the baseline is given in terms of the following equation:

$$\int g(s)d\Lambda_0(s) = \sum g(T_i)\Delta\Lambda_0(T_i) = \theta , \tag{7}$$

where θ is a given constant, and $g(\cdot)$ is a given function. The second expression above assumes a discrete hazard that only have possible jumps at the observed survival times, T_i (like the Nelson-Aalen estimator). This type of information includes many familiar cases. For example, if $g(s) = I_{[s \le 45]}$ and $\theta = -\log 0.5$, then the extra information corresponds to "median equal to 45".

For simplicity, we assume $T_1 < T_2 < \cdots < T_n$ for the rest of this paper. The modified Cox estimator is defined via the empirical likelihood. Let

$w_i^0 = \Delta\Lambda_0(T_i)$ for $i = 1, 2, \cdots, n$. We rewrite the log empirical likelihood (6) in terms of w_i^0:

$$\log \mathcal{A}L^c(\beta, \Lambda_0) = \sum_{i=1}^{n} \delta_i \log w_i^0 + \sum_{i=1}^{n} \delta_i \beta z_i - \sum_{i=1}^{n} w_i^0 \sum_{j=i}^{n} \exp(\beta z_j).$$

To maximize the above with respect to β and w_i^0, at the same time keep in mind of the extra information (7) imposed on the baseline hazard, we form the target function to be used by Lagrange multiplier method

$$G = \sum_{i=1}^{n} \delta_i \log w_i^0 + \sum_{i=1}^{n} \delta_i \beta z_i - \sum_{i=1}^{n} w_i^0 \sum_{j=i}^{n} \exp(\beta z_j) - n\lambda[\sum_{i=1}^{n} g(T_i)\delta_i w_i^0 - \theta] \; .$$

Taking partial derivatives of G with respect to β and w_i^0, and letting them equal to zero, we obtain

$$\frac{\partial G}{\partial w_i^0} = \frac{\delta_i}{w_i^0} - \sum_{j=i}^{n} \exp(\beta z_j) - n\lambda g(T_i)\delta_i = 0, \quad i = 1, 2, \cdots, n \tag{8}$$

and

$$\frac{\partial G}{\partial \beta} = \sum_{i=1}^{n} \delta_i z_i - \sum_{i=1}^{n} w_i^0 \sum_{j=i}^{n} z_j \exp(\beta z_j) = 0. \tag{9}$$

Solving (8), we have

$$w_i^0 = \frac{\delta_i}{\sum_{j=i}^{n} \exp(\beta z_j) + n\lambda g(T_i)\delta_i} \quad i = 1, 2, \cdots, n. \tag{10}$$

The λ in the above equation is the solution of

$$m(\beta, \lambda) = \sum_{i=1}^{n} \frac{\delta_i g(T_i)}{\sum_{j=i}^{n} \exp(\beta z_j) + n\lambda g(T_i)\delta_i} - \theta = 0. \tag{11}$$

Substituting (10) into (9), we get the equation

$$\ell^*(\beta, \lambda) = \sum_{i=1}^{n} \delta_i z_i - \sum_{i=1}^{n} \delta_i \frac{\sum_{j \in \Re_i} z_j \exp(\beta z_j)}{\sum_{j \in \Re_i} \exp(\beta z_j) + n\lambda g(T_i)\delta_i} = 0. \tag{12}$$

Solving (12) and (11) for β and λ simultaneously requires iterative methods. We notice that the computation for the regular Cox estimator of β needs iteration too. We shall discuss the computation in more detail in the next section. Let us use $\hat{\beta}$, $\hat{\lambda}$ to denote the solution of (12) and (11). The solution $\hat{\beta}$ is also the modified estimator of the regression coefficient.

Theorem 3 *As $n \to \infty$, the regression estimator, $\hat{\beta}$, incorporating the additional information (7) on the baseline has the following limiting distribution:*

$$\sqrt{n}(\hat{\beta} - \beta_0) \xrightarrow{\mathcal{D}} N(0, (\Sigma^*)^{-1}) ,$$

where $\Sigma^* = \Sigma + BA^{-1}B$ and

$$A = \lim A_n = \lim \sum_{i=1}^{n} \frac{\delta_i g^2(T_i)}{\left[\sum_{j \in \Re_i} \exp(\beta_0 z_j)\right]^2} ;$$

$$B = \lim B_n = \lim \sum_{i=1}^{n} \frac{\delta_i g(T_i) \sum_{j \in \Re_i} z_j \exp(\beta_0 z_j)}{\left[\sum_{j \in \Re_i} \exp(\beta_0 z_j)\right]^2} .$$

Notice the variance is smaller than that of the regular Cox estimator.

The proof of Theorem 3 is deferred to appendix.

We also have the Empirical Likelihood Theorem (Wilks) for the modified regression estimator.

Theorem 4 *Assume all the conditions of Theorem 1. In addition we assume $g(\cdot)$ is integrable. Finally assume the true baseline hazard satisfy (7). Then we have, as $n \to \infty$,*

$$-2 \log \mathcal{A}LR^c(\beta_0, \Lambda_0) = -2 \log \frac{\sup \mathcal{A}L^c(\beta_0, \Lambda_0)}{\sup \mathcal{A}L^c(\beta, \Lambda_0)} \xrightarrow{\mathcal{D}} \chi^2_{(1)} ,$$

where the numerator $\mathcal{A}L^c$ is maximized with β fixed at β_0 and Λ_0 satisfy (11). The denominator $\mathcal{A}L^c$ is maximized with Λ_0 satisfy (11) but β may change freely.

See appendix for proof.

Remark 2.1. If the regression coefficient β is a vector, then a similar proof will show that the limiting distribution in Theorem 4 becomes a $\chi^2_{(p)}$ where $p = \dim(\beta)$.

We may also consider the situation where the additional information is not given by (7) but by an interval type constraint,

$$C_1 \leq \sum w_i^0 g(T_i) \leq C_2$$

or equivalently replace equation (11) by $k_1 \leq m(\beta, \lambda) \leq k_2$. This is probably more practical since people are often reluctant to assume a precise value for the baseline feature, but a range is much more reasonable. As sample size tends to infinity this type of information may not yield any improvements in estimation/testing since $\sum w_i^0 g(T_i) \to \theta$ and thus $k_1 \leq m(\beta, \lambda) \leq k_2$ holds with probability approach one (assuming k_1 and k_2 are fixed). But for finite samples, there is always some probability that the inequalities will be violated and the adjustment that forces the summation value into the interval $[C_1, C_2]$ will lead to improvements of the estimation for β. This means we only need to adjust the estimator when the feature of the (unadjusted) baseline falls outside the interval and when it does, we only do minimal adjustment to pull the feature to the boundary of the interval. We call this "finite sample adjustment" in simulation section next.

The above discussion assumes the true value θ satisfy $C_1 < \theta < C_2$. If however θ equals to one of the boundaries, then the asymptotics are more complicated. If θ is outside the interval $[C_1, C_2]$, then the modified estimate/test will not be consistent.

3. Computation of the Modified Estimator and Simulations

We have modified the programs for the regular Cox model in R language (Gentleman and Ihaka 1996) to do the computations for the Cox estimator with additional information on the baseline hazard, (available at `http://www.ms.uky.edu/~mai/splus/library/coxEL`). These programs solve (by iterative method) equation (12) for a given λ value and $g(\cdot)$ function. It does not solve (11) but rather, it will give the value (13) in the output. If you happen to pick θ equal to this value then this λ also solves (11). In general, we need to solve another equation in terms of λ to obtain the λ for a given θ.

The package is called `coxEL`. The relevant function is `coxphEL()`. This function is similar to the function `coxph()` in the `survival` package. But you need to supply an extra value `lam` and a function $g(\cdot)$ when calling the function `coxphEL()`. The program will output, among other things, $\hat{\beta}$, the modified regression coefficient estimator, and the value

$$\sum_{i=1}^{n} \frac{\delta_i g(T_i)}{\sum_{j \in \Re_i} e^{\hat{\beta} z_j^*} + n\lambda g(T_i)\delta_i} = \sum \delta_i g(T_i) w_i = \int g(t) d\Lambda_0(t) \ , \tag{13}$$

where $z_j^* = z_j - \bar{z}$.

R/Splus (also SAS) re-centers the covariates, z, automatically, therefore the baseline hazard is actually the hazard for a subject with $z = \bar{z}$ instead of $z = 0$. If you would rather recover the constraint value for the hazard at $z = 0$, you need to multiply the value obtained in (13) by $\exp(-\hat{\beta}\bar{z})$. This is because we are in a proportional hazards model and the ratio of hazards for $z = \bar{z}$ and $z = 0$ is $\exp(\hat{\beta}\bar{z})$.

If the constraint value (13) in the output is not what you wanted, then you should adjust the value of `lam`. Keep changing the input `lam` value until you get the desired output value. This is relatively easy due to the fact that the value (13) is monotone in λ and one dimensional. In the simulation, we achieve that by calling the `uniroot( )` function in R.

For `lam`$= 0$, you get the regular Cox estimator, $\hat{\beta}_c$, and the constraint value (13) is the NPMLE of the integral $\int g d\Lambda_0$.

3.1. *Some simulation results*

We use a two sample setup and both samples are exponentially distributed, and having the same sample size. Sample one $\sim \exp(0.2)$. Sample two $\sim \exp(0.3)$. We use a binary covariate, z, to indicate the samples: if $z_i = 0$ then y_i is from sample one; if $z_i = 1$ then y_i is from sample two.

The risk ratio or hazard ratio is 0.3/0.2. In Cox model, this imply the true β should be $\log(0.3/0.2) = 0.4054651$.

We did not impose censoring in this simulation. The extra information we suppose we have on the baseline hazard is

$$\int \exp(-t) d\Lambda_0(t) = \theta = 0.2 \ .$$

We generated 400 such samples (each of size 400) and for each sample we computed the Cox estimator of the regression coefficient, $\hat{\beta}_c$, and the modified estimator, $\hat{\beta}$. The sample means and sample variances below are based on 400 simulation runs.

	sample mean	sample variance
Regular Cox estimator $\hat{\beta}_c$	0.4160447	0.009736113
Modified estimator $\hat{\beta}$	0.4129862	0.008310867

Table 1. Comparison of two estimators.

The simulation show $\hat{\beta}$ has smaller bias and smaller variance compared to that of $\hat{\beta}_c$. We expect the improvements for smaller samples to be more visible.

3.2. *Additional information given as interval*

The extra information about the baseline may take the form $\int g(t) d\Lambda_0(t) \in [C_1, C_2]$, instead of assume we know its exact value.

In the next two simulations, we only adjust the regular Cox estimator when the value of the integration $\int g(t)d\hat{\Lambda}_0(t)$ falls outside the interval $[\theta-\epsilon, \theta+\epsilon]$ where θ is the true theoretical value of the integration (=0.2 in this case). For sample size $n = 400$, $\epsilon = 0.05$ we obtained the following results:

	sample mean	sample variance
Regular Cox estimator $\hat{\beta}_c$	0.4160447	0.009736113
Modified estimator $\hat{\beta}$	0.4085578	0.009332247

Table 2. Sample size 400. 400 simulations.

For sample size $n = 180$ (equal sample size of 90 each), $\epsilon = 0.1$, the results of 500 simulation runs gave the following table 3:

	sample mean	sample variance
Regular Cox estimator $\hat{\beta}_c$	0.4194698	0.02715997
Modified estimator $\hat{\beta}$	0.4187548	0.02708653

Table 3. Sample size 180. 500 simulations.

We see that the adjusted estimator is again having smaller bias and smaller variance. Although as ϵ increases the improvement diminished.

4. Growing Information about $\Lambda_0(t)$

In this section we suppose there are many more information available about the nuisance parameter $\Lambda_0(t)$. If the additional information is given in the form of *several* equations like (7),

$$\text{for } i = 1, 2, \cdots, k; \qquad \int g_i(t)d\Lambda_0(t) = \theta_i \ , \tag{14}$$

then analysis similar to those in section two leads to

Theorem 5 *Let the maximum empirical likelihood estimator in the Cox model with additinal information (14) be denoted by $\hat{\beta}$. Under some mild regularity conditions the asymptotic distribution of $\sqrt{n}(\hat{\beta}-\beta_0)$ is normal with zero mean and variance given by*

$$[\Sigma^*]^{-1} = [\Sigma + B^T A^{-1} B]^{-1} \ ,$$

where the vector B and matrix A is defined by

$$A = (A_{rs}), \quad A_{rs} = \lim \sum_{i=1}^{n} \frac{\delta_i g_r(T_i) g_s(T_i)}{\left[\sum_{j\in\Re_i} \exp(\beta_0 z_j)\right]^2} \ ;$$

$$B_m = \lim \sum_{i=1}^{n} \frac{\delta_i g_m(T_i) \sum_{j\in\Re_i} z_j \exp(\beta_0 z_j)}{\left[\sum_{j\in\Re_i} \exp(\beta_0 z_j)\right]^2} \; .$$

Next we take a closer look at the asymptotic variance of $\hat{\beta}$. Let us call $\Sigma^* = [\Sigma + B^T A^{-1} B]$ the Fisher information for β in the Cox model with additional information (14) on the baseline hazard. In view of Theorem 1, we see that the quantity $B^T A^{-1} B$ is the increment of the Fisher information due to the additional information (14) on the baseline.

When $g_i(t)$ are the indicator functions: $g_i(t) = I_{[t\le u_i]}$ for some constants u_i, $i = 1, 2, \cdots, k$, the increment in the Fisher information, $B^T A^{-1} B$, takes a particularly simple form (see Kim and Zhou (2002) for a proof):

$$B^T A^{-1} B = \sum_{i=1}^{k} \frac{[h(u_i) - h(u_{i-1})]^2}{V(u_i) - V(u_{i-1})} \; ,$$

where $h(u_i) = B_i$ and $V(\min(u_i, u_j)) = A_{ij}$. When $k \to \infty$ and u_i become dense, this summation will approach from below the integral

$$\int \frac{[h'(t)]^2}{V'(t)} dt \; .$$

This integration can also be written as

$$\int \frac{[h'(t)]^2}{V'(t)} dt = \lim \int \frac{[\sum z_j e^{\beta z_j} / \sum e^{\beta z_j}]^2}{n / \sum e^{\beta z_j}} d\Lambda_0(t) = \lim \sum_{i=1}^{n} \left(\frac{\sum z_j e^{\beta z_j}}{\sum e^{\beta z_j}} \right)^2 \frac{\delta_i}{n} .$$

In view of the expression of $I(\beta_0)$ in (4), we see that the Fisher information, Σ^*, in Theorem 5 can approach but never exceed the upper bound

$$\Sigma^* = [\Sigma + B^T A^{-1} B] \le \Sigma^{**}$$

with

$$\Sigma^{**} = \lim \frac{1}{n} \sum_{i=1}^{n} \delta_i \frac{\sum_{j\in\Re_i} z_j^2 \exp(\beta_0 z_j)}{\sum_{j\in\Re_i} \exp(\beta_0 z_j)} \; .$$

The relation between Σ and Σ^{**} is like that of a variance and a second moment.

As the next lemma and theorem reveals, the expectation of this information upper bound, Σ^{**}, is identical to that of the parametric model Fisher information.

Lemma 1 *In the parametric proportional hazards model where the baseline is completely specified, the expected information for β (when there is no censoring) is*

$$I_{para}(\beta) = \sum_{i=1}^{n} z_i^2 \ .$$

With censoring, the information is

$$I_{para}(\beta) = E\sum z_i^2 H_i(\min(Y_i, C_i)) = \sum z_i^2 E H_i(\min(Y_i, C_i)) = \sum_{i=1}^{n} z_i^2 E\delta_i.$$

The proof of the lemma is straight forward and is omitted.

Theorem 6 *We have the following equality concerning the expected informations*

$$E\Sigma^{**} = \frac{1}{n} I_{para} \ ,$$

where the expectation is over all the possible ordering of the observations.

PROOF: This can be proved by induction. Assuming no censoring and for two observations, notice the probability

$$p = 1 - q = P(Y_1 < Y_2) = \frac{e^{\beta_0 z_1}}{e^{\beta_0 z_1} + e^{\beta_0 z_2}} \ .$$

We can now compute the expectation:

$$p(z_1^2 p + z_2^2 q + z_2^2) + q(z_2^2 q + z_1^2 p + z_1^2) = z_1^2 + z_2^2 \ .$$

Assume the Theorem is true for $(n-1)$ observations, then for n observations the expectation is

$$\sum p_{i1} p_{i2} \cdots p_{in} \left[z_{i1}^2 p_{i1} + \cdots + z_{in}^2 p_{in} + \sum_{i=1}^{n-1} \frac{\sum z_j^2 \exp(\beta_0 z_j)}{\sum \exp(\beta_0 z_j)} \right]$$

$$= \sum_i (z_i^2 p_i + p_i \sum_{j \neq i} z_j^2) = \sum_{i=1}^{n} z_i^2 \ .$$

For details of the induction proof in the censored data cases, see Luan (2004). ◇

This theorem gives us the following picture: additional information on the baseline hazard increases the Fisher information of β. These Fisher informations form a *continuous spectrum* from the completely unspecified baseline model (i.e. Cox model with Fisher information Σ) to completely specified baseline model (parametric proportional hazards model with Fisher information Σ^{**}).

The fact that the maximum empirical likelihood estimators achieve all these Fisher informations in the spectrum reinforces the view that empirical likelihood is the extension of parametric likelihood.

Appendix

Lemma 2 (Joint CLT): *Assume the same conditions as in Theorem 1. In addition, assume that $g(\cdot)$ is square integrable with respect to $\Lambda_0(\cdot)$, then we have*

$$\left[\frac{\ell(\beta_0)}{\sqrt{n}}, \sqrt{n}\cdot m(\beta_0, 0)\right] \xrightarrow{\mathcal{D}} N(0, V),$$

as $n \to \infty$ where m is defined as

$$m(\beta, 0) = \sum_{i=1}^{n} \frac{\delta_i g(T_i)}{\sum_{j\in\Re_i} e^{\beta z_j}} - \int g(t) d\Lambda_0(t) .$$

The variance-covariance matrix V is diagonal, $V = \text{diag}(\Sigma, V_{22})$, Σ is defined in Theorem 1 and

$$V_{22} = \lim_{n\to\infty} \int \frac{g^2(s)d\Lambda_0(s)}{\frac{1}{n}\sum_j \exp(\beta_0 z_j) I_{[T_j \geq s]}} = \lim \sum_{i=1}^{n} \frac{n g^2(T_i)\delta_i}{\left[\sum_{j\in\Re_i} \exp(\beta_0 z_j)\right]^2} .$$

PROOF: It is now a standard result that we have the following martingale representation:

$$\begin{aligned} m(\beta_0, 0) &= \sum_{i=1}^{n} \frac{\delta_i g(T_i)}{\sum_{T_j \geq T_i} e^{\beta_0 z_j}} - \int g(s) d\Lambda_0(s) \\ &= \sum_{i=1}^{n} \int \frac{g(s)}{\sum_{j=1}^{n} e^{\beta_0 z_j} I_{[T_j \geq s]}} dM_i(s) \end{aligned}$$

and

$$\ell(\beta_0) = \sum_{i=1}^{n} \int \left(z_i - \frac{\sum_{j=1}^{n} z_j e^{\beta_0 z_j} I_{[T_j\geq s]}}{\sum_{j=1}^{n} e^{\beta_0 z_j} I_{[T_j \geq s]}} \right) dM_i(s)$$

whcro

$$M_i(t) = I_{[T_i \leq t, \delta_i = 1]} - \int^t I_{[T_i \geq s]} \exp(\beta_0 z_i) d\Lambda_0(s)$$

with

$$\langle M_i(t) \rangle = \int_0^t I_{[T_i \geq s]} \exp(\beta_0 z_i) d\Lambda_0(s) .$$

Standard computation of the predictable quadratic variation process for the martingales yields

$$\langle \sqrt{n}m(\beta_0,0), \ell(\beta_0)/\sqrt{n}\rangle = 0, \ \langle \sqrt{n}m(\beta_0,0)\rangle = \int \frac{g^2(s)d\Lambda_0(s)}{\frac{1}{n}\sum_j \exp(\beta_0 z_j) I_{[T_j \ge s]}}.$$

By the martingale central limit theorem, we see the lemma is proved. ◇

Lemma 3 *The simultaneous solution of equations (12) and (11) has the following representation:*

$$\sqrt{n}[\hat\beta - \beta_0, \hat\lambda] = -\sqrt{n}\left[\frac{\ell(\beta_0)}{\sqrt{n}}, \sqrt{n}m(\beta_0,0)\right] D^{-1} + o_p(1)$$

where D is a matrix

$$D = \begin{pmatrix} -I(\beta_0)/\sqrt{n} & -\sqrt{n}B \\ \sqrt{n}B & -\sqrt{n}A \end{pmatrix},$$

with the quantity A and B defined in Theorem 3.

Proof: The $\hat\beta$ and $\hat\lambda$ are the solutions of $(0,0) = [\frac{\ell^*(\beta,\lambda)}{\sqrt{n}}, m(\beta,\lambda)\sqrt{n}]$. By Taylor expansion

$$\begin{aligned}&[\frac{\ell^*(\beta,\lambda)}{\sqrt{n}}, m(\beta,\lambda)\sqrt{n}] \\ &= [\frac{\ell^*(\beta_0,0)}{\sqrt{n}}, m(\beta_0,0)\sqrt{n}] + (\beta-\beta_0,\lambda)\cdot D + o(|\beta-\beta_0| + |\lambda|)\end{aligned}$$

where D is the matrix of the first derivatives of the vector. We let $\beta = \hat\beta$ and $\lambda = \hat\lambda$ in the above to get

$$(0,0) = [\frac{\ell^*(\beta_0,0)}{\sqrt{n}}, m(\beta_0,0)\sqrt{n}] + (\hat\beta-\beta_0,\hat\lambda)\cdot D + o(|\hat\beta-\beta_0| + |\lambda|)\ .$$

Notice $\ell^*(\beta_0,0) = \ell(\beta_0)$, which gives

$$[\hat\beta - \beta_0, \hat\lambda] = -[\frac{\ell(\beta_0)}{\sqrt{n}}, \sqrt{n}m(\beta_0,0)] * D^{-1} + o_p(1/\sqrt{n})\ . \ ◇$$

PROOF OF THEOREM 3: From Lemma 3 we have

$$\sqrt{n}(\hat\beta - \beta_0) = \frac{\frac{\ell(\beta_0)}{\sqrt{n}}A + \sqrt{n}m(\beta_0,0)B}{AI(\beta_0)/n + B^2} + o_p(1)\ .$$

The asymptotic normality is immediate from Lemma 2. We need to compute the asymptotic variance. Since ℓ and m are asymptotically independence (Lemma 2), we compute

$$\lim Var(\sqrt{n}(\hat\beta - \beta_0)) = \lim \frac{A^2\Sigma + B^2 V_{22}}{(\Sigma A + B^2)^2}\ .$$

Since $\lim V_{22} = \lim A$, we have

$$= \frac{A}{\Sigma A + B^2} = \frac{1}{\Sigma + B^2/A} = \frac{1}{\Sigma^*} .$$

Notice $A > 0$, therefore we have

$$\lim Var(\sqrt{n}(\hat{\beta} - \beta_0)) = (\Sigma^*)^{-1} < \Sigma^{-1} = \lim Var\sqrt{n}(\hat{\beta}_c - \beta_0) \ . \ \diamond$$

Lemma G (Graybill 1976) *Suppose* $Y \xrightarrow{\mathcal{D}} N(0,V)$ *and* M *is a symmetric matrix. Then* $YMY^T \xrightarrow{\mathcal{D}} \chi^2_p$ *if and only if* MV *is idempotent and* $rank(MV) = p$.

PROOF OF THEOREM 4: In the Wilks theorem, the log of the empirical likelihood ratio becomes the difference of two terms. We shall compute each term separately:

Step I: We first compute the maximum of the log empirical likelihood (6) when β is fixed at β_0, and with the additional information (7).

Let $w_i^0 = \Delta\Lambda_0(T_i)$ for $i = 1, 2, \cdots, n$. We write the logarithm of $\mathcal{A}L^c(\beta_0, \Lambda_0)$ in terms of w_i^0's as follows

$$\begin{aligned}\log\mathcal{A}L^c(\beta_0, \Lambda_0) &= \sum_{i=1}^n \delta_i \log w_i^0 + \sum_{i=1}^n \delta_i \beta_0 z_i - \sum_{i=1}^n \sum_{j=1}^i w_j^0 \exp(\beta_0 z_i) \\ &= \sum_{i=1}^n \delta_i \log w_i^0 + \sum_{i=1}^n \delta_i \beta_0 z_i - \sum_{i=1}^n w_i^0 \sum_{j=i}^n \exp(\beta_0 z_j).\end{aligned}$$

To maximize the above empirical likelihood under the constraint (7) via Lagrange multiplier, we form the target function:

$$G = \sum_{i=1}^n \delta_i \log w_i^0 + \sum_{i=1}^n \delta_i \beta_0 z_i - \sum_{i=1}^n w_i^0 \sum_{j=i}^n \exp(\beta_0 z_j) - n\gamma[\sum g(T_i) w_i^0 - \theta]$$

Taking derivatives of G with respect to w_i^0 for $i = 1, 2, \cdots, n$, and letting them equal to zero, we obtain the equations

$$\frac{\partial G}{\partial w_i^0} = \frac{\delta_i}{w_i^0} - \sum_{j=i}^n \exp(\beta_0 z_j) - n\gamma g(T_i)\delta_i = 0.$$

It follows that

$$w_i^0 = \frac{\delta_i}{\sum_{j=i}^n \exp(\beta_0 z_j) + n\gamma g(T_i)\delta_i}$$

for $i = 1, 2, \cdots, n$. The value of the γ in the above can be obtained as the solution of the equation

$$0 = m(\beta_0, \gamma) = \sum_{i=1}^{n} \frac{g(T_i)\delta_i}{\sum_{j=i}^{n} \exp(\beta_0 z_j) + n\gamma g(T_i)\delta_i} - \theta \, . \tag{15}$$

The derivative of $m(\beta_0, \gamma)$ with respect to γ is always negative, so there is a unique γ solution, for the feasible values of θ.

By using the Taylor expansion on (15), it is easy to see that the solution, $\hat{\gamma}$, of (15) with $\theta = \int g(s)d\Lambda(s)$ has the following representation

$$\hat{\gamma} = m(\beta_0, 0) \times A^{-1} + o_p(1/\sqrt{n})$$

where $A = \lim A_n$ is defined in Theorem 3. We notice that $A = \lim A_n = V_{22}$.

The Hessian matrix of $\log \mathcal{A}L^c(\beta_0, \Lambda_0)$ is negative-definite so the w_i^0's provide the maximum of the log likelihood. Thus the maximized log likelihood under the extra baseline constraint is: (maximized over baseline, with β fixed at β_0)

$$\begin{aligned}
&\log \mathcal{A}L^c(\beta_0, \hat{\Lambda}(\beta_0)) \\
&= \sum_{i=1}^{n} \delta_i \log \frac{\delta_i}{\sum_{j=i}^{n} e^{\beta_0 z_j} + n\hat{\gamma} g(T_i)\delta_i} + \sum_{i=1}^{n} \delta_i \beta_0 z_i \\
&\quad - \sum_{i=1}^{n} \frac{\delta_i \sum_{j=i}^{n} \exp(\beta_0 z_j)}{\sum_{j=i}^{n} e^{\beta_0 z_j} + n\hat{\gamma} g(T_i)\delta_i} \\
&= \sum_{i=1}^{n} \delta_i \beta_0 z_i - \sum_{i=1}^{n} \delta_i \log \left(\sum_{j=i}^{n} e^{\beta_0 z_j} + n\hat{\gamma} g(T_i)\delta_i \right) \\
&\quad - \sum_{i=1}^{n} \frac{\delta_i \sum_{j=i}^{n} \exp(\beta_0 z_j)}{\sum_{j=i}^{n} e^{\beta_0 z_j} + n\hat{\gamma} g(T_i)\delta_i} \, ,
\end{aligned}$$

where $\hat{\gamma}$ is the solution of the equation (15).

Step II: We now compute the maximum of the log empirical likelihood without fixing the β. The extra information on the baseline hazard, (7), shall remain in effect. Recall that the maximum is achieved at $(\beta = \hat{\beta}, \lambda = \hat{\lambda})$.

Substituting β in (10) by $\hat{\beta}$, λ by $\hat{\lambda}$, we get the expression of w_i^0:

$$w_i^0 = \frac{\delta_i}{\sum_{j \in \Re_i} \exp(\hat{\beta} z_j) + n\hat{\lambda} g(T_i)\delta_i}, \quad i = 1, 2, \cdots, n. \tag{16}$$

The Hessian matrix of log$\mathcal{A}L^c(\beta, \Lambda_0)$ is negative-definite so the stationary point of log$\mathcal{A}L^c(\beta, \Lambda_0)$ is a maximum point. Therefore we obtain the expression for the maximum of the log likelihood

$$\log \max_{\{\beta,\Lambda_0 \ll \hat{\Lambda}_{NA}\}} \mathcal{A}L^c(\beta, \Lambda_0) = \sum_{i=1}^{n} \delta_i \hat{\beta} z_i - \sum_{i=1}^{n} \delta_i \log\left(\sum_{j=i}^{n} e^{\hat{\beta} z_j} + n\hat{\lambda} g(T_i)\delta_i\right) - \sum_{i=1}^{n} \frac{\delta_i \sum_{j=i}^{n} \exp(\hat{\beta} z_j)}{\sum_{j=i}^{n} e^{\hat{\beta} z_j} + n\hat{\lambda} g(T_i)\delta_i}.$$

If we let

$$C(\beta, \lambda) = \sum_{i=1}^{n} \delta_i \beta z_i - \sum_{i=1}^{n} \delta_i \text{log}\left(\sum_{j=i}^{n} \exp(\beta z_j) + n\lambda g(T_i)\delta_i\right) - \sum_{i=1}^{n} \frac{\delta_i \sum_{j=i}^{n} \exp(\beta z_j)}{\sum_{j=i}^{n} e^{\beta z_j} + n\lambda g(T_i)\delta_i},$$

and combine step I and II, we have the Wilks statistic

$$-2\text{log}\mathcal{A}LR^c(\beta_0, \Lambda_0) = -2\log \frac{\max_{\{\Lambda_0 \ll \hat{\Lambda}_{NA}, \Lambda_0 \text{satisfy}(7)\}} \mathcal{A}L^c(\beta_0, \Lambda_0)}{\max_{\{\beta, \Lambda_0 \ll \hat{\Lambda}_{NA}, \Lambda_0 \text{satisfy}(7)\}} \mathcal{A}L^c(\beta, \Lambda_0)}$$

$$= 2(C(\hat{\beta}, \hat{\lambda}) - C(\beta_0, \hat{\gamma})) = U_1 - U_2. \qquad (say)$$

We can verify easily that for any β value we have

$$\frac{\partial C(\beta, \lambda)}{\partial \lambda}|_{\lambda=0} \tag{17}$$

$$= -\sum \frac{n\delta_i g(T_i)}{\sum e^{\beta z_j} + n\lambda g(T_i)}|_{\lambda=0} + \sum \frac{n\delta_i g(T_i) \sum e^{\beta z_j}}{(\sum e^{\beta z_j} + n\lambda g(T_i))^2}|_{\lambda=0} = 0 \tag{18}$$

and

$$\frac{\partial^2 C(\beta, \lambda)}{\partial \lambda^2}|_{\lambda=0,\beta=\beta_0}$$

$$= \sum \frac{\delta_i n^2 g^2(T_i)}{(\sum)^2} - 2\sum \frac{\delta_i n^2 g^2(T_i)}{(\sum)^2} = -n^2 \sum \frac{\delta_i g^2(T_i)}{[\sum e^{\beta_0 z_j}]^2} = -nA < 0 .$$

We have the following Taylor expansion

$$U_2 = 2\{C(\beta_0, 0) + \hat{\gamma} C'(\beta_0, 0) + 1/2C''(\beta_0, 0)\hat{\gamma}^2 + o(1)\}$$
$$= 2C(\beta_0, 0) - nA\hat{\gamma}^2 + o(1) .$$

On the other hand

$$\begin{aligned}U_1 &= 2\{C(\beta_0,0)+(\hat\beta-\beta_0,\hat\lambda)(C'_\beta(\beta_0,0),C'_\lambda(\beta_0,0))^T\\ &\qquad +(\hat\beta-\beta_0,\hat\lambda)Q/2(\hat\beta-\beta_0,\hat\lambda)^T+o(1)\}\\ &= 2C(\beta_0,0)+2(\hat\beta-\beta_0)C'_\beta(\beta_0,0)+(\hat\beta-\beta_0,\hat\lambda)Q(\hat\beta-\beta_0,\hat\lambda)^T+o(1)\end{aligned}$$

where Q is the second derivative matrix of $C(\beta,\lambda)$ at $\beta=\beta_0,\lambda=0$.

Now we compute Q. Notice also that

$$\frac{\partial C(\beta,\lambda)}{\partial\beta}|_{\lambda=0}=\sum\delta_i z_i-\sum\delta_i\frac{\sum z_j e^{\beta z_j}}{\sum e^{\beta z_j}}=\ell(\beta)\ .$$

and

$$\frac{\partial^2 C(\beta,\lambda)}{\partial\beta^2}|_{\lambda=0}=-I(\beta).$$

Also

$$\frac{\partial^2 C(\beta,\lambda)}{\partial\lambda\partial\beta}|_{\lambda=0}=\frac{\partial^2 C(\beta,\lambda)}{\partial\beta\partial\lambda}|_{\lambda=0}=0.$$

Thus we have a diagonal matrix Q:

$$Q=\mathrm{diag}[-I(\beta_0),-nA].$$

Putting these all together we have

$$\begin{aligned}&-2\log ALR^c(\beta_0,\Lambda_0)\\ &=\{nA\hat\gamma^2+2(\hat\beta-\beta_0)C'_\beta(\beta_0,0)+(\hat\beta-\beta_0,\hat\lambda)Q(\hat\beta-\beta_0,\hat\lambda)^T+o(1)\}\\ &=(\sqrt{n}\frac{m(\beta_0,0)}{\sqrt{A}}+o_p(1))^2+2((\hat\beta-\beta_0)C'_\beta(\beta_0,0))\\ &\qquad+[\frac{\ell(\beta_0)}{\sqrt{n}},\sqrt{n}m(\beta_0,0)]D^{-1}\sqrt{n}\frac{Q}{n}(\sqrt{n}D^{-1})^T[\frac{\ell(\beta_0)}{\sqrt{n}},\sqrt{n}m(\beta_0,0)]^T\\ &=[\frac{\ell(\beta_0)}{\sqrt{n}},\sqrt{n}m(\beta_0,0)]\cdot(I+II+III)\cdot[\frac{\ell(\beta_0)}{\sqrt{n}},\sqrt{n}m(\beta_0,0)]^T+o_p(1)\\ &=[\frac{\ell(\beta_0)}{\sqrt{n}},\sqrt{n}m(\beta_0,0)]\cdot M\cdot[\frac{\ell(\beta_0)}{\sqrt{n}},\sqrt{n}m(\beta_0,0)]^T+o_p(1)\end{aligned}$$

where

$$I=\begin{pmatrix}0&0\\0&1/A\end{pmatrix}\ ,\quad II=\frac{1}{A\Sigma+B^2}\begin{pmatrix}2A&B\\B&0\end{pmatrix}$$

and

$$III=\frac{1}{A\Sigma+B^2}\begin{pmatrix}-A&0\\0&-\Sigma\end{pmatrix}\ ,\qquad M=\frac{1}{A\Sigma+B^2}\begin{pmatrix}A&B\\B&B^2/A\end{pmatrix}\ .$$

In view of the Lemma G and Lemma 2, we only need to verify two matrix properties. To this end we compute

$$M \cdot V = \frac{1}{A\Sigma + B^2} \begin{pmatrix} A\Sigma & AB \\ B\Sigma & B^2 \end{pmatrix} .$$

It is easy to verify that the above matrix is idempotent and has rank 1. By Lemma 2 and Lemma G we have the desired result. ◇

References

1. Andersen, P.K., Borgan, O., Gill, R. and Keiding, N. (1993), *Statistical Models Based on Counting Processes.* Springer, New York.
2. Chen, M. (2005). Some contributions to the empirical likelihood method. Ph.D. Thesis. Department of Statistics, University of Kentucky.
3. Cox, D. R. (1972). Regression Models and Life Tables (with discussion) *J. Roy. Statist. Soc B.*, **34**, 187-220.
4. Cox, D. R. (1975). Partial Likelihood. *Biometrika*, **62**, 269-276.
5. Gentleman, R. and Ihaka, R. (1996). R: A Language for data analysis and graphics. *J. of Computational and Graphical Statistics*, **5**, 299-314.
6. Gill, R. (1980), *Censoring and Stochastic Integrals.* Mathematical Centre Tracts 124. Mathematisch Centrum, Amsterdam.
7. Graybill, F. (1976). *Theory and Application of the Linear Model.* Wadsworth Publishing Company Inc., Belmont, California.
8. Kim, K. and Zhou, M. (2004). Symmetric location estimation/testing by empirical likelihood *Communications in Statistics: Theory and Methods* **33**, 2233-2243.
9. Luan, J. (2004). Empirical Likelihood and Right-censoring and Left-truncation Data. Ph.D. Thesis. Department of Statistics, University of Kentucky.
10. Owen, A. (1988). Empirical Likelihood Ratio Confidence Intervals for a Single Functional. *Biometrika*, **75**, 237-249.
11. Owen, A. (2001). *Empirical Likelihood.* Chapman and Hall, London.
12. Pan, X.R. (1997). *Empirical Likelihood Ratio Method for Censored Data. Ph.D. Thesis, Univ. of Kentucky, Dept. of Statist.*
13. Thomas, D. R. and Grunkemeier, G. L. (1975). Confidence interval estimation of survival probabilities for censored data. *J. Amer. Statist. Assoc.* **70**, 865-871.

Nonparametric Statistics

THE LOGISTIC DISTRIBUTION AND A RANK TEST FOR NON-TRANSITIVITY

B.M. BROWN

Department of Statistics and Applied Probability, National University of Singapore
Singapore 117546
stabbm@nus.edu.sg

R.H. HETTMANSPERGER

graphics@redcushion.com, Louisville, CO 80027, USA

T.P. HETTMANSPERGER

Department of Statistics, The Pennsylvania State University
State College, PA 16802, USA
tph@stat.psu.edu

Dedicated to Yuan Shih Chow, a peerless scholar and teacher

A circularity statistic, based upon pairwise Mann-Whitney statistics, and measuring the non-transitivity effect $A > B > C > A$, was introduced in Brown & Hettmansperger (2002). In the present paper, its large sample null distribution is shown to be logistic. To test for non-transitivity, possibly indicating the presence of mixture terms, one of the components of the logistic limit variable is used as a regulator to prevent the circularity statistic being inflated by large transitive rather than non-transitive effects. An example is discussed.

Some key words: Brownian bridge; Circularity; Cramer-von Mises distribution; Efron mixtures; Location shifts; Mann-Whitney; Mixtures; Stochastic ordering.

1. Introduction

In k-sample testing between treatments, the most familiar paradigm is a model of *stochastic ordering*, where symbolically denoting treatments by A, B etc, $A > B$ and $B > C$ implies $A > C$. Among such stochastically ordered or *transitive* models, the simplest is that of *location shifts* between treatments. The location shift parameters are interpretable, and produce estimates and confidence intervals using either parametric or nonparametric error assumptions.

Signals of *non-transitivity*, where pairwise comparisons might yield $A > B > C > A$, or *circularity*, are of interest because they may indicate the presence of mixtures with different components. This phenomenon was examined in Brown & Hettmansperger (2002), in the context of rank tests. It was shown that for three treatments A, B and C, with sample sizes n_1, n_2, n_3, in testing H_0 : *A, B and C observations are from the same distribution*, the three pairwise Mann-Whitney statistics T_{12}, T_{23} and T_{31} had a rank 3 covariance matrix. Two degrees of freedom were used up by the Kruskal-Wallis test statistic (see Hettmansperger & McKean, 1998 p 240), while the remaining degree of freedom was attributable to an independent 'circularity' statistic

$$C_{123} = \frac{T_{12}}{n_1 n_2} + \frac{T_{23}}{n_2 n_3} + \frac{T_{31}}{n_3 n_1}. \tag{1}$$

The Mann-Whitney statistics are defined for example by $T_{12} = \sum_{i,j} sgn(B_j - A_i)$, where A_i is the ith A-observation and B_j is the jth B-observation.

The potential use of the circularity statistic C_{123} for testing non-transitivity was discussed. It was shown that this statistic was inflated by mixture models based upon the well-known *Efron dice*; see Gardner (1970). However, two issues were left outstanding:

(i) the limit null distribution of C_{123} was non-Gaussian, of unknown form, and;

(ii) values of C_{123} are inflated not only by non-transitive effects, but also by large location shifts, meaning that C_{123} alone would not be a suitable test for non-transitivity.

The present paper addresses both of these problems.

There are several hypotheses of interest about the distributions of treatments A, B and C:

H_0: the location shift model holds, and all location shifts are zero; this is the same as H_0 above;

H_1: the location shift model holds, ie location shifts can be non-zero; and

H_2: there is non-transitivity; the distributions of A, B and C are not stochastically ordered.

For detecting H_2, we propose test statistics based on the circularity statistic C_{123} and other measures, all with null distributions based on H_0, but which are not inflated if H_1 is true.

In Section 2 it is shown that C_{123} is a *swept area functional* on a random walk which converges to a Brownian process. An internet link is provided to executable code for simulating these random walks and illustrating the swept area functional. The limit distribution of C_{123} is that of the swept area functional on the Brownian process, which is shown in Section 3 to have a logistic distribution.

One of the components of the limit logistic random variable is the square root of a *second order Cramer-von Mises variable*, derived from the radial part of the random walk. Applying location shifts to keep this 'radial' random variable near its mean value enables it to be used as a regulator to prevent the circularity statistic becoming inflated by H_1 rather than H_2. This procedure is outlined in Section 4 and illustrated by an example.

2. The Swept Area Functional

In this Section, the circularity statistic C_{123} is shown to be a *swept area functional* on a certain random walk, defined from the random order of the A, B and C observations.

The random walk is $X = \{X_j, j = 0, 1, \ldots, N\}$, where $X_0 = X_N = 0$, and the increment $Y_j = X_j - X_{j-1}$ depends on whether an A, B or C observation is in the jth position of the combined sample order; thus $N = n_1 + n_2 + n_3$. Let $Y_j = a$ if the jth element is an A, $= b$ if the jth element is a B, or $= c$ if the jth element is a C, where $a, b, c \in R^2$, and are as follows:

$$a^T = \left(0, \quad \sqrt{\frac{n_2 + n_3}{Nn_1}}\right)$$

$$b^T = \left(-\sqrt{\frac{n_3}{n_2(n_2 + n_3)}}, \quad -\sqrt{\frac{n_1}{N(n_2 + n_3)}}\right),$$

$$c^T = \left(\sqrt{\frac{n_2}{n_3(n_2 + n_3)}}, \quad -\sqrt{\frac{n_1}{N(n_2 + n_3)}}\right)$$

From this definition it follows that

$$n_1 a + n_2 b + n_3 c = (0, 0)^T, \tag{2}$$

and

$$n_1 aa^T + n_2 bb^T + n_3 cc^T = I_2, \tag{3}$$

the 2×2 identity matrix. Now the swept area functional is

$$S_N = \frac{1}{2} \sum_{j=1}^{N} det(X_{j-1}, Y_j).$$

Note that increments to this functional have sign depending on the direction of rotation, so that S_N is a *signed swept-area* functional. Suppose that $X_{j-1} = j_1 a + j_2 b + j_3 c$. Some detailed calculations using the definitions of a, b and c now show that

$$\begin{aligned} det(X_{j-1}, Y_j) &= \frac{1}{n_1}\sqrt{\frac{n_1 n_2 n_3}{N}}\{\frac{j_3}{n_3} - \frac{j_2}{n_2}\}, \qquad \text{if} \quad Y_j = a, \\ &= \frac{1}{n_2}\sqrt{\frac{n_1 n_2 n_3}{N}}\{\frac{j_1}{n_1} - \frac{j_3}{n_3}\}, \qquad \text{if} \quad Y_j = b, \\ &= \frac{1}{n_3}\sqrt{\frac{n_1 n_2 n_3}{N}}\{\frac{j_2}{n_2} - \frac{j_1}{n_1}\}, \qquad \text{if} \quad Y_j = c. \end{aligned}$$

These terms can be compared to the jth contribution to the circulant

$$C_{123} = \frac{T_{12}}{n_1 n_2} + \frac{T_{23}}{n_2 n_3} + \frac{T_{31}}{n_3 n_1}.$$

For example, if $Y_j = a$, the contribution to C_{123} is

$$\frac{1}{n_1}(\frac{j_2}{n_2} - \frac{j_3}{n_3}).$$

The contributions if $Y_j = b$ or $Y_j = c$ are similar. Therefore, the swept area functional is

$$S_N = \frac{1}{2}\sqrt{\frac{n_1 n_2 n_3}{N}} C_{123}.$$

In Brown & Hettmansperger (2002) it was shown that $E(S_N) = 0$ and $var(S_N) = 1/12$. These values will correspond to the moments of the limit distribution to be derived in Section 3.

Note from the definitions of a^T, b^T and c^T that the vector a points up, the vector b points down to the left, and the vector c points down to the right. As the random walk steps through the sequence of As, Bs and Cs it sweeps out a signed area. Area swept in a clockwise direction is negative, and in a counter-clockwise direction positive. The net area is proportional to C_{123}. When each sample size is equal to n, C_{123} is proportional to $T = T_{12} + T_{23} + T_{31}$. Then $E(T) = 0$ and the standard deviation of T is simply n. Based on the logistic approximation in Section 3, $P(|T| > 2n)$ is approximately 0.052. This provides a quick two-sided, roughly 5% test for nontransitivity. See Section 4 for more discussion, and an illustration.

A simple example is related to the Efron dice in Fig 1 of Brown and Hettmansperger (2002): AACCBBBBAACCCCBBAA. This is a nontransitive sequence with sample sizes 6 for each sample. The link http://www.stat.psu/~tph/Efron provides a graphical visualizer for a random walk based on equal sample sizes. The above sequence results in a display with $T = -12$, implying, at the 5% level, a significant nontransitive effect. There is a readme file on the website that describes how to use the visualizer. Data can be either read into the application or copied and pasted.

Another simple example consists of the sequence AAACBBCACBC-CBABCAB, which results in $T = -2$ with $n = 6$, and is not close to significance. There is no evidence, at a reasonable level, of nontransitivity.

3. The Logistic Limit Distribution

In this Section, the following result, about the large sample null distribution of the circulant statistic, is stated and proved.

Theorem. Assume that for $i = 1, 2, 3$

$$lim_{N \to \infty} \frac{n_i}{N} = \lambda_i > 0. \tag{4}$$

Then as $N \to \infty$,

$$L_N = 2\pi S_N = \pi \sqrt{\frac{n_1 n_2 n_3}{N}} C_{123} \to^{\mathcal{D}} L,$$

where L has a standard logistic distribution, with mean 0 and variance $\pi^2/3$.

Proof. The proof takes place through a series of lemmas. The stages are

(i) show that the random walk X converges in distribution as $N \to \infty$ to a two-dimensional Brownian bridge process W with independent components, and hence that the limit distribution of the swept-area functional S_N is the distribution of S, the swept-area of W;

(ii) show that W is representable in terms of two independent processes, a *radial part* R and a *turned distance part* D, and that D is standard Brownian motion;

(iii) hence show that $S = ZV/2$, where $Z \sim N(0, 1)$ and V^2, the average radius-squared of the process W, are independent; and

(iv) then use the fact that V^2 has a *second order Cramer-von Mises distribution*, representable as a weighted sum of independent χ^2_2 rvs, to show that the product $S = ZV/2$ has a logistic distribution.

Lemma 1. As $N \to \infty$, $X_N \to^{\mathcal{D}} W$, a standard two-dimensional Brownian Bridge process.

Proof. The process W is defined in Brown (1982) by all linear combinations of components being the appropriate one-dimensional Brownian Bridge processes. The conditions (2), (3) and (4) provide convergence in distribution of all these one-dimensional linear combinations, from Billingsley(1968, page 209); see also Lemma A1 of Brown (1982).

Let $W = \{W_1(t), W_2(t), 0 \leq t \leq 1\}$. Define the *radial* and *turned distance* processes R and D by $W_1(t) = R(t)\cos(\theta_t), W_2(t) = R(t)\sin(\theta_t), R^2(t) = W_1^2(t) + W_2^2(t)$, and $dD(t) = R(t)d\theta_t$.

Lemma 2. The processes R and D are independent, and D is standard Brownian motion.

Proof. Standard calculus gives

$$dR(t) = \cos(\theta_t)dW_1(t) + \sin(\theta_t)dW_2(t)$$

and

$$dD(t) = R(t)d\theta_t = -\sin(\theta_t)dW_1(t) + \cos(\theta_t)dW_2(t).$$

These two equations show that the increments $dR(t)$ and $dD(t)$ are independent of each other. Let $\mathcal{F}_t$ be the σ-field generated by $\{W(s), 0 \leq s \leq t\}$, and write $E_t = E(.|\mathcal{F}_t)$. Each of the independent Brownian bridge processes W_1 and W_2 has exchangeable increments summing to zero, so that for $i = 1, 2$

$$E_t\{dW_i(t)\} = -\frac{W_i(t)}{(1-t)}dt.$$

Thus it follows that

$$\begin{aligned} E_t[dD(t)] &= -\sin(\theta_t)\{-\frac{W_1(t)}{(1-t)}\}dt + \cos(\theta_t)\{-\frac{W_2(t)}{(1-t)}\}dt \\ &= 0 \qquad \text{for all } t, \end{aligned}$$

and

$$E_t[dD(t)]^2 = \{\sin^2(\theta_t) + \cos^2(\theta_t)\}dt = dt.$$

Therefore $\{D(t), 0 \leq t \leq 1\}$ is standard Brownian motion.

Lemma 3. Conditional on the radial process R, S is representable as $2S/V = Z$, a standard $N(0,1)$ rv. Hence Z, V are independent, and $S = ZV/2$.

Proof. Clearly,

$$S = \frac{1}{2}\int_0^1 R^2(t)d\theta_t = \frac{1}{2}\int_0^1 R(t)dD_t.$$

Therefore, by Lemma 2, conditional on the radial process R, $S \sim N(0, \frac{1}{4}\int_0^1 R^2(t)dt)$. Write $\int_0^1 R^2(t)dt = V^2$. It then follows that $S = ZV/2$, where $Z \sim N(0,1)$ and V are independent.

Lemma 4. The distribution of V^2 is *second-order Cramer-von Mises*, ie $V^2 = U_1 + U_2$, where U_1, U_2 are independent Cramer-von Mises rvs, and

$$V^2 = \sum_{j=1}^{\infty} \frac{Y_j}{j^2\pi^2},$$

where $\{Y_j\}$ are independent χ_2^2 rvs. Hence $E(V^2) = 1/3$ and $var(V^2) = 2/45$.

Proof. This result is a special case for $r = 2$ of the Corollary to Proposition A2 of Brown (1982). Alternatively, it may be deduced directly from the representation $V^2 = \int_0^1 \{W_1^2(t) + W_2^2(t)\}dt = U_1 + U_2$. See Durbin (1973) for more background details. The expressions for mean and variance come from utilising the representation of V^2 as a weighted sum of χ_2^2 rvs.

Lemma 5. Let Q have an extreme-value distribution, ie $\Pr(Q \leq x) = exp\{-e^{-x}\}$. Then the characteristic function of Q is ϕ_Q, where $\phi_Q(t) = Ee^{itQ} = \Gamma(1 - it)$.

Proof. The result comes directly from definitions.

Lemma 6. Let Q_1, Q_2 be independent extreme-value rvs, and let $L = Q_1 - Q_2$. Then L has a standard logistic distribution, with distribution function $\Pr(L \leq x) = (1 + e^{-x})^{-1}$.

Proof. The result comes from elementary calculations.

Lemma 7. The characteristic function of a standard logistic rv L is ϕ_L, where

$$\phi_L(t) = \Gamma(1 + it)\Gamma(1 - it) = \prod_{j=1}^{\infty}(1 + \frac{t^2}{j^2})^{-1}$$

Proof. Apply Lemmas 5 and 6, and the infinite product form of the gamma function,

$$\Gamma(1 - it) = e^{i\gamma t}\prod_{j=1}^{\infty}\frac{e^{-it/j}}{(1 - it/j)},$$

where γ is Euler's constant.

Lemma 8. Let ψ be the Laplace transform of the distribution of V^2. Then

$$\psi(\theta) = \prod_{j=1}^{\infty}(1 + \frac{2\theta}{j^2\pi^2})^{-1}.$$

Proof. From Lemma 4,

$$\psi(\theta) = E[exp(-\theta V^2)] = E\prod_{j=1}^{\infty} exp(-\theta j^{-2}\pi^{-2}Y_j),$$

and the result follows after applying the form of the Laplace transform of a χ_2^2 distribution.

Completion of proof of the Theorem. From Lemma 3, the characteristic function of S is

$$\begin{aligned}\phi_S(t) &= E(e^{itZV/2}) \\ &= E\{E[exp(itZV/2)|V]\} \\ &= E\{exp(-\frac{t^2}{2}\frac{V^2}{4})\} \\ &= \psi(t^2/8) \\ &= \prod_{j=1}^{\infty}(1+\frac{t^2}{4j^2\pi^2})\end{aligned}$$

from Lemma 8. From Lemma 7, this is the characteristic function of $L/(2\pi)$. But from Lemma 1, $S_N = -\frac{1}{2}\sqrt{n_1 n_2 n_3 N^{-1}}C_{123} \to^{\mathcal{D}} S$, which $=^{\mathcal{D}} L/(2\pi)$. This gives the result.

Remark. The standard logistic rv L has variance $\pi^2/3$. Thus $L_N = 2\pi S_N$ and $L = 2\pi S$ both have variance $= \pi^2/3$. The two-sided P-value from observing $L_N = l$ is approximately

$$P(|L| \geq |l|) = \frac{2}{1+e^{-|l|}}.$$

4. Statistical Application of Circularity Results

As outlined in Section 1, the aim is to find a test for non-transitivity which is not inflated by location shifts between the three treatment distributions A, B and C; ie we seek a test statistic with null distribution under H_0, which is inflated if H_2 is true, but not inflated if H_1 is true. Using the circularity statistic C_{123} alone creates a difficulty, because the swept area S is inflated both by circularity effects, and by significant location shifts between the treatments. Simple demonstrations of these facts are available from considering orderings such as $AAAABBBBCCCC$, produced by large location shifts, or by the prototype Efron dice ordering $ABCBCACAB$.

The key to controlling the effects of location shifts is the variable V^2, the average squared radial distance of the Brownian process W from the origin. Routine calculations show that the sample version of V^2, ie based on the sample random walk X instead of W, is

$$V_N^2 = \frac{1}{N}\sum_{j=1}^{n-1}\Big(\frac{j_1^2}{n_1}+\frac{j_2^2}{n_2}+\frac{j_3^2}{n_3}-\frac{j^2}{N}\Big),$$

where after j steps, there are j_1, j_2, and j_3 occurrences of treatments A, B and C respectively. The value of V_n^2 is easily calculated from the rank ordering of A, B and C.

The null asymptotic structure of the swept-area functional S, given in Lemma 3, is $S = ZV/2$, where $Z \sim N(0,1)$ is independent of V^2, and is the total *turned-distance* of the process W. It is necessary to separate the effects on S of *large radial values* and *turning.* Both are inflated by large location shifts, which 'separate' the treatments, and make the random walk X swing around through large radial values. But if the treatment distributions are equally centered in some sense, ie there are no significant radial effects of location shifts, then V^2 will not be large. But if at the same time there is non-transitivity in the form of genuine circularity effects, the turned-distance rv Z will become inflated due to the extra turning that occurs. This is the effect which we want to detect.

Thus, an effective approach may be to adjust the location shifts until V^2 is not large. Informally, apply location shifts to A, B and C samples until V^2 is $\leq$ its expected value of $1/3$. This can be done by using sample means, or robustly, with sample medians.

Then, considering the representation $S_N = -\frac{1}{2}\sqrt{n_1 n_2 n_3 N^{-1}}C_{123} = Z_N V_N/2$, it is possible to use either S_N or Z_N as test statistic for non-transitivity. It will be conservative to use the circularity statistic $L_N = 2\pi S_N$, and refer it to a logistic null distribution, because V_N^2 has been constrained to be less than its mean value. If $Z_N = 2V_N^{-1}S_N$ is used, its value can be referred to a null standard normal distribution.

5. An Example

We consider the data from Anderson (2000) that consist of responses to a survey conducted among expatriate employees returning after a period working abroad. They were scored on their opinions concerning adequacy of preparation, training and support that they received. There are three groups of employer organisations, A: private enterprise, with $n_1 = 47$, B: government, with $n_2 = 41$, and C: religious, with $n_3 = 35$. The data was analyzed in Brown and Hettmansperger (2002). The Kruskal-Wallis test yields a value of 10.7 and a P-value = 0.0048. The follow-up analysis sugggests the transitive pattern $A < B < C$.

However, as we will see, there is also strong evidence of circularity in the data, which is obscured by the transitive pattern. First, we align the data by translating the samples so that $V_N^2 = 0.393$ (the expected value is 0.333), and the Kruskal-Wallis value is now 0.14, with P-value of 0.933. Note that $V_N^2 = 2.38$ (with expected value = 0.333) for the raw, unaligned data. Hence, the transitive, or location shift effects have been removed.

On the aligned data $C_{123} = 0.1257$ and $L_N = 9.25$, with a two-sided logistic P-value = 0.0002. The less conservative $Z_N = 4.7$, and provides a two-sided P-value that is essentially zero. Thus, there is a strong circularity or nontransitive effect in the data. Inspection shows that $A > B > C > A$. This reverses the circular order concluded in the initial analysis. The message in this example is that the true circularity cannot be measured and interpreted in the presence of strong transitive effects.

References

1. ANDERSON, B. (2000). *The scope, effectiveness and importance of Australian private, public and non-government sector expatriate management policy and practices.* Ph.D. dissertation, University of South Australia.
2. BILLINGSLEY, P. (1968). *Convergence of Probability Measures.* New York: Wiley.
3. BROWN, B.M. (1982). Cramer-von Mises distributions and permutation tests. *Biometrika* **69**, 619-624.
4. BROWN, B.M. & HETTMANSPERGER, T.P. (2002). Kruskal-Wallis, multiple comparisons and Efron dice . *ANZ J Statist* **44**, 110-114.
5. DURBIN, J. (1973). *Distribution Theory for Tests based on the Sample Distribution Function.* Philadelphia: Society for Industrial and Applied Mathematics.
6. GARDNER, M. (1970). The paradox of the nontransitive dice and the elusive principle of indifference . *Scientific American* **223**, 427-438.
7. HETTMANSPERGER, T. P. & MCKEAN, J.W. (1998). *Robust Nonparametric Statistical Methods.* London: Arnold.

A NOTE ON THE CONSISTENCY IN BAYESIAN SHAPE-RESTRICTED REGRESSION WITH RANDOM BERNSTEIN POLYNOMIALS

I-SHOU CHANG[1], CHAO A. HSIUNG[2], CHI-CHUNG WEN[2], and YUH-JENN WU[1]

[1] *President's Laboratory,* [2] *Div. of Biostatistics and Bioinformatics*
National Health Research Institutes
35 Keyan Road, Zhunan Town, Miaoli County 350, Taiwan

We describe a Bayesian framework for shape-restricted regression in which the prior is given by Bernstein polynomials. We present consistency theorems concerning the posterior distribution in this Bayesian approach. This study includes monotone regression and a few other shape-restricted regressions.

Some key words: Bayesian inference; Bernstein Polynomial; Consistency; Shape-restricted Regression.

1. Introduction

Search for a simple, smooth and efficient estimate of a monotone smooth regression function is of considerable interest. There exists a large literature in this area and an excellent review of it is given by Gijbles (2004); more recent developments include the interesting paper by Dette et al. (2005).

We proposed a nonparametric Bayesian approach to monotone regression where the prior is introduced by Bernstein polynomials and the posterior distribution is simulated by reversible jump Metropolis-Hasting algorithm (Green 1995). This prior features the properties that it has a large support, selects only smooth functions, can easily incorporate geometric information into the prior, and can be generated without computational difficulty; simulation studies indicate that the numerical performance of this approach is quite satisfactory (Chang, Chien, Hsiung, Wen and Wu 2006). The purpose of this note is to establish a consistency property of this Bayes estimator.

For integers $0 \le i \le n$, let $\varphi_{i,n}(t) = C_i^n t^i (1-t)^{n-i}$ for $t \in [0,1]$, where $C_i^n = n!/(i!(n-i)!)$. $\{\varphi_{i,n} \mid i = 0, \cdots, n\}$ is called the Bernstein basis for polynomials of order n. Let $B_n = \mathbb{R}^{n+1}$ and $\mathcal{B} = \cup_{n=1}^{\infty}(\{n\} \times B_n)$. Let π be a probability measure on $\mathcal{B}$. For $\tau > 0$, we define $\mathbf{F} : \mathcal{B} \times [0, \tau] \to \mathbb{R}^1$ by

$$\mathbf{F}(n, b_{0,n}, \cdots, b_{n,n}, t) = \sum_{i=0}^{n} b_{i,n}\varphi_{i,n}(\frac{t}{\tau}), \tag{1.1}$$

where $(n, b_{0,n}, \cdots, b_{n,n}) \in \mathcal{B}$ and $t \in [0, \tau]$. We also denote (1.1) by $F_{b_n}(t)$ if $b_n = (b_{0,n}, \cdots, b_{n,n})$. π is called a Bernstein prior, and $\mathbf{F}$ is called the random Bernstein polynomial for the prior π. It is a stochastic process on $[0, \tau]$ with smooth sample paths. Important references for Bernstein polynomials include Lorentz (1986) and Altomare and Campiti (1994), among others.

Assume that on a probability space $(\mathcal{B} \times \mathbb{R}^{\infty}, \mathcal{F}, \mathcal{P})$, there are random variables $\{Y_k \mid k = 1, \cdots, K\}$ satisfying the property that, conditional on $\mathbf{B} = (n, b_n)$,

$$Y_k = F_{b_n}(X_k) + \epsilon_k,$$

with $\{\epsilon_k \mid k = 1, \cdots, K\}$ being independent random variables, ϵ_k having known density g, $\mathbf{B}$ being the projection from $\mathcal{B} \times \mathbb{R}^{\infty}$ to $\mathcal{B}$, F_{b_n} being the function on $[0, \tau]$ associated with $(n, b_n) \in \mathcal{B}$ defined in (1.1), $X_1, \cdots, X_K$ being constant design points and $\mathcal{F}$ being the Borel σ−field on $\mathcal{B} \times \mathbb{R}^{\infty}$. We also assume the marginal distribution of $\mathcal{P}$ on $\mathcal{B}$ is the prior π. Instead of explaining the existence of the probability space $(\mathcal{B} \times \mathbb{R}^{\infty}, \mathcal{F}, \mathcal{P})$, which is straightforward, we would like to remark that the above mathematical formulation is only meant to facilitate a simple formal presentation. In fact, $\mathcal{P}$ is constructed after the prior and the likelihood are specified. A natural way to introduce the prior π is to specify $p(n) = \pi(\{n\} \times B_n)$ and the conditional density of π on $\{n\} \times B_n$, denoted by $\pi(\cdot \mid \{n\} \times B_n) = \pi_n(\cdot)$. Given $\mathbf{B} = (n, b_n)$, the likelihood for the data $\{(X_k, Y_k) \mid k = 1, \cdots, K\}$ is

$$\prod_{k=1}^{K} g(Y_k - F_{b_n}(X_k)).$$

Thus the posterior density ν of the parameter (n, b_n) given the data is proportional to

$$\prod_{k=1}^{K} g(Y_k - F_{b_n}(X_k))\pi_n(b_n)p(n). \tag{1.2}$$

We note that Chang, Chien, Hsiung, Wen and Wu (2006) considered the situation that $\pi_n(\cdot)$ is introduced in terms of the coefficients $b_n = (b_{0,n}, \cdots, b_{n,n})$. In this note, Section 2 presents a few statements regarding consistency for (1.2) and Section 3 contains the proofs.

2. Asymptotic Behavior when n Is Truncated

Here we provide two statements regarding the asymptotic behavior of the above Bayes estimate of the shape-restricted regression function. For practical reasons, the order of the Bernstein polynomial may be truncated to a maximum value n_0 so that the prior has support on the set of polynomials of order less than n_0; in addition, the value of the regression function may be assumed in a certain range. We study the asymptotic behavior of the posterior in this situation. There are a few papers on the asymptotic behavior of Bayesian estimators in the case of a possibly incorrect parametric model. See, for example, Berk (1966; 1970), Bunke and Milhaud (1998), Petrone and Wasserman (2002), and Chang, Hsiung, Wu and Yang (2005). These results provide regularity conditions under which the posterior almost surely concentrates on the subset Θ_0 of the parameter space Θ on which Kullback-Leibler divergence of the true distribution function against the distribution in the model is minimal.

2.1. *Monotone regression*

Let the true regression F be an increasing continuous function on $[0, \tau]$. Assume that $X_1, X_2, \cdots$ is an i.i.d. sequence having density h whose support is $[0, \tau]$. Let $\mathcal{K}(\cdot,\cdot)$ denote the Kullback-Leibler divergence. Then we have

Theorem 2.1. *Suppose that the prior distribution π has support $\mathcal{B}_{n_0}^M = \cup_{n=2}^{n_0}\{n\} \times B_n^M$, where $B_n^M = \{b_n = (b_{0,n}, \cdots, b_{n,n}) : -M \le b_{i,n} \le b_{i+1,n} \le M,\ i = 0, 1, \cdots, n-1\}$ for some $M \in (0, \infty)$. Assume g is the normal density function with mean 0 and known variance $\sigma^2 > 0$. Fix $\epsilon > 0$, let*

$$U_\epsilon = \{(n, b_n) \in \mathcal{B}_{n_0}^M : \mathcal{K}(h(X)g(Y - F(X)), h(X)g(Y - F_{b_n}(X)) < \eta + \epsilon\},$$

where $\eta = \inf_{(n,b_n)\in\mathcal{B}_{n_0}^M}\{\mathcal{K}(h(X)g(Y - F(X)), h(X)g(Y - F_{b_n}(X))\}$. Then $\pi(U_\epsilon \mid (X_k, Y_k),\ k = 1, 2, \cdots, K)$ converges to 1 almost surely as K goes to infinity.

The following corollary expresses the convergence in terms of the L_2-norm of the regression function, which is more relevant than the Kullback-Leibler divergence used in the theorem.

Corollary 2.1. *Let*

$$Q_\epsilon = \{(n, b_n) \in \mathcal{B}_{n_0}^M : c\int_0^\tau \mid F(x) - F_{b_n}(x) \mid^2 h(x)dx \le \eta + \epsilon\}.$$

There exists a constant $c > 0$, dependent only on M, σ and F, such that $\pi(Q_\epsilon \mid (X_k, Y_k),\ k = 1, 2, \cdots, K)$ converges to 1 almost surely as K goes to infinity.

It is easy to see that elements in $\mathcal{B}_{n_0}^M$ represent non-decreasing function on $[0, \tau]$. We can use Bernstein-Weierstrass approximation theorem to show that an increasing continuous function on $[0, \tau]$ can be approximated in uniform norm by elements in $\mathcal{B}_n^M$ if both n and M are large. This indicates that the support of the Bernstein prior can be quite large.

We note that, although we assume $\mathbf{g}$ is a known normal density, the above results can be extended to other densities or even to the case that $\mathbf{g}$ has certain parametric form with priors on the parameters.

2.2. *General shape-restricted regression*

The above method can be used to study other shape-restricted regression like concave regression, unimodal regression, etc. As an example, we present the following theorem regarding the case that the regression function F is continuously differentiable on $[0, \tau]$ and as t increases, $F(t)$ has value zero initially, starts to increase after a while, and reaches its maximum before starting to decrease eventually. This type of regression function can be used to model the time course gene expression level of viruses, because viruses do not have cells and they start to express only after they get into a cell (see Chang, Chien, Gupta, Wen, Wu, Jiang, Juang, Hsiung 2006 for details). Let $t_0 \in (0, \tau)$ and $F_{t_0, b_n}(t) = \sum_{i=0}^{n} b_{i,n}\varphi_{i,n}(\frac{t-t_0}{\tau-t_0})1_{(t_0,\tau]}(t)$. Then we have

Theorem 2.2. *Suppose that the prior distribution π has support $\mathcal{B}_{n_0}^{M_\nu} = \cup_{n=2}^{n_0}\{n\} \times B_n^{M_\nu}$, where $B_n^{M_\nu} = \{(t_0, b_n) = (t_0, b_{0,n}, \cdots, b_{n,n}) : t_0 \in (0, t_1), 0 = b_{0,n} = b_{1,n} \leq b_{2,n} \leq \cdots \leq b_{l,n}, b_{l,n} \geq b_{l+1,n} \geq \cdots \geq b_{n-1,n} > b_{n,n}$ for some $l = 2, 3, \cdots, n-1; -M \leq b_{i,n} \leq M,\ i = 0, 1, 2, \cdots, n\}$ for some $M \in (0, \infty)$ and $t_1 \in (0, \tau)$. Assume $\mathbf{g}$ is the normal density function with mean 0 and known variance $\sigma^2 > 0$. Fix $\epsilon > 0$, let*

$$U_\epsilon^\nu = \{(n, t_0, b_n) \in \mathcal{B}_{n_0}^{M_\nu} : \mathcal{K}(h(X)\mathbf{g}(Y - F(X)), h(X)\mathbf{g}(Y - F_{t_0,b_n}(X)) < \eta + \epsilon\},$$

where $\eta = \inf_{(n,t_0,b_n) \in \mathcal{B}_{n_0}^{M_\nu}}\{\mathcal{K}(h(X)\mathbf{g}(Y - F(X)), h(X)\mathbf{g}(Y - F_{t_0,b_n}(X))\}$. Then $\pi(U_\epsilon^\nu \mid (X_k, Y_k),\ k = 1, 2, \cdots, K)$ converges to 1 almost surely as K goes to infinity.

Corollary 2.2. *Let*

$$Q_\epsilon^\nu = \{(n, t_0, b_n) \in \mathcal{B}_{n_0}^{M_\nu} : c\int_0^\tau \mid F(x) - F_{t_0,b_n}(x) \mid^2 h(x)dx \leq \eta + \epsilon\}.$$

There exists a constant $c > 0$, dependent only on t_1, M, σ and F, such that $\pi(Q_\epsilon^\nu \mid (X_k, Y_k),\ k = 1, 2, \cdots, K)$ converges to 1 almost surely as K goes to infinity.

It is easy to see, by straightforward calculation, that elements in $\mathcal{B}_{n_0}^{M_\nu}$ are continuously differentiable functions on $[0, \tau]$ having value zero initially, starting to increase after a while, and reaching the maximum before starting to decrease eventually. With suitably chosen norm, we can show that a function satisfying the property in the last sentence can be approximated by elements in $\mathcal{B}_n^{M_\nu}$ if both n and M are large, which shows that the support of the prior is large enough for the purpose of inference.

3. Proofs

Since the proofs for Theorem 2 and Corollary 2 are similar to those for Theorem 1 and Corollary 1, although technically more involved, they are omitted. Since Corollary 1 follows from Theorem 1 and Proposition 1 immediately, we prove Theorem 1 and Proposition 1 only. Let (X, Y) be a random vector having the same distribution as (X_1, Y_1). Then the joint density of (X, Y) is $h(X)g(Y - F(X))$ on $[0, \tau] \times (-\infty, \infty)$. Let $\tilde{F}$ be a real-valued continuous function on $[0, \tau]$. Then we have

Proposition 3.1. *There exists $\tilde{C} > 0$, dependent only on F and $\tilde{F}$, such that*

$$\mathcal{K}(h(X)g(Y - F(X)), h(X)g(Y - \tilde{F}(X))) \geq \tilde{C} \int_0^\tau \mid F(x) - \tilde{F}(x) \mid^2 h(x)dx.$$

Proof. Observe that

$$\begin{aligned}
&\int_{-\infty}^{\infty} \mid g(y - F(x)) - g(y - \tilde{F}(x)) \mid dy \\
= &\ \frac{1}{\sqrt{2\pi}\sigma} \int_{-\infty}^{\infty} \mid e^{-\frac{1}{2\sigma^2}(y - F(x))^2} - e^{-\frac{1}{2\sigma^2}(y - \tilde{F}(x))^2} \mid dy \\
= &\ \frac{1}{\sqrt{2\pi}\sigma} \int_{-\infty}^{\infty} e^{-\frac{1}{2\sigma^2}(\xi(x,y))^2} \mid \frac{1}{2\sigma^2}(y - F(x))^2 - \frac{1}{2\sigma^2}(y - \tilde{F}(x))^2 \mid dy \\
= &\ \frac{1}{2\sqrt{2\pi}\sigma^3} \int_{-\infty}^{\infty} e^{-\frac{1}{2\sigma^2}(\xi(x,y))^2} \mid 2y(\tilde{F}(x) - F(x)) + F^2(x) - \tilde{F}^2(x) \mid dy \\
= &\ \frac{1}{2\sqrt{2\pi}\sigma^3} \mid F(x) - \tilde{F}(x) \mid \int_{-\infty}^{\infty} e^{-\frac{1}{2\sigma^2}(\xi(x,y))^2} \mid 2y - F(x) - \tilde{F}(x) \mid dy \\
\geq &\ \frac{1}{2\sqrt{2\pi}\sigma^3} \mid F(x) - \tilde{F}(x) \mid \int_{-1}^{1} e^{-\frac{1}{2\sigma^2}(\xi(x,y))^2} \mid 2y - F(x) - \tilde{F}(x) \mid dy
\end{aligned}$$

$$\geq \ C_1 \mid F(x) - \tilde{F}(x) \mid \int_{-1}^{1} \mid 2y - F(x) - \tilde{F}(x) \mid dy$$

$$\geq \ C_2 \mid F(x) - \tilde{F}(x) \mid, \tag{3.1}$$

where $\xi(x,y)$ lies between $y - F(x)$ and $y - \tilde{F}(x)$ and C_2 depends only on F, $\tilde{F}$ and σ^2. We note that the second equality in (3.1) follows from the mean-value theorem. It follows form (3.1) and the inequality of Kemperman (1969) that

$$\begin{aligned}
&\mathcal{K}(h(X)g(Y - F(X)), h(X)g(Y - \tilde{F}(X))) \\
= \ & \int_0^{\tau} \int_{-\infty}^{\infty} h(x)g(y - F(x)) \log \frac{g(y - F(x))}{g(y - \tilde{F}(x))} dydx \\
\geq \ & \frac{1}{4} \int_0^{\tau} (\int_{-\infty}^{\infty} \mid g(y - F(x)) - g(y - \tilde{F}(x)) \mid dy)^2 h(x) dx \\
\geq \ & \frac{C_2^2}{4} \int_0^{\tau} \mid F(x) - \tilde{F}(x) \mid^2 h(x) dx.
\end{aligned}$$

This completes the proof.

Proof of Theorem 1. Let P^* denote the outer measure of P, the true probability measure on the probability space $[0,\tau]^\infty \times \mathbb{R}^\infty$, determined by g, h and F. The proof of the theorem consists of establishing the following three statements, namely

(i) $\pi(U_{\frac{\epsilon}{2}}) > 0$,

(ii) The class $\mathcal{F} = \{\log \frac{g(Y - F(X))}{g(Y - F_{b_n}(X))} \mid (n, b_n) \in \mathcal{B}_{n_0}^M\}$ is a Glivenko-Cantelli class, which means $\sup_{(n,b_n) \in \mathcal{B}_{n_0}^M} \mid \frac{1}{K} \sum_{j=1}^{K} \log \frac{g(Y_j - F(X_j))}{g(Y_j - F_{b_n}(X_j))} - \mathcal{K}(h(X)g(Y - F(X)),$
$h(X)g(Y - F_{b_n}(X))) \mid$ converges to 0 almost surely relative to P^*.

(iii) $\pi(U_\epsilon \mid (X_k, Y_k),\ k = 1, 2, \cdots, K)$ converges to 1 almost surely as K goes to infinity.

We first show that (i) and (ii) imply (iii).

Let $D_K = \frac{1}{K} \sum_{k=1}^{K} \log \frac{g(Y_k - F(X_k))}{g(Y_k - F_{b_n}(X_k))}$. It follows from (ii) that for any given $0 < s < \frac{\epsilon}{6}$, there exists K_1 such that for every $K > K_1$,

$$\mathcal{K}(h(X)g(Y - F(X)), h(X)g(Y - F_{b_n}(X))) - s < D_k$$

$$< \mathcal{K}(h(X)g(Y - F(X)), h(X)g(Y - F_{b_n}(X))) + s$$

for every $(n, b_n) \in \mathcal{B}_{n_0}^M$. Thus, for $K > K_1$, $D_k < \eta + \frac{\epsilon}{2} + s$ for $(n, b_n) \in U_{\frac{\epsilon}{2}}$ and $D_k > \eta + \frac{\epsilon}{2} + 2s$ for $(n, b_n) \in E$, where $E = \{(n, b_n) \in \mathcal{B}_{n_0}^M \mid$

$\mathcal{K}(h(X)g(Y-F(X)), h(X)g(Y-F_{b_n}(X))) > \eta + \frac{\epsilon}{2} + 3s\}$. Using these, we know for $K > K_1$,

$$\begin{aligned}
& \frac{\pi(E \mid (X_k, Y_k),\ k=1,2,\cdots,K)}{\pi(E^c \mid (X_k, Y_k),\ k=1,2,\cdots,K)} \\
\leq & \frac{\pi(E \mid (X_k, Y_k),\ k=1,2,\cdots,K)}{\pi(U_{\frac{\epsilon}{2}} \mid (X_k, Y_k),\ k=1,2,\cdots,K)} \\
= & \frac{\int_E \prod_{k=1}^K h(X_k)g(Y_k - F_{b_n}(X_k))\pi(db)}{\int_{U_{\frac{\epsilon}{2}}} \prod_{k=1}^n h(X_k)g(Y_k - F_{b_n}(X_k))\pi(db)} \\
= & \frac{\int_E e^{-KD_K}\pi(db)}{\int_{U_{\frac{\epsilon}{2}}} e^{-KD_K}\pi(db)} \\
\leq & \frac{e^{-K(\eta+\frac{\epsilon}{2}+2s)}\pi(E)}{e^{-K(\eta+\frac{\epsilon}{2}+s)}\pi(U_{\frac{\epsilon}{2}})} \\
= & e^{-Ks}\frac{\pi(E)}{\pi(U_{\frac{\epsilon}{2}})}.
\end{aligned} \tag{3.2}$$

Using (i), we know (3.2) converges to 0 , as K goes to infinity. This, in turn, implies that $\pi(U_\epsilon \mid (X_k, Y_k),\ k=1,2,\cdots,K)$ also converges to 1 almost surely as K goes to infinity.

This shows that (iii) follows from (i) and (ii).

Now we prove (ii).

Using $\log \frac{g(Y-F(X))}{g(Y-F_b(X))} = \log g(Y-F(X)) - \log \frac{1}{\sqrt{2\pi}\sigma} + \frac{1}{2\sigma^2}y^2 - \frac{1}{2\sigma^2}2yF_{b_n}(X) + \frac{1}{2\sigma^2}F_{b_n}^2(X)$ and the stability property of Gliverko-Cantelli (see, for examples, Exercise 3, p.125, van der Vaart and Wellner 1996), we know it suffices to prove that every one of $\mathcal{F}_0$, $\mathcal{F}_1$ and $\mathcal{F}_2$ is Glivenko-Cantelli. Here

$$\begin{aligned}
\mathcal{F}_0 &= \{\log g(Y-F(X)) - \log \frac{1}{\sqrt{2\pi}\sigma} + \frac{1}{2\sigma^2}y^2\}, \\
\mathcal{F}_1 &= \{\frac{1}{2\sigma^2}2yF_{b_n}(X) \mid (n, b_n) \in \mathcal{B}_{n_0}^M\}, \\
\mathcal{F}_2 &= \{\frac{1}{2\sigma^2}F_{b_n}^2(X) \mid (n, b_n) \in \mathcal{B}_{n_0}^M\}.
\end{aligned}$$

We first show that they are Vapnik-Cervonenkis (henceforth V-C) classes. It is obvious that $\mathcal{F}_0$ is a V-C class, because $\mathcal{F}_0$ consists of one element. Since $\{F_{b_n} \mid (n, b_n) \in \mathcal{B}_{n_0}^M\}$ and $\{F_{b_n}^2 \mid (n, b_n) \in \mathcal{B}_{n_0}^M\}$ are subsets of polynomials having an upper bound on their orders, we know from Example 19.17 in van der Vaart (1998) that they are V-C classes.

Using this argument and the permanence propertis of V-C class (see, for example, Lemma 2.6.18, van der Vaart and Wellner 1996), we know both $\mathcal{F}_1$ and $\mathcal{F}_2$ are V-C classes. It is easy to see that $\mathcal{F}_0$,$\mathcal{F}_1$ and $\mathcal{F}_2$ have an integral envelope. These together with the fact that they are V-C classes imply that $\mathcal{F}_0$,$\mathcal{F}_1$ and $\mathcal{F}_2$ are Glivenko-Cantelli (see, for example, Theorem 2.4.3 and Theorem 2.6.7 in van der Vaart and Wellner 1996). This shows (ii).

Finally we prove (i). We can apply the dominated convergence theorem to show that $\mathcal{K}(h(X)g(Y - F(X)), h(X)g(Y - F_{b_n}(X)))$ is continuous as a function of b_n. From the definition of η, it follows that there exists $(n_1, b_{n_1}) \in \{n_1\} \times B_{n_1}^M$ such that $\mathcal{K}(h(X)g(Y - F(X)), h(X)g(Y - F_{b_{n_1}}(X))) < \eta + \frac{\epsilon}{4}$. Since $\mathcal{K}$ is continuous at b_{n_1}, there exists an open neighborhood $N_\epsilon \subset B_{n_1}^M$ of b_{n_1} such that $\mathcal{K}(h(X)g(Y - F(X)), h(X)g(Y - F_w(X))) < \eta + \frac{\epsilon}{2}$ for every w in N_ϵ. Hence $\pi(U_{\frac{\epsilon}{2}}) \geq \pi(N_\epsilon) \geq p(n_1)\int_{N_\epsilon} \pi_{n_1}(b)db > 0$. Thus, the proof is complete.

References

1. ALTOMARE, F. & CAMPITI, M. (1994). *Korovkin-type Approximation Theory and its Application.* Walter Gruyter, Berlin.
2. BERK, R. H. (1996). Limiting behavior of posterior distributions when the model is incorrect. *Ann. Math. Statist.* **37**, 51-58.
3. BERK, R. H. (1970). Consistency of a posteriori. *Ann. Math. Statist.* **41**, 894-906.
4. BUNKE, O. & MILHAUD, X. (1998). Asymptotic behavior of Bayes estimates under possibly incorrect models. *Ann. Statist.* **26**, 617-644.
5. CHANG, I. S., CHIEN, L. C., HSIUNG, C. A., WEN, C. C. & WU, Y.J. (2006). Shape-restricted regression with random Bernstein polynomials. (Preprint)
6. CHANG, I. S., HSIUNG, C. A., WU, Y. J. & YANG, C. C. (2005). Bayesian survival analysis using Bernstein polynomials. *Scand. J. of Statist.* **32**, 447-466.
7. CHANG, I. S., CHIEN, L. C., GUPTA, PRAMOD K., WEN, C. C., WU, Y. J., JIANG, S. S., JUANG, J. L. & HSIUNG, C. A. (2006). A Bayes approach to virus-gene time-course expression data. (In preparation)
8. DETTE, H., NEUMEYER, N. & PILZ, K. F. (2005). A simple nonparametric estimator of a monotone regression function. (To appear in Bernoulli)
9. GIJBLES, I. (2004). Monotone regression. Discussion paper 0334, Institute de Statistique, Université Catholique de Louvain. http://www.stat.ucl.ac.be
10. GREEN, P. J. (1995). Reversible jump Markov chain Monte Carlo computation and Bayesian model determination. *Biometrika* **82**, 711-732.
11. KEMPERMAN, J. H. B. (1969). On the optimum rate of transmitting information. *Ann. Math. Statist.* **40**, 2056-2177.

12. LORENTZ, G. G. (1986). *Bernstein Polynomials.* Chelsea, New York.
13. PETRONE, S. & WASSERMAN, L. (2002). Consistency of Bernstein polynomial posteriors. *J. Roy. Statist. Soc. Ser. B* **64**, 79-100.
14. VAN DER VAART, A. W. & WELLNER, J. A. (1996). *Weak Convergence and Empirical Processes.* Springer, New York.
15. VAN DER VAART, A. W. (1998). *Asymptotic Statistics.* Cambridge Univ. Press, Cambridge.

Corresponding Author:
I-Shou Chang
Institute of Cancer Research
National Health Research Institutes
35 Keyan Road, Zhunan
Miaoli 350
Taiwan
E-mail: ischang@nhri.org.tw
Fax: 886-37-586-410

A STABILIZED UNIFORM Q-Q PLOT TO DETECT NON-MULTINORMALITY

KAI-TAI FANG

Department of Mathematics, Hong Kong Baptist University
Kowloon Tong, Hong Kong, China

JIAJUAN LIANG

University of New Haven, School of Business
300 Boston Post Road, West Haven, CT 06516, U.S.A.

FRED J. HICKERNELL

Department of Applied Mathematics, Illinois Institute of Technology
Chicago, IL 60616, U.S.A.

RUNZE LI

Department of Statistics and The Methodology Center
The Pennsylvania State University, University Park, PA 16802, U.S.A.

By using the theory of spherical distributions and some properties of invariant statistics, we develop a stabilized uniform Q-Q plot for checking the multinormality assumption in high-dimensional data analysis. Acceptance regions associated with the plot are given. Empirical performance of the acceptance regions is studied by Monte Carlo simulation. Application of the Q-Q plot is illustrated by a real data set.

Some key words: Multinormality; Q-Q plot; Spherical distribution.

1. Introduction

Methodologies for testing multivariate normality or multinormality have been the continuous interest to statisticians because of the imposed normal assumption in many statistical models for high-dimensional data analysis. The properties and empirical performance of many existing statistics for testing multinormality have been studied or reviewed since the past two decades (see, e.g., Mardia, 1980; Horswell & Looney, 1992; Romeu & Ozturk, 1993; Henze, 2002; Srivastava & Muholkar, 2003; and Mecklin

& Mundfrom, 2004). Compared to analytical methods, graphical methods for finding evidence of non-multinormality are appreciated by practising statisticians who are less sophisticated in statistical theory. For inexperienced data analysts, graphical methods can help them accept the analytical results more readily after seeing the plots (Cleveland, 1993). Among the existing graphical methods for detecting non-normality, the quantile-quantile (Q-Q) plot is one of the convenient and effective plotting techniques in the univariate case (see, e.g., Brown & Hettmansperger, 1996).

To detect non-multinormality by the Q-Q plot technique, it is usually necessary to find a transformation to project high-dimensional data into one-dimensional ones (see, e.g., Healy, 1968; Small, 1978; Ahn, 1992; Koziol, 1993; Li, Fang & Zhu, 1997; Liang & Bentler 1999; and Liang, Pan & Yang, 2004). Easton & McCulloch (1990) generalized the Q-Q plot technique to the multivariate case, but one may meet some difficulty in finding "matching samples" when implementing the multivariate Q-Q plot in high-dimensional case.

In this paper we will develop a stabilized uniform Q-Q plot to detect non-multinormality by using the theory of spherical distributions and the property of the univariate stabilized probability plot (Michael, 1983). The theoretical derivation of the Q-Q plot is given in Section 2. Section 3 presents the power performance and application of the Q-Q plot associated with the acceptance regions. Some concluding remarks are given in the last section.

2. Theoretical Development

2.1. *Preliminaries and a basic theorem*

Let $\boldsymbol{x}_1, \ldots, \boldsymbol{x}_n$ be an i.i.d. (independently identically distributed) sample with a distribution function (d.f.) $F(\boldsymbol{x})$ ($\boldsymbol{x} \in R^p$). A general framework for testing multinormality of $F(\boldsymbol{x})$ can be stated as follows:

$$H_0: \ F(\boldsymbol{x}) \text{ is the d.f. of a multinormal distribution } N_p(\boldsymbol{\mu}, \boldsymbol{\Sigma}), \qquad (2.1)$$

versus H_1: $F(\boldsymbol{x})$ is not a normal d.f., where $\boldsymbol{\mu} \in R^p$ and $\boldsymbol{\Sigma} > \mathbf{0}$ (positive definite) are unknown parameters. Denote by $\boldsymbol{X}_{n\times p} = (\boldsymbol{x}_1, \cdots, \boldsymbol{x}_n)'$ the observation matrix. Using the vector space approach in Eaton (1986), we have $\boldsymbol{X} \sim N_{n\times p}(\mathbf{1}_n\boldsymbol{\mu}', \boldsymbol{I}_n \otimes \boldsymbol{\Sigma})$ under the null hypothesis (2.1), where $\mathbf{1}_n$ is a column vector of ones, and "$\otimes$" stands for the Kronecker product of matrices. In developing the Q-Q plot technique, we will project an observation matrix $\boldsymbol{X}_{n\times p}$ into a random vector $\boldsymbol{z}_{(n-1)\times 1}$ which acts as the role of a one-dimensional observation vector. Then the Q-Q plot technique will be based on a univariate statistics $t(\boldsymbol{z})$. Under the null hypothesis (2.1), the projection of $\boldsymbol{X}$ leads to $\boldsymbol{z}$ having a spherical distribution. A random vector $\boldsymbol{z}_{(n-1)\times 1}$ is said to have a spherical distribution if for any constant orthogonal matrix $\boldsymbol{\Gamma}_{(n-1)\times(n-1)}$, $\boldsymbol{\Gamma z} \stackrel{d}{=} \boldsymbol{z}$, where the sign "$\stackrel{d}{=}$" means that the two sides of the equality have the same distribution. Using the notation in Fang, Kotz & Ng (1990), we denote by $\boldsymbol{z} \sim S_{n-1}^{+}(\phi)$ if $\boldsymbol{z}$ has an $(n-1)$-dimensional spherical distribution and $P(\boldsymbol{z} = \boldsymbol{0}) = 0$, where $\phi(\cdot)$ stands for the characteristic function (c.f.) of a spherical distribution. It is a scale function with the form $\phi(\boldsymbol{t}'\boldsymbol{t}) = \phi(\|\boldsymbol{t}\|^2)$ ($\boldsymbol{t} \in R^{n-1}$), that is, $\phi(\boldsymbol{t})$ depends on $\boldsymbol{t} \in R^{n-1}$ only through its Euclidean norm $\|\boldsymbol{t}\|$. The following lemmas are useful for our basic theorem.

Lemma 2.1. (*Theorem 2.22* of Fang, Kotz & Ng, 1990, p. 51) Let $\boldsymbol{z}$ be a random vector in R^{n-1}. Assume that $t(\boldsymbol{z})$ is a statistic based on $\boldsymbol{z}$. If $\boldsymbol{z} \sim S_{n-1}^{+}(\phi)$ and

$$t(a\boldsymbol{z}) \stackrel{d}{=} t(\boldsymbol{z}), \quad \text{for any constant } a > 0, \tag{2.2}$$

then $t(\boldsymbol{z}) \stackrel{d}{=} t(\boldsymbol{z}_0)$, where $\boldsymbol{z}_0 \sim N_{n-1}(\boldsymbol{0}, \boldsymbol{I}_{n-1})$.

Lemma 2.2. Let $\boldsymbol{Y}_{(n-1)\times p} \sim N_{(n-1)\times p}(\boldsymbol{0}, \boldsymbol{I}_{n-1}\otimes\boldsymbol{\Sigma})$ ($\boldsymbol{\Sigma} > \boldsymbol{0}$) and random vector $\boldsymbol{d} = f(\boldsymbol{Y}'\boldsymbol{Y})$ ($p \times 1$) be a function of $\boldsymbol{Y}'\boldsymbol{Y}$. Define a random vector $\boldsymbol{z}_{(n-1)\times 1}$ by $\boldsymbol{z} = \boldsymbol{Yd}$. Then $\boldsymbol{z} \sim S_{n-1}^{+}(\phi)$ for some scale function $\phi(\cdot)$.

If a statistic $t(\boldsymbol{z})$ satisfies (2.2), we will call it an invariant statistic hereafter. Lemma 2.2 is due to the theory of spherical matrix distribution (Fang & Zhang, 1990; and Läuter, 1996). The random vector $\boldsymbol{d}$ in Lemma 2.2 acts as a projection direction. Läuter, Glimm & Kropf (1996) applied Lemma 2.2 to some invariant statistics for testing the mean value of a multivariate normal population. They recommended several choices of $\boldsymbol{d}$ for some practical problems.

Theorem 2.1. Let $\boldsymbol{X} = (\boldsymbol{x}_1, \ldots, \boldsymbol{x}_n)' \sim N_{n\times p}(\boldsymbol{1}_n\boldsymbol{\mu}', \boldsymbol{I}_n\otimes\boldsymbol{\Sigma})$ ($\boldsymbol{\Sigma} > \boldsymbol{0}$) and $\boldsymbol{Y} = (\boldsymbol{y}_1, \ldots, \boldsymbol{y}_{n-1})'$ with $\boldsymbol{y}_i$'s ($i = 1, \ldots, n-1,\ n \geq p+2$) given by

$$\boldsymbol{y}_i = (\boldsymbol{x}_1 + \cdots + \boldsymbol{x}_i - i\boldsymbol{x}_{i+1})/\sqrt{i(i+1)}. \tag{2.3}$$

Assume that random vector $\boldsymbol{d} = f(\boldsymbol{Y}'\boldsymbol{Y})$ ($p \times 1$), which is a function of $\boldsymbol{Y}'\boldsymbol{Y}$ and uniquely determined by $\boldsymbol{Y}'\boldsymbol{Y}$. Let

$$\boldsymbol{z} = \boldsymbol{Yd}. \tag{2.4}$$

Then

(1) $\boldsymbol{Y} \sim N_{(n-1)\times p}(\boldsymbol{0}, \boldsymbol{I}_{n-1}\otimes\boldsymbol{\Sigma})$ and $\boldsymbol{z} \sim S_{n-1}^{+}(\phi)$ for some scale function $\phi(\cdot)$;

(2) If a statistic $t(\boldsymbol{z})$ is invariant in the sense of (2.2), then the distribution of $t(\boldsymbol{z})$ remains unchanged under the linear transformation

$$\boldsymbol{x}_i \rightarrow \boldsymbol{B}\boldsymbol{x}_i + \boldsymbol{b}, \quad i = 1, \ldots, n, \tag{2.5}$$

for each nonsingular constant matrix $\boldsymbol{B}_{p\times p}$ and a constant vector $\boldsymbol{b} \in R^p$, in particular, $t(\boldsymbol{z}) \stackrel{\mathrm{d}}{=} t(\boldsymbol{z}_0)$, where $\boldsymbol{z}_0 \sim N_{n-1}(\boldsymbol{0}, \boldsymbol{I}_{n-1})$.

Proof: (1) Let $\boldsymbol{A}_{(n-1)\times n} = \left(a_{ij}\right)$ be defined by

$$a_{ij} = \begin{cases} \frac{1}{\sqrt{i(i+1)}}, & j = 1, \ldots, i, \\ \frac{-i}{\sqrt{i(i+1)}}, & j = i+1, \\ 0, & \text{otherwise}, \end{cases} \tag{2.6}$$

for $i = 1, \ldots, n-1$. Then $\boldsymbol{A}$ is a row-orthogonal matrix, i.e., $\boldsymbol{A}\boldsymbol{A}' = \boldsymbol{I}_{n-1}$ and the matrix $\boldsymbol{Y}$ can be written as

$$\boldsymbol{Y} = \boldsymbol{A}\boldsymbol{X}. \tag{2.7}$$

It is easy to verify:

$$\boldsymbol{X} \sim N_{n\times p}(\boldsymbol{1}_n\boldsymbol{\mu}', \boldsymbol{I}_n \otimes \boldsymbol{\Sigma}) \Rightarrow \boldsymbol{Y} \sim N_{(n-1)\times p}(\boldsymbol{0}, \boldsymbol{I}_{n-1} \otimes \boldsymbol{\Sigma}). \tag{2.8}$$

Then assertion (1) holds by Lemma 2.2.

(2) Under the linear transformation (2.5), denoting by $\boldsymbol{U} = \boldsymbol{X}\boldsymbol{B}' + \boldsymbol{1}_n\boldsymbol{b}'$, we have the following assertions:

$$\begin{aligned} &\boldsymbol{X} \sim N_{n\times p}(\boldsymbol{1}_n\boldsymbol{\mu}', \boldsymbol{I}_n \otimes \boldsymbol{\Sigma}), \\ \Rightarrow &\boldsymbol{U} \sim N_{n\times p}(\boldsymbol{1}_n\boldsymbol{\mu}'\boldsymbol{B}' + \boldsymbol{1}_n\boldsymbol{b}', \boldsymbol{I}_n \otimes \boldsymbol{B}\boldsymbol{\Sigma}\boldsymbol{B}'), \\ \Rightarrow &\boldsymbol{A}\boldsymbol{U} \sim N_{(n-1)\times p}(\boldsymbol{0}, \boldsymbol{I}_{n-1} \otimes \boldsymbol{B}\boldsymbol{\Sigma}\boldsymbol{B}'). \end{aligned} \tag{2.9}$$

Under transformation (2.5), choose a projection direction $\boldsymbol{d}_a = g(\boldsymbol{U}'\boldsymbol{A}'\boldsymbol{A}\boldsymbol{U})$, which is a function of $\boldsymbol{U}'\boldsymbol{A}'\boldsymbol{A}\boldsymbol{U}$ and is uniquely determined by $\boldsymbol{U}'\boldsymbol{A}'\boldsymbol{A}\boldsymbol{U}$. Similarly, the random vector $\boldsymbol{z}$ given by (2.4) is transformed into $\boldsymbol{z}_a = \boldsymbol{A}\boldsymbol{U}\boldsymbol{d}_a$, which still has a spherical distribution $S_{n-1}^+(\phi)$ by Lemma 2.2. Then $t(\boldsymbol{z}_a) \stackrel{\mathrm{d}}{=} t(\boldsymbol{z}) \stackrel{\mathrm{d}}{=} t(\boldsymbol{z}_0)$ with $\boldsymbol{z}_0 \sim N_{n-1}(\boldsymbol{0}, \boldsymbol{I}_{n-1})$ by Lemma 2.1. This completes the proof.

Note: Equation (2.3) in Theorem 2.1 establishes a transformation from an i.i.d. sample $\{\boldsymbol{x}_1, \ldots, \boldsymbol{x}_n\} \sim N_p(\boldsymbol{\mu}, \boldsymbol{\Sigma})$ to another i.i.d. sample $\{\boldsymbol{y}_1, \ldots, \boldsymbol{y}_{n-1}\} \sim N_p(\boldsymbol{0}, \boldsymbol{\Sigma})$ with a zero mean. As a result, one sample size is lost by avoiding the unknown mean vector. The value of the new sample $\{\boldsymbol{y}_1, \cdots, \boldsymbol{y}_{n-1}\}$ may depend on the ordering of the original sample $\{\boldsymbol{x}_1, \ldots, \boldsymbol{x}_n\}$, but the distribution of the random vector $\boldsymbol{z}$ resulted from equation (2.4) does not depend on the ordering. As a result, any test statistic based on $t(\boldsymbol{z})$ satisfying the condition (2.2) in Lemma 2.1 will have the same distribution as $t(\boldsymbol{z}_0)$ with $\boldsymbol{z}_0 \sim N_{n-1}(\mathbf{0}, \boldsymbol{I}_{n-1})$ if the original sample $\{\boldsymbol{x}_1, \ldots, \boldsymbol{x}_n\}$ is from a normal population $N_p(\boldsymbol{\mu}, \boldsymbol{\Sigma})$.

In Theorem 2.1, the conclusion

$$\boldsymbol{X} \sim N_{n\times p}(\mathbf{1}_n\boldsymbol{\mu}', \boldsymbol{I}_n \otimes \boldsymbol{\Sigma}) \Rightarrow \boldsymbol{Y}_{(n-1)\times p}(\mathbf{0}, \boldsymbol{I}_{n-1} \otimes \boldsymbol{\Sigma}) \Rightarrow \boldsymbol{z} = \boldsymbol{Y}\boldsymbol{d} \sim S^+_{n-1}(\phi) \tag{2.10}$$

(for some c.f. ϕ of the form $\phi(\boldsymbol{t}'\boldsymbol{t})$, $\boldsymbol{t} \in R^{n-1}$) provides the basis for transforming a test for hypothesis H_0 in (2.1) into a test for the following hypothesis

$$H_{0d}: \ \boldsymbol{z} \text{ has a spherical distribution,} \tag{2.11}$$

versus H_{1d}: $\boldsymbol{z}$ is not spherically distributed, where $\boldsymbol{z} = \boldsymbol{Y}\boldsymbol{d}$ is computed by (2.10) for any given direction $\boldsymbol{d}$ satisfying the condition in Theorem 2.1. We will call a test based on an invariant statistic $t(\boldsymbol{z})$ for (2.11) a necessary test for hypothesis (2.1). It is obvious that rejection of (2.11) implies rejection of (2.1), but acceptance of (2.11) generally does not lead to acceptance of (2.1).

The determination of the projection direction $\boldsymbol{d}$ in transformation (2.10) plays an important role in obtaining desirable power performance. It may be infeasible to search for a universally optimal projection direction. Instead, we can follow the idea in principal component analysis and the discussion on some feasible choices of $\boldsymbol{d}$ in Läuter, Glimm & Kropf (1996). Let $\boldsymbol{d}$ be obtained by one of the following ways:

(1) Solutions to the eigenvalue problem

$$\boldsymbol{Y}'\boldsymbol{Y}\boldsymbol{D} = \boldsymbol{D}\boldsymbol{\Lambda}, \tag{2.12}$$

where $\boldsymbol{D}_{p\times p} = [\boldsymbol{d}_1, \ldots, \boldsymbol{d}_p]$ consists of the eigenvectors corresponding to the eigenvalues $\lambda_1 \geq \ldots \geq \lambda_p > 0$, and $\boldsymbol{\Lambda} = \text{diag}(\lambda_1, \ldots, \lambda_p)$ (a diagonal matrix with the specified numbers as its diagonal elements). To ensure the unique solution, the random matrix $\boldsymbol{D}$ is assumed to have positive diagonal elements. The three directions $\boldsymbol{d}_1$, $\boldsymbol{d}_m$, $\boldsymbol{d}_p$ will be chosen for a Monte Carlo study, where $m = [p/2]$, the integer part of $p/2$.

(2) Let $\boldsymbol{d} = \boldsymbol{d}_{ss} = [\text{diag}(\boldsymbol{Y}'\boldsymbol{Y})]^{-1/2}\mathbf{1}_p$, where $\text{diag}(\boldsymbol{Y}'\boldsymbol{Y})$ denotes the diagonal matrix with the same diagonal elements as those of $\boldsymbol{Y}'\boldsymbol{Y}$. A statistic $t(\boldsymbol{z})$ based on $\boldsymbol{d}_{ss}$ is called the SS test (standardized sum test) by Läuter, Glimm & Kropf (1996). $\boldsymbol{z} = \boldsymbol{Y}\boldsymbol{d}_{ss}$ does not depend on the measurement scale of the original sample.

(3) Solutions to the eigenvalue problem

$$\boldsymbol{Y}'\boldsymbol{Y}\,\widetilde{\boldsymbol{D}} = \text{diag}(\boldsymbol{Y}'\boldsymbol{Y})\,\widetilde{\boldsymbol{D}}\,\boldsymbol{\Lambda}, \quad \widetilde{\boldsymbol{D}}'\,\text{diag}(\boldsymbol{Y}'\boldsymbol{Y})\,\widetilde{\boldsymbol{D}} = \boldsymbol{I}_p, \tag{2.13}$$

where $\widetilde{\boldsymbol{D}}= [\widetilde{\boldsymbol{d}}_1, \ldots, \widetilde{\boldsymbol{d}}_p]$ consists of the eigenvectors corresponding to the eigenvalues $\widetilde{\lambda}_1 \geq \ldots \geq \widetilde{\lambda}_p > 0$, and $\boldsymbol{\Lambda} = \mathrm{diag}(\widetilde{\lambda}_1, \ldots, \widetilde{\lambda}_p)$. Similar conditions to those on the $\boldsymbol{d}_i$'s determined by (2.12) are imposed on the $\widetilde{\boldsymbol{d}}_i$'s to ensure the uniqueness. A statistic $t(\boldsymbol{z})$ based on $\widetilde{\boldsymbol{d}}_1$ is called the PC test by Läuter, Glimm & Kropf (1996). The three directions $\widetilde{\boldsymbol{d}}_1$, $\widetilde{\boldsymbol{d}}_m$ ($m = [p/2]$) and $\widetilde{\boldsymbol{d}}_p$ will be chosen for a Monte Carlo study. For convenience, we list the above-mentioned projection directions for a Monte Carlo study as follows:

$$\boldsymbol{d}_1, \quad \boldsymbol{d}_m, \quad \boldsymbol{d}_p, \quad \boldsymbol{d}_{ss}, \quad \widetilde{\boldsymbol{d}}_1, \quad \widetilde{\boldsymbol{d}}_m, \quad \widetilde{\boldsymbol{d}}_p \,. \tag{2.14}$$

These choices of $\boldsymbol{d}$ in transformation (2.10) will be used for the Monte Carlo study in Section 3.

2.2. *The stabilized uniform Q-Q plot*

The stabilized uniform Q-Q plot is based on the following corollary.

Corollary 2.1. Assume that $\boldsymbol{X} \sim N_{n\times p}(\mathbf{1}_n\boldsymbol{\mu}', \boldsymbol{I}_n \otimes \boldsymbol{\Sigma})$ and $\boldsymbol{z} = (z_1, \cdots, z_{n-1})'$ is obtained by transformation (2.10). Let $k_0 = [(n-1)/2]-1$ and

$$\begin{gathered} v_1 = z_1^2 + z_2^2, \quad v_2 = z_3^2 + z_4^2, \quad \ldots, \quad v_{k_0+1} = z_{2k_0+1}^2 + z_{2k_0+2}^2, \\ s = \sum_{i=1}^{k_0+1} v_i \quad \text{and} \quad u_r = (\sum_{i=1}^{r} v_i)/s, \quad r = 1, \ldots, k_0. \end{gathered} \tag{2.15}$$

Then u_r has a beta distribution $\mathrm{Be}(r, k_0 - r + 1)$.

Proof. Note that $\boldsymbol{z} \sim S_{n-1}^{+}(\phi)$ by Theorem 2.1. It is obvious that the u_r's given by (2.15) are invariant statistics in the sense of (2.2). If we denote the u_r's by $u_r(\boldsymbol{z})$'s, then $u_r(\boldsymbol{z}) \stackrel{\mathrm{d}}{=} u_r(\boldsymbol{z}_0)$ by Lemma 2.1, where $\boldsymbol{z}_0 \sim N_{n-1}(\mathbf{0}, \boldsymbol{I}_{n-1})$. The assertion of Corollary 2.1 holds by the definition of the beta distribution.

Let $U_1, \ldots, U_{k_0}$ ($k_0 = [(n-1)/2] - 1$) be i.i.d. random variables with a uniform distribution $U(0,1)$ and $U_{(1)} \leq \cdots \leq U_{(k_0)}$ the associated order statistics. It is easy to verify that $U_{(r)} \sim \mathrm{Be}(r, k_0 - r + 1)$ (Csörgo, Seshadri, & Yalovsky, 1973). Therefore, under the null hypothesis (2.1), we have

$$u_r \stackrel{\mathrm{d}}{=} U_{(r)}, \qquad r = 1, \ldots, k_0, \tag{2.16}$$

where u_r's are given by (2.15). This implies that the u_r's act as the role of k_0 order statistics obtained from k_0 i.i.d. $U(0,1)$-variates. The $[(2r-1)/(2k_0)]$-quantiles ($r = 1, \ldots, k_0$) of $U(0,1)$ can be used as estimates for the u_r's given by (2.15) (Ahn, 1992). When given an observation matrix $\boldsymbol{X}_{n\times p}$, we perform transformation (2.10) to get $\boldsymbol{z}$ and obtain the u_r's by (2.15). A Q-Q plot for detecting non-spherical distribution of the $\boldsymbol{z}$-vector can be summarized as: plot the u_r calculated by (2.15) versus the $U(0,1)$-quantiles $\{(2r-1)/(2k_0): r=1,\dots,k_0\}$ in the unit square

$$S_{uq} = \{(x,y): 0 \le x \le 1, \ 0 \le y \le 1\}. \tag{2.17}$$

If the plot has a "significant" departure from the equiangular line $y=x$ in the square (2.17), the distribution of $\boldsymbol{z}$ shows evidence of non-spherical symmetry. That is, hypothesis (2.11) is rejected and consequently, $\boldsymbol{X}$ shows evidence of non-multinormality. We will call this Q-Q plot the *uniform Q-Q plot* in the subsequent context.

In order to measure the "significant" departure from the line $y=x$ for the uniform Q-Q plot, we need to define a statistic for testing the deviation resulting from using $(2r-1)/(2k_0)$ to estimate u_r $(r=1,\dots,k_0)$. The idea of stabilized probability plot proposed by Michael (1983) can be applied to the uniform Q-Q plot. Let

$$S = \frac{2}{\pi}\arcsin(U^{1/2}), \tag{2.18}$$

where $U \sim U(0,1)$. Then S has a density function $(\pi/2)\sin(\pi s)$ for $0 \le s \le 1$. If $S_1 \le \cdots \le S_n$ is an ordered sample from the distribution given by (2.18), then as $n \to \infty$ and $i/n \to a$ (a is a constant, $1 \le i \le n$), the asymptotic variance of nS_i is $1/\pi^2$, which is independent of the constant a. Now we let

$$S_{(r)} = \frac{2}{\pi}\arcsin(u_r^{1/2}), \quad r=1,\dots,k_0, \tag{2.19}$$

where the u_r's are given by (2.15). Then the $S_{(r)}$'s act as the role of k_0 ($k_0 = [(n-1)/2]-1$) ordered random sample from the distribution (2.18). Let

$$b_r = \frac{2}{\pi}\arcsin\{((2r-1)/(2k_0))^{1/2}\}, \quad r=1,\dots,k_0. \tag{2.20}$$

Then b_r can be used as an estimate of $S_{(r)}$ given by (2.19). The uniform Q-Q plot can be substituted by plotting the $S_{(r)}$'s versus the b_r's in the unit square (2.17). If the plot has a "significant" departure from the equiangular line $y=x$ in the square (2.17), the original observation matrix $\boldsymbol{X}$ shows evidence of non-multinormality. We will call this Q-Q plot the *stabilized uniform Q-Q plot* and denote it by SUQ-plot for convenience. Michael (1983) proposed the following Kolmogorov-Smirnov (K-S) type statistic to test goodness-of-fit of the SUQ-plot:

$$D_{sp} = \max_{1 \le r \le k_0} |S_{(r)} - b_r|. \tag{2.21}$$

Large values of D_{sp} imply a "significant" departure from the line $y = x$ for the SUQ-plot. Percentiles for D_{sp} ($k_0 \leq 100$) are listed in Table 2 of Michael (1983). Based on the statistic D_{sp}, we perform Monte Carlo simulation on the type I error rates for the SUQ-plot when the null hypothesis (2.1) is true. Since the null distribution of the $\boldsymbol{z}$-vector calculated from (2.10) is independent of the unknown parameters $\boldsymbol{\mu}$ and $\boldsymbol{\Sigma}$ in the normal distribution $N_p(\boldsymbol{\mu}, \boldsymbol{\Sigma})$, the null distribution of D_{sp} is also independent of $\boldsymbol{\mu}$ and $\boldsymbol{\Sigma}$. In the simulation, we let $\boldsymbol{\mu} = \mathbf{0}$ and $\boldsymbol{\Sigma} = I_n$. Table 1 presents the empirical type I error rates by using the statistic D_{sp}, where the projection directions are given by (2.14). The results in Table 1 imply that the statistic D_{sp} has acceptable control of the type I error rates.

Table 1. Simulated type I error rates for the SUQ-plot for the nominal levels $\alpha = 1\%$, 5% and 10%

Sample Dimension $p = 5$								
		Projection Directions						
level	n	$\boldsymbol{d}_1$	$\boldsymbol{d}_m$	$\boldsymbol{d}_p$	$\boldsymbol{d}_{ss}$	$\tilde{\boldsymbol{d}}_1$	$\tilde{\boldsymbol{d}}_m$	$\tilde{\boldsymbol{d}}_p$
$\alpha = 1\%$	51	0.0095	0.0110	0.0085	0.0165	0.0070	0.0080	0.0070
	103	0.0120	0.0125	0.0090	0.0105	0.0105	0.0050	0.0085
	203	0.0120	0.0120	0.0115	0.0080	0.0140	0.0060	0.0125
$\alpha = 5\%$	51	0.0455	0.0500	0.0430	0.0595	0.0385	0.0550	0.0475
	103	0.0475	0.0460	0.0435	0.0505	0.0485	0.0495	0.0515
	203	0.0530	0.0465	0.0615	0.0495	0.0565	0.0485	0.0585
$\alpha = 10\%$	51	0.0990	0.1070	0.1070	0.1085	0.0945	0.1095	0.1080
	103	0.0940	0.0900	0.0880	0.0975	0.1010	0.1000	0.0975
	203	0.0965	0.0915	0.1095	0.0930	0.0975	0.1030	0.1105
Sample Dimension $p = 10$								
		Projection Directions						
level	n	$\boldsymbol{d}_1$	$\boldsymbol{d}_m$	$\boldsymbol{d}_p$	$\boldsymbol{d}_{ss}$	$\tilde{\boldsymbol{d}}_1$	$\tilde{\boldsymbol{d}}_m$	$\tilde{\boldsymbol{d}}_p$
$\alpha = 1\%$	51	0.0110	0.0120	0.0130	0.0065	0.0060	0.0080	0.0110
	103	0.0105	0.0095	0.0125	0.0125	0.0100	0.0090	0.0115
	203	0.0090	0.0115	0.0125	0.0100	0.0125	0.0095	0.0105
$\alpha = 5\%$	51	0.0510	0.0575	0.0420	0.0420	0.0495	0.0430	0.0475
	103	0.0575	0.0510	0.0515	0.0455	0.0480	0.0470	0.0555
	203	0.0590	0.0565	0.0580	0.0470	0.0545	0.0525	0.0545
$\alpha = 10\%$	51	0.1035	0.1175	0.0995	0.0965	0.1020	0.0970	0.0900
	103	0.0920	0.0985	0.1045	0.0885	0.0975	0.0965	0.1020
	203	0.1115	0.1060	0.1115	0.0955	0.0975	0.0930	0.1090

Based on the K-S type statistic D_{sp} in (2.21), we propose the acceptance regions for the SUQ-plot:

$$b_r \pm d_{sp}, \quad r = 1, \ldots, k_0, \tag{2.22}$$

where the b_r's are given by (2.20) and d_{sp} is the $(1-\alpha)$-percentile of D_{sp}, which is given by Table 2 of Michael (1983).

3. Power Study and Application

3.1. *Power study*

In this subsection we will study the power performance of the SUQ-plot by Monte Carlo simulation. In order to see the efficiency of the SUQ-plot for detecting non-multinormality of "minor" non-normal and "severe" non-normal distributions, two groups of alternative distributions are selected for generating empirical samples. The first group of alternatives consists of five elliptical distributions. By the notation in Chapter 3 of Fang, Kotz & Ng (1990), these distributions are: 1) multivariate t-distribution (Mul-t) with $m = 5$; 2) Kotz type distribution with $N = 2$, $s = 1$ and $r = 0.5$; 3) Pearson type VII distribution with $m = 2$ and $N = 10$; 4) Pearson type II distribution with $m = 3/2$ and 5) multivariate Cauchy distribution. The TFWW algorithm (Tashiro, 1977; and Fang & Wang 1994, pp. 166–170) is employed to generate the empirical samples from these distributions. The power of the SUQ-plot is calculated by using the acceptance regions (2.22). Since the null distribution (i.e., H_0 in (2.1) is true) of the K-S type statistic D_{sp} in (2.21) is location and scale invariant (its null distribution does not depend on the mean vector and the covariance matrix of the null distribution), without loss of generality, we can choose the five elliptical distributions with location parameter $\boldsymbol{\mu} = \mathbf{0}$ and the scale parameter $\boldsymbol{\Sigma} = \boldsymbol{I}_p$.

In the second group of alternatives the following four multivariate non-normal distributions are considered:

(1) the multivariate double-exponential distribution comprises of i.i.d. univariate double-exponential (D-Exp) distribution, each of which has a density $f(x) = \exp(-|x|)/2$, $x \in R$;

(2) the multivariate exponential (Exp) distribution comprises of i.i.d. univariate exponential distribution with a density function $\exp(-x)$, $x > 0$;

(3) the multivariate chi-squared distribution comprises of i.i.d. univariate χ_2^2;

(4) N+χ^2: the multivariate distribution with i.i.d. marginals, $[p/2]$ (the integer part of $p/2$) marginals have a normal distribution and $p - [p/2]$ marginals have a chi-square distribution χ_2^2.

In generating the non-elliptical samples, a p-dimensional empirical observation matrix $X_{n\times p} = (\boldsymbol{x}_1, \ldots, \boldsymbol{x}_n)'$ is generated from each of the second

Table 2. Power comparisons of the SUQ-plot by using different directions as the one-dimensional projection (based on simulation with 2,000 replications, $p = 5$ and $\alpha = 5\%$)

	Mul-t	Kotz	PVII	PII	Cauchy	D-Exp	Exp	χ^2_2	N+χ^2_2
d_1, $n = 51$	0.4585	0.1190	0.1330	0.0065	0.9925	0.2390	0.3975	0.6045	0.3825
$n = 103$	0.5540	0.1350	0.1650	0.5540	0.9990	0.2825	0.4895	0.7080	0.4845
$n = 203$	0.6280	0.1245	0.1340	0.0080	1.0000	0.3070	0.5365	0.7885	0.5395
d_m, $n = 51$	0.2095	0.0750	0.0960	0.0085	0.9355	0.1595	0.2420	0.4035	0.2450
$n = 103$	0.3270	0.0905	0.1015	0.3270	0.9960	0.2090	0.3145	0.5325	0.3240
$n = 203$	0.4120	0.1195	0.1285	0.0120	1.0000	0.2230	0.4020	0.6160	0.4095
d_p, $n = 51$	0.0700	0.0615	0.0565	0.0345	0.4095	0.0935	0.1145	0.2365	0.0590
$n = 103$	0.1070	0.0680	0.0650	0.1070	0.7630	0.1135	0.1675	0.3580	0.0565
$n = 203$	0.1495	0.0785	0.0725	0.0215	0.9570	0.1460	0.2425	0.4795	0.0550
d_{ss}, $n = 51$	0.2065	0.0765	0.0790	0.0185	0.8805	0.0845	0.1050	0.1415	0.0820
$n = 103$	0.2975	0.0930	0.1060	0.2975	0.9865	0.0995	0.1285	0.1960	0.0930
$n = 203$	0.3570	0.1000	0.1025	0.0100	0.9975	0.1100	0.1635	0.2585	0.1040
$\tilde{d}_1$, $n = 51$	0.4155	0.1130	0.1140	0.0070	0.9860	0.1520	0.2145	0.3490	0.1460
$n = 103$	0.4960	0.1245	0.1440	0.4960	1.0000	0.1690	0.2855	0.4285	0.1795
$n = 203$	0.5765	0.1245	0.1420	0.0070	1.0000	0.1920	0.3030	0.5000	0.1985
$\tilde{d}_m$, $n = 51$	0.1010	0.0595	0.0670	0.0285	0.6185	0.0930	0.1005	0.1850	0.0790
$n = 103$	0.1690	0.0795	0.0925	0.1690	0.9060	0.1195	0.1715	0.3045	0.1230
$n = 203$	0.2135	0.0835	0.0915	0.0150	0.9910	0.1415	0.2315	0.3975	0.1430
$\tilde{d}_p$, $n = 51$	0.0710	0.0640	0.0625	0.0375	0.4415	0.0685	0.0985	0.1475	0.0665
$n = 103$	0.1205	0.0670	0.0670	0.1205	0.8120	0.0900	0.1325	0.2240	0.0960
$n = 203$	0.1595	0.0755	0.0840	0.0230	0.9610	0.1190	0.1785	0.3230	0.1120

Table 2. (Continued for $p = 10$ and $\alpha = 5\%$)

	Mul-t	Kotz	PVII	PII	Cauchy	D-Exp	Exp	χ^2_2	N+χ^2_2
d_1, $n = 51$	0.5970	0.1205	0.3000	0.0090	0.9970	0.2400	0.4000	0.5930	0.4245
$n = 103$	0.7070	0.1360	0.3340	0.7070	1.0000	0.2520	0.4610	0.6835	0.4870
$n = 203$	0.7700	0.1175	0.3435	0.0110	1.0000	0.2645	0.4815	0.7380	0.5445
d_m, $n = 51$	0.1045	0.0690	0.0740	0.0320	0.7245	0.0980	0.1180	0.1965	0.0995
$n = 103$	0.1670	0.0770	0.1080	0.1670	0.9495	0.1025	0.1505	0.2905	0.1890
$n = 203$	0.2595	0.0845	0.1285	0.0230	0.9995	0.1160	0.2070	0.3570	0.2380
d_p, $n = 51$	0.0635	0.0530	0.0550	0.0575	0.1900	0.0670	0.0635	0.1225	0.0520
$n = 103$	0.0835	0.0575	0.0640	0.0835	0.4575	0.0715	0.1025	0.1930	0.0510
$n = 203$	0.1025	0.0635	0.0665	0.0395	0.8010	0.0990	0.1420	0.3105	0.0520
d_{ss}, $n = 51$	0.1885	0.0775	0.1135	0.0325	0.8700	0.0555	0.0735	0.0890	0.0680
$n = 103$	0.2610	0.0760	0.1285	0.2610	0.9860	0.0670	0.0890	0.1225	0.0715
$n = 203$	0.3340	0.0885	0.1375	0.0190	1.0000	0.0735	0.0970	0.1490	0.0715
$\tilde{d}_1$ $n = 51$	0.5605	0.1265	0.2610	0.0110	0.9945	0.1330	0.2060	0.3455	0.1180
$n = 103$	0.6645	0.1315	0.3040	0.6645	1.0000	0.1570	0.2505	0.4105	0.1310
$n = 203$	0.7405	0.1050	0.3080	0.0150	1.0000	0.1530	0.2705	0.4235	0.1535
$\tilde{d}_m$, $n = 51$	0.0930	0.0695	0.0645	0.0310	0.5405	0.0750	0.0960	0.1435	0.0765
$n = 103$	0.1420	0.0855	0.0840	0.1420	0.8645	0.1045	0.1205	0.2230	0.0960
$n = 203$	0.2265	0.0805	0.1120	0.0270	0.9855	0.1190	0.1830	0.2880	0.1130
$\tilde{d}_p$, $n = 51$	0.0595	0.0460	0.0500	0.0585	0.2235	0.0495	0.0530	0.0845	0.0595
$n = 103$	0.0800	0.0560	0.0715	0.0800	0.5325	0.0660	0.0870	0.1425	0.0585
$n = 203$	0.1140	0.0675	0.0645	0.0405	0.8605	0.0840	0.0995	0.1620	0.0690

group of alternative distributions in each Monte Carlo simulation, then the generated data are centered by $X - E(X)$ ($E(X)$ is a constant matrix by taking the expcctation for each element of X). In this way we obtain a sample distributed in R^p and it is comparable to that of a normal sample. Table 2 presents the empirical power of the the SUQ-plot, where the projection directions are given by (2.14). From the simulation results in Table 2, we can summarize the following conclusions: 1) the projection direction $\boldsymbol{d}_1$ performs the best in each case of sample size n for almost all of the selected alternative distributions; 2) the power of the SUQ-plot increases as the sample size n increases; and 3) for the seven selected projection directions, the SUQ-plot has poor power against the two distributions: Kotz type and the Pearson type II (PII). For small sample sizes, it also has poor power against the the Pearson type VII and the double exponential distributions. Based on the Monte Carlo study on the power of the SUQ-plot, we suggest using the projection direction $\boldsymbol{d}_1$ as the first choice for a general purpose.

3.2. *Application*

Now we apply the SUQ-plot to a real data set. We will compare the conclusions drawn by the SUQ-plot with those drawn by some existing statistics. The multivariate skewness (b_{1p}) and kurtosis (b_{2p}) statistics proposed by Mardia (1970) have been found to have better power performance against many types of alternative non-normal distributions based on Monte Carlo studies (Horswell & Looney, 1992; and Romeu & Ozturk 1993). In applying the Mardia's (1970) skewness and kurtosis statistics, their standardized forms (whose limiting distributions are known) are usually used and their critical values for some finite sample sizes and given dimensions were tabulated by Romeu & Ozturk (1993). When simulating the p-values in the following example, we employ the standardized Mardia's skewness and kurtosis statistics and choose the critical values from Romeu & Ozturk (1993) if provided, or obtain them by simulation with 10,000 replications if not provided. The following data set has been used for testing multinormality in the literature (Small, 1980; and Royston, 1983).

Table 3. Empirical p-values of the three types of data sets and the mixed data sets in Example 1

(X_1, X_2, X_3, X_4)		
data set	p-value of b_{1p}	p-value of b_{2p}
setosa data	0.1276	0.0297
versicolor data	0.1416	0.4993
virginica data	0.1129	0.2009
(X_1, X_2, X_3, X_4)		
setosa+versicolor	0.1386	0.0797
setosa+virginica	0.0012	0.2103
versicolor+virginica	0.0539	0.4815
setosa+versicolor+virginica	0.0000	0.4449

Example 1. Three types of variates (setosa, versicolor and virginica) are studied. For each type of variate, the following variates were measured on $n = 50$ plants.

X_1: sepal length, X_2: sepal width, X_3: petal length, X_4: petal width.

There are 150 observations for the three types of variates. The full data are given by Table 3.4 of Kendall (1975) (p. 40). For the setosa data, the variate X_4 was found to be markedly non-normal while (X_1, X_2, X_3) shows no evidence of departure from multinormality (Royston, 1983). For all of these three sets of data (setosa, versicolor and virginica), we use the standardized Mardia's skewness and kurtosis statistics to test the 4-dimensional joint normality. The p-values of b_{1p} and b_{2p} obtained by Monte Carlo simulation with 10,000 replications for each data set are presented in Table 3, where "setosa+versicolor" means that the data set is composed of the setosa data and versicolor data (sample size $n = 50 + 50 = 100$), other mixed data sets have a similar meaning.

The simulation results on the p-values in Table 3 imply that the setosa data show evidence of non-multinormality by b_{2p} at $\alpha = 5\%$ significance level while both of the versicolor data and the virginica data show no evidence of departure from multinormality by both of b_{1p} and b_{2p} at significance levels $\alpha = 1\%$, 5% and 10%. All of the four mixed data sets show evidence of departure from multinormality by b_{1p} or b_{2p} at the significance level $\alpha = 10\%$, where the mixed data set by the three types of variates (setosa, versicolor and virginica) (sample size $n = 150$) shows severe non-multinormality by b_{1p} at significance levels $\alpha = 1\%$, 5% and 10%. The SUQ-plots for the data sets in Table 3 are presented in Figure 1.

Based on Figure 1 with the significance levels $\alpha = 1\%$, 5% and 10%, we can draw the following conclusions corresponding to the p-values in Table 3:

(1) the SUQ-plot (a1) shows that the setosa data set with the first three variates (X_1, X_2, X_3) shows no evidence of departure from multinormality;

(2) the SUQ-plot (a2) shows that the setosa data set with the full four variates (X_1, X_2, X_3, X_4) shows no evidence of departure from multinormalit. This fails to detect severe non-normality of the variates X_4 in the setosa data set. The SUQ-plot (a2) implies that, the non-multinormality of (X_1, X_2, X_3, X_4) in the setosa data set is "masked" by multinormality of its subset (X_1, X_2, X_3);

(3) the versicolor and the virginica data sets show no evidence of departure from multinormality by their SUQ-plots: SUQ (b) and SUQ (c) in Figure 1, these two plots do not exceed the acceptance regions;

(4) all of the four mixed data sets listed in Table 3 show obvious evidence of departure from multinormality by their SUQ-plots: SUQ (ab), SUQ (ac), SUQ (bc) and SUQ (abc) in Figure 1. Most of these conclusions are consistent with those drawn by Mardia's skewness and kurtosis statistics given in Table 3.

4. Concluding Remarks

The SUQ-plot has the property: given a fixed sample size, the increase of sample dimension has little influence on the efficiency. This property results from the fact that the null distribution of the K-S type statistic (2.21) on which the SUQ-plot associated with its acceptance regions is based is independent of the sample dimension. The Monte Carlo study on the power performance of the SUQ-plot demonstrated by Table 2 also verifies this "power stability" in the sense that the SUQ-plot keeps relatively stable power when the dimension increases and the sample size is fixed. Therefore, it might be an advantage to apply the SUQ-plot to high-dimensional data analysis when the sample size is limited due to certain conditions. Concerning about the choice of the projection direction $\boldsymbol{d}$ in (2.4), we only emphasize the idea in principal component analysis as in (2.12) and (2.13). There are certainly many other choices (satisfying the condition in Theorem 2.1) which may perform better. In practical application of the SUQ-plot, it can enhance the conclusion if providing the SUQ-plot by choosing different projection directions, such as those used in the Monte Carlo study in Section 3. It is pointed out that SUQ-plot only makes use of $k_0 = [(n-1)/2]-1$ plotting positions as can be seen from Example 1. This is a weakness if the sample size is not large enough. As a complement to testing multinormality, the SUQ-plot is able to enhance some analytical conclusions if combining with other plotting methods and analytical statistics in applications.

Acknowledgment

Liang's research was partially supported by an SRCC Research Grant from Hong Kong Baptist University, and a University of New Haven 2005 Summer Faculty Fellowship. Li's research was supported by NSF grant DMS-0348869 and National Institute on Drug Abuse grant P50 DA10075.

References

1. Ahn, S.K. (1992). *F*-probability plot and its application to multivariate normality. *Commun. Statist. – Theory & Methods* **21**, 997-1023.

2. Brown, B.M. & Hettmansperger, T.P. (1996). Normal scores, normal plots and tests for normality. *J. Amer. Statist. Assoc.* **91**, 1668-1675.
3. Cleveland, W.S. (1993). *Visualizing Data*, AT & T Bell Lab., Murray Hill, New Jersey.
4. Csörgo, S., Seshadri, V. & Yalovsky, M. (1973). Some exact tests for normality in the presence of unknown parameters. *J. Roy. Statist. Soc.* (B) **35**, 507-522.
5. Easton, G.S. & McCulloch, R.E. (1990). A multivariate generalization of quantile-quantile plot. *J. Amer. Statist. Assoc.* **85**, 376-386.
6. Eaton, M.L. (1986). *Multivariate Analysis: – A Vector Space Approach*, John Wiley & Sons, New York.
7. Fang, K.-T., Kotz, S. & Ng, K.W. (1990). *Symmetric Multivariate and Related Distributions*, Chapman and Hall, London and New York.
8. Fang, K.-T. & Wang, Y. (1994). *Number-Theoretic Methods in Statistics*, Chapman and Hall, London.
9. Fang, K.-T. & Zhang, Y.-T. (1990). *Generalized Multivariate Analysis*, Science Press and Springer-Verlag, Beijing and Berlin.
10. Healy, M.J.R. (1968). Multivariate normal plotting. *Applied Statistics* **17**, 157-161.
11. Henze, N. (2002). Invariant tests for multivariate normality: A critical review. *Statist. Papers* **43**, 467-506.
12. Horswell, R.L. & Looney, S.W. (1992). A comparison of tests for multivariate normality that are based on measures of multivariate skewness and kurtosis. *J. Statist. Comput. & Simul.* **42**, 21-38.
13. Kendall, M.G. (1975). *Multivariate Analysis*, London: Griffin.
14. Koziol, J.A. (1993). Probability plots for assessing multivariate normality. *The Statistician* **42**, 161-173.
15. Läuter, J. (1996). Exact t and F tests for analyzing studies with multiple endpoints. *Biometrics* **52**, 964-970.
16. Läuter, J., Glimm, E. & Kropf, S. (1996). New multivariate tests for data with an inherent structure. *Biometrical Journal* **38**, 5-23.
17. Li, R., Fang, K.-T. & Zhu, L.-X. (1997). Some Q-Q probability pots to test spherical and elliptical symmetry. *J. Comput. & Graph. Statist.* **6**, 435-450.
18. Liang, J. & Bentler, P.M. (1999). A t-Ddistribution plot to detect non-multinormality. *Comput. Statist. & Data Anal.* **30**, 31-44.
19. Liang, J., Pan, W. & Yang, Z.-H. (2004). Characterization-based Q-Q plots for testing multinormality. *Statist. & Prob. Letters* **70**, 183-190.
20. Mardia, K.V. (1970). Measures of multivariate skewness and kurtosis with applications. *Biometrika* **57**, 519-530.
21. Mardia, K.V. (1980). Tests of univariate and multivariate normality. In: Krishnaiah, P.R. (Ed.), *Handbook of Statistics*, Vol. **1**, North-Holland Publishing Company, pp. 279-320.
22. Mecklin, C.J. & Mundfrom, D.J. (2004). An appraisal and bibliography of tests for multivariate normality. *Internat. Statist. Rev.* **72**, 123-138.
23. Michael, J.R. (1983). The stabilized probability plot. *Biometrika* **70**, 11-17.

24. Romeu, J.L. & Ozturk, A. (1993). A Comparative study of goodness-of-fit tests for multivariate normality. *J. Multivar. Anal.* **46**, 309-334.
25. Royston, J.P. (1983). Some techniques for assessing multivariate normality based on the Shapiro-Wilk W. *Appl. Statist.* **32**, 121-133.
26. Small, N.J.H. (1978). Plotting squared radii. *Biometrika* **65**, 657-658.
27. Srivastava, D.K. & Muholkar, G.S. (2003). Goodness-of-fit tests for univariate and multivariate normal models. In: Khattree, R. & Rao, C.R. (Eds.), *Handbook of Statistics*, Vol. **22**, Elsevier, North-Holland, Amsterdam, 2003, pp. 869-906.
28. Tashiro, D. (1977). On methods for generating uniform points on the surface of a sphere, *Ann. Inst. Statist. Math.* **29**, 295-300.

BAYESIAN STOCHASTIC ESTIMATION OF THE MAXIMUM OF A REGRESSION FUNCTION

CHENG-DER FUH *

Institute of Statistical Science, Academia Sinica, Taipei, Taiwan, R.O.C.

INCHI HU †

Department of Information and Systems Management
Hong Kong University of Science and Technology
Clear Water Bay, Kowloon, Hong Kong

In this paper we consider the problem of finding the maximum of a regression function using the Bayesian approach. We first express the solution as a function of unknown parameters of a model. We then invoke the plug-in principle to substitute Bayes estimates for unknown parameters at each stage. The resulting scheme differs from the classic Kiefer-Wolfowitz scheme (Kiefer and Wolfowitz, 1952) in several important aspects. In particular it does not depend on a sequence of constants $a_n \to 0$ to dampen the effect of random errors as in Robbins and Monro (1951). The convergence of the scheme to the desired value with probability one is established under mild conditions. In the case of several independent variables, our model is the same as the second order model considered in response surface methods. Hence the convergence result of this paper would be of interest to response surface methods.

1. Introduction

Consider the problem in which a response of interest is influenced by several variables and the objective is to optimize this response. For example, suppose that a chemical engineer wishes to find the levels of temperature and pressure that maximize the yield of a process. The process yield is a function of the levels of temperature and pressure, say,

$$y = M(x) + \epsilon,$$

where ϵ represents the random error contained in the response y. Thus the goal is to find the levels of independent variables such that the regression

*Research partially supported by NSC 94-2118-M-001-028.
†Research partially supported by Hong Kong RGC Grant HKUST6212/04H.

function $M(\cdot)$ is maximized when responses are contaminated by errors. Partly motivated by a paper of Robbins and Monro (1951), Kiefer and Wolfowitz (1952) made their seminal contribution to the aforementioned problem by proposing what is now known as the Kiefer-Wolfowitz scheme for finding the maximum of a regression function.

The Kiefer-Wolfowitz scheme takes the form

$$x_{n+1} = x_n + a_n \Delta(x_n),$$

where at the nth stage a pair of observations y_n' and y_n'' are taken at the design levels $x_n' = x_n - c_n$ and $x_n'' = x_n + c_n$, respectively, a_n and c_n are positive constants, and

$$\Delta(x_n) = \frac{(y_n'' - y_n')}{2c_n}.$$

To dampen the effect of the errors, Kiefer and Wolfowitz (1952) assumed

$$\sum_{n=1}^{\infty} (a_n/c_n)^2 < \infty. \tag{1}$$

This implies that $\sum_{n=1}^{\infty} a_n(\epsilon_n'' - \epsilon_n')/c_n$ converges almost surely, where ϵ_n' and ϵ_n'' denote the random errors contained in y_n' and y_n'', respectively. To ensure convergence to the maximized value, they also required

$$\sum_{n=1}^{\infty} a_n = \infty, \quad a_n \to 0, \tag{2}$$

and

$$\sum_{n=1}^{\infty} a_n c_n < \infty. \tag{3}$$

Under certain conditions on M and ϵ, Kiefer and Wolfowitz (1952) showed that the scheme converges in probability to the maximizing value. These assumptions were subsequently weakened by Blum (1954) and Dvoretzky (1956), who removed the requirement (1.3) and also established its almost sure convergence.

In what follows we will consider a Bayesian version of Kiefer-Wolfowitz scheme. Specifically, we consider the following model.

$$y_n = \alpha + \beta x_n + \gamma x_n^2 + \epsilon_n, \quad n = 1, 2, \cdots, \tag{4}$$

where the random errors $\{\epsilon_n\}_{n=1}^{\infty}$ are martingale differences with bounded conditional variances. We shall assume that the unknown parameters α, β, γ have a prior distribution μ, which is independent of the errors, $\epsilon_1, \epsilon_2, \cdots$.

After taking n observations $y_1, \cdots, y_n$, from the model (1.4), the posterior distribution of α, β, γ can be calculated as the conditional distribution given $y_1, \cdots, y_n$. Let $\mathcal{F}_n$ be the σ-field generated by $y_1, \cdots, y_n$. The expectation taken with respect to the posterior distribution given the data $y_1, \cdots, y_n$, is a version of the conditional expectation $E(\cdot|\mathcal{F}_n)$. Then the Bayes estimates of (α, β, γ) given the observations $y_1, \cdots, y_n$ is $(\alpha_n, \beta_n, \gamma_n) = E[(\alpha, \beta, \gamma)|\mathcal{F}_n]$. If (α, β, γ) were known, the maximum of $E(y)$ is achieved at $x = -\beta/(2\gamma)$. When β, γ are unknown, we can replace them by Bayes estimates $x_n = -\beta_n/(2\gamma_n)$. Following a similar line of thoughts, we propose the following Bayesian version of Kiefer-Wolfowitz scheme (BKWS).

At stage n, take a pair of observations y'_n, y''_n at $x'_n = x_n - c_n$ and $x''_n = x_n + c_n$, respectively, where c_n is a sequence of positive constants satisfying

$$c_n \to 0 \text{ and } \sum_{n=1}^{\infty} c_n^2 = \infty. \tag{5}$$

Let $\tilde{y}_n = y''_n - y'_n$ and let $\tilde{\mathcal{F}}_n$ be the σ-field generated by $\tilde{y}_1, \cdots, \tilde{y}_n$. The Bayes estimates $\gamma_n = E(\gamma|\tilde{\mathcal{F}}_n)$ and $\beta_n = E(\beta|\tilde{\mathcal{F}}_n)$ will be employed to calculate the input x_{n+1} at stage $n+1$ according to

$$x_{n+1} = -\frac{\beta_n}{2\gamma_n}. \tag{6}$$

Our main result is encapsulated in the following theorem.

Theorem 1.1. *Suppose observations are generated sequentially from the model (4) according to BKWS described in the preceding paragraph which satisfies (1.5) and (1.6). The unknown parameters α, β, and γ have a prior distribution μ such that*

$$\int (\alpha^2 + \beta^2 + \gamma^2) d\mu < \infty. \tag{7}$$

Furthermore we assume $\mu\{\gamma < 0\} = 1$. Let the errors be such that with probability one

$$E(\epsilon_n|\epsilon_1, \cdots, \epsilon_{n-1}) = 0; \ \sup_{1 \le n} E(\epsilon_i^2|\epsilon_1, \cdots, \epsilon_{n-1}) < \infty. \tag{8}$$

Then $\lim_{n\to\infty} x_n = -\beta/2\gamma$ almost surely. In other words, BKWS converges to the value that maximizes the regression function with probability one.

Remark 1. We can replace the condition $\mu\{\gamma < 0\} = 1$ by $\mu\{\gamma > 0\} = 1$. The proof will be almost the same. The point is that as long as we know we are searching for the maximum ($\mu\{\gamma < 0\} = 1$) or the minimum ($\mu\{\gamma > 0\} = 1$), BKWS will converges to the optimizing value.

Remark 2. The almost sure convergence of $\lim_{n\to\infty} x_n = -\beta/(2\gamma)$ in Theorem 1.1 is with respect to the product measure $\mu \times Q$, where μ is the prior probability measure and Q is the probability measure induced by random errors $\{\epsilon_n\}$. Unless the context clearly indicates otherwise, this interpretation holds for all almost sure statements in this paper.

The rest of the article is organized as follows. In Section 2, we will document some results on posterior covariance matrices for stochastic regression models. These results will be used to prove Theorem 1.1. We will consider the case of several independent variables and its relationship with response surface methods in Section 3. We compare the proposed scheme with the Kiefer-Wolfowitz scheme in Section 4, where some discussions and concluding remarks are also given.

2. Posterior Covariance Matrices and the Proof of Theorem 1.1

The model (4) with x_n generated from BKWS is a special case of stochastic regression models, which we now introduce. Consider the following multiple regression model

$$y_n = \theta^t \mathbf{x}_n + \epsilon_n, \tag{1}$$

where y_n is the response and $\theta^t = (\theta_1, \cdots, \theta_k)$ is the unknown parameters and $\mathbf{x}_n = (x_{n1}, \cdots, x_{nk})^t$ is the design vector. Here and in the sequel, we will use $\mathbf{v}^t$ to denote the transpose of a vector or a matrix $\mathbf{v}$. The random errors $\{\epsilon_n\}$ satisfy (1.8). We also assume that the design vector is stochastic and satisfying the following measurable condition: $\mathbf{x}_n$ is measurable with respect to the σ-field generated by $y_1, \cdots, y_{n-1}$. This measurability condition implies that the design vector at the nth stage can be computed from observations $y_1, \cdots, y_{n-1}$. This stochastic regression model is very general and has been studied extensively in several fields of statistical inquiry including time series analysis, dynamic input-output systems, stochastic approximation schemes, adaptive control and sequential design. We will show that the BKWS scheme applying to model (4) satisfies the conditions for stochastic regression models.

Let

$$C_n = E[\{\theta - E(\theta|\mathcal{F}_n)\}\{\theta - E(\theta|\mathcal{F}_n)\}^t|\mathcal{F}_n]$$

be the covariance matrix calculate under the posterior distribution of θ given $y_1, \cdots, y_n$. For any two $p \times p$ matrices A, B, we write $A \geq B$ if $\mathbf{v}^t A \mathbf{v} \geq \mathbf{v}^t B \mathbf{v}$ for all $\mathbf{v} \in \mathbf{R}^p$. The posterior covariance matrices of the stochastic regression model (2.1) obeys a recursive relation given in the following lemma.

Lemma 2.1. *The posterior covariance matrices in model (2.1) C_n $n \geq 1$ satisfy the following recursive relation*

$$E(C_n|\mathcal{F}_{n-1}) \leq C_{n-1} - \frac{C_{n-1}\mathbf{x}_n\mathbf{x}_n^t C_{n-1}}{\sigma_n^2 + \mathbf{x}_n^t C_{n-1}\mathbf{x}_n},$$

where $\sigma_n^2 = E(\epsilon_n^2|\mathcal{F}_{n-1})$.

Proof. The proof can be found in Hu (1997).

Proof of Theorem 1.1. By Lemma 2.1, it follows that, for any $\mathbf{v} \in \mathbf{R}^p$, $V_n = \mathbf{v}^t C_n \mathbf{v}$ is a nonnegative supermartingale, which converges almost surely. By a result in Robbins and Siegmund (1971) concerning nonnegative almost supermartingale, we obtain from Lemma 2.1 that for all $\mathbf{v} \in \mathbf{R}^p$

$$\sum_{n=1}^{\infty}(\sigma_n^2 + \mathbf{x}_n^t C_{n-1}\mathbf{x}_n)^{-1}(\mathbf{v}^t C_{n-1}\mathbf{x}_n)^2 < \infty \text{ almost surely.} \qquad (2)$$

We can rewrite (2.2) as

$$\sum_{n=1}^{\infty}\{(\frac{\sigma_n}{\|\mathbf{x}_n\|})^2 + \tilde{\mathbf{x}}_n^t C_{n-1}\tilde{\mathbf{x}}_n\}^{-1}(\mathbf{v}^t C_{n-1}\tilde{\mathbf{x}}_n)^2 < \infty \text{ almost surely,} \qquad (3)$$

where $\tilde{\mathbf{x}}_n = \mathbf{x}_n/\|\mathbf{x}_n\|$.

For BKWS, it is not hard to see that

$$\begin{aligned}\tilde{y}_n &= [\alpha + \beta(x_n + c_n) + \gamma(x_n + c_n)^2 + \epsilon_n''] \\ &\qquad -[\alpha + \beta(x_n - c_n) + \gamma(x_n - c_n)^2 + \epsilon_n'] \\ &= 2c_n(\beta + 2\gamma x_n) + \epsilon_n'' - \epsilon_n'. \end{aligned} \qquad (4)$$

Hence $\tilde{y}_n$ obeys model (2.1) with $\mathbf{x}_n = 2c_n(1, 2x_n)$ and $\tilde{\mathbf{x}}_n = \frac{1}{1+4x_n^2}(1, 2x_n)$. By (1.8), $\sup_{n\geq 1}\sigma_n^2 < \infty$ almost surely . From (1.5), $||\tilde{\mathbf{x}}_n|| = 1$, and the fact that $\{C_n\}$ converges to a finite limit, we have $\sup_{n\geq 1} c_n^2 \tilde{\mathbf{x}}_n^t C_{n-1}\tilde{\mathbf{x}}_n < \infty$ almost surely. Thus there exists a constant K such that for all $n \geq 1$

$$\{(\frac{\sigma_n}{\|\mathbf{x}_n\|})^2 + \tilde{\mathbf{x}}_n^t C_{n-1}\tilde{\mathbf{x}}_n\}^{-1} \geq K c_n^2. \qquad (5)$$

From (1.5), (2.3), and (2.5), it follows that $\lim_{n\to\infty} \mathbf{v}^t C_{n-1}\tilde{\mathbf{x}}_n = 0$ almost surely, for all $\mathbf{v} \in \mathbf{R}^p$. This implies that

$$\lim_{n\to\infty} \tilde{\mathbf{x}}_n^t C_{n-1}\tilde{\mathbf{x}}_n = 0 \text{ almost surely} \tag{6}$$

because $||\tilde{\mathbf{x}}_n|| = 1$.

Next we show that $x_n = -\beta_n/(2\gamma_n)$ converges to a finite limit with probability one. It is clear that $\{\beta_n, \tilde{\mathcal{F}}_n\}$ and $\{\gamma_n, \tilde{\mathcal{F}}_n\}$ are both uniformly integrable martingales. Hence both β_n and γ_n converge almost surely to the limits β_∞ and γ_∞, respectively. Therefore it is sufficient to show that $\lim_{n\to\infty} \gamma_n = \gamma_\infty > 0$ with probability one. Let $A = \{\omega : \gamma_\infty(\omega) \leq 0\}$. Clearly $A \in \tilde{\mathcal{F}}_\infty$. Assume $P(A) > 0$. On the one hand $\lim_{n\to\infty} \int_A \gamma_n = \int_A \gamma_\infty \leq 0$. On the other hand, since we assume that $\mu\{\gamma > 0\} = 1$ (The case of $\mu\{\gamma < 0\} = 1$ can be treated similarly) and $\gamma_\infty = E(\gamma|\tilde{\mathcal{F}}_\infty)$, we have $\int_A \gamma_\infty = \int_A \gamma > 0$. This is a contradiction and so we conclude that γ_∞ is positive with probability one. Consequently, $x_n = -\beta_n/(2\gamma_n)$ converges to the finite limit $\beta_\infty/(2\gamma_\infty)$ with probability one.

Since the sequence $\{x_n\}$ converges to a finite limit with probability one and that $\tilde{\mathbf{x}}_n = (1, 2x_n)(1 + 4x_n^2)^{-1/2}$, from (2.6), it follows that $\mathrm{Var}(\beta + 2x_\infty\gamma|\tilde{\mathcal{F}}_\infty) = 0$. Or equivalently, the posterior distribution of $\beta + 2x_\infty\gamma$ is a point mass. The point mass must be

$$E(\beta + 2\gamma x_\infty|\mathcal{F}_\infty) = E(\beta|\tilde{\mathcal{F}}_\infty) + E(\gamma|\tilde{\mathcal{F}}_\infty)x_\infty = \beta_\infty + 2\gamma_\infty(-\frac{\beta_\infty}{2\gamma_\infty}) = 0.$$

Thus $\beta + 2x_\infty\gamma = 0$ almost surely, or equivalently $\lim_{n\to\infty} x_n = x_\infty = -\beta/(2\gamma)$ almost surely. The proof is complete.

3. Response Surface Methods and the Case of Several Independent Variables

The BKWS described in Section 2 can be extended to solve the problem of maximizing a regression function of several independent variables. Consider the following model of k independent variables.

$$y = M(\mathbf{x}) + \epsilon = \sum_{i=1}^{k} \gamma_i x_i^2 + \sum_{1\leq i<j\leq k} \beta_{ij} x_i x_i + \sum_{i=1}^{k} \delta_i x_i + \alpha + \epsilon, \tag{1}$$

where $\mathbf{x} = (x_1, \cdots, x_k)^t$ denotes the levels of k independent variables and $\gamma_i, \beta_{ij}, \delta_i, \alpha$ are unknown parameters. Note that the preceding model is the same as the second order model considered in response surface methods (RSM). Our goal here is also the same as that of RSM, to lead the experimenter rapidly and efficiently along a path of improvement toward the general vicinity of the optimum (see e.g. Montgomery, 1997). In our setting this goal is achieved by sequentially producing a sequence of design vectors $\{(x_1^{(n)}, \cdots, x_k^{(n)})\}$ that converges to the value that optimizes $E(y)$ of (3.1). Thus the convergence result described below can be viewed as an attempt to develop a theory relevant to response surface methods.

Let

$$D = \begin{pmatrix} 2\gamma_1 & \beta_{12} & \cdots & \beta_{1k} \\ \beta_{12} & 2\gamma_2 & \cdots & \beta_{2k} \\ \vdots & \vdots & \ddots & \vdots \\ \beta_{1k} & \beta_{2k} & \cdots & 2\gamma_k \end{pmatrix}.$$

We shall assume that $\mu\{D$ is negative definite$\} = 1$. This is to ensure that the regression function $M(\mathbf{x})$ of (3.1) has a maximum. The maximum is attained at the solution of the following system of equations

$$\begin{cases} 2\gamma_1 x_1 + \sum_{i\neq 1} \beta_{1i} x_i + \delta_1 = 0; \\ \qquad\vdots \\ 2\gamma_k x_k + \sum_{i\neq k} \beta_{ki} x_i + \delta_k = 0. \end{cases} \tag{2}$$

Suppose that at the nth stage the scheme produces an approximation, $\mathbf{x}_n = (x_1^{(n)}, \cdots, x_k^{(n)})$, to the solution of (3.2). Then take k pairs of observations y'_{in}, y''_{in}, $i = 1, \cdots, k$, at $(x_1^{(n)}, \cdots, x_{i-1}^{(n)}, x_i^{(n)} + c_n, x_{i+1}^{(n)}, \cdots, x_k^{(n)})$ and $(x_1^{(n)}, \cdots, x_{i-1}^{(n)}, x_i^{(n)} - c_n, x_{i+1}^{(n)}, \cdots, x_k^{(n)})$, respectively. Define $\tilde{y}_{in} = y'_{in} - y''_{in}$. Let $\tilde{\mathbf{y}}_n = (\tilde{y}_{1n}, \cdots, \tilde{y}_{kn})$. Let $\tilde{\mathcal{F}}_n$ be the σ-field generated by $\{\tilde{\mathbf{y}}_1, \tilde{\mathbf{y}}_2, \cdots, \tilde{\mathbf{y}}_n\}$. That is, we update the σ-field $\tilde{\mathcal{F}}_i$ to $\tilde{\mathcal{F}}_{i+1}$ only after we have observed the whole vector $\tilde{\mathbf{y}}_i$. Let $(\gamma_i^{(n)}, \beta_{ij}^{(n)}, \delta_i^{(n)}) = E(\gamma_i, \beta_{ij}, \delta_i | \tilde{\mathcal{F}}_n)$, $1 \le i < j \le k$, be the Bayes estimates for the respective unknown parameters. At stage $(n+1)$ the scheme produces an approximation, $\mathbf{x}_{n+1} = (x_1^{(n+1)}, \cdots, x_k^{(n+1)})$ as the solution to the following system of k equations

$$\begin{cases} 2\gamma_1^{(n)} x_1^{(n+1)} + \sum_{i\neq 1} \beta_{1i}^{(n)} x_i^{(n+1)} + \delta_1^{(n)} = 0; \\ \qquad\vdots \\ 2\gamma_k^{(n)} x_k^{(n+1)} + \sum_{i\neq k} \beta_{ki}^{(n)} x_i^{(n+1)} + \delta_k^{(n)} = 0. \end{cases} \tag{3}$$

To rule out the possibility of nonexistence of the solution $\mathbf{x}_{n+1}$ to (3.3), we need the following result.

Lemma 3.1. *Let $X \in R^m$ be a random vector with $E(|X|) < \infty$. Let A be a convex set of R^m. If $P\{X \in A\} = 1$, then $E(X) \in A$.*

Proof. The lemma is easily seen to be true when the distribution of X is discrete. For the general case, approximate the distribution of X by a discrete distribution. Some care is needed for the boundary of A. The details being straightforward is omitted here. $\square$

Since we have assumed that $\mu\{D \text{ is negative definite}\} = 1$, it is clear that

$$P\{D \text{ is negative definite } |\tilde{\mathcal{F}}_n\} = 1$$

for all $n \geq 1$. Since the set $\{D \text{ is negative definite}\}$ is also a convex set of $R^{k(k+1)/2}$, by the preceding lemma, we can conclude that the matrix

$$E(D|\tilde{\mathcal{F}}_n) = \begin{pmatrix} 2\gamma_1^{(n)} & \beta_{12}^{(n)} & \cdots & \beta_{1k}^{(n)} \\ \beta_{12}^{(n)} & 2\gamma_2^{(n)} & \cdots & \beta_{2k}^{(n)} \\ \vdots & \vdots & \ddots & \vdots \\ \beta_{1k}^{(n)} & \beta_{2k}^{(n)} & \cdots & 2\gamma_k^{(n)} \end{pmatrix}$$

is negative definite with probability one. This guarantees the existence of the solution to (3.3).

Since all coefficients of (3.3) are uniformly integrable martingales, they converge to a finite limit almost surely. Furthermore, by martingale convergence theorem, the limiting matrix $E(D|\tilde{\mathcal{F}}_\infty)$ is negative definite with probability one. Consequently, the sequence $\{\mathbf{x}_n\}$ converges to a finite limit $\mathbf{x}_\infty = (x_1^{(\infty)}, \cdots, x_k^{(\infty)})$ with probability one. We now proceed to characterize $\mathbf{x}_\infty$.

Clearly

$$\begin{cases} \tilde{y}_{1n} = y'_{1n} - y''_{1n} = 2c_n(2\gamma_1 x_1^{(n)} + \sum_{i\neq 1} \beta_{1i} x_i^{(n)} + \delta_1) + \tilde{\epsilon}_{1n}; \\ \vdots \\ \tilde{y}_{kn} = y'_{kn} - y''_{kn} = 2c_n(2\gamma_k x_k^{(n)} + \sum_{i\neq k} \beta_{ik} x_i^{(n)} + \delta_k) + \tilde{\epsilon}_{kn}, \end{cases} \tag{4}$$

where $\tilde{\epsilon}_{in} = \epsilon'_{in} - \epsilon''_{in}$, $i = 1, \cdots, k$, with ϵ'_{in} and ϵ''_{in} being the random errors contained in y'_{in} and y''_{in}, respectively. From (3.4) and recalling the definition of $\tilde{\mathcal{F}}_n$, it is not hard to see that the model (2.1) prevails for the sequence of observations $\tilde{y}_{11}, \cdots, \tilde{y}_{k1}, \tilde{y}_{12}, \cdots, \tilde{y}_{k2}, \cdots, \tilde{y}_{1n}, \cdots, \tilde{y}_{kn}, \cdots$.

Applying the same argument as in the one-dimensional case, which establishes (2.2)-(2.6), to each of the k equations of (3.4), along with the fact that $\{\mathbf{x}_n\}$ converges to a finite limit $\mathbf{x}_\infty = (x_1^{(\infty)}, \cdots, x_k^{(\infty)})$ with probability one, we arrive at

$$\begin{cases} \text{Var}(2\gamma_1 x_1^{(\infty)} + \sum_{i\neq 1} \beta_{1i} x_i^{(\infty)} + \delta_1 | \tilde{\mathcal{F}}_\infty) = 0; \\ \vdots \\ \text{Var}(2\gamma_k x_k^{(\infty)} + \sum_{i\neq k} \beta_{ki} x_i^{(\infty)} + \delta_k | \tilde{\mathcal{F}}_\infty) = 0. \end{cases} \tag{5}$$

Or equivalently, the posterior distributions of $2\gamma_l x_l^{(\infty)} + \sum_{i\neq l} \beta_{li} x_i^{(\infty)} + \delta_l$, $l = 1, \cdots, k$, are all point masses.

From (3.3), it follows that

$$\begin{cases} E(2\gamma_1 x_1^{(\infty)} + \sum_{i\neq 1} \beta_{1i} x_i^{(\infty)} + \delta_1 | \tilde{\mathcal{F}}_\infty) = 0; \\ \vdots \\ E(2\gamma_k x_k^{(\infty)} + \sum_{i\neq k} \beta_{ki} x_i^{(\infty)} + \delta_k | \tilde{\mathcal{F}}_\infty) = 0. \end{cases} \tag{6}$$

From (3.5) and (3.6), it follows that the limit $\mathbf{x}_\infty$ of $\{\mathbf{x}_n\}$ solves (3.2). As a summary, we state the following theorem to conclude the discussion on the case of several independent variables.

Theorem 3.1. *Suppose observations are taken sequentially from the model (3.1) with random errors satisfying (1.8). Consider the scheme which generates a sequence of approximations $\{\mathbf{x}_n\}$ to the solution of (3.2) and satisfies (1.5) and (3.3). The unknown parameters α, β_{ij}, γ_i, δ_i have a prior distribution μ such that*

$$\int (\alpha^2 + \sum_{1\leq i<j\leq k} \beta_{ij}^2 + \sum_{i=1}^{k} \gamma_i^2) d\mu < \infty. \tag{7}$$

Assume that $\mu\{D$ is negative definite $\} = 1$. Then the sequence $\{\mathbf{x}_n\}$ converges to the solution of (3.2) almost surely.

Remark. We can change the assumption, $\mu\{D$ is negative definite$\} = 1$, to $\mu\{D$ is positive definite$\} = 1$. This corresponds to finding the minimum of the regression function. The proof can be carried out in much the same way.

4. Concluding Remarks

Following the plug-in principle (see e.g. Efron and Tibshirani, 1993), the most natural scheme for finding the maximum (minimum) of the regression function under model (1.4) would proceed as follows. Under model (1.4) the regression function is maximized at $-\beta/(2\gamma)$. Given $n-1$ observations from model (1.4) one obtains an estimate $(\hat{\beta}_{n-1}, \hat{\gamma}_{n-1})$ for (β, γ), then take another observation at $x_n = -\hat{\beta}_{n-1}/(2\hat{\gamma}_{n-1})$. With this additional observation one can update the estimate to $(\hat{\beta}_n, \hat{\gamma}_n)$ and the next observation will be taken at $x_{n+1} = -\hat{\beta}_n/(2\hat{\gamma}_n)$. Repeating the same step, one obtains a sequence of approximations $\{x_n\}$ to the solution $-\beta/(2\gamma)$. The convergence of the preceding scheme with well-known estimators (such as MLE, Bayes, least-squares etc.) to the desirable value $-\beta/(2\gamma)$ is an open problem.

The scheme described in Section 2 employs the plug-in principle on model (2.4), which is the "derivative" of model (1.4). In model (2.4), $-\beta/(2\gamma)$ is the root of the regression function. By working with model (2.4) instead of (1.4), the problem of finding the maximum of the regression function is transformed to that of finding the root of the regression function. The latter is the stochastic approximation problem proposed in Robbins and Monro (1951). The convergence of schemes employing the plug-in principle in stochastic approximation and sequential design has been obtained by Hu (1997) and Hu (1998).

In stochastic approximation, to dampen the effect of random errors, one considers the sequence

$$x_{n+1} = x_n - a_n y_n,$$

where the sequence of positive constants $\{a_n\}$ satisfies, $\sum_{n=1}^{\infty} a_n = \infty$, and $\sum_{n=1}^{\infty} a_n^2 < \infty$. The optimal choice of $\{a_n\}$ that yields the best convergence rate is $1/(bn)$, where b equals the derivative of the regression function at the root. Hence the efficiency of the stochastic approximation algorithm depends on the consistency of the slope estimator which is rather difficult to establish (see, e.g. Lai and Robbins, 1981). In our approach using the plug-in principle, the proposed scheme converges to the desired value without relying on the sequence $\{a_n\}$.

The convergence results of Theorem 1.1 and Theorem 3.1 together with the fact that we have used the best estimates (Bayes estimates) for unknown parameters at each stage suggest that the convergence may be efficient. Indeed, the efficiency of schemes employing plug-in principle was established in some cases (see e.g. Chen and Hu, 1998). It is also interesting to note that when one adopts the optimal choice of $a_n = O(1/n)$, then $c_n = O(n^{-1/2})$ is not allowed for the Kiefer-Wolfowitz scheme, but it is permissible under our setting.

At the first look, the model (3.1) seems to be more restrictive than models considered in the Kiefer-Wolfowitz scheme literature. However, model (3.1) is only intended to be a local approximation to the true model. That is, if the true model is smooth and unimodal with a unique maximum and near the maximum the regression function is well approximated by the model (3.1) then it seems that the convergence result of Theorem 3.1 still holds. One advantage of model (3.1) is that in the process of computing the sequence of approximations to the solution, one needs to obtain estimates for the parameters in model (3.1). These estimates can be used as diagnostic tools to check whether we are near a global/local maximum or other types of stationary points (local minimum and saddle points). For example, we can use parameter estimates to estimate the matrix D. When we are searching for the maximum and the estimated D is not negative definite, it is likely that we are in the wrong region and no where near the maximum.

Let's discuss briefly the computation issue related to BKWS given by (1.4) -(1.6). By (2.4), if the random errors $\epsilon_1, \epsilon_2, \cdots$, are i.i.d. normally distributed and the prior distributions for β and γ are both normal, then the posterior distributions are also normal and the posterior means β_n and γ_n can be expressed explicitly as functions of the previous designs and observations, a special case of *Kalman filter*. See e.g. (16) and (17) of Chen and Hu (1998) for details. However, here we require that the prior distribution of γ has support on the positive real line, which is obviously non-Gaussian. Therefore the posterior means do not have a close form. Actually, very few cases apart from Gaussianity and linearity the analytical solution exists. We largely rely on simulation method for online computation of posterior distributions.

Starting from Gordon et al (1993), the method *bootstrap filter*, also called *particle filter*, has been developed for on-line posterior inference when observations arrive sequentially in time. This method designed to cope with non-Gaussianity and nonlinearity is particularly suitable for BKWS. The numerical performance of the bootstrap filter in BKWS and related issues will be pursued in another paper.

References

1. BLUM, J. R. (1954). Approximation methods which converge with probability one. *Ann. Math. Statist.* **25,** 382-386.
2. CHEN, K. AND HU, I. (1998). On consistency of Bayes estimates in a certainty equivalent adaptive system. *IEEE Trans. Aut. Control* **AC-43.** 943-947.
3. DVORETZKY (1956). On stochastic approximation. *Proc. Third Berkeley Symp. Math. Statist. and Probability* **1,** 39-55.
4. EFRON, B. AND TIBSHIRANI R. J. (1993). *An Introduction to the Bootstrap.* Chapman & Hall, New York.
5. GORDON, N. J., SALMOND, D. J. AND SMITH, A. F. M. (1993). Novel approach to nonlinear non-Gaussian Bayesian state estimation.. *IEE Pro-*

ceedings on Radar and Signal Processing **140,** 107-113.
6. HU, I. (1997). Posterior covariance matrices in stochastic regression models with applications to strong consistency of Bayes estimates. *Biometrika* **84,** 744-749.
7. HU, I. (1998). On sequential design in nonlinear problems. *Biometrika* **85,** 496-503.
8. KIEFER, J. AND WOLFOOWITZ, J. (1952). Stochastic estimation of the maximum of a regression function. *Ann. Math. Statist.* **23,** 462-466.
9. LAI, T. L. AND ROBBINS, H. (1981). Consistency and asymptotic efficiency of slope estimates in stochastic approximation schemes. *Z. Wahrsch. verw. Gebiete* **56,** 329-360.
10. MONTGOMERY, D. C. (1997). *Design and Analysis of Experiments. 4th edition.* John Wiley & Sons, New York.
11. ROBBINS, H. AND MONRO, S. (1951). A stochastic approximation method. *Ann. Math. Statist.* **22,** 400-407.
12. ROBBINS, H. AND SIEGMUND, D. (1971). A convergence theorem for nonnegative almost supermartingales and some applications. In: *Optimizing Methods in Statistics.* (J.S. Rstagi, ed.), Academic Press, New York, 233-257.

THE L_1-NORM KERNEL ESTIMATOR OF CONDITIONAL MEDIAN FOR STATIONARY PROCESSES*

ZHENGYAN LIN

Department of Mathematics, Zhejiang University
Xixi Campus, Hangzhou, Zhejiang 310028, China
zlin@zju.edu.cn

DEGUI LI

Department of Mathematics, Zhejiang University
Yuquan Campus, Hangzhou, Zhejiang 310027, China
ldgofzju@tom.com

We consider L_1-norm kernel estimator of the conditional median $\theta(\mathbf{x})$, $\mathbf{x} \in \mathcal{R}^d$ for a wide class of stationary processes. Asymptotic normality of the resulting estimator $\theta_n(\mathbf{x})$ is established under different regularity conditions on bandwidths. Applications of our main results are also given.

Some key words: Asymptotic normality; Conditional median; L_1-norm kernel estimation; Stationary process.

1. Introduction

Over the last three decades, conditional quantile, also called quantile regression or regression quantile, introduced by Koenker and Bassett (1978), has been used widely in various disciplines, such as finance, economics and so on. For the stationary process $\{\mathbf{X}_i, Y_i\}$ with $\mathbf{X}_i$ being $\mathcal{R}^d$-valued and Y_i being real-valued, denote $G(y|\mathbf{x})$ the conditional distribution Y_n given $\mathbf{X}_n = \mathbf{x}$, where $\mathbf{x} = (x_1, \cdots, x_d) \in \mathcal{R}^d$. The conditional quantile function of $G(y|\mathbf{x})$ at $\mathbf{x}$ is defined as for $0 < \tau < 1$,

$$G^{-1}(\tau|\mathbf{x}) = \inf\{y : P(Y_n \leq y|\mathbf{X}_n = \mathbf{x}) \geq \tau\}. \tag{1}$$

*Supported by National Natural Science Foundation of China (Grant No. 10131040) and Specialized Research Fund for the Doctor Program of Higher Education (Grant No. 2002335090)

For more details about the recent developments of conditional quantile, we refer to Xiang (1996), Koenker and Hollock (2001), Cai (2002) and the references therein. It is easy to check that the conditional median $\theta_1(x)$ corresponding to $G^{-1}(\frac{1}{2}|\mathbf{x})$ is a kind of special conditional quantiles. When the process $\{\mathbf{X}_i, Y_i\}$ is strong mixing, Zhou and Liang (2000) studied asymptotic normality for the L_1-norm kernel estimator of $\theta_1(\mathbf{x})$. In this paper, we shall establish analogous result for a wide class of stationary processes which include many useful models.

Consider the process

$$X_n = Q(\cdots, \varepsilon_{n-1}, \varepsilon_n), \tag{2}$$

where Q is a measurable function and $\{\varepsilon_n\}_{n\in\mathcal{Z}}$ is a sequence of independent and identically distributed (i.i.d.) random variables. Clearly $\{X_n\}_{n\in\mathcal{Z}}$ is a stationary process which contains a wide class of time series models such as ARMA models, fractional ARIMA models and ARCH models. Therefore, it is interesting to study the process $\{X_n\}_{n\in\mathcal{Z}}$ defined by (2). To formulate the regression problem, we consider the model

$$Y_n = M(X_{n-1}, \cdots, X_{n-d}, \eta_n), \tag{3}$$

where $\{\eta_n\}_{n\in\mathcal{Z}}$ is a sequence of i.i.d. random variables and η_n is independent of $(\cdots, \varepsilon_{n-1}, \varepsilon_n)$, M is a measurable function.

Assume that (2) and (3) hold. Let $F(y|\mathbf{x})$ be the conditional distribution of Y_n given $\mathbf{X}_{n-1} = \mathbf{x}$, where $\mathbf{X}_{n-1} = (X_{n-1}, \cdots, X_{n-d})$. The conditional quantile function of $F(y|\mathbf{x})$ at $\mathbf{x}$ is defined as, for any $0 < \tau < 1$,

$$F^{-1}(\tau|\mathbf{x}) = \inf\{y : P(Y_n \leq y|\mathbf{X}_{n-1} = \mathbf{x}) \geq \tau\}. \tag{4}$$

In this paper, we shall consider the L_1-norm kernel estimator of conditional median $\theta(\mathbf{x})$ corresponding $F^{-1}(\frac{1}{2}|\mathbf{x})$. Let $J_n = J_n(\mathbf{x})$ be the set $\{1 \leq i \leq n : |X_{i-j} - x_j| \leq h_n$ for all $1 \leq j \leq d\}$, where $h_n \to 0$, and $N_n = N_n(\mathbf{x})$ be the number of the points in J_n. The estimator $\theta_n(\mathbf{x})$ is defined as follows

$$\theta_n(\mathbf{x}) = \theta_{h_n}(\mathbf{x}) = F_n^{-1}(\frac{1}{2}|\mathbf{x}), \tag{5}$$

where $F_n(y|\mathbf{x}) = \frac{1}{N_n} \sum_{i\in J_n} I(Y_i \leq y)$. We can find that $\theta_n(\mathbf{x})$ is just the solution of the following problem

$$\min_{a\in\mathcal{R}} \sum_{i=1}^{n} K_{n,i}(\mathbf{x})|Y_i - a|, \tag{6}$$

where $K_{n,i}(\mathbf{x}) = I\{|X_{i-j} - x_j| \leq h_n$ for all $1 \leq j \leq d\}$(cf. Truong (1989)). Here and in the sequel, $I(A)$ denotes the indicator function of the set A.

For the stationary process defined by (2) and the model defined by (3), Wu (2003) estimated the conditional mean of Y_n given $\mathbf{X}_{n-1} = \mathbf{x}$ based on L_2-norm. However, estimators based on L_2-norm are sensitive to outliers and do not perform well when the error distribution is heavy-tailed. Estimators based on L_1-norm are better than that based on L_2-norm from the point of robustness. They have several other advantages. For example, we can study asymptotic normality for the L_1-norm kernel estimator of conditional median without any moment conditions.

In this paper, the proposed method to establish asymptotic normality is based on martingale approximation, which is first clearly employed by Ho and Hsing (1996, 1997) in the context of nonlinear functionals of linear processes. The main idea is to approximate sums of stationary processes by martingales. Such approximation method acts as a bridge that connects stationary processes and martingales. Therefore, some results for martingales can be applied. For some applications of the martingale approximations, we refer to Wu and Mielniczuk (2002) and Wu (2003, 2005).

The rest of the paper is structured as follows. Some assumptions and the main results of our paper are provided in Section 2. The proofs of the main results are given in Section 3. Some applications to linear processes are given in Section 4.

2. Assumptions and Main Results

Let $\mathcal{C}_n = (\cdots, \varepsilon_{n-2}, \varepsilon_{n-1})$ be the shift process. For $\mathbf{x} = (x_1, \cdots, x_d)$, denote $g(\mathbf{x}|\mathcal{C}_j)$ the conditional density function of $\mathbf{X}_{j+d-1}$ at $\mathbf{x}$ given $\mathcal{C}_j$ and denote $f(y|\mathbf{x})$ the conditional density function of Y_n at y given $\mathbf{X}_{n-1} = \mathbf{x}$. Let $F(\cdot)$ be the distribution function of $\mathbf{X}_n$. Throughout the paper, convergence in distribution and convergence in probability are denoted by $\xrightarrow{d}$ and $\xrightarrow{p}$, respectively. C denotes a positive constant which may change from one place to another. Before stating the main results, we first give the following assumptions.

A1. For the neighborhood $U_{\mathbf{x}} := U_{\mathbf{x}}(\xi) = \{\mathbf{z} = (z_1, \cdots, z_d) \in \mathcal{R}^d : |z_j - x_j| \le \xi, 1 \le j \le d\}$, $\xi > 0$ and $\mathbf{x} = (x_1, \cdots, x_d) \in \mathcal{R}^d$, the distribution of $\mathbf{X}_n$ is absolutely continuous and its density function $f(\cdot)$ is bounded away from zero. That is,

$$M_1^{-1} \le f(\mathbf{z}) \le M_1 \quad \text{for } \mathbf{z} \in U_{\mathbf{x}}, \tag{1}$$

where M_1 may depend on $\mathbf{x}$. Furthermore, $f(\mathbf{z})$ has bounded partial derivatives for $\mathbf{z} \in U_{\mathbf{x}}$.

A2. $0 < \frac{\partial F(y|\mathbf{x})}{\partial y}|_{y=\theta(\mathbf{x})} \le M_2$ and $|\frac{\partial F(y|\mathbf{z})}{\partial z_j}| \le M_3$, for $j = 1, \cdots, d$, and

$(\mathbf{z}, y) \in B(\mathbf{x})$, where $B(\mathbf{x}) = U_{\mathbf{x}} \times U_{\theta(\mathbf{x})}$, $U_{\theta(\mathbf{x})} := U_{\theta(\mathbf{x})}(\xi) = \{y \in \mathcal{R} : |y - \theta(\mathbf{x})| \leq \xi\}$, $\xi > 0$, M_2 and M_3 are positive constants.

A3. Let $h_n > 0$ be bandwidths such that

$$h_n \to 0 \ , nh_n^d \to \infty \text{ and } nh_n^{d+2} \to 0 \text{ as } n \to \infty. \tag{2}$$

A3$'$. Let $h_n > 0$ be bandwidths such that

$$h_n \to 0 \ , nh_n^d \to \infty \text{ and } nh_n^{d+4} \to \Theta \text{ as } n \to \infty, \tag{3}$$

for some positive constant Θ.

A4. Define the projection operator $\mathcal{P}_j\zeta = E(\zeta|\mathcal{C}_j) - E(\zeta|\mathcal{C}_{j-1})$.

$$\sup_{\mathbf{z}} \|\mathcal{P}_0 g(\mathbf{z}|\mathcal{C}_j)\| \leq M_4 j^{-\alpha} l(j), \ \ j \geq 1, \tag{4}$$

for some $\frac{1}{2} < \alpha < 1$ and some positive constant M_4;

$$\sup_{\mathbf{z}} g(\mathbf{z}|\mathcal{C}_0) < M_5 \tag{5}$$

for some positive constant M_5. Here and in the sequel, $\|\cdot\|$ denotes the L_2-norm, $l(x)$ is a positive slowly varying function.

A4$'$. (5) holds and

$$\sup_{\mathbf{z}} \sum_{j=1}^{\infty} \|\mathcal{P}_0 g(\mathbf{z}|\mathcal{C}_j)\| < \infty. \tag{6}$$

Remark 1. (*Discussion of Conditions*) Conditions A1 and A2 are the same as Condition 1.1 in Zhou and Liang (2000). Conditions A3 and A3$'$ give some limits on bandwidths. Later, we shall establish asymptotic normality under Conditions A3 (A3$'$, *resp.*). Conditions A4 and A4$'$ are useful when applying the martingale approximation method. Furthermore, it is easy to check that Condition A4$'$ implies Condition A4.

Theorem 1. Suppose that (2) and (3) hold and Conditions A1-A4 are satisfied. Then

$$(nh_n^d)^{\frac{1}{2}}(\theta_n(\mathbf{x}) - \theta(\mathbf{x})) \xrightarrow{d} N(0, \sigma^2), \tag{7}$$

where $\sigma^2 = [2^{d+2} f^2(\theta(\mathbf{x})|\mathbf{x}) f(\mathbf{x})]^{-1}$, provided $f(\theta(\mathbf{x})|\mathbf{x}) > 0$ and $f(\mathbf{x}) > 0$.

Theorem 2. Assume that Conditions A3 and A4 are replaced by A3$'$ and A4$'$ respectively in Theorem 1. For $(\mathbf{z}, y) \in B(\mathbf{x})$, $f(\mathbf{z})$ and $F(y|\mathbf{z})$ have bounded twice partial derivatives. Then

$$(nh_n^d)^{\frac{1}{2}}(\theta_n(\mathbf{x}) - \theta(\mathbf{x})) \xrightarrow{d} N(-\mu, \sigma^2), \tag{8}$$

where $\mu = \frac{\sqrt{\Theta}\nu}{6f(\theta(\mathbf{x})|\mathbf{x})}$, $\nu = \sum_{j=1}^{d}\{\frac{\partial^2 F(\theta(\mathbf{x})|\mathbf{z})}{\partial z_j^2} + \frac{2}{f(\mathbf{z})}\frac{\partial f(\mathbf{z})}{\partial z_j}\frac{\partial F(\theta(\mathbf{x})|\mathbf{z})}{\partial z_j}\}_{|\mathbf{z}=\mathbf{x}}$ and σ^2 was defined in Theorem 1.

Remark 2. Wu (2003) consider the regression estimation problem for (1) and (2). He established asymptotic normality for the Nadaraya-Watson estimator of the regression function $E\{Y_n|\mathbf{X}_{n-1} = \mathbf{x}\}$ under some moment conditions on Y_n. We study the estimator of the conditional median of Y_n given $\mathbf{X}_{n-1} = \mathbf{x}$ without any moment conditions on Y_n.

Remark 3. For the stationary strong mixing process $\{\mathbf{X}_i, Y_i\}$, Zhou and Liang (2000) studied the L_1-norm estimator $\theta_n(\mathbf{x})$ of the conditional median $\theta_1(\mathbf{x})$. They proved asymptotic normality for $\theta_n(\mathbf{x})$ under more rigid condition on bandwidth, i.e.

$$\frac{nh_n^d}{(\log n)^3} \to \infty. \tag{9}$$

We establish analogous results for stationary processes defined by (2) and (3) and remove the restriction (9).

3. Proofs of the Main Results

Lemma 1. Define $\Delta_n(\mathbf{z}) = \sum_{j=1}^{n} g(\mathbf{z}|\mathcal{C}_j) - nf(\mathbf{z})$ for $\mathbf{z} \in \mathcal{R}^d$. If Condition A4 is satisfied, we have

$$\sup_{\mathbf{z}} \|\Delta_n(\mathbf{z})\|^2 = O(n^{3-2\alpha}l^2(n)); \tag{1}$$

if Condition A4$'$ is satisfied, we have

$$\sup_{\mathbf{z}} \|\Delta_n(\mathbf{z})\|^2 = O(n). \tag{2}$$

Proof. We only prove (1) since the proof of (2) can be found in Wu (2003). Note that $\|\mathcal{P}_k g(\mathbf{x}|\mathcal{C}_j)\|_2 = 0$ for $k > j$. By stationarity, we have

$$\begin{aligned}
\|\Delta_n(\mathbf{x})\|^2 &= \sum_{k=-\infty}^{n} \|\mathcal{P}_k \Delta_n(\mathbf{x})\|^2 \\
&\le \sum_{k=-\infty}^{-n-1} [\sum_{j=1}^{n} \|\mathcal{P}_k g(\mathbf{x}|\mathcal{C}_j)\|]^2 + \sum_{k=-n}^{0} [\sum_{j=1}^{n} \|\mathcal{P}_k g(\mathbf{x}|\mathcal{C}_j)\|]^2 \\
&\quad + \sum_{k=1}^{n} [\sum_{j=k}^{n} \|\mathcal{P}_k g(\mathbf{x}|\mathcal{C}_j)\|]^2 \\
&\le C \sum_{k=n+1}^{\infty} [\sum_{j=k+1}^{k+n} j^{-\alpha} l(j)]^2 + C \sum_{k=0}^{n} [\sum_{j=k+1}^{k+n} j^{-\alpha} l(j)]^2 \\
&\quad + C \sum_{k=1}^{n} [\sum_{j=1}^{n} j^{-\alpha} l(j)]^2 \\
&= O(n^{3-2\alpha} l^2(n)).
\end{aligned}$$

The proof of Lemma 1 is completed.

Lemma 2. Suppose that Conditions A1 and A4 are satisfied. If $h_n \to 0$ and $nh_n^d \to \infty$, then we have

$$(nh_n^d)^{-1} N_n - 2^d f(\mathbf{x}) = o_p(1), \tag{3}$$

where N_n is defined as before.
Proof. By the definition of N_n, it is easy to see that

$$N_n = \sum_{i=1}^{n} I\{|X_{i-j} - x_j| \le h_n \text{ for all } 1 \le j \le d\} = \sum_{i=1}^{n} K_{n,i}. \tag{4}$$

By Taylor's expansion, it follows that for $\mathbf{z} \in U_{\mathbf{x}}$,

$$f(\mathbf{z}) = f(\mathbf{x}) + \sum_{j=1}^{d} \frac{\partial f(\mathbf{z})}{\partial z_j}|_{\mathbf{z}=\phi}(z_j - x_j), \tag{5}$$

where $\mathbf{z} = (z_1, z_2, \cdots, z_d)$, $\mathbf{x} = (x_1, x_2, \cdots, x_d)$ and ϕ is some real-valued vector between $\mathbf{z}$ and $\mathbf{x}$. Therefore, by (5) and Condition A1, we have, for

n large enough,

$$\begin{aligned} EK_{n,1} &= f(\mathbf{x})(2h_n)^d + \sum_{j=1}^{d} \int \cdots \int_{\mathbf{z} \in U_{\mathbf{x}}(h_n)} \frac{\partial f(\mathbf{z})}{\partial z_j} |_{\mathbf{z}=\phi}(z_j - x_j) dz_1 \cdots dz_d \\ &= f(\mathbf{x})(2h_n)^d + O(h_n^{d+1}) \\ &= f(\mathbf{x})(2h_n)^d + o(h_n^d). \end{aligned}$$

In order to show (3), it suffices to prove that

$$\frac{1}{n}\sum_{i=1}^{n} K_{n,i} - EK_{n,1} = o_p(h_n^d). \tag{6}$$

It is enough to show that

$$\frac{1}{n}\sum_{i=1}^{n}(K_{n,i} - E(K_{n,i}|\mathcal{C}_{i-d})) = o_p(h_n^d) \tag{7}$$

and

$$\frac{1}{n}\sum_{i=1}^{n}(E(K_{n,i}|\mathcal{C}_{i-d}) - EK_{n,i}) = o_p(h_n^d). \tag{8}$$

By the definition of $\mathcal{C}_i$ and $K_{n,i}$, we can find that $K_{n,i}$ is $\mathcal{C}_i$-measurable. Hence, we have

$$\frac{1}{n}\sum_{i=1}^{n}(K_{n,i} - E(K_{n,i}|\mathcal{C}_{i-d})) = \frac{1}{n}\sum_{j=0}^{d-1}\sum_{i=1}^{n}(E(K_{n,i}|\mathcal{C}_{i-j}) - E(K_{n,i}|\mathcal{C}_{i-j-1})). \tag{9}$$

Therefore, in order to prove (7), it is enough to show that for $0 \le j \le d-1$,

$$\frac{1}{n}\sum_{i=1}^{n}(E(K_{n,i}|\mathcal{C}_{i-j}) - E(K_{n,i}|\mathcal{C}_{i-j-1})) = o_p(h_n^d).$$

We only prove the case of $j = 0$, i.e.

$$\frac{1}{n}\sum_{i=1}^{n}(K_{n,i} - E(K_{n,i}|\mathcal{C}_{i-1})) = o_p(h_n^d), \tag{10}$$

the case of $1 < j < d - 1$ can be proved similarly. Define

$$K_n = \sum_{i=1}^{n}(K_{n,i} - E(K_{n,i}|\mathcal{C}_{i-1})).$$

Noting that $EK_{n,1}^2 = O(h_n^d)$, by the Markov inequality, we have

$$P(|K_n| > \kappa n h_n^d) \le \frac{EK_n^2}{(\kappa n h_n^d)^2} \le \frac{CnEK_{n,1}^2}{(\kappa n h_n^d)^2} = O(\frac{1}{nh_n^d}) = o(1).$$

Hence, (7) is true. Next, we shall show that (8) holds. By Conditions A1, A4, Lemma 1 and the Hölder inequality, we have

$$
\begin{aligned}
&\sup_{\mathbf{z}\in U_{\mathbf{x}}(h_n)} E|\sum_{j=1}^{n} g(\mathbf{z}|\mathcal{C}_{j-d}) - nf(\mathbf{z})| \\
&\leq \sup_{\mathbf{z}\in U_{\mathbf{x}}(h_n)} E|\sum_{j=1}^{d} g(\mathbf{z}|\mathcal{C}_{j-d}) - df(\mathbf{z})| \\
&\qquad + \sup_{\mathbf{z}\in U_{\mathbf{x}}(h_n)} E|\sum_{j=d+1}^{n} g(\mathbf{z}|\mathcal{C}_{j-d}) - (n-d)f(\mathbf{z})| \\
&\leq d(M_5 + M_1) + O((n-d)^{\frac{3}{2}-\alpha} l(n-d)) \\
&= O(n^{\frac{3}{2}-\alpha} l(n)).
\end{aligned}
\tag{11}
$$

Therefore, we have

$$
\begin{aligned}
&E|\tfrac{1}{n}\sum_{i=1}^{n}(E(K_{n,i}|\mathcal{C}_{i-d}) - EK_{n,i})| \\
&\leq \tfrac{1}{n}\int\cdots\int_{\mathbf{z}\in U_{\mathbf{x}}(h_n)} |\sum_{j=1}^{n} g(\mathbf{z}|\mathcal{C}_{j-d}) - nf(\mathbf{z})| dz_1\cdots dz_d \\
&= O(n^{\frac{1}{2}-\alpha} l(n) h_n^d) = o(h_n^d).
\end{aligned}
\tag{12}
$$

Hence, (8) is true. The proof of Lemma 3 is completed.

Proof of Theorem 1. Define $\rho = \sigma t(nh_n^d)^{-\frac{1}{2}}$. By simplifying, we have

$$
\begin{aligned}
&P\{(nh_n^d)^{\frac{1}{2}}(\theta_n(\mathbf{x}) - \theta(\mathbf{x})) < t\sigma\} \\
&= P\{\tfrac{1}{2} < F_n(\theta(\mathbf{x}) + \rho|\mathbf{x})\} \\
&= P\{\tfrac{1}{N_n}\sum_{i\in J_n} I(Y_i > \theta(\mathbf{x}) + \rho) < \tfrac{1}{2}\} \\
&= P\{\tfrac{1}{N_n}\sum_{i\in J_n} (I(Y_i > \theta(\mathbf{x}) + \rho) - [1 - F(\theta(\mathbf{x}) + \rho|\mathbf{X}_{i-1})]) \\
&\qquad < \tfrac{1}{N_n}\sum_{i\in J_n} F(\theta(\mathbf{x}) + \rho|\mathbf{X}_{i-1}) - \tfrac{1}{2}\}.
\end{aligned}
\tag{13}
$$

Recall that $nh_n^{d+2} \to 0$. By Condition A2, we have

$$
\begin{aligned}
&\frac{1}{N_n}\sum_{i\in J_n} F(\theta(\mathbf{x}) + \rho|\mathbf{X}_{i-1}) - \frac{1}{2} \\
&= \frac{1}{N_n}\sum_{i\in J_n}\{F(\theta(\mathbf{x}) + \rho|\mathbf{X}_{i-1}) - F(\theta(\mathbf{x}) + \rho|\mathbf{x})\} + \{F(\theta(\mathbf{x}) + \rho|\mathbf{x}) - \frac{1}{2}\} \\
&= O(h_n) + \rho f(\theta(\mathbf{x})|\mathbf{x}) + o((nh_n^d)^{-\frac{1}{2}}) \\
&= \rho f(\theta(\mathbf{x})|\mathbf{x}) + o((nh_n^d)^{-\frac{1}{2}}).
\end{aligned}
$$

Hence, the last term of (13) equals

$$
P\{\frac{1}{N_n}\sum_{i\in J_n}(I(Y_i > \theta(\mathbf{x}) + \rho) - [1 - F(\theta(\mathbf{x}) + \rho|\mathbf{X}_{i-1})])
$$

$$< \rho f(\theta(\mathbf{x})|\mathbf{x}) + o((nh_n^d)^{-\frac{1}{2}})\}. \tag{14}$$

By Lemma 2, we have

$$\frac{1}{nh_n^d} N_n \xrightarrow{p} 2^d f(\mathbf{x}). \tag{15}$$

Therefore, in order to show that (7) holds, it suffices to prove that

$$\frac{1}{\sqrt{n}} \sum_{i=1}^{n} Z_{n,i} \xrightarrow{d} N(0,1), \tag{16}$$

where $Z_{n,i} = (\frac{4}{f(\mathbf{x})(2h_n)^d})^{\frac{1}{2}} K_{n,i}(\mathbf{x})\{I(Y_i > \theta(\mathbf{x}) + \rho) - [1 - F(\theta(\mathbf{x}) + \rho|\mathbf{X}_{i-1})]\}$. Define $\mathcal{F}_i = \sigma(\cdots, \eta_{i-1}, \varepsilon_{i-1}, \eta_i, \varepsilon_i)$. It is easy to see that $\{Z_{n,i}\}$ is $\mathcal{F}_i$-measurable. Furthermore,

$$\begin{aligned} E(Z_{n,i}|\mathcal{F}_{i-1}) = (\frac{4}{f(\mathbf{x})(2h_n)^d})^{\frac{1}{2}} K_{n,i}(\mathbf{x})\{E[I(Y_i > \theta(\mathbf{x}) + \rho)|\mathcal{F}_{i-1}] \\ -[1 - F(\theta(\mathbf{x}) + \rho|\mathbf{X}_{i-1})]\} = 0 \quad \text{a.s.} \end{aligned}$$

Therefore, $\{Z_{n,i}\}$ is a stationary sequence of martingale differences. By Lemma 2, we have

$$\frac{1}{nh_n^d} \sum_{i=1}^{n} K_{n,i}(\mathbf{x}) \xrightarrow{p} 2^d f(\mathbf{x}). \tag{17}$$

Noting that $K_{n,i}^2(\mathbf{x}) = K_{n,i}(\mathbf{x})$, by (17), we have

$$\begin{aligned} \sum_{i=1}^{n} E(Z_{n,i}^2|\mathcal{F}_{i-1}) &= \sum_{i=1}^{n} \tfrac{4}{f(\mathbf{x})(2h_n)^d} K_{n,i}^2(\mathbf{x})(\tfrac{1}{4} + O(\rho + h_n)) \\ &= \sum_{i=1}^{n} \tfrac{1}{f(\mathbf{x})(2h_n)^d} K_{n,i}(\mathbf{x})(1 + O(\rho + h_n)) \\ &= n + o_p(n). \end{aligned} \tag{18}$$

Therefore, we have

$$\frac{1}{n} \sum_{i=1}^{n} E(Z_{n,i}^2|\mathcal{F}_{i-1}) \xrightarrow{p} 1. \tag{19}$$

By the martingale central limit theorem (cf. Hall and Heyde (1980)), we have to check the Lindeberg condition, which follows from

$$\frac{1}{\sqrt{n}} E|Z_{n,1}|^3 \le \frac{C}{\sqrt{nh_n^{3d}}} EK_{n,1}^3(\mathbf{x}) = O(\frac{1}{\sqrt{nh_n^d}}) = o(1). \tag{20}$$

In view of (19) and (20), (16) holds. The proof of Theorem 1 is completed.

Next, we turn to the proof of Theorem 2.

Lemma 3. Define $V_{n,i} = K_{n,i}(\mathbf{x})(I(Y_i > \theta(\mathbf{x})+\rho)-\frac{1}{2})$. Then, if Conditions A1, A2 and A3$'$ hold, we have

$$EV_{n,1} = -(\frac{f(\mathbf{x})(2h_n)^d}{4n})^{\frac{1}{2}}t - \frac{2^d h_n^{d+2} f(\mathbf{x})}{6}\nu + o(\rho h_n^d) + o(h_n^{d+2}), \tag{21}$$

where ν and ρ are defined as above.

Proof. Notice that

$$EV_{n,1} = EK_{n,1}(\mathbf{x})I(Y_1 > \theta(\mathbf{x}) + \rho) - \frac{1}{2}EK_{n,1}(\mathbf{x}). \tag{22}$$

Using Taylor's expansion, we have

$$EK_{n,1}(\mathbf{x}) = f(\mathbf{x})(2h_n)^d + \frac{1}{2}\sum_{i=1}^{d}\sum_{j=1}^{d}\frac{\partial^2 f(\mathbf{z})}{\partial z_i \partial z_j}|_{\mathbf{z}=\mathbf{x}} w_{i,j}(h_n) + o(h_n^{d+2}), \tag{23}$$

where $w_{i,j}(h_n) = \int\cdots\int_{\mathbf{z}\in U_{\mathbf{x}}(h_n)}(z_i - x_i)(z_j - x_j)dz_1\cdots dz_d$; on the other hand,

$$\begin{aligned}&EK_{n,1}(\mathbf{x})I(Y_1 > \theta(\mathbf{x}) + \rho)\\ &= \tfrac{1}{2}f(\mathbf{x})(2h_n)^d - \rho f(\mathbf{x})f(\theta(\mathbf{x})|\mathbf{x})(2h_n)^d\\ &+\tfrac{1}{2}\sum_{i=1}^{d}\sum_{j=1}^{d}[\tfrac{\partial^2 f(\mathbf{z})}{\partial z_i \partial z_j} - \tfrac{\partial^2 (f(\mathbf{z})F(\theta(\mathbf{z})|\mathbf{z}))}{\partial z_i \partial z_j}]|_{\mathbf{z}=\mathbf{x}} w_{i,j}(h_n) + o(\rho h_n^d) + o(h_n^{d+2}).\end{aligned} \tag{24}$$

Note that $w_{i,j}(h_n) = 0$ $(i \neq j)$ and $w_{i,i}(h_n) = \frac{2^d}{3}h_n^{d+2}$. By (22), (23) and (24), we have

$$\begin{aligned}EV_{n,1} &= -\rho f(\mathbf{x})f(\theta(\mathbf{x})|\mathbf{x})(2h_n)^d - \frac{1}{2}\sum_{j=1}^{d}[\frac{\partial^2 F(\theta(\mathbf{x})|\mathbf{z})}{\partial z_j^2}f(\mathbf{z})\\ &\quad +2\frac{\partial f(\mathbf{z})}{\partial z_j}\frac{\partial F(\theta(\mathbf{x})|\mathbf{z})}{\partial z_j}]|_{\mathbf{z}=\mathbf{x}} w_{j,j}(h_n) + o(\rho h_n^d) + o(h_n^{d+2})\\ &= -(\frac{f(\mathbf{x})(2h_n)^d}{4n})^{\frac{1}{2}}t - \frac{2^d h_n^{d+2} f(\mathbf{x})}{6}\nu + o(\rho h_n^d) + o(h_n^{d+2}).\end{aligned}$$

Proof of Theorem 2. Similarly as (13), we have

$$\begin{aligned}P\{(nh_n^d)^{\frac{1}{2}}(\theta_n(\mathbf{x}) - \theta(\mathbf{x})) < t\sigma\} &= P\{\tfrac{1}{N_n}\sum_{i\in J_n} I(Y_i > \theta(\mathbf{x}) + \rho) < \tfrac{1}{2}\}\\ &= P\{\sum_{i=1}^{n} V_{n,i} < 0\}\end{aligned} \tag{25}$$

where $V_{n,i}$ is defined as in Lemma 3. Denote $U_{n,i} = (\frac{4}{f(\mathbf{x})(2h_n)^d})^{\frac{1}{2}} V_{n,i}$. By Lemma 3 and $nh_n^{d+4} \to \Theta$,

$$\sqrt{n}EU_{n,1} = -t - \frac{(2^{d+2}f(\mathbf{x})\Theta)^{\frac{1}{2}}}{6}\nu + o(1) = -t - \frac{\mu}{\sigma} + o(1). \tag{26}$$

Therefore, the last term in (25) equals

$$P\{\frac{1}{\sqrt{n}}\sum_{i=1}^{n}(U_{n,i} - EU_{n,i}) < t + \frac{\mu}{\sigma} + o(1)\}.$$

Hence, (8) is equivalent to

$$\frac{1}{\sqrt{n}}\sum_{i=1}^{n}(U_{n,i} - EU_{n,i}) \xrightarrow{d} N(0,1). \tag{27}$$

We apply the martingale approximation method to prove (27), i.e.

$$\frac{1}{\sqrt{n}}\sum_{i=1}^{n}[U_{n,i} - E(U_{n,i}|\mathcal{F}_{i-1})] \xrightarrow{d} N(0,1), \tag{28}$$

$$\frac{1}{\sqrt{n}}\sum_{i=1}^{n}[E(U_{n,i}|\mathcal{F}_{i-1}) - EU_{n,i}] \xrightarrow{p} 0, \tag{29}$$

where $\mathcal{F}_i$ was defined in the proof of Theorem 1. The proof of (28) is the same as that of (16). Hence, we only show that (29) holds. It suffices to prove

$$\sum_{i=1}^{n}[E(U_{n,i}|\mathcal{F}_{i-1}) - E(U_{n,i}|\mathcal{F}_{i-d-1})] = o_p(\sqrt{n}), \tag{30}$$

$$\sum_{i=1}^{n}[E(U_{n,i}|\mathcal{F}_{i-d-1}) - EU_{n,i}] = o_p(\sqrt{n}). \tag{31}$$

By (2), the proof of (31) is analogous as that of (12). Noting that $F(\theta(\mathbf{x})|\mathbf{x}) = \frac{1}{2}$, by Condition A2, we have

$$\begin{aligned}
&\|E(U_{n,1}|\mathcal{F}_0)\|^2 \\
&= \tfrac{2^{d+2}}{f(\mathbf{x})h_n^d}E\{K_{n,1}(\mathbf{x})[F(\theta(\mathbf{x}+\rho)|\mathbf{X}_0) - \tfrac{1}{2}]\}^2 \\
&\le \tfrac{C}{h_n^d}E\{K_{n,1}(\mathbf{x})[F(\theta(\mathbf{x}+\rho)|\mathbf{X}_0) - F(\theta(\mathbf{x})|\mathbf{X}_0) + F(\theta(\mathbf{x})|\mathbf{X}_0) - F(\theta(\mathbf{x})|\mathbf{x})]\}^2 \\
&\le \tfrac{C(\rho+h_n)^2}{h_n^d}EK_{n,1}^2(\mathbf{x}) \\
&\le O((nh_n^d)^{-1} + h_n^2) = o(1).
\end{aligned} \tag{32}$$

Following the argument in Wu (2003) and by (32), we have for $j = 1, \cdots, d$,

$$\begin{aligned}
&\| \sum_{i=0}^{[n/d]-1} [E(U_{n,id+j}|\mathcal{F}_{id-1+j}) - E(U_{n,id+j}|\mathcal{F}_{id-d-1+j})]\|^2 \\
&= [n/d]\|E(U_{n,0}|\mathcal{F}_{-1}) - E(U_{n,0}|\mathcal{F}_{-d-1})\|^2 \\
&\le [n/d]\|E(U_{n,0}|\mathcal{F}_{-1}\|^2 = o(n).
\end{aligned}$$

Hence, (30) follows from

$$\begin{aligned}
&E|\sum_{i=1}^{n}[E(U_{n,i}|\mathcal{F}_{i-1}) - E(U_{n,i}|\mathcal{F}_{i-d-1})]| \\
&\le \sum_{j=1}^{d} E| \sum_{i=0}^{[\frac{n}{d}]-1} [E(U_{n,id+j}|\mathcal{F}_{id-1+j}) - E(U_{n,id+j}|\mathcal{F}_{id-d-1+j})]| \\
&\le \sum_{j=1}^{d} \| \sum_{i=0}^{[n/d]-1} [E(U_{n,id+j}|\mathcal{F}_{id-1+j}) - E(U_{n,id+j}|\mathcal{F}_{id-d-1+j})]\| \\
&= o(\sqrt{n}).
\end{aligned}$$

Therefore, the proof of Theorem 2 is completed.

4. Applications

If the measurable function Q has the linear form

$$X_n = Q(\cdots, \varepsilon_{n-1}, \varepsilon_n) = \sum_{j=0}^{\infty} a_j \varepsilon_{n-j}, \tag{1}$$

where $\sum_{i=0}^{\infty} a_i^2 < \infty$ and $\{\varepsilon_i\}$ is a sequence of i.i.d. random variables, $E\varepsilon_1 = 0$, $E\varepsilon_1^2 < \infty$, then X_n is well defined and includes many useful processes such as ARMA and fractional ARIMA models. When $\sum_{i=0}^{\infty} |a_i| < \infty$, it is easy to check that the covariances of X_n are absolutely summable and the sequence is short-range dependent. When the covariances of X_n are not absolutely summable, the sequence is long-range dependent. Many authors studied the nonparametric estimation for the linear processes. For instance, Tran (1992), Wu and Mielniczuk (2002) discussed the kernel density estimator for linear processes. Honda (2000) studied nonparametric estimation of the conditional median for long-range processes. Throughout the section, we assume that for the linear process (1), there exists a positive integer d such that

$$f(x|\mathcal{C}_i) = f(x|\mathbf{X}_{i-1}), \quad \text{a.s.} \tag{2}$$

where $f(x|\mathcal{C}_i)$ is the conditional density of X_i at x given $\mathcal{C}_i = (\cdots, \varepsilon_{i-2}, \varepsilon_{i-1})$, and $f(x|\mathbf{X}_{i-1})$ is the conditional density of X_i at x given $\mathbf{X}_{i-1} = (X_{i-1}, \cdots, X_{i-d})$. It is easy to check that (2) is satisfied if $X_n = R(X_{n-1}, \cdots, X_{n-d}, \varepsilon_n)$, where R is a multivariate measurable function. As an application of our main results, we shall consider the L_1-norm estimator $\theta_n(\mathbf{x})$ of the conditional median of X_n given $\mathbf{X}_{i-1} = \mathbf{x}$.

Proposition 1. Let linear process X_n be defined as (1), where $|a_j| \le Cj^{-\alpha}l(j)$ for some $\frac{1}{2} < \alpha < 1$. Assume that Conditions A2, A3 and (2) are satisfied. If there exists the density function $f_\varepsilon(\cdot)$ of ε_1 such that

$$\sup_x f_\varepsilon(x) < \infty \text{ and } \sup_x |\frac{df_\varepsilon(x)}{dx}| < \infty, \tag{3}$$

then (7) holds.

Proof. Without loss of generality, let $d = 2$ and $a_0 = 1$. First we show that (3) implies Conditions A1 and A4. By (3), the density function $f_\varepsilon(\cdot)$ is Lipschitz continuous with bounded partial derivatives. Using Lemma 1 in Wu and Mielniczuk (2002), the density function of X_n exists and has bounded partial derivatives. Hence, Condition A1 is satisfied. On the other hand, it is easy to check that

$$g(\mathbf{x}|\mathcal{C}_j) = f_\varepsilon(x_1 - Y_j - a_1(x_2 - Z_j))f_\varepsilon(x_2 - Z_j), \tag{4}$$

where $\mathbf{x} = (x_1, x_2)$, $Y_j = \sum_{i=2}^{\infty} a_i\varepsilon_{j+1-i}$ and $Z_j = \sum_{i=1}^{\infty} a_i\varepsilon_{j-i}$. By (4), $\sup_x f_\varepsilon(x) < \infty$ implies (5). By the analogous proof as that of Theorem 4 in Wu (2003), we can show that

$$\|E[g(\mathbf{x}|\mathcal{C}_j)|\mathcal{C}_0] - E[g(\mathbf{x}|\mathcal{C}_j)|\mathcal{C}_{-1}]\| = O(|a_{j+2}| + |a_{j+3}|). \tag{5}$$

Therefore, (5) and $|a_j| \le Cj^{-\alpha}l(j)$ imply that (4) is satisfied. Therefore, Conditions A1-A4 are satisfied. Using (2), the proof is analogous as that of Theorem 1 with some modifications.

Remark 4. The condition $|a_j| \le Cj^{-\alpha}l(j)$, $\frac{1}{2} < \alpha < 1$, implies that X_n can be either a short-range linear process or a long-range linear process. Therefore, Proposition 1 gives asymptotic normality for L_1-norm estimator of the conditional median of X_n given $\mathbf{X_{n-1}} = \mathbf{x}$ under both short-range dependence and long-range dependence.

Remark 5. Next, we shall show that Condition A2 is satisfied under certain conditions. For example, consider the nonlinear AR (d) model

$$X_n = \Phi(X_{n-1}, \cdots, X_{n-d}) + \Psi(X_{n-1}, \cdots, X_{n-d})\varepsilon_n, \tag{6}$$

where ε_n is a sequence of i.i.d. random variables. Consider the conditional median estimator of X_n, given $(X_{n-1}, \cdots, X_{n-d})$. If

$$\sup_{(z_{n-1},\cdots,z_{n-d})\in\mathcal{R}^d} \{|\Phi(z_{n-1}, \cdots, z_{n-d})| + |\Psi(z_{n-1}, \cdots, z_{n-d})|^{-1}\} < \infty, \tag{7}$$

$$\sup_{(z_{n-1},\cdots,z_{n-d})\in\mathcal{R}^d} \{|\frac{\partial\Phi(z_{n-1}, \cdots, z_{n-d})}{\partial z_{n-i}}| + |\frac{\partial(\Psi(z_{n-1}, \cdots, z_{n-d}))^{-1}}{\partial z_{n-j}}|\} < \infty \tag{8}$$

for any $1 \leq i, j \leq d$ and $\sup_x f_\varepsilon(x) < \infty$, then Condition A2 is satisfied. Details are omitted.

Example. Consider AR (d) model

$$\Lambda(B)X_n = \varepsilon_n, \tag{9}$$

where $\Lambda(z) = 1 - a_1 z - \cdots - a_d z^d$, B is a lag operator and $\{\varepsilon_i\}$ is defined as before. For the conditional median estimator of X_n, given $(X_{n-1}, \cdots, X_{n-d})$, we have the following result. Suppose that Condition A3 and (3) holds. Then, (7) holds if

$$\Lambda(z) \neq 0 \text{ for all } |z| \leq 1. \tag{10}$$

Recall that (10) is equivalent to that X_n is causal. Hence, by the proof of Proposition 1, (3) implies that Conditions A1 and A4 are satisfied. On the other hand, it is easy to check that (7) and (8) are satisfied. Therefore, by Remark 5, Condition A2 holds. Furthermore, it is easy to check that (2) holds. Finally, by Proposition 1, we obtain the result.

References

1. Cai Z.W. (2002). Regression quantile for time series. *Econometric Theory* **18**, 169-192.
2. Hall P. and Heyde C.C. (1980). *Martingale Limit Theory and Its Application.* Academic Press, New York.
3. Ho H.C. and Hsing T. (1996). On the asymptotic expansion of the empirical process of long-memory moving averages. *Ann. Statist.* **24**, 992-1024.

4. Ho H.C. and Hsing T. (1997). Limit theorems for functionals of moving averages. *Ann. Probab.* **25**, 1636-1669.
5. Honda T. (2000). Nonparametric estimation of a conditional median function for long-range dependent processes. *J. Japan Statist. Soc.* **30**, 185-198.
6. Koenker R. and Bassett G.W. (1978). Regression quantiles. *Econometrica* **46**, 33-50.
7. Koenker R. and Hallock K.F. (2001). Quantile regression: An introduction. *J. Economic Perspectives* **15**, 143-157.
8. Tran L.T. (1992). Kernel density estimation for linear processes. *Stochastic Process Appl.* **26**, 281-296.
9. Truong Y.K. (1989). Asymptotic properties of kernel estimators based on local medians. *Ann. Statist.* **17**, 606-617.
10. Wu W.B. (2003). Nonparametric estimation for stationary processes. (preprint)
11. Wu W.B. (2005). On the Bahadur representation of sample quantile for dependent sequences. *Ann. Statist.* **33**, 1934-1963.
12. Wu W.B. and Mielniczuk J. (2002). Kernel density estimation for linear processes. *Ann. Statist.* **30**, 1441-1459.
13. Xiang X.J. (1996). A kernel estimate of a conditional quantile. *J. Multi. Anal.* **59**, 206-216.
14. Zhou Y. and Liang H. (2000). Asymptotic normality for L_1-norm kernel estimator of conditional median under α-mixing dependence. *J. Multi. Anal.* **73**, 136-154.

Statistical Applications

FROM MICROSTRUCTURES TO PROPERTIES: STATISTICAL ASPECTS OF COMPUTATIONAL MATERIALS SCIENCE

CHUANSHU JI *

Department of Statistics and Operations Research, University of North Carolina
Chapel Hill, NC 27599, U.S.A.
cji@email.unc.edu

We discuss some statistical aspects in materials science that involve microstructures and materials properties. Materials scientists have applied homogenization theory and the finite element method (FEM) for justification and computation of certain effective properties, which are considered as expectations with respect to the distributions of underlying microstructures. However, many important issues remain unresolved, including the implementation of more efficient Monte Carlo computational methods, quantification of statistical variability of FEM, etc. Through several examples, we illustrate stochastic geometry models for microstructural features, Markov chain Monte Carlo (MCMC) methods in computation of elastic moduli and thermal (or electrical) conductivity, and MCMC based confidence intervals related to FEM.

Some key Words: Materials science; Microstructure, Effective property; Homogenization; Finite element; Markov chain Monte Carlo.

1. Introduction

Materials scientists wish to design processes to produce components with desired *performance* from materials with certain *properties* by controlling *microstructure*. Many research problems along the line "performance-property-microstructure" are statistical in nature, because they often raise difficult issues regarding different levels of uncertainty. Standard statistical methods are used in the analysis of material performance data. Statistical reliability models have played a role in studies of material properties. In contrast, statisticians have done very little work in modelling and computation related to material microstructure. This is mainly due to the enormous

*Research supported in part by AFOSR Grant F49620-96-1-0133.

complexity involved in microstructure data, which hinders many traditional statistical techniques. Therefore, further studies of microstructure, related fracture mechanics, and various macroscopic properties require novel statistical approaches.

The research in statistical aspects of computational materials science has a clear interdisciplinary nature. There are many existing theories and methods in computational materials science, successful in various degrees, noticeably finite element methods (FEM). But their connections to modern statistical methodology have not been articulated as they should be. The main goal of this work is to illustrate some important statistical issues in several examples, and attract more attentions from statisticians to develop relevant statistical methods in computational materials science.

2. Effective and Critical Properties

Computation of macroscopic properties always takes a center stage in materials science. The use of microstructural information in such computation is widely recognized as a necessary but daunting task. A basic question: why not measure those properties in lab tests? Experimental materials scientists do that on daily basis. But it need not always be easy. The undertaking to prepare the required conditions for a certain desirable lab test might be far greater than to compute a property. Even if lab tests are convenient, it would still be scientifically sound to verify an experimental result or predict other possibilities via a computer program. In short, computational materials science and experimental materials science complement each other.

From the statistical point of view, properties of interest can be classified in two groups: effective properties and critical properties. In principle, members of the first group can be expressed as expectations (ensemble averages) with respect to the probability distributions for underlying microstructures, thus can naturally be approximated by ergodic averages through Monte Carlo simulation. Examples in this group include effective conductivity (thermal or electrical), permeability, and various elastic moduli. However, derivation of the expectation expressions need not be obvious, and it is the core of homogenization theory for random media. See Papanicolaou (1995) for an excellent survey. Moreover, homogenization theory only provides a mathematical formulation to justify the definitions of effective properties, but not tools to compute them, which is a more useful but difficult problem in materials science. There have been many techniques introduced for computation in different situations.

Members of the second group are not averages, but extreme values.

Examples include breakdown strength, fracture toughness, etc. Compared to effective properties, this is a much less developed area. An exception is the fiber-composite materials, for which 1D statistical reliability models, based on the weakest length theory, were studied in Phoenix and Taylor (1973). A multidimensional extension is the class of spring networks [Curtin and Scher (1990)] which also serves as a computational alternative to FEM. It is not used by materials scientists as extensively as FEM, and sometimes criticized for introducing an unnatural scaling effect [Jagota and Bennison (1994a) (1994b)]. A major distinction lies in the way of discretization. A spring network is defined on a lattice — the discretized material; while FEM discretize the partial differential equations (PDEs) governing the physical law defined on a continuum material. One could study stochastic PDEs in statistical modelling. But until now, the development is still in its infancy, especially the numerical analysis aspect.

A general idea is hierarchical modelling. There are at least two basic layers in modelling the microstructure-property relationships, denoted by L_m (for microstructure) and L_p (for property) respectively. The Bayesian paradigm would rely on the conditional probability measures $P(L_p|L_m)$ and $P(L_m|L_p)$. Both of them are complex due to the variety of microstructure and the involvement of underlying physics. In fact, very little is known about $P(L_m|L_p)$, because it is not even clear how to define and formulate it. We will mainly focus on $P(L_p|L_m)$ which falls on the same line as the estimation of properties based on microstructural information.

3. Computation of Effective Properties

In this section, we describe two problems which illustrate the impact of statistics in the studies of effective properties. We also discuss a link between statistics and FEM.

3.1. *Effective conductivity σ^* of two-phase random media*

Materials scientists are often interested in the effective behavior of heterogeneous media. In this set-up, *if* there is a homogeneous medium, called the effective medium S^*, whose conductivity σ^* is equal to the conductivity of the original medium S when measured on a large space scale, then σ^* is called the effective conductivity. Consider a simple 1D deterministic conductor S in Figure 1:

Fig. 1.

Clearly,

$$\sigma^* = \frac{1}{\sigma_1^{-1} + \sigma_2^{-1}}.$$

If we extend it to a 1D random conductor, consisting of iid Bernoulli random variables

$$\sigma(x) = \begin{cases} \sigma_1, \, x \in \text{Phase 1 (with probability } p) \\ \sigma_2, \, x \in \text{Phase 2 (with probability } q), \end{cases}$$

then we have

$$\sigma^* = \frac{1}{p\,\sigma_1^{-1} + q\,\sigma_2^{-1}} = \frac{1}{E\left[\frac{1}{\sigma(x)}\right]}.$$

The previous deterministic conductor, after normalization, can be treated as a special case with $p = q = 1/2$. Unfortunately, such a neat harmonic mean representation of the effective conductivity, $\sigma^* = \left[E\frac{1}{\sigma(x)}\right]^{-1}$, does not hold for general random media.

Consider a two-phase composite material represented by the 2D domain S. Each phase i itself is homogeneous with constant (electrical or thermal) conductivity $\sigma_i > 0$, $i = 1, 2$. Assume the phase 1 region, $R_1 \subset S$, is precisely covered by the configuration ξ_n with n nonoverlapping ellipses, and $R_2 = S \backslash R_1$ is the phase 2 region. Define the local conductivity

$$\sigma(x) = \sigma_1 I_{\{x \in R_1\}} + \sigma_2 I_{\{x \in R_2\}}, \quad x \in S. \tag{1}$$

Hence $\{\sigma(x),\, x \in S\}$ is a random field due to the random distribution of ξ_n.

There are two important issues. The first one is to justify the existence of σ^*, i.e. under what conditions the averaging, i.e. homogenization, takes place so that the complex small scale random structure can be replaced by an asymptotically equivalent homogeneous deterministic structure. Here is a summary.

Start with a stochastic PDE

$$\nabla \cdot [\sigma(x)\ \nabla u(x)] = 0 \tag{2}$$

with a boundary condition, where u is the temperature distribution function. A general formulation for random media is to attach multiple scales to (2) by introducing a small parameter $\epsilon > 0$:

$$\nabla \cdot [\sigma(x/\epsilon, \omega)\ \nabla u(x/\epsilon, \omega)] + a\ u(x/\epsilon, \omega) = g(x) \tag{3}$$

with a boundary condition. If the solution u "converges" to u_0 in some sense as $\epsilon \to 0$, so that the limit is governed by a deterministic PDE

$$\nabla \cdot [\sigma^* \nabla u_0(x)] + a\, u_0(x) = g(x) \tag{4}$$

with a boundary condition and for some constant σ^*, then σ^* is referred to as the effective conductivity. The effective conductivity σ^* is defined as the conductivity of a macroscopically homogeneous "equivalent" (virtual) material. σ^* depends on σ_1, σ_2 and how the two phases are distributed (i.e. the microstructure).

The second issue is the computation of σ^*, which is more relevant in computational materials science. There are several methods that apply to different situations. One is the perturbation expansion argument, originally developed by Brown (1955) and amplified by Torquato and his collaborators in a sequence of papers, e.g. Torquato (1985), Sen and Torquato (1989), Torquato and Sen (1990), and Torquato (1991).

The perturbation expansion for d-dimensional media (in particular, $d =$ 2 or 3) is expressed as

$$(q_i\beta_{ij})^2(\sigma^* - \sigma_j \cup)^{-1}[\sigma^* + (d-1)\sigma_j \cup] \;=\; q_i\beta_{ij} \cup - \sum_{k=2}^{\infty} A_k^{(i)} \beta_{ij}^k, \tag{5}$$

$1 \leq i \neq j \leq 2$, where the tensor coefficients $A_k^{(i)}$ are represented as integrals:

$$A_2^{(i)} \;=\; \frac{d}{2\pi(d-1)} \int T_{12} \cdot (P_{1,2}^i - q_i^2)\, dz_2, \tag{6}$$

and

$$A_k^{(i)} = -q_i^{2-k} \left[\frac{-d}{2\pi(d-1)}\right]^{k-1} \int \cdots \int (T_{12} \cdots T_{k-1,k} \cdot D_k^i)\, dz_2 \cdots dz_k, \tag{7}$$

$k \geq 3$; with the notation:
q_i is the volume (area) fraction of phase i; $\beta_{ij} = \frac{\sigma_i - \sigma_j}{\sigma_i + (d-1)\sigma_j}$; for each $m = 1, \ldots, k$, $z_m \in S$ is a d-dimensional vector, and $T_{m-1,m}$ is the dipole-dipole interaction tensor dependent upon z_{m-1} and z_m; D_k^i is a $(k-1) \times (k-1)$ determinant with entries P_{m_1,m_2}^i — the probability that all $(m_2 - m_1 + 1)$ points $z_{m_1}, z_{m_1+1}, \ldots, z_{m_2}$ are contained in the phase i region R_i, where $m_1, m_2 = 1, \ldots, k$ and $m_1 \leq m_2$.

A major problem is to tackle the massive computation of $A_k^{(i)}$, especially for large k. For many cases of interest, the only available results in the literature are lower-order bounds (for $k \leq 5$), and even those bounds are not always sharp. Further progress in this direction requires good approximations to the family of k-point probability functions (not just for small k)

as a crucial step. The family of k-point probability functions defined in a multiphase composite is in a core of computing effective properties of the composite, because it characterizes the uncertainty of microstructure distribution. A k-point probability is the probability that *given* k points in the composite are contained in a certain configuration of those phases. When the number k and the locations of the k points vary, the corresponding probabilities constitute the family of k-point probability functions.

A new MCMC-based method, introduced in Derr (1999), shed some light on this problem. The basic idea seems clear: given k points in S, generate a long run of MCMC samples and use the empirical relative frequency in the MCMC sequence as an estimate of the k-point probability. For the family of k-point probability functions, the procedure needs to be repeated many times with different k and varying locations of the k points. For the MCMC sample X_t at each time t, the empirical frequency can be counted by translating the k points in the window S under the stationarity assumption. Here arises the challenge: with large k and spread patterns of k points, the count in each X_t becomes very low, thus more MCMC samples are needed to come up with a reasonable estimate. To alleviate the computational burden, a two-stage procedure was proposed in Derr (1999). The first stage is to estimate the 2-point probability function via MCMC; while the second stage is to compute the k-point probability functions for $k \geq 3$ iteratively based on the estimated 2-point probability function obtained in the first stage. The advantage of this procedure is that the empirical relative frequency can be easily obtained in the first stage since only two points are involved; and no MCMC simulation is involved in the second stage.

Denote the probability that k points $z_1, \ldots, z_k$ fall in the phase 1 region by $P_k(z_1, \ldots, z_k)$ and their estimates by $\hat{P}_k$. The second stage relies on the iterative formula:

$$\begin{aligned}&\hat{P}_k(z_1, \ldots, z_k) \qquad (8)\\ &= \hat{P}_{k-1}(z_1, \ldots, z_{k-1}) \left[\frac{(1-\hat{q}) \sum_{i=1}^{k-1} \hat{P}_2(z_i, z_k) - (k-1)\hat{q}^2}{\hat{q} + \sum_{i=1}^{k-1} w_i \sum_{j \neq i}^{k-1} \hat{P}_2(z_i, z_j) - (k-1)\hat{q}^2} + \hat{q} \right],\end{aligned}$$

where q is the volume fraction of phase 1, and the weights $w_1, \ldots, w_{k-1}$ are defined as follows:

Let $r_{ij} = |z_i - z_j|$ be the (Euclidean) distance between two points z_i and z_j. For each $i = 1, \ldots, k-1$, let

$$W_i = \frac{r_{1k}}{r_{1k} + r_{ik}} \frac{r_{2k}}{r_{2k} + r_{ik}} \cdots \frac{r_{i-1,k}}{r_{i-1,k} + r_{ik}} \frac{r_{i+1,k}}{r_{i+1,k} + r_{ik}} \cdots \frac{r_{k-1,k}}{r_{k-1,k} + r_{ik}}$$

and

$$w_i = \frac{W_i}{\sum_{j=1}^{k-1} W_j}. \tag{9}$$

In practice, the numerical value in [ ] (denoted by [ ]) in (8) should be constrained in $[0, 1]$ by taking $\min(\max([\ \ ], 0), 1)$.

In comparison to direct MCMC, there is a tremendous reduction in computational time by using the above two-stage procedure. In Derr (1999), the comparison was demonstrated in an example of computing the determinant D_5^1, for which the direct MCMC took 21 minutes while the two-stage procedure only took 15 seconds. However, there is a noticeable bias incurred when using the two-stage procedure in estimation.

Bias correction in this situation is a tricky problem. The bias is due to using an oversimplified "lower bound" q for all conditional probabilities $P(z_k \mid z_1, \ldots, z_{k-1})$ in the iterative formula (8). q would be greater than the actual lower bound when z_k is neither very far from nor very close to $\{z_1, \ldots, z_{k-1}\}$, and an elliptical boundary passes between z_k and $\{z_1, \ldots, z_{k-1}\}$. The correct lower bound is just as difficult to estimate as $P(z_k \mid z_1, \ldots, z_{k-1})$. A simple fix is to replace q by the minimum of $P(z_2 \mid z_1)$ (easily obtained from the estimate of $P(z_1, z_2)$) as the lower bound, but this can obviously be off again for large k. For further improvement, one needs to check the relative positions among $z_1, \ldots, z_{k-1}, z_k$ in each set, especially to see how $z_1, \ldots, z_{k-1}$ are distributed in the surrounding of z_k. For instance, partition the window into three or four cones of equal angle with the common vertex z_k, and retain one representative point closest to z_k among $z_1, \ldots, z_{k-1}$ in each cone (if any). This is based on the idea of spatial Markov property and gives rise to approximations to higher-order k-point probability functions by estimates of lower-order, e.g. $k \leq 5$. More "trial and error" will be carried out to see whether the approximations work well.

3.2. *Effective elastic moduli of solids with cavities/cracks*

Most works in computation of effective elastic moduli are deterministic in nature, and no statistical models involved. For instance, in Budiansky and O'Connell (1976), Thorpe and Sen (1985), Day et al. (1992), the only statistical parameter related to microstructure is the phase 1 volume fraction q, assuming a two-phase material.

Kachanov and his collaborators [Kachanov et al. (1994a), and Kachanov et al. (1994b), denoted by KTS1 and KTS2 respectively in what follows], added the orientation parameter of the statistical distribution of cavities into the expression of effective elastic moduli.

The authors provided a number of examples of the *marginal* distribution for orientation to which their analysis applied, ranging from fully parallel to fully random. But they did not address how to specify the *entire* statistical distribution of microstructure, i.e. the *joint* distribution of the cavities/cracks, to back up their analysis.

These works started with the expression of elastic potential energy in stress tensor σ:

$$f(\sigma) = f_0(\sigma) + \Delta f,$$

where $f_0(\sigma)$ is the potential in a 2D homogeneous solid without cavities, and Δf is the change in the potential due to cavities.

Taking derivatives $\epsilon_{ij} = \frac{\partial f}{\partial \sigma_{ij}} = S_{ijkl}\sigma_{kl}$ yields the effective compliances S_{ijkl} as a sum of the compliance of the homogeneous matrix S^0_{ijkl} and the increments ΔS_{ijkl} due to cavities.

In the case of *non-interacting* elliptical cavities, suppose a heterogeneous material with area A contains n nonoverlapping elliptical cavities with major and minor half-axes $\{(a_k, b_k),\ k = 1, \dots, n\}$. Then

$$\Delta f = \frac{1}{2E_0} q[4\text{tr}(\sigma \cdot \sigma) - (\text{tr}\sigma)^2] + 2\sigma \cdot \sigma : (\beta - qI), \tag{10}$$

where E_0 is the Young's modulus and $\beta = \frac{1}{A}\sum_{k=1}^{n} \pi(a_k^2\ a_k^{\perp}a_k^{\perp} + b_k^2\ b_k^{\perp}b_k^{\perp})$ is the second rank cavity density tensor in which for the kth cavity, $a_k^{\perp}$ and $b_k^{\perp}$ are unit normal vectors to the half-axes a_k and b_k respectively. Or equivalently, if we let $\alpha = \left(\sum_{k=1}^{n} b_k^2\right)\left(\sum_{k=1}^{n} a_k^2\right)^{-1}$ (interpreted as a sort of average squared aspect ratio) and $\rho = \frac{1}{A}\sum_{k=1}^{n} a_k^2$ (interpreted as the scalar crack density of "fictitious" cracks associated with the major half-axes), then we have

$$\beta = \pi\rho\{[\alpha + (1-\alpha)M_1]e_1e_1 - M_2(1-\alpha)(e_1e_2 + e_2e_1) + [1-(1-\alpha)M_1]e_2e_2\}, \tag{11}$$

where e_1 and e_2 are the basis vectors of the corresponding coordinate system, and the expectations

$$M_1 = \int \sin^2\theta\ h(\theta)d\theta \quad \text{and} \quad M_2 = \int \sin\theta\cos\theta\ h(\theta)d\theta \tag{12}$$

are taken with respect to the marginal density $h(\theta)$ of the orientation θ between the major axis of an ellipse and e_1. (11) and (12) demonstrate the connection between property β and microstructural features M_1 and M_2.

The case of interacting elliptical cavities is more interesting and also more challenging. KTS1 and KTS2 recommended using the Mori-Tanaka approximation scheme (MTS) in terms of a simple adjustment in stress $\tilde{\sigma} = \frac{1}{1-q}\,\sigma$ and in the potential increment $\widetilde{\Delta f} = \frac{1}{1-q}\,\Delta f$, where the symbol $\tilde{}$ is attached to the quantities with interaction. In particular, for cracks the porosity $q = 0$ and MTS gives the same result as in the case of no interaction. Although the mechanical analysis carried out in KTS1, KTS2 is elegant, several issues remain unresolved. It would be particularly useful to specify a certain range for parameters in the statistical distribution of microstructure which makes the MTS (or alike) valid. Here are some important issues in this direction:

(i) Kachanov et al. emphasized that their mechanical analysis relies on the specification of statistical distribution for cavities/cracks, and the ability of generating Monte Carlo samples of these defects. The hard-core elliptical processes studied in Derr and Ji (2004) just fit such a need. The assumption of non-interacting defects is clearly unrealistic. Its validity could be justified in the case of dilute limit (small porosity q). A more specific description of dilute limit is to use the Dobrushin condition given in Jensen (1993). Furthermore, the marginal distribution $h(\theta)$ of orientation can be easily built in the hard-core elliptical processes to maintain tremendous modelling flexibility. This has been demonstrated by various synthetic microstructures in Derr and Ji (2004). See Figure 2 for one such example.

(ii) In the case of interacting cracks, that the MTS gives rise to the same result as in the case of no interaction looks dubious. Notice that this is not a dilute limit. The stress field at a crack tip becomes strong and nearly singular. Hence the stress fields around nearby cracks ought to interact. Even if the crack orientation is fully random (i.e. uniformly distributed over $[0, \pi)$) to cause overall "cancellation" of the stress field, it is unclear how such cancellation takes place. The problem is that no explicit connection has been established between the statistical distribution of microstructure and the effective elastic moduli to be calculated. Important quantities, such as k-point probability functions, have not shown up in the MTS or other approximation schemes. Beran (1968) and Willis (1981) provided some general mathematical framework in this direction, but hard analysis and calculation need to be carried out for the interesting cases considered in KTS1 and KTS2.

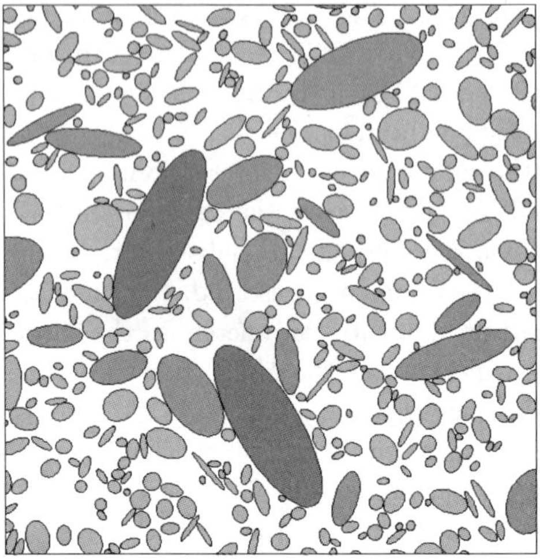

Fig. 2.

3.3. *Confidence intervals for effective elastic moduli in connection with FEM software*

A selling point of the perturbation expansion is that it ties k-point probability functions with effective properties. But the argument goes a long way through several demanding steps. For computation of material properties, FEM still enjoy the most popularity in materials science community. One can connect statistics of microstructure to FEM through construction of confidence intervals for effective properties. FEM are deterministic. For each given microstructure X_k as an input, a property of interest, say μ_k, can be calculated as an output of FEM. If μ is a property of interest, then repeating this procedure n times will yield the average $\bar{\mu} = n^{-1}(\mu_1 + \cdots + \mu_n)$ as an estimate for μ. The issue of sampling variability should be addressed, i.e. how values of the same property differ when they are calculated based on similar microstructures sampled from the same material? Confidence intervals for μ will provide such information. With sufficient computational power, we could obtain the sample variance of the property and use it to construct a confidence interval. But this simple procedure would not enable us to answer any questions related to microstructures. Materials scientists are interested in quantification of error propagation, i.e. how should the variance of an effective property be affected by various microstructural features and their variability.

An object oriented finite element solver "OOF" developed at NIST can read a microstructure into it then carry out a variety of virtual mechanical tests, thus calculate mechanical properties. One useful task will be to produce a statistical "toolbox" for OOF which has some functionality of quantifying error propagation.

References

1. Beran, M.J. (1968). *Statistical Continuum Theories.* Wiley, New York.
2. Brown, W.F.J. (1955). Solid mixture permittivities. *J. Chem. Phys.* **23**, 1514-1517.
3. Budiansky, B. and O'Connell, R.J. (1976). Elastic moduli of a cracked solid. *Int. J. Solids Structures 12*, 81-97.
4. Curtin, W.A. and Scher, H. (1990). Brittle fracture in disordered materials: A spring network model. *J. Mater. Res.* **5**, 535-553.
5. Day, A.R., Snyder, K.A., Garboczi, E.J., and Thorpe, M.F. (1992). The elastic moduli of a sheet containing circular holes. *J. Mech. Phys. Solids* **40**, 1031-1051.
6. Derr, R.E. (1999). *Statistical Modelling of Microstructure with Applications to Effective Property Computation in Materials Science.* Ph.D. dissertation, UNC-Chapel Hill
(available on the site **www.stat.unc.edu/faculty/ji.html**).
7. Derr, R.E. and Ji, C. (2004). Fitting microstructural models in materials science. Submitted and available on the site
www.stat.unc.edu/faculty/ji.html.
8. Herrmann, H.J. and Roux, S. (1990). *Statistical Models for the Fracture of Disordered Media.* North-Holland.
9. Jagota, A. and Bennison, S.J. (1994a). Spring-network and finite-element models for elasticity and fracture. To appear in the proceedings of a workshop on *Breakdown and Non-linearity in Soft Condensed Matter*, K.K. Bardhan et al. (ed.) Springer.
10. Jagota, A. and Bennison, S.J. (1994b). Element breaking rules in computational models for brittle fracture. Submitted to *Modelling Simul. Mater. Sci. Eng.*
11. Jensen, J.L. (1993). Asymptotic normality of estimates in spatial point processes. *Scand. J. Stat.* **20**, 97-109.
12. Kachanov, M., Tsukrov, I. and Shafiro, B. (1994a). Effective moduli of solids with cavities of various shapes. *Appl. Mech. Rev.* **47**, S151-S174.
13. Kachanov, M., Tsukrov, I. and Shafiro, B. (1994b). On anisotropy of solids with non-randomly oriented cavities. *Fracture and Damage in Quasibrittle Structures*, Z. P. Bažant et al. (ed.), E&FN, 19-24.
14. Papanicolaou, G.C. (1995). Diffusion in random media. *Surveys in Applied Mathematics*, Keller, J.P. (ed.) **1**, 205-253. Plenum Press, New York.
15. Phoenix, S.L. and Taylor, H.M. (1973). The asymptotic strength distribution of a general fiber bundle. *Adv. Appl. Prob.* **5**, 200-216.

16. Sen, A.K. and Torquato, S. (1989). Effective conductivity of anisotropic two-phase composite media. *Physical Review B 39*, 4504-4515.
17. Snyder, K.A, Garboczi, E.J. and Day, A.R. (1992). The elastic moduli of simple two-dimensional isotropic composites: computer simulation and effective medium theory. *J. Appl. Phys.* **72**, 5948-5955.
18. Sridhar, N., Yang, W., Srolovitz, D.J. and Fuller, E.R. (1994). A microstructural mechanics model of anisotropic thermal expansion induced microcracking. *J. Amer. Ceramics Soc.* **77**, 1123-1138.
19. Thorpe, M.F. and Sen, P.N. (1985). Elastic moduli of two-dimensional composite continua with elliptical inclusions. *J. Acoust. Soc. Am.* **77**, 1674-1680.
20. Torquato, S. (1985). Effective electrical conductivity of two-phase disordered composite media. *J. Appl. Phys.* **58**, 3790-3797.
21. Torquato, S. (1991). Random heterogeneous media: microstructure and improved bounds on effective properties. *Appl. Mech. Rev.* **44**, 37-76.
22. Torquato, S. and Sen, A.K. (1990). Conductivity tensor of anisotropic composite media from the microstructure. *J. Appl. Phys.* **67**, 1145-1155.
23. Willis, J.R. (1981). Variational and related methods for the overall properties of composites. *Adv. Appl. Mech.* **21**, 1-78.

THE IMPACT OF POPULATION STRATIFICATION ON COMMONLY USED STATISTICAL PROCEDURES IN POPULATION-BASED QTL ASSOCIATION STUDIES

HONG QIN*

Department of Statistics, Central China Normal University, Wuhan, Hubei, China

HONG ZHANG*

Department of Statistics, University of Science and Technology of China Hefei, Anhui, China

ZHAOHAI LI†

Department of Statistics, George Washington University, Washington, DC, U.S.A.

Population-based association study using unrelated individuals is a powerful strategy (Risch and Merikangas, 1996; Risch, 2000) for detecting association between markers and quantitative trait loci (QTLs). However, association test using unrelated individuals may suffer from confounding due to population structure. In this paper, we examine the impact of confounding due to population substructure on commonly used statistical procedures. Two study designs for genetic association study are considered: 1) retrospective sampling of cases and controls according to two cutoff points of the quantitative trait values (high or low) with allele frequency based test statistic; 2) random sample of individuals with regression analysis. For the first design, we consider the impact of confounder on association analysis between markers and QTLs. It is found that the false positive rate (or type I error) could be inflated substantially in the presence of both population stratification and trait heterogeneity; under other situations (e.g., in the presence of population stratification without trait heterogeneity), the inflated false positive rate is relatively minor. For the second design, we consider candidate marker association analysis. It is found that the inflated false positive rate could also be considerable in the presence of both population stratification and trait heterogeneity. Simulation studies confirm the theoretical results.

Some key words: Association study; Case control design; False positive rate;

*Also at Department of Statistics, George Washington University, Washington, DC

†Also at Biostatistics Branch, Division of Cancer Epidemiology and Genetics, National Cancer Institute, Bethesda, MD, U.S.A.

Population structure; Quantitative trait locus (QTL); Regression analysis.

1. Introduction

Genetic association study has received a great deal of attentions recently (Risch and Merikangas, 1996; Risch, 2000). Disease association with a genetic marker is often taken as a preliminary indication of linkage between the genetic marker and disease locus. However, population stratification and admixture may lead to disease association even in the absence of linkage. In order to eliminate spurious associations due to population stratification, family-based association designs have been developed, both for qualitative traits (Spielman et al., 1993; Ewens and Spielman, 1995; Morton and Collins, 1998; Risch and Teng, 1998; Teng and Risch, 1999; Li et al., 2001) and for quantitative traits (Rabinowitz, 1997; Thompson and Neel, 1997; Xiong and Guo, 1997; Rannala and Slatkin, 1998; Van den Oord, 1999; Xiong et al., 2002). Population samples consisting of unrelated individuals, however, may be easier and less expensive to collect, and such designs are, in general, more powerful compared to family-based association designs. The main shortcoming of population-based designs is that spurious associations due to population substructure may occur.

Li et al. (2005) examined the impact of confounding due to population substructure on the population-based case control studies. They only studied the impact in case control association studies of qualitative traits. In the present study, the impact of confounding on commonly used statistical procedures in population-based association studies of quantitative traits are investigated. We consider two popular designs and their corresponding methods for analysis. For the first design, one ascertains cases by selecting individuals with trait values exceed pre-specified threshold and controls with trait values below a value. The method of analysis is based on allele frequency comparison for this design. We consider the impact of confounding on association between markers and QTLs. It is found that the false positive rate could be inflated substantially in the presence of both population stratification and trait heterogeneity; the inflated false positive rate is minor under other situations. For the second design, a random sample of individuals are obtained and regression analysis is applied. We consider association analysis of a candidate QTL for this design. It is found that the inflated false positive rate could also be considerable in the presence of both population stratification and trait heterogeneity. Simulation studies demonstrate that the theoretical conclusions are valid in practical situations.

2. Allele Frequency-Based Test

2.1. *Model and a naise test*

We assume a biallelic QTL with alleles D and D', and a biallelic marker locus with alleles M and M'. The population frequencies of D and M are denoted by p and m, respectively. We are interested in testing the association between the two loci. Denote by y the quantitative trait value of an individual. The quantitative trait value is commonly assumed to have the form

$$y = \mu + g + \epsilon, \tag{2.1}$$

where

$$g = \begin{cases} a, & \text{for an } DD \text{ individual} \\ d, & \text{for an } DD' \text{ individual} \\ -a, & \text{for an } D'D' \text{ individual} \end{cases}$$

is the genotype effect, μ is the intercept and ϵ is the error term, which has mean 0 and variance σ^2. Define an individual with trait value greater than τ_2 as a *case*, and an individual with trait value less than τ_1 as a *control*, where τ_1 and τ_2 are given threshold values. Suppose that we sample n_1 cases and n_2 controls from the population. Let the numbers of cases with genotypes MM, MM' and $M'M'$ be D_2, D_1 and D_0, respectively. Likewise, C_2, C_1 and C_0 are the numbers for controls. The data are summarized in Table 1.

Table 1. Observed numbers of marker genotypes in cases and controls.

	MM	MM'	$M'M'$	Sum
Case	D_2	D_1	D_0	n_1
Control	C_2	C_1	C_0	n_2
Sum	E_2	E_1	E_0	n

Unlike the candidate marker case, our marker-QTL association analysis is to test that the two loci are in linkage equilibrium (H_{0A} defined in Appendix A). While in the candidate marker case, one tests $a = d = 0$ (no genetic effect) instead. A commonly used allele frequency-based test statistic for association is (Ohashi et al., 2001)

$$T_A = \frac{\hat{m}_D - \hat{m}_C}{\sqrt{(\frac{1}{2n_1} + \frac{1}{2n_2})\hat{m}(1-\hat{m})}}, \tag{2.2}$$

where $\hat{m}_D = \frac{2D_2+D_1}{2n_1}$, $\hat{m}_C = \frac{2C_2+C_1}{2n_2}$ and $\hat{m} = \frac{2E_2+E_1}{2n}$ are the MLEs of frequencies of allele M in the cases, controls and combined population, respectively. If there is no population structure and Hardy-Weinburg equilibrium holds in the population, the expectation and variance of $\hat{m}_D - \hat{m}_C$ under the null hypothesis of no association are 0 and $(\frac{1}{2n_1} + \frac{1}{2n_2})m(1-m)$, respectively. Thus, T_A has asymptotic standard normal distribution under the null hypothesis, since $\hat{m}$ is a consistent estimate of m. The rejection region of the test at α level of significance is taken as

$$|T_A| > u_{\alpha/2} \tag{2.3}$$

in favor of two-sided alternative of association, where $u_{\alpha/2}$ is the $\alpha/2$-upper quantile of the standard normal distribution. This test is valid if the population is homogeneous and Hardy-Weinburg equilibrium holds in the population. However, this test may be invalid in the presence of population structure, as discussed in the next subsection.

2.2. *Bias and overdispersion due to population structure*

We assume that the overall population consists of K subpopulations. First we introduce some further notation. For subpopulation k, denote by w_k the probability that a randomly selected individual is from the kth subpopulation, p_k the frequency of allele D for the kth subpopulation at trait locus, and m_k the frequency of allele M at marker locus. Population stratification (or marker allele frequency heterogeneity) exists at marker locus if at least two m_k's are unequal. When all m_k's are the same there is no population stratification at marker locus. The values of μ, a and d in subpopulation k are denoted by μ_k, a_k and d_k, respectively. In addition, we assume that population structure affects the error distribution only on its scale, and that the variance of ϵ in subpopulation k is σ_k^2. This assumption is satisfied naturally when the error follows the normal distribution. When μ_k, a_k, d_k, and σ_k^2 are the same for all $k = 1, ..., K$, the trait value structure has no heterogeneity among subpopulations. The frequencies of D and M in the overall population are $p = \sum_{k=1}^K w_k p_k$ and $m = \sum_{k=1}^K w_k m_k$, respectively.

Let $f_{2k} = P(y > \tau_2|DD, S_k)$, $f_{1k} = P(y > \tau_2|DD', S_k)$ and $f_{0k} = P(y > \tau_2|D'D', S_k)$ be the proportions of cases with genotypes DD,

DD' and $D'D'$ in the kth subpopulation, respectively. Here, S_k stands for the event that a randomly sampled individual is from subpopulation k. Likewise, $g_{2k} = P(y < \tau_1|DD, S_k)$, $g_{1k} = P(y < \tau_1|DD', S_k)$ and $g_{0k} = P(y < \tau_1|D'D', S_k)$ are for controls. Since we have assumed that the population structure affects the error distribution only on its scale, for the kth subpopulation, ϵ/σ_k follows a distribution that is independent of k. Denote by F the common cumulative distribution function of ϵ/σ_k. Then,

$$f_{2k} = 1 - F(\frac{\tau_2 - \mu_k - a_k}{\sigma_k}), \quad g_{2k} = F(\frac{\tau_1 - \mu_k - a_k}{\sigma_k}), \tag{2.4}$$

$$f_{1k} = 1 - F(\frac{\tau_2 - \mu_k - d_k}{\sigma_k}), \quad g_{1k} = F(\frac{\tau_1 - \mu_k - d_k}{\sigma_k}), \tag{2.5}$$

$$f_{0k} = 1 - F(\frac{\tau_2 - \mu_k + a_k}{\sigma_k}), \quad g_{0k} = F(\frac{\tau_1 - \mu_k + a_k}{\sigma_k}). \tag{2.6}$$

It is shown in Appendix A that the null expectations of $\hat{m}_D$ and $\hat{m}_C$ are

$$\bar{m}_D \equiv E_{H_0}(\hat{m}_D) = \sum_{k=1}^{K} m_k \alpha_k \text{ and } \bar{m}_C \equiv E_{H_0}(\hat{m}_C) = \sum_{k=1}^{K} m_k \gamma_k, \tag{2.7}$$

respectively, where

$$\alpha_k = \frac{[f_{2k}p_k^2 + f_{1k}2p_k(1-p_k) + f_{0k}(1-p_k)^2]w_k}{\sum_{j=1}^{K}[f_{2j}p_j^2 + f_{1j}2p_j(1-p_j) + f_{0j}(1-p_j)^2]w_j}$$

and

$$\gamma_k = \frac{[g_{2k}p_k^2 + g_{1k}2p_k(1-p_k) + g_{0k}(1-p_k)^2]w_k}{\sum_{j=1}^{K}[g_{2j}p_j^2 + g_{1j}2p_j(1-p_j) + g_{0j}(1-p_j)^2]w_j}$$

are the proportions of subpopulation k in the cases and controls, respectively. Meanwhile, the null variances of $\hat{m}_D$ and $\hat{m}_C$ are

$$Var_{H_{0A}}(\hat{m}_D) = \frac{1}{2n_1}[\bar{m}_D(1-\bar{m}_D) + \sigma_D^2]$$

and

$$Var_{H_{0A}}(\hat{m}_C) = \frac{1}{2n_2}[\bar{m}_C(1-\bar{m}_C) + \sigma_C^2],$$

respectively, where

$$\sigma_D^2 = \sum_{k=1}^{K} \alpha_k (m_k - \bar{m}_D)^2 \text{ and } \sigma_C^2 = \sum_{k=1}^{K} \gamma_k (m_k - \bar{m}_C)^2.$$

These expressions are similar to those in Li (1969). Thus, the null expectation and variance of $\hat{m}_D - \hat{m}_C$ are

$$\delta_A \equiv \bar{m}_D - \bar{m}_C \tag{2.8}$$

and

$$\sigma_A^2 \equiv \frac{1}{2n_1}[\bar{m}_D(1-\bar{m}_D)+\sigma_D^2] + \frac{1}{2n_2}[\bar{m}_C(1-\bar{m}_C)+\sigma_C^2], \tag{2.9}$$

respectively. The law of large number theory gives that the estimated variance used in T_A is asymptotically equal to

$$(\sigma_A^*)^2 = (\frac{1}{2n_1}+\frac{1}{2n_2})\bar{m}(1-\bar{m}), \tag{2.10}$$

where $\bar{m} = \frac{n_1}{n}\bar{m}_D + \frac{n_2}{n}\bar{m}_C$ is the expectation of frequency of M in the combined population. Thus, the bias is δ_A and the overdispersion $\sigma_A^2 - (\sigma_A^*)^2$ is equal to

$$\frac{1}{2n_1}[\bar{m}_D(1-\bar{m}_D)-\bar{m}(1-\bar{m})+\sigma_D^2] + \frac{1}{2n_2}[\bar{m}_C(1-\bar{m}_C)-\bar{m}(1-\bar{m})+\sigma_C^2].$$

As in Li et al. (2005), we can estimate σ_A^2 consistently and thus the impact due to overdispersion could be eliminated. However, the bias is hard to estimate, which has much bigger impact on the false positive rate, as discussed in the end of this subsection.

Remark 1. (I) If there is no marker allele frequency heterogeneity, that is, $m_k = m$ for $k = 1, \cdots, K$. Then $\bar{m}_D = \bar{m}_C = m$ and $\sigma_D^2 = \sigma_C^2 = 0$. In this case, both the bias and the overdispersion vanish. (II) If there is no trait heterogeneity but the marker allele heterogeneity presents (i.e., $\sigma_k = \sigma$, $\mu_k = \mu$, $a_k = a$ and $d_k = d$, $k = 1, \cdots, K$, but $m_k \neq m$ for some k), then, $f_{2k} = f_{1k} = f_{0k}$ and $g_{2k} = g_{1k} = g_{0k}$ for $k = 1, \cdots, K$. Hence, $\alpha_k = \gamma_k = w_k$ for $k = 1, \cdots, K$ and bias vanishes. However, overdispersion always exist as σ_D^2 and σ_C^2 are nonzero. (III) If both trait and the marker allele frequency heterogeneities exist, then, both the bias and the overdispersion are nonzero.

For the two-sided test with rejection region (2.3), the excessive false positive rate (EFPR) is approximately

$$1 - \Phi((u_{\alpha/2}\sigma_A^* - \delta_A)/\sigma_A) + \Phi((-u_{\alpha/2}\sigma_A^* - \delta_A)/\sigma_A) - \alpha, \tag{2.11}$$

where $\Phi(\cdot)$ is the standard normal distribution function.

Remark 2. Note that both $(\sigma_A^*)^2$ and σ_A^2 are of order $1/n$, the excessive false positive rate tends to $1-\alpha$ as n tends to infinity, provided that δ_A is positive. Thus, the bias has much bigger impact on the false positive rate than the overdispersion.

3. Regression-Based Test

3.1. *Model and a naive test*

We consider a candidate marker with alleles A and A'. We assume a linear relationship between the quantitative trait value and the candidate marker genotype:

$$y = \beta_0 + \beta_1 x + \epsilon, \tag{3.1}$$

where ϵ is the error term with mean 0 and variance σ^2, $x = I_{AA} + dI_{AA'} - I_{A'A'}$. Here I_{AA} is the indicator function such that $I_{AA} = 1$ if the genotype is AA and 0 otherwise. The dominance parameter $d = -1$, 0 and 1 corresponds to the recessive, additive and dominant models, respectively. For the candidate marker, we are interested in testing $\beta_1 = 0$.

Suppose that we observe n unrelated individuals, with the values of y and x for individual i being y_i and x_i, respectively. The least squares (LS) estimate of β_1 is

$$\hat{\beta}_1 = \sum_{i=1}^{n} b_i y_i,$$

where

$$b_i = \frac{x_i - \bar{x}}{\sum_{i=1}^{n}(x_i - \bar{x})^2} \quad \text{with } \bar{x} = \frac{1}{n}\sum_{i=1}^{n} x_i.$$

The conditional variance of $\hat{\beta}_1$ given the genotypes is $Var(\hat{\beta}_1) = \sum_i b_i^2 \sigma^2$. A Wald type test for $\beta_1 = 0$ is

$$T = \frac{\sum_{i=1}^{n} b_i y_i}{\sqrt{\sum_{i=1}^{n} b_i^2 \tilde{\sigma}^2}}, \tag{3.2}$$

where $\tilde{\sigma}^2$ is an estimate of σ^2. A widely used estimate is using residuals as follows

$$\hat{\sigma}^2 = \frac{1}{n-2}\mathbf{y}'[I_n - \mathbf{x}(\mathbf{x}'\mathbf{x})^{-1}\mathbf{x}']\mathbf{y}, \tag{3.3}$$

where I_n is the n-identity matrix, $\mathbf{y}' = (y_1, \cdots, y_n)$ and

$$\mathbf{x}' = \begin{pmatrix} 1 & 1 & \cdots & 1 \\ x_1 & x_2 & \cdots & x_n \end{pmatrix}.$$

Thus, a natural estimate of the variance of $\hat{\beta}_1$ is

$$\widehat{Var(\hat{\beta}_1)} = \sum_{i=1}^{n} b_i^2 \hat{\sigma}^2.$$

3.2. *Bias and overdispersion due to population structure*

Suppose that the overall population consists of K subpopulations. For subpopulation k, the parameters β_0, β_1 and σ^2 become β_{0k}, β_{1k} and σ_k^2, respectively. Then, the null hypothesis is

$$H_{0R} : \beta_{1k} = 0, \quad k = 1, \cdots, K. \tag{3.4}$$

Without loss of generality, we assume that individuals $n_{k-1}+1, \cdots, n_k$ are from subpopulation k, where $n_0 = 0$ and $\sum_{k=1}^K n_k = n$. Thus, the trait value of individual $n_1 + \cdots + n_{k-1} + j$ is equal to

$$y_{kj} = \beta_{0k} + \beta_{1k}x_{kj} + \epsilon_{kj}, \quad j = 1, \cdots, n_k, \; k = 1, \cdots, K,$$

where $y_{kj} = y_{n_1+\cdots+n_{k-1}+j}$, $x_{kj} = x_{n_1+\cdots+n_{k-1}+j}$, and $\epsilon_{kj} = \epsilon_{n_1+\cdots+n_{k-1}+j}$; the expectation and variance of ϵ_{kj} are 0 and σ_k^2, respectively. Then, the LS estimate $\hat{\beta}_1$ obtained under model (3.1) can be rewritten as

$$\hat{\beta}_1 = \sum_{k=1}^{K}\sum_{j=1}^{n_k} b_{kj}y_{kj}, \text{ with } b_{kj} = \frac{x_{kj} - \bar{x}}{\sum_{k=1}^K \sum_{j=1}^{n_k}(x_{kj} - \bar{x})^2}.$$

Let $w_k = n_k/n$, then, for large sample size,

$$\begin{aligned}\sum_{k=1}^{K}\sum_{j=1}^{n_k}(x_{kj} - \bar{x})^2 &= \sum_{k=1}^{K}\sum_{j=1}^{n_k} x_{kj}^2 - n(\bar{x})^2 \\ &\approx n\Bigg(\sum_{k=1}^{K} w_k E(x_{k1}^2) - (E\bar{x})^2\Bigg) = n\sum_{k=1}^{K} w_k E(x_{k1} - E\bar{x})^2,\end{aligned}$$

which follows that

$$b_{kj} \approx \frac{1}{n}\frac{x_{kj} - \bar{x}}{\sum_{k=1}^K w_k E(x_{k1} - E\bar{x})^2}. \tag{3.5}$$

Here and in the sequel, we say $U_1 \approx U_2$ if U_1 is equal to U_2 plus a negligible term, i.e., $U_1 = U_2(1 + o_p(1))$. Since the null expectation of y_{kj} is β_{0k}, the null expectation of the estimate $\hat{\beta}_1$ is approximated by

$$\delta_R \equiv \frac{\sum_{k=1}^K w_k(Ex_{k1} - E\bar{x})\beta_{0k}}{\sum_{k=1}^K w_k E(x_{k1} - Ex)^2}. \tag{3.6}$$

The above δ_R can be regarded as the bias due to using the regression model (3.1) when population structure exits. If the Hardy-Weinberg equilibrium holds in each of the K subpopulations, then, no allele frequency heterogeneity implies $E[x_{k1}] = E[\bar{x}]$ and hence the bias is approximately 0, since the expectations of x_{k1} and $\bar{x}$ depend only on allele frequencies and d; also, the bias vanishes if the intercepts $\beta_{01}, \cdots, \beta_{0K}$ are equivalent, since $E[\bar{x}] = \sum_{k=1}^{K} w_k E[x_{k1}]$, which implies that $\sum_{k=1}^{K} w_k E(x_{k1} - \bar{x}) = 0$; otherwise, the null expectation is nonzero.

The variance of the estimator $\hat{\beta}_1$ under the null hypothesis, $Var_{H_{0R}}(\hat{\beta}_1)$, is approximated by (Appendix B)

$$\sigma_R^2 \equiv \frac{\sum_{k=1}^{K} w_k [E(x_{k1} - E\bar{x})^2 \sigma_k^2 + Var(x_{k1})(\beta_{0k} - \bar{\beta})^2]}{n[\sum_{k=1}^{K} w_k E(x_{k1} - E\bar{x})^2]^2}. \tag{3.7}$$

It is easy to see that the estimator defined in (3.3) is equal to

$$\hat{\sigma}^2 = \frac{1}{n-2}\epsilon'[I_n - \mathbf{x}(\mathbf{x}'\mathbf{x})^{-1}\mathbf{x}']\epsilon + \frac{1}{n-2}\beta'[I_n - \mathbf{x}(\mathbf{x}'\mathbf{x})^{-1}\mathbf{x}]\beta = I_1 + I_2,$$

with $\epsilon' = (\epsilon_1, \cdots, \epsilon_n)$ and $\beta' = (\beta_{01} 1'_{n_1}, \cdots, \beta_{0K} 1'_{n_K})$. By standard asymptotic expansion, we obtain that

$$I_1 \approx \sum_{k=1}^{K} w_k \sigma_k^2$$

and

$$\begin{aligned} I_2 \approx & \sum_{k=1}^{K} w_k \beta_{0k}^2 - (\sum_{k=1}^{K} w_k \beta_{0k})^2 \\ & + \left[(\sum_{k=1}^{K} w_k \beta_{0k} E x_{k1})^2 - \sum_{k=1}^{K} w_k \beta_{0k} \sum_{k=1}^{K} w_k E x_{k1} \right] \\ & \div \left[\sum_{k=1}^{K} w_k E[x_{k1}^2] - (\sum_{k=1}^{K} w_k E x_{k1})^2 \right]. \end{aligned}$$

We see that I_2 vanishes if $\beta_{01} = \cdots = \beta_{0K}$. It follows that the estimated variance of $\hat{\beta}_1$, $\widehat{Var(\hat{\beta}_1)} = \sum_{i=1}^{n} b_i^2 \hat{\sigma}^2$, is approximated by

$$(\sigma_R^*)^2 \equiv \frac{1}{n \sum_{k=1}^{K} w_k E(x_{k1} - E\bar{x})^2} \left[\sum_{k=1}^{K} w_k \sigma_k^2 + I_2 \right]. \tag{3.8}$$

Write

$$w'_k = \frac{w_k E(x_{k1} \quad E\bar{x})^2}{\sum_{l=1}^{K} w_l E(x_{l1} - E\bar{x})^2},$$

then, the overdispersion is approximately

$$\sigma_R^2 - (\sigma_R^*)^2 = \frac{\sum_{k=1}^{K} (w'_k - w_k)\sigma_k^2 + \sum_{k=1}^{K} w_k Var(x_{k1})(\beta_{0k} - \bar{\beta}_0)^2 + I_2}{n \sum_{k=1}^{K} w_k E(x_{k1} - E\bar{x})^2}. \tag{3.9}$$

The overdispersion arises from either trait heterogeneity or allele frequency heterogeneities. It is also interesting to see that the variance could be overestimated, we say there exists *underdispersion* in this case. While for the allele frequency-based method, the variance can not be overestimated.

Remark 3. (I) If there is no allele frequency heterogeneity, then bias vanishes. (II) If there is neither allele frequency heterogeneity nor trait heterogeneity, then overdispersion or underdispersion vanish. (III) The bias exists in the presence of both allele frequency heterogeneity and trait heterogeneity. Also, overdispersion or underdispersion exists in the presence of either allele heterogeneity or trait heterogeneity.

The EFPR for the two-sided test is equal to

$$1 - \Phi((u_{\alpha/2}\sigma_R^* - \delta_R)/\sigma_R) + \Phi((-u_{\alpha/2}\sigma_R^* - \delta_R)/\sigma_R) - \alpha. \tag{3.10}$$

Remark 4. From (3.7) and (3.8), we see that both the variance and its estimate are of order $1/n$. Thus, the excessive false positive rate could be considerable in the presence of positive bias, and the bias has much bigger impact on false positive rate than the overdispersion.

4. Simulation Studies

First, we considered the allele frequency-base test. Define the threshold values τ_1 and τ_2 as the q_1 and q_2 quantiles of quantitative value in the overall population, respectively. We assumed that the overall population consists of two subpopulations with equal proportions and that the errors were normally distributed. To study the impact of all factors on the T_A test, we consider various combinations of parameters m_k, p_k, μ_k, a_k, σ_k for $k = 1, 2$ and q_1 and q_2. We set $m_2 = 0.5$, $m_1 = m_2$ or $m_2/2$; $p_2 = 0.5$, $p_1 = p_2$ or $p_2/2$; $\mu_2 = 0$, $\mu_1 = 0, -1$ or 1; $a_2 = 1$, $a_1 = a_2$ or $2a_2$; $\sigma_2 = 1$, $\sigma_1 = \sigma_2$ or $2\sigma_2$; $q_2 = 0.9$, $q_1 = 0.1$ or 0.9. To verify the null expectation and variance of $\hat{p}_D - \hat{p}_C$, we define a "test statistic"

$$T_A^* = \frac{\hat{p}_D - \hat{p}_C - \delta_A}{\sqrt{\sigma_A^2}},$$

where δ_A and σ_A^2 are defined in (2.8) and (2.9), respectively. For each of parameter combinations, we simulated 10,000 replicates of 100 cases (with trait value greater than q_2 quantile) and 100 controls (with trait value less than q_1 quantile) under null hypothesis of linkage equilibrium between marker locus and QTL. The resulting false positive rates of the two-sided tests at 0.05 level of significance (with rejection regions $\{|T_A| > 1.96\}$ and $\{|T_A^*| > 1.96\}$) are presented in Table 2 and Table 3. Table 2

Table 2. Type I error rates of T_A and T_A^* tests with homogenous marker allele frequencies.

p_1	μ_1	a_1	σ_1	$q_1 = 0.1$		$q_1 = 0.9$	
				T_A	T_A^*	T_A	T_A^*
0.5	0	1	1	0.051	0.051	0.054	0.054
			2	0.049	0.049	0.048	0.048
		2	1	0.055	0.055	0.049	0.049
			2	0.050	0.050	0.049	0.049
	-1	1	1	0.047	0.047	0.052	0.052
			2	0.051	0.051	0.051	0.051
		2	1	0.053	0.053	0.051	0.051
			2	0.049	0.049	0.052	0.052
	1	1	1	0.051	0.051	0.052	0.052
			2	0.051	0.051	0.050	0.050
		2	1	0.053	0.053	0.054	0.054
			2	0.051	0.051	0.052	0.052
0.25	0	1	1	0.051	0.051	0.052	0.052
			2	0.052	0.052	0.053	0.053
		2	1	0.047	0.047	0.048	0.048
			2	0.053	0.053	0.056	0.056
	-1	1	1	0.051	0.051	0.050	0.050
			2	0.053	0.053	0.053	0.053
		2	1	0.050	0.050	0.051	0.051
			2	0.052	0.052	0.053	0.053
	1	1	1	0.053	0.053	0.051	0.051
			2	0.050	0.050	0.051	0.051
		2	1	0.053	0.053	0.050	0.050
			2	0.053	0.053	0.050	0.050

$m_1 = 0.5$, $m_2 = 0.5$, $\mu_2 = p_2 = 0.5$, $\mu_2 = 0$, $a_2 = 1$, $d_1 = d_2 = 0$, $\sigma_2 = 1$, $q_2 = 0.9$.

is for homogenous marker allele frequencies ($m_1 = m_2$). We see that both of the two tests have almost correct false positive rates, since both the bias and the overdispersion vanish, as asserted in Remark I. Table 3 is for heterogenous marker allele frequencies ($m_1 = m_2/2$). In this case, the false positive rate is always inflated, except that the QTL has nothing to do with the quantitative trait value ($p_1 = p_2 = 0.5$, $\mu_1 = \mu_2$, $a_1 = a_2 = 1$,

$\sigma_1 = \sigma_2 = 1$). The excessive false positive rate varies a lot, which could be considerable (as large as $0.939 - 0.05 = 0.889$) or very minor (as small as $0.054 - 0.05 = 0.004$). For each of the parameter combinations, the modified T_A^* test has almost correct false positive rate, which varies from 0.042 to 0.057. This verifies the null expectation and variance of $\hat{p}_D - \hat{p}_C$ defined in (2.8) and (2.9).

Table 3. Type I error rates of T_A and T_A^* tests with heterogenous marker allele frquencies.

p_1	μ	a_1	σ_1	$q_1 = 0.1$		$q_1 = 0.9$	
				T_A	T_A^*	T_A	T_A^*
0.5	0	1	1	0.059	0.052	0.056	0.050
			2	0.060	0.050	0.506	0.052
		2	1	0.057	0.046	0.085	0.049
			2	0.054	0.044	0.557	0.048
	-1	1	1	0.538	0.048	0.208	0.053
			2	0.225	0.055	0.250	0.049
		2	1	0.484	0.052	0.095	0.052
			2	0.201	0.044	0.294	0.057
	1	1	1	0.535	0.050	0.223	0.046
			2	0.225	0.056	0.719	0.049
		2	1	0.501	0.054	0.336	0.051
			2	0.196	0.042	0.740	0.049
0.25	0	1	1	0.191	0.048	0.097	0.048
			2	0.101	0.054	0.378	0.047
		2	1	0.477	0.045	0.115	0.053
			2	0.206	0.054	0.290	0.051
	-1	1	1	0.831	0.056	0.375	0.051
			2	0.418	0.048	0.140	0.050
		2	1	0.939	0.050	0.367	0.050
			2	0.560	0.053	0.092	0.051
	1	1	1	0.189	0.048	0.090	0.048
			2	0.100	0.055	0.631	0.050
		2	1	0.060	0.050	0.074	0.045
			2	0.054	0.043	0.543	0.044

$m_1 = 0.1$, $m_2 = 0.5$, $\mu_2 = p_2 = 0.5$, $\mu_2 = 0$, $a_2 = 1$, $d_1 = d_2 = 0$, $\sigma_2 = 1$, $q_2 = 0.9$.

Table 4. Type I error rates of T_R test and T_R^* test

			$d=-1$		$d=0$		$d=1$	
p_1	β_{01}	σ_1^2	T_R	T_R^*	T_R	T_R^*	T_R	T_R^*
0.5	0	1	0.055	0.056	0.056	0.056	0.054	0.055
		0.5	0.056	0.056	0.052	0.054	0.052	0.053
		2	0.054	0.057	0.055	0.054	0.052	0.054
	-1	1	0.050	0.050	0.049	0.050	0.054	0.055
		0.5	0.056	0.058	0.055	0.056	0.060	0.060
		2	0.055	0.054	0.049	0.048	0.054	0.056
	1	1	0.054	0.055	0.053	0.054	0.050	0.050
		0.5	0.057	0.057	0.052	0.052	0.052	0.054
		2	0.050	0.053	0.052	0.054	0.050	0.052
0.25	0	1	0.055	0.059	0.049	0.052	0.052	0.049
		0.5	0.090	0.055	0.061	0.051	0.040	0.046
		2	0.023	0.057	0.044	0.052	0.059	0.051
	-1	1	0.210	0.047	0.361	0.050	0.304	0.053
		0.5	0.301	0.048	0.490	0.048	0.409	0.051
		2	0.082	0.050	0.177	0.052	0.179	0.050
	1	1	0.211	0.049	0.366	0.053	0.301	0.051
		0.5	0.303	0.051	0.481	0.045	0.410	0.050
		2	0.085	0.052	0.186	0.056	0.177	0.053

$p_2 = 0.5$, $\beta_{02} = 0$, $\sigma_2^2 = 1$

Second, we study the behavior of the regression-based test. To verify the null expectation and variance defined in (3.6) and (3.7), define a "test statistic"

$$T_R^* = \frac{\hat{\beta}_1 - \delta_R}{\sqrt{\sigma_R^2}}.$$

Also, we assumed that the overall population consisted of two subpopulations with equal proportions and that the errors were normally distributed. To study the impact of allele frequencies, intercepts and error variances on the false positive rates of the test, let $p_2 = 0.5$, $p_1 = p_2$ or $p_2/2$; $\beta_{02} = 0$, $\beta_{01} = 0, -1$ or 1; $\sigma_2 = 1$, $\sigma_1 = \sigma_2$ or $2\sigma_2$. For each of parameter combinations, 100 individuals were randomly drawn from the overall population under null hypothesis of no association between candidate QTL and trait value. For $d = -1$, 0 or 1, the two-sided tests (with rejection regions $\{|T_R| > 1.96\}$ and $\{|T_R^*| > 1.96\}$) at 0.05 level of significance were conducted with $10,000$ replicates. The resulting false positive rates are presented in Table 4. When $p_1 = p_2$, the T_R test has almost correct false positive rates, since the bias and overdispersion vanish. It is seen that very minor inflation in false positive rate is induced, which varies from -0.001 to 0.01. This is due to the randomness of the genotype effect. When the allele frequencies are heterogenous, the false positive rate of the T_R test is inflated for most of the parameter combinations. In our simulation situations, although the heterogeneities in allele frequencies, intercepts and error variances are moderate or zero, the false positive rate could be as large as 0.490. It is interesting to find that the false positive rate could be deflated. For example, when $p_1 = p_2/2 = 0.25$, $\beta_{01} = \beta_{02} = 0$, $d = -1$ and $\sigma_1 = 2\sigma_2 = 2$, the false positive rate is 0.023. This is not surprising since the overdipersion defined in (3.9) could be negative. In all of the cases, the T_R^* test has almost correct false positive rate, which varies from 0.045 to 0.060. This verifies the null expectation and variance defined in (3.6) and (3.7).

5. Discussion

In this paper, we studied the impact of population stratification on the QTL association analysis. It is found from both the theoretical studies and the simulation studies that the false positive rate could be inflated considerably in the presence of both population stratification and trait heterogeneity. Because the population-based association design is in general more powerful than the family-based association design (Risch and Merikangas, 1996; Risch, 2000), there is need to derive powerful tests with the desired false positive rate (type I error) based on unrelated individuals. To control for false positive rate arises from population stratification, several methods for population-based association studies have been proposed (Devlin and Roeder, 1999; Bacanu et al., 2000; Pritchard et al., 2000a; Pritchard et al., 2000b; Devlin et al., 2001; Satten et al., 2001; Zhang and Zhao, 2001; Zhang et al., 2002). However, some of them need to put strong mathematical conditions which may not be satisfied in practice, some of them have to know or estimate the population structure, which may not be reliable in reality. Most of the proposed methods focused on adjustment of overdispersion. How to correct bias might be more important. The issue investigated here also occurs in many other medical research areas, the finding may have a broad application.

Appendices

A. Derivations of the expectations and variances of $\hat{m}_D$ and $\hat{m}_C$

Let

$$I_{MM} = \begin{cases} 1, & \text{if the case individual's genotype is } MM; \\ 0, & \text{otherwise.} \end{cases}$$

and

$$I_{MM'} = \begin{cases} 1, & \text{if the case individual's genotype is } MM'; \\ 0, & \text{otherwise.} \end{cases}$$

Then

$$\hat{m}_D = \frac{1}{2n_1} \sum (2I_{MM} + I_{MM'}),$$

where the summation is taken for all case individuals.

The expectation of $\hat{m}_D$ is equal to

$$E(\hat{m}_D) = \frac{1}{2}[2E(I_{MM}) + E(I_{MM'})] = \frac{1}{2}[2P(MM|y > \tau_2) + P(MM'|y > \tau_2)] \tag{A.1}$$

The variance of $\hat{m}_D$ is equal to

$$Var(\hat{p}_D) = \frac{1}{(2n_1)^2}\left[4n_1 Var(I_{MM}) + n_1 Var(I_{MM'}) + 4n_1 Cov(I_{MM}, I_{MM'})\right], \tag{A.2}$$

where

$$Var(I_{MM}) = P(MM|y > \tau_2)(1 - P(MM|y > \tau_2)), \tag{A.3}$$

$$Var(I_{MM'}) = P(MM'|y > \tau_2)(1 - P(MM'|y > \tau_2)), \tag{A.4}$$

$$Cov(I_{MM}, I_{MM'}) = -P(MM|y > \tau_2)P(MM'|y > \tau_2). \tag{A.5}$$

To calculate the expectation and variance, we need only calculate $P(MM|y > \tau_2)$ and $P(MM'|y > \tau_2)$.

The proportion of subpopulation k in the case individuals is

$$\begin{aligned} \alpha_k &= P(S_k|y > \tau_2) \\ &= \frac{\sum_{G_T} P(y > \tau_2|G_T, S_k)P(G_T|S_k)P(S_k)}{\sum_{j=1}^{K}\left[\sum_{G_T} P(y > \tau_2|G_T, S_j)P(G_T|S_j)P(S_j)\right]} \\ &= \frac{[f_{2k}p_k^2 + f_{1k}2p_k(1-p_k) + f_{0k}(1-p_k)^2]w_k}{\sum_{j=1}^{K}[f_{2j}p_j^2 + f_{1j}2p_j(1-p_j) + f_{0j}(1-p_j)^2]w_j}, \end{aligned} \tag{A.6}$$

where $f_{2k} = P(y > \tau_2|DD, S_k)$, $f_{1k} = P(y > \tau_2|DD', S_k)$, $f_{0k} = P(y > \tau_2|D'D', S_k)$ and the summation $\sum_{G_T}$ is taken for DD, DD' and $D'D'$. Similarly, The proportion of subpopulation k in the control individuals is

$$\begin{aligned}\gamma_k &= P(S_k|y < \tau_1) \\ &= \frac{[g_{2k}p_k^2 + g_{1k}2p_k(1-p_k) + g_{0k}(1-p_k)^2]w_k}{\sum_{j=1}^{K}[g_{2j}p_j^2 + g_{1j}2p_j(1-p_j) + g_{0j}(1-p_j)^2]w_j},\end{aligned} \tag{A.7}$$

where $g_{2k} = P(y < \tau_1|DD, S_k)$, $g_{1k} = P(y < \tau_1|DD', S_k)$, $g_{0k} = P(y < \tau_1|D'D', S_k)$.

Suppose that HWE holds in each of the K subpopulations. Then, under the null hypothesis that linkage equilibrium between the marker and QTL holds in all the subpopulation, marker genotype and quantitative trait locus genotype are independent. In mathematical notation,

$$P_{H_{0A}}(G_M, G_T|S_k) = P(G_M|S_k)P(G_T|S_k), \tag{A.8}$$

where the null hypothesis H_{0A} is the null hypothesis that will be defined in (A.9) and G_M and G_T are any marker genotype and QTL genotype, respectively. We show this fact as follows.

Let the frequencies of the four gametes MD, MD', $M'D$ and $M'D'$ in subpopulation k are denoted by

$$P(MD|S_i) = x_{k1},\ P(MD'|S_k) = x_{k2},\ P(M'D|S_k) = x_{k3},\ P(M'D'|S_k) = x_{k4}.$$

Then, the frequencies of alleles D and M are

$$p_k = x_{k1} + x_{k3} \quad \text{and} \quad m_k = x_{k1} + x_{k2},$$

respectively. The linkage disequilibrium δ_k within subpopulation k is defined as

$$\delta_k = x_{k1}x_{k4} - x_{k2}x_{k3},$$

then, $x_{k1} = p_k m_k + \delta_k$. The null hypothesis of linkage equilibrium is

$$H_{0A} : \delta_k = 0, \quad k = 1, \cdots, K. \tag{A.9}$$

Suppose that random mating in each of the K subpopulations holds, then the genotype pair $MMDD$ in the next generation is equal to

$$P(MMDD|S_k) = [p_k m_k + (1-\theta)\delta_k]^2,$$

where θ is the recombination fraction between marker and trait loci. Thus, under the null hypothesis H_{0A}, the marker genotype MM and trait genotype DD are independent, since $P(DD|S_k) = p_k^2$ and $P(MM|S_k) = m_k^2$. Similarly, we can show (A.8) for any marker genotype and trait genotype.

By (A.8), we have

$$\begin{aligned}
&P_{H_{0A}}(MM|S_k, y > \tau_2)\\
&= \sum_{G_T} P_{H_{0A}}(MM, G_T|S_k, y > \tau_2)\\
&= \sum_{G_T} P_{H_{0A}}(MM, G_T, y > \tau_2|S_k)/P_{H_{0A}}(y > \tau_2|S_k)\\
&= \sum_{G_T} P_{H_{0A}}(y > \tau_2|MM, G_T, S_k)P_{H_{0A}}(MM, G_T|S_k)/P(y > \tau_2|S_k)\\
&= P(MM|S_k)\sum_{G_T} P(y > \tau_2|G_T, S_k)P(G_T|S_k)/P(y > \tau_2|S_k)\\
&= P(MM|S_k).
\end{aligned}$$

Thus,

$$\begin{aligned}
P_{H_{0A}}(MM|y > \tau_2) &= \sum_{k=1}^{K} P_{H_{0A}}(MM|y > \tau_2, S_k)P(S_k|y > \tau_2)\\
&= \sum_{k=1}^{K} P(MM|S_k)P(S_k|y > \tau_2)\\
&= \sum_{k=1}^{K} m_k^2\alpha_k. \qquad \text{(A.10)}
\end{aligned}$$

Similarly, we have

$$P_{H_{0A}}(MM'|y > \tau_2) = \sum_{k=1}^{K} 2m_k(1 - m_k)\alpha_k. \qquad \text{(A.11)}$$

By (A.10) and (A.11), we have

$$\begin{aligned}
E_{H_{0A}}(\hat{m}_D) = P_{H_{0A}}(M|y > \tau_2) &= P_{H_{0A}}(MM|y > \tau_2) + \frac{1}{2}P_{H_{0A}}(MM'|y > \tau_2)\\
&= \sum_{k=1}^{K} m_k\alpha_k \equiv \bar{m}_D. \qquad \text{(A.12)}
\end{aligned}$$

Similarly,

$$E_{H_{0A}}(\hat{m}_C) = \sum_{k=1}^{K} m_k\gamma_k \equiv \bar{m}_C. \qquad \text{(A.13)}$$

Using the presentation of Li (1969), we have

$$P_{H_{0A}}(MM|y > \tau_2) = \bar{m}_D^2 + \sigma_D^2 \quad \text{with} \quad \sigma_D^2 = \sum_{k=1}^{K} \alpha_k(m_k - \bar{m}_D)^2, \text{ (A.14)}$$

and

$$P_{H_{0A}}(MM'|y > \tau_2) = 2\bar{m}_C(1-\bar{m}_C)-2\sigma_C^2 \quad \text{with} \quad \sigma_C^2 = \sum_{k=1}^{K} \gamma_k(m_k-\bar{m}_C)^2. \tag{A.15}$$

By (A.2)-(A.5), (A.14) and (A.15), we have

$$Var_{H_{0A}}(\hat{m}_D) = \frac{1}{2n_1}[\bar{m}_D(1-\bar{m}_D)+\sigma_D^2]. \tag{A.16}$$

Similarly,

$$Var_{H_{0A}}(\hat{m}_C) = \frac{1}{2n_2}[\bar{m}_C(1-\bar{m}_C)+\sigma_C^2]. \tag{A.17}$$

Equations (A.12), (A.13), (A.16) and (A.17) are what we need.

B. Derivation of the variance of LS estimator

Under the null hypothesis, $(x_{kj}-\bar{x})$ is independent of y_{kj}, hence,

$$\begin{aligned} Var_{H_{0R}}[(x_{kj}-\bar{x})y_{kj}] &= Var(x_{kj}-\bar{x})E_{H_{0R}}(y_{kj}^2) + Var_{H_{0R}}(y_{kj})E(x_{kj}-\bar{x})^2 \\ &= Var[(x_{kj}-\bar{x})\beta_{0k}] + E(x_{kj}-\bar{x})^2\sigma_k^2 \end{aligned} \tag{B.1}$$

For $j_1 \neq j_2$, y_{kj_1} and y_{kj_2} are independent, also, both of them are independent of x_{kj_1} and x_{kj_2} under the null hypothesis, thus,

$$\begin{aligned} & Cov_{H_{0R}}[(x_{kj_1}-\bar{x})y_{kj_1}, (x_{kj_2}-\bar{x})y_{kj_2}] \\ &= Cov(x_{kj_1}-\bar{x}, x_{kj_2}-\bar{x})E_{H_{0R}}[y_{kj_1}]E_{H_{0R}}[y_{kj_2}] \\ &= Cov[(x_{kj_1}-\bar{x})\beta_{0k}, (x_{kj_2}-\bar{x})\beta_{0k}]. \end{aligned} \tag{B.2}$$

By (B.1) and (B.2), we have

$$\begin{aligned} & Var_{H_{0R}}[\sum_{j=1}^{n_k}(x_{kj}-\bar{x})y_{kj}] \\ &= \sum_{j=1}^{n_k} Var_{H_{0R}}[(x_{kj}-\bar{x})y_{kj}] + \sum_{j_1\neq j_2} Cov_{H_{0R}}[(x_{kj_1}-\bar{x})y_{kj_1}, (x_{kj_2}-\bar{x})y_{kj_2}] \\ &= n_k E(x_{k1}-\bar{x})^2\sigma_k^2 + Var[\sum_{j=1}^{n_k}(x_{kj}-\bar{x})\beta_{0k}]. \end{aligned} \tag{B.3}$$

Similarly, for $k_1 \neq k_2$, we have

$$Cov_{H_{0R}}[\sum_{j=1}^{n_{k_1}}(x_{k_1 j} - \bar{x})y_{k_1 j}, \sum_{j=1}^{n_{k_2}}(x_{k_2 j} - \bar{x})y_{k_2 j}]$$

$$= Cov[\sum_{j=1}^{n_{k_1}}(x_{k_1 j} - \bar{x})\beta_{0k_1}, \sum_{j=1}^{n_{k_2}}(x_{k_2 j} - \bar{x})\beta_{0k_2}]. \tag{B.4}$$

It follows from (B.3) and (B.4) that

$$Var_{H_{0R}}[\sum_{k=1}^{K}\sum_{j=1}^{n_k}(x_{kj} - \bar{x})y_{kj}]$$

$$= \sum_{k=1}^{K} Var_{H_{0R}}[\sum_{j=1}^{n_k}(x_{kj} - \bar{x})y_{kj}]$$

$$+ \sum_{k_1 \neq k_2} Cov_{H_{0R}}[\sum_{j=1}^{n_k}(x_{k_1 j} - \bar{x})y_{k_1 j}, \sum_{j=1}^{n_k}(x_{k_2 j} - \bar{x})y_{k_2 j}]$$

$$= n\sum_{k=1}^{K} w_k \sigma_k^2 E(x_{k1} - \bar{x})^2 + Var[\sum_{k=1}^{K}\sum_{j=1}^{n_k}(x_{kj} - \bar{x})\beta_{0k}].$$

Thus,

$$Var_{H_{0R}}(\hat{\beta}_1) = Var_{H_{0R}}(\sum_{k=1}^{K}\sum_{j=1}^{n_k} b_{kj}y_{kj}) \approx \frac{Var_{H_{0R}}[\sum_{k=1}^{K}\sum_{j=1}^{n_k}(x_{kj} - \bar{x})y_{kj}]}{\left[n\sum_{k=1}^{K} w_k E(x_{kj} - E\bar{x})^2\right]^2}$$

$$= \frac{n\sum_{k=1}^{K} w_k E(x_{k1} - \bar{x})^2\sigma_k^2 + Var[\sum_{k=1}^{K}\sum_{j=1}^{n_k}(x_{kj} - \bar{x})\beta_{0k}]}{\left[n\sum_{k=1}^{K} w_k E(x_{k1} - E\bar{x})^2\right]^2}. \tag{B.5}$$

In what follows, we give the expressions of $\sum_{k=1}^{K} w_k E(x_{k1} - \bar{x})^2\sigma_k^2$ and $Var(\sum_{k=1}^{K}\sum_{j=1}^{n_k}(x_{kj} - \bar{x})\beta_{0k})$. By the independence of the observations, we have

$$E(x_{kj}\bar{x}) = \frac{1}{n}[Ex_{k1}^2 + \sum_{i=1}^{n} Ex_{k1}Ex_i - (Ex_{k1})^2] = \frac{1}{n}Var(x_{k1}) + Ex_{k1}E\bar{x} \tag{B.6}$$

and

$$E(\bar{x})^2 = \frac{1}{n}\sum_{k=1}^{K}\sum_{j=1}^{n_k} E(x_{kj}\bar{x}) = \frac{1}{n}\sum_{k} w_k Var(x_{k1}) + (E\bar{x})^2. \tag{B.7}$$

By (B.6) and (B.7), we have

$$\begin{aligned}
&\sum_k w_k\sigma_k^2 E(x_{k1}-\bar{x})^2\\
&=\sum_k w_k\sigma_k^2\left(Ex_{k1}^2-2E(x_{k1}\bar{x})+E(\bar{x})^2)\right)\\
&=\sum_{k=1} w_k\sigma_k^2\left(Ex_{k1}^2-\frac{2}{n}Var(x_{k1})-2Ex_{k1}E\bar{x}+\frac{1}{n}\sum_k w_k Var(x_{k1})+(E\bar{x})^2\right)\\
&\approx\sum_k w_k\sigma_k^2 E(x_{k1}-E\bar{x})^2. \qquad\qquad (B.8)
\end{aligned}$$

By the independence of the observations, we have

$$\begin{aligned}
Var(x_{kj}-\bar{x}) &= Var\left(\frac{n-1}{n}x_{kj}-\frac{1}{n}\sum_{i\neq n_1+\cdots+n_{k-1}+j} x_i\right)\\
&=\left(\frac{n-1}{n}\right)^2 Var(x_{k1})+\frac{1}{n^2}\left(\sum_i Var(x_i)-Var(x_{k1})\right)\\
&=\frac{n-2}{n}Var(x_{k1})+\frac{1}{n}\sum_k w_k Var(x_{k1}).
\end{aligned}$$

By (B.6) and (B.7), we have

$$\begin{aligned}
Cov(x_{kj_1}-\bar{x},x_{kj_2}-\bar{x}) &= Ex_{kj_1}Ex_{kj_2}-E(x_{kj_1}\bar{x})-E(x_{kj_2}\bar{x})+E(\bar{x})^2\\
&\quad -[Ex_{kj_1}Ex_{kj_2}-Ex_{kj_1}E\bar{x}-Ex_{kj_2}E\bar{x}+(E\bar{x})^2]\\
&=-\frac{2}{n}Var(x_{k1})+\frac{1}{n}\sum_k w_k Var(x_{k1}).
\end{aligned}$$

It follows from the above two expressions that

$$\begin{aligned}
&Var\Big[\sum_j (x_{kj}-\bar{x})\Big]\\
&=\sum_j Var(x_{kj}-\bar{x})+\sum_{j_1\neq j_2} Cov(x_{kj_1}-\bar{x},x_{kj_2}-\bar{x})\\
&=\frac{1}{n}\Bigg\{(n-2)n_k Var(x_{k1})+n_k\sum_k w_k Var(x_{k1})\\
&\quad +n_k(n_k-1)\Big[-2Var(x_{k1})+\sum_k w_k Var(x_{k1})\Big]\Bigg\}\\
&=\frac{1}{n}\Bigg\{nn_k Var(x_{k1})+n_k^2\Big[-2Var(x_{k1})+\sum_k w_k Var(x_{k1})\Big]\Bigg\}.
\end{aligned}$$

Again, by (B.6) and (B.7),

$$
\begin{aligned}
&Cov[\sum_j (x_{k_1 j} - \bar{x}), \sum_j (x_{k_2 j} - \bar{x})] \\
&= \sum_{j_1} \sum_{j_2} Cov(x_{k_1 j_1} - \bar{x}, x_{k_2 j_2} - \bar{x}) \\
&= \sum_{j_1} \sum_{j_2} \Big(Ex_{k_1 1} Ex_{k_2 1} - E(x_{k_1 1}\bar{x}) - E(x_{k_2 1}\bar{x}) + E(\bar{x})^2 \\
&\quad - Ex_{k_1 1} Ex_{k_2 1} + Ex_{k_1 1} E\bar{x} + Ex_{k_2 1} E\bar{x} - (E\bar{x})^2 \Big) \\
&= \frac{1}{n} n_{k_1} n_{k_2} \Big(- Var(x_{k_1 1}) - Var(x_{k_2 1}) + \sum_k w_k Var(x_{k1}) \Big)
\end{aligned}
$$

By the above two expressions, we have

$$
\begin{aligned}
&Var[\sum_{k=1}^{K} \sum_{j=1}^{n_k} (x_{kj} - \bar{x}) \beta_{0k}] \\
&= \sum_{k=1} \beta_{0k}^2 Var(\sum_{j=1}^{n_k} (x_{kj} - \bar{x})) + \sum_{k_1 \neq k_2} \beta_{k_1} \beta_{k_2} Cov[\sum_j (x_{k_1 j} - \bar{x}), \sum_j (x_{k_2 j} - \bar{x})] \\
&= n \Big(\sum w_k \beta_{0k}^2 Var(x_{k1}) - 2 \sum_k w_k \beta_{0k} \sum_k w_k \beta_{0k} Var(x_{k1}) \\
&\quad + (\sum_k w_k \beta_{0k})^2 \sum_k w_k Var(x_{k1}) \Big) \\
&= n \sum_k w_k Var(x_{k1}) (\beta_{0k} - \bar{\beta}_0)^2, \quad \text{with } \bar{\beta}_0 = \sum_k w_k \beta_{0k}. \qquad \text{(B.9)}
\end{aligned}
$$

It follows from (B.5), (B.8) and (B.9) that the variance of $\hat{\beta}_1$ can be approximated by (3.7).

Acknowledgments

This work was supported in part by an NIH grant EY014478 (HQ, HZ and ZL), National Natural Science Foundation of China (10471136), Ph. D. Program Foundation of Ministry of Education of China, and Special Foundations of the Chinese Academy of Science and USTC.

References

1. Bacanu, S. A., Devlin, B. and Roeder, K. (2000). The power of genomic control. *American Journal of Human Genetics* **28**, 782-793.

2. Devlin, B. and Roeder, K. (1999). Genomic control for association studies, *Biometrics* **55**, 997-1004.
3. Devlin, B., Roeder, K. and Wasserman, L. (2001). Genomic control, a new approach to genetic-based association studies. *Theoretical Population Biology* **60**, 155-166.
4. Ewens, W. J. and Spielman, R. (1995). The Transmission/ Disequilibrium test: history, subdivision, and admixture. *American Journal of Human Genetics* **57**, 455-464.
5. Li, C. C. (1969). Population subdivision with respect to multiple alleles. *Annals of Human Genetic* **33**, 23-29.
6. Li, Z., Gail, M. H., Pee, D. and Gastwirth, J. L. (2001). Statistical properties of Teng and Risch's sibship type tests for detecting an association between diseases and a candidate allele. *Human Heredity* **53**, 114-129.
7. Li, Z., Zhang, H., Gastwirth J. L. and Gail, M. H. (2005). Excessive false positive rate caused by population stratification and disease rate heterogeneity in case-control association studies. (Personal communication)
8. Morton, N. E. and Collins, A. (1998). Tests and estimates of allelic association in complex inheritance. *Proceedings of the National Academy of Sciences of the United States of America* **95**, 11389-11393.
9. Ohashi, J., Yamamoto, S., Tsuchiya, N., Hatta, Y., Komata, T., Matsushita, M. and Tokunaga, K. (2001). Comparison of statistical power between 2×2 allele frequency and allele positivity tables in case-control studies of complex disease genes. *Annals of Human Genetic* **65**, 197-206.
10. Pritchard, J. K., Stephens, M., Donnelly, P. (2000a). Inference of population structure using multilocus genotype data. *Genetics* **65**, 220-228.
11. Pritchard, J. K., Stephens, M., Rosenberg, N. A., Donnelly, P. (2000b). Association mapping in structured population. *American Journal of Human Genetics* **67**, 170-181.
12. Rabinowitz, D. (1997). A transmission disequilibrium test for qauntitative trait loci. *Human Heredity* **47**, 342-350.
13. Rannala, B. and Slatkin, M. (1998). Likelihood analysis of disequilibrium mapping, and related problems. *American Journal of Human Genetics* **62**, 459-473.
14. Risch, N. (2000). Searching for genetic determinants in the new millennium. *Nature* **405**, 847-856.
15. Risch, N. and Merikangas, K. (1996). The future of genetic studies of complex human diseases. *Science* **273**, 1516-1517.
16. Risch, N. and Teng, J. (1998). The relative power of family-based and case-control designs for linkage disequilibrium studies of complex human diseases. I. DNA pooling, *Genome Research* **8**, 1273-1288.
17. Satten, G. A., Flanders, W. D. and Yang, Q. (2001). Accounting for unmeasured population substructure in case-control studies of genetic association using a novel latent-class model, *American Journal of Human Genetics* **68**, 466-477.
18. Spielman, R. S., McGinnis, R. E. and Ewens, W. J. (1993). Transmission test for linkage disequilibrium: the insulin gene region and insulin-dependent

diabetes mellitus (IDDM), *American Journal of Human Genetics*, **52**:506-513.

19. Teng, J. and Risch, N. (1999). The relative power of family-based and case-control designs for linkage disequilibrium studies of complex human diseases. II. individual genotyping. *Genome Research* **9**, 234-241.
20. Thompson, E. A. and Neel, J. V. (1997). Allelic disequilibrium and allele frequency distribution as a function of social and demographic history. *American Journal of Human Genetics* **60**, 197-204.
21. van den Oord, E. J. C. G. (1999). A comparison between different designs and tests to detect QTLs in association studies. *Behavior Genetics* **29**, 245-256.
22. Xiong, M. M., Fan, R. Z. and Jin, L. (2002). Linkage disequilibrium mapping of quantitative trait loci under truncation selection. *Human Heredity* **53**, 158-172.
23. Xiong, M. M. and Guo, S. W. (1997). Fine-scale genetic mapping based on linkage disequilibrium: theory and application. *American Journal of Human Genetics* **60**, 1513-1531.
24. Zhang, S. L., Kidd, K. K. and Zhao, H. Y. (2002). Detecting genetic association in case-control studies using similarity-based association tests. *Statistica Sinica* **12**, 337-359.
25. Zhang, S. L. and Zhao, H. Y. (2001). Quantitative similarity-based association tests using population samples. *American Journal of Human Genetics* **69**, 601-614.

PROBABILITY MODELING AND STATISTICAL ANALYSIS OF STANDARDIZED TEST DATA FOR PURPOSES OF UNDERSTANDING AND MEASURING TEST EQUITY, LATENT ABILITY SPACE MULTI-DIMENSIONALITY, AND SKILLS DIAGNOSTIC ASSESSMENT

WILLIAM STOUT*

Department of Statistics, University of Illinois

The paper surveys almost two decades of progress by me and colleagues in three psychometric research areas involving the probability modeling and statistical analysis of standardized ability test data: nonparametric multidimensional latent ability structure modeling and assessment, test fairness modeling and assessment, and modeling and assessment of skills diagnosis via educational testing. In the process, it is suggested that the unidimensional scoring testing paradigm that has driven standardized ability testing research for over half a century is giving way to a new multidimensional latent ability modeling and multiple scoring paradigm that in particular explains and allows the effective detection of test bias and embraces skills-level formative assessment, opening up a plethora of challenging, exciting, and societally important research problems for psychometricians. It is hoped that this light-stepping history will interest probabilists and statisticians in exploring the field of psychometrics.

Informally, test bias occurs when an examinee is under or over evaluated by his test score in terms of the purpose of the test. Also informally, skills diagnosis refers to evaluating examinee levels of mastery (usually done dichotomously as master versus nonmaster of each skill) on a moderate number of carefully selected skills for which having student skills profiles can greatly help individual student learning and classroom level.

My strategy, strongly influenced by his probabilistic background, for producing interesting and effective psychometric research is to choose psychometric research questions from practical challenges facing educational testing. Then, I and colleagues bring to bear sophisticated probability modeling and modern statistical thought to solve these questions, making effectiveness of the resulting research in meeting the educational testing challenges the ultimate criterion for judging its worth.

It is somebody's ancient proverb that the acorn sometimes falls far from the oak tree. Y. S. Chow taught me the tools of probability limit theorem research,

*I wish to thank all my former Ph.D. students: Without their contributions, the content of this paper would have been quite different and much less interesting!

taught me to approach research with enthusiasm and tenacity, and provided a very supportive environment for me and his other graduate students. Although psychometrics/educational measurement is far from the probabilistic oak tree, whatever success I've had as a psychometrician has been strongly influenced by the supportive, demanding, and creative environment Y. S. creates for his students. By now I have had many Ph.D. students in psychometrics: it was the just described model of Y. S.'s for mentoring Ph.D. students that I followed with all of them.

Some key words: Nonparametric IRT; NIRT; Latent unidimensionality; Latent multidimensionality; Essential unidimensionality; Monotone locally independent unidimensional IRT model; MLI1; Item pair conditional covariances; DIMTEST; HCA/CCPROX; DETECT; Generalized compensatory IRT model; Approximate simple structure; DIF; Differential item functioning; Differential bundle functioning; DBF; Valid subtest; Multidimensional model for DIF; MMD; SIBTEST; MultiSIB; Mantel-Haenszel; PolySIB; CrossingSIB; Skills diagnosis; Formative assessment; Unified Model; Bayes Unified Model; MCMC; PSAT Score Report *Plus*; University of Illinois Department of Statistics Statistical Laboratory for Educational and Psychological Measurement.

1. Introductory Remarks

A festschrift (in my opinion) is an occasion where one has the right to present broad survey papers and to present one's personal perspective about the past and future of the area of the paper, especially how it relates to the person being honored. This paper summarizes my efforts together with those of a large group of colleagues to individually and collectively use the tools of probability and statistics to address three core issues currently confronting educational and psychological testing. Most of these colleagues got their Ph.D.s under my direction and, as such, have at one time been members of the Statistical Laboratory for Educational and Psychological Measurement (the "Lab"), which I founded in the Department of Statistics at the University of Illinois in the late 1980s and which is currently co-directed by Jeff Douglas and Louis Roussos, former Ph.D. student Lab members and now faculty members at the University of Illinois.

During the first half of my research career, spent as a mathematics professor at the University of Illinois, I solved a few probability limit problems, excited by their intrinsic intellectual appeal. In doing so, I was acting as artist rather than engineer. This pure-research-driven aesthetic motivation, which works so well for many researchers and worked very well for me for the first third of my career, was becoming in the late '70s insufficient to sustain my research drive. My response was to switch fields from probability limit theorems to psychometrics/educational measurement. I became determined for the remainder of my career to do research that, in addition to being intrinsically intellectually interesting, would bear fruit that could be effectively *applied* to important societal problems.

Thus, my personal and enthusiastic entry into the field of psychometrics in the early 1980s, after I had already become a full professor of mathematics, was my private resolution of my professional "midlife crisis." In resolving this crisis, I discovered two important things about the process of doing psychometric research that are worth stressing to other probabilists and statisticians possibly drawn to the field. First, issues and problems arising out of the *actual practice* of educational testing often provide fertile ground for the generation of excellent research problems, for which a deeply theoretical probabilistic modeling approach often produces effective and valuable solutions from the applications perspective. Second, enormous excitement and intellectual power can flow forth when a carefully selected group of researchers dedicates itself to collaboratively and supportively solving important and interesting psychometric research problems. In fact, after switching fields I was determined to intensify the collaborative research style I had learned while doing mathematical research, especially as I had learned from observing Y.S. (such as his well-known collaboration with Herbert Robbins) and while working jointly and extensively with my mathematical colleague and friend Walter Philipp (see, for example, Philipp and Stout, 1975).

Psychometrics is a field that is both intellectually interesting and relevant for a society in which effective education and training are vital prerequisites to progress. Indeed, psychometrics' intellectual appeal and great societal importance were what most motivated my choosing it, noting that so many fields of application can have the same appeal for others of course.

I decided the way to maximize the probability that the Laboratory's psychometric research would be usefully applied to societally important educational measurement problems (my personal goal) was to choose research problems wisely from the vast hurly-burly of practical problems and issues flowing out of actual educational measurement practice as conducted by testing companies and other practitioners of the standardized testing art. From this applications perspective, I attempted to choose research problems for which successful solutions would clearly have a widespread impact on bringing about improved educational testing and assessment, my thus accepting the basic premise that, done right, standardized testing can be of great benefit to the educational enterprise.

In the case of the Laboratory that I directed in the '80s and '90s over the almost two decades of its existence, our focus was on three core applied issues growing out of the practice of standardized testing: assessing the multidimensional structure of the latent ability that is modeled to stochastically drive test performance, assessing test fairness, and diagnosing examinee latent skills as a means of accomplishing formative assessment. Here, "latent" refers to the cognitive psychologist's distinguishing an examinee's observed performance such as his/her responses to test questions from the model-postulated but unobservable inside-the-brain "latent" variables the examinee possesses and that monotonically relate to the probability of successful observed performance. By formative assessment, I mean the assessment of students while they are still learning, with the purpose of facilitating both future teaching and the students' future learning.

Because of my Ph.D. mentoring under Y.S. and the fact that my early field of research had been probability theory, my tendency as a psychometrician has always been to stress the importance of sophisticated probability modeling in addressing psychometric research. Thus, in each of the three identified research areas - latent structure, fairness, and skills diagnosis - the challenge of developing appropriate probability models, as well as then deriving important mathematical implications of these developed models, became very important. Another tendency has been to stress the importance of bringing modern statistical thought and methodologies to bear on psychometric research problems. This viewpoint has had a major influence on my group's progress in the three research areas and in addition has from time to time drawn research statisticians to the field of psychometrics. In this regard, I strongly believe, and have seen first hand, that psychometrics can powerfully appeal to research statisticians (probabilists too) that are casting about for interesting applied areas of research and applications to ply their trade.

As a sequel to the probability modeling and associated study of the mathematical properties of these psychometric models, the development of specialized statistical procedures was often required. It was judged vital to return to the applied setting that had precipitated the research problem by providing a practical and easy-to-implement statistical solution to the practitioner's problem that hopefully will be widely applicable to other similar applied problems. That is, the goal has never been just the publishing of papers; it has also been to produce successful applications played out in actual educational measurement practice. For the research to have an important and wide impact on educational measurement, it must have easy and wide transferability to similar measurement settings. This transferability has been facilitated by making methodological software available to practitioners and researchers, as discussed below.

Below this paper describes progress in the three applied research areas I and my colleagues have focused on, with occasional suggested directions for future research given.

2. Nonparametric Latent Ability Structure Assessment

We consider the assessment of latent structure unidimensionality first. Then more generally we consider the assessment of multidimensional latent structure.

2.1. *Unidimensionality from the weak, or conditional covariance, perspective*

The psychometric modeling of examinee performance on standardized tests has for the last two decades been dominated by what is referred to as Item Response Theory (IRT) modeling. An IRT model specifies the probability of successfully answering a test question (called an "item") by a monotone increasing function of a real valued (but sometimes n-dimensional) latent ability variable. Conditional on the value of the latent ability, the examinee's responses to different items are independent (often called "local independence" or LI). The responses of the different test-taking examinees are assumed independent. When the latent variable producing LI of item responses is real valued, the IRT model is called a unidimensional IRT model. Unidimensional IRT models have dominated the field in applications and are interpreted as a single underlying latent variable (e.g., algebra ability) "explaining" the responses of the examinees to the items of the test. From this perspective, it is natural to refer to an algebra test, etc.

In the influential monograph *Applications of Item Response Theory to Practical Testing Problems* (Lord, 1980), Fred Lord stated, "There is a great need for a statistical significance test for the unidimensionality of a set of items." This strong statement, made when practical applications of unidimensional IRT modeling to testing was in its relative infancy, reminded the testing community of the great need to have an effective statistical test of unidimensionality. Hypothesis test based acceptance of unidimensionality would help to legitimize applying the universally used unidimensional logistic IRT-based calibration and prediction methodologies such as BILOG.

Further, one important way to address the issue of whether to summarize examinee test performance via a single unidimensional latent scale is by asking the psychometric question of whether the test data is unidimensional. Thus, our two decades long quest into nonparametric latent structure assessment began by addressing the question of how to model and to statistically assess important departures from latent unidimensionality.

The question of when a unidimensional IRT model is adequate provides a nice example of the principle that sound theory is a prerequisite for good psychometric data analysis practice. My probabilistic limit theorem approach to unidimensionality was both simple and nonparametric (Stout, 1987). By taking a nonparametric approach, one does not confound lack of model fit exhibited by a particular unidimensional parametric family of models with the data having been generated by an intrinsically multidimensional latent-trait model.

Rigorously defining latent unidimensionality, an essential step, requires one to first ask the foundational question of how best to define latent unidimensionality. The primary insight is that the issue of unidimensionality must be framed as whether the inferred manifest test distribution that describes examinee test responding can be represented by a unidimensional, locally independent, and monotone IRT model appearing in a de Finetti-like integral to produce the manifest distribution. (In principle, by observing enough examinees, the manifest examinee test response distribution can be statistically inferred from test data with any desired accuracy and hence becomes "manifest." Because of these considerations, the unidimensionality problem is posed by presuming that a specific manifest test distribution has been specified.) It is interesting and important to note the obvious logical conclusion that being able to exhibit a substantively and mathematically appealing two-dimensional locally independent, monotone IRT model that fits the given manifest distribution *does not* prove that unidimensionality fails for the given manifest distribution.

By contrast, providing a unidimensional model fitting the manifest distribution does prove unidimensionality holds for the manifest distribution. Further, the issue of unidimensionality cannot be addressed by showing the lack of fit of an appealing unidimensional IRT model for the given manifest distribution. Rather, the issue of unidimensionality is the issue of whether any unidimensional IRT model *exists* that fits the manifest test distribution.

For simplicity, we assume dichotomous item responses (right or score 1 versus wrong or score 0) throughout the paper, even though most of the issues discussed have polytomous (e.g., one could score 0, 1, 2, or 3 on an item) versions too. Although it is the test that is described as unidimensional, the dimensionality is determined by the interaction between the test structure (given by the item response functions; denoted IRFs) and the latent ability structure (given by the latent ability examinee population distribution: possibly multidimensional). In fact, the same test could easily be unidimensional for one examinee population and multidimensional for another, noting that changing the examinee population alters the manifest distribution of examinee test responding.

We now define test unidimensionality rigorously. To avoid unneeded notational complexity, we make the easily removable assumption that all latent random variables are of continuous type.

Definition 1: A length n test $\mathsf{U}' = (U_1, U_2, \mathrm{K}, U_n)$ with manifest distribution $P(\boldsymbol{U} = \boldsymbol{u})$ with the components of $\boldsymbol{u}$ being 0 or 1 is said to be *unidimensional* if there exists a real-valued random variable Θ with latent density denoted by $f(\theta)$ such that for all possible response patterns $\boldsymbol{u}$

$$P(\mathsf{U} = \mathsf{u}) = \int_{-\infty}^{\infty} P(\mathsf{U} = \mathsf{u} | \Theta = \theta) f(\theta) d\theta \tag{1}$$

for which local independence (LI) and IRF monotonicity (M) relative to θ holds. Any such IRT model satisfying LI and M for a unidimensional latent trait Θ is called a *monotone locally independent unidimensional model* and herein is denoted by MLI1. For simplicity, we will sometimes refer to such a model as a unidimensional model. □

The widely studied MLI1 model goes back at least to Mokken (1971). What is being denoted in this paper as a MLI1 model has many names in the literature including "monotone unidimensional latent trait model," "monotone homogeneity model," "monotone latent variable model," and "monotone IRT model." Thus, great care is required in reading and interpreting the psychometric literature concerning the concept of a unidimensional IRT model.

It is worth noting that if one drops M as a requirement, the conceptual idea behind the intuitively startling fact that one can always exhibit a unidimensional LI latent model for any given manifest test distribution can be easily explained. This result was formally shown by Suppes and Zanotti (1981). Assuming a deterministically-responding but random-sampled examinee perspective (see Holland, 1990a) as the source of the examinee responses being random , the basic idea of the proof of a unidimensional LI IRT model always being possible to construct is to assign to each examinee, as the value of his or her unidimensional latent variable, the binary expansion corresponding to his or her true deterministic knowledge state for all of the items (that is, 1 in ith place of the binary expansion if item i answer is "known" by the examinee, 0 if not known). The probability assigned to each such binary-expansion-represented latent state is the probability assigned by the manifest test distribution to the corresponding examinee item response pattern $\boldsymbol{u}$ producing the "latent" binary expansion. It is then easy to see that LI holds and M fails for the resulting unidimensional model.

Conceptually useful generalizations of Definition 1 are possible by appropriate weakening of either the concept of local independence or of monotonicity. One weakening, important from the practitioner perspective, is to replace the traditional definition of local independence by that of pairwise local independence or what is termed weak local independence (see McDonald, 1994), as provided in Definition 3.

Definition 2: A test $\boldsymbol{U}$ is said to be *(strongly) locally independent* (denoted LI, or SLI when it is necessary to contrast LI of this definition with weak LI of Definition 3 below) with respect to a latent variable Θ if for all $\boldsymbol{u}$ and θ,

$$P(\mathsf{U} = \mathsf{u}|\Theta = \theta) = \prod_{i=1}^{n} P(U_i = u_i|\Theta = \theta). \tag{2}$$

□

Definition 3: A test $\boldsymbol{U}$ is said to be *weakly locally independent* (*WLI*) with respect to a latent variable Θ if for all item pairs i, i' and all θ,

$$\mathrm{Cov}(U_i, U_{i'}|\Theta = \theta) = 0. \tag{3}$$

□

In the binary (0, 1) item-scoring case, it is trivial that WLI is equivalent to pairwise LI. That is, WLI holds if and only if pairwise LI holds for all item pairs i, i'. That is, for all θ WLI is equivalent to

$$P(U_i = u_i, U_{i'} = u_{i'}|\Theta = \theta) = P(U_i = u_i|\Theta = \theta)P(U_{i'} = u_{i'}|\Theta = \theta) \tag{4}$$

for all item pairs i, i' and all θ. □

From the practitioner's perspective, it seems a reasonable practical approximation that a dichotomous item test could be viewed as unidimensional, or, equivalently, a latent model viewed as MLI1, even if statistical evidence only shows that Definition 1 holds with local independence replaced by weak local independence. The perceptive reader will note that by making this replacement, an assumption has (sneakily) inserted itself: If a test is declared unidimensional according to Definition 1 but with WLI (equivalently, pairwise LI) used as the definition of local independence, then this presumes it is also unidimensional (it supports a MLI1 IRT model) with SLI replacing pairwise LI. I agree with the assumption from the empirical viewpoint: From the measurement practitioner's perspective, much of the time WLI seems for practical purposes empirically equivalent (trivially, WLI and SLI are *not* mathematically equivalent) to the stronger and more complex (and much harder to verify statistically) reality of SLI.

In this regard it is useful to quote the well known psychometrician Rod McDonald (1994) when he argues that the insertion of WLI in place of SLI amounts to "not changing our definition or our substantive conceptualization of latent traits." McDonald goes on to state that it is:

> unlikely that investigators seriously imagine that the conditional covariances of the items vanish while they still possess higher order dependencies in probability. That is, we are unlikely to believe that while every pair of items gives statistically independent responses [conditional on a latent variable], responses to some items [conditional on the latent variable] are dependent on responses to two or more other items.

Our research group's acceptance of the empirical equivalence of WLI and SLI from the educational test practitioner's perspective undergirds my statistical approach to latent dimensionality assessment. That is, in order to study and statistically assess latent dimensional structure, by taking into account the just-discussed practitioner's perspective, it is reasonable that we merely statistically investigate conditional covariances given an appropriately chosen latent trait θ, thereby statistically investigating WLI instead of investigating the full SLI.

Thus, the practitioner-driven need for a useful and effective statistical test of the unidimensionality of an educational test led me to formulate a theory of latent space dimensionality structure assessment based on conditional covariances. This, as will be discussed below, in turn led to conditional covariance based theoretically-defensible statistical procedures such as DIMTEST, DETECT, and HCA/CCPROX. The use of these procedures has been justified empirically via both large-scale simulation studies and real-data-based standardized test applications.

In addition to a focus on conditional covariances as a basic latent dimensionality assessment tool, our efforts had refocused us on the importance of the fundamental MLI1 model, which in turn has helped invigorate the recent psychometric research emphasis on foundational and applied nonparametric item response theory (NIRT) research: see especially Junker and Sijtsma, 2001. Indeed, our conditional covariance approach to latent dimensionality assessment has meshed nicely with the general resurgence of interest in NIRT, propelled in part on the European side by a long-standing focus on Mokken scaling. The interested reader is in particular referred to the September 2001 *Applied Psychological Measurement* "Special Issue on NIRT," edited by Brian Junker and Klaas Sijtsma, for a superb survey of modern NIRT research, and to a slightly earlier methodological NIRT survey paper by Sijtsma (1998).

2.2. *Infinite test length unidimensional asymptotic MLI1 modeling*

A flurry of foundational work on NIRT flowed from the first DIMTEST paper (Stout, 1987), all aimed at developing a useful practitioner-oriented foundational understanding of multidimensional latent trait modeling, especially as revealed by conditional covariances. Much of this work then formed the basis for the further development of conditional-covariance-based statistical tools developed for practitioners in order to estimate important characteristics of the multidimensional latent ability space, especially the statistical procedures HCA/CCPROX and DETECT.

Early on I had realized that the finite-length-test grounded notion of unidimensionality, with either WLI or SLI as its underpinning, was too stringent a concept from the practical perspective of wanting to model and count only the important or influential latent dimensions and in particular wanting to theoretically characterize a test consisting of only one important dimension, the construction of a unidimensional test very often being the practitioner's goal. In particular, from the practitioner perspective, it is useful to partition the latent dimensions of a LI M multidimensional IRT model into the important (called "essential," "dominant," "major," etc.) dimensions and the unimportant (called "inessential," "weak," "nuisance," "minor," etc.) dimensions, which can be ignored from the practitioner's perspective.

Thus, we wanted a theoretical construction that separates essential from inessential dimensions, counting only the number of essential dimensions and, in particular, defining in what sense a test can have just one "essential" dimension with possibly numerous inessential dimensions. This idea has factor analytic roots: see, for example, Tucker, Koopman, and Linn (1969) for a factor analytic model distinguishing between minor (and hence inessential) factors and major factors. The resulting definition of essential unidimensionality (Stout, 1990) with respect to a latent variable θ uses conditional covariances and is based on modeling the dimensionality properties of a finite-length test $\mathsf{U}'_n = (U_1, U_2, \mathrm{K}, U_n)$ asymptotically by representing U'_n as naturally and appropriately embedded in an infinite length test

$$\mathsf{U}'_\infty = (U_1, U_2, \mathrm{K}, U_n, \mathrm{K}) = (\mathsf{U}'_n, U_{n+1}, \mathrm{K}).$$

A philosophical (foundations) remark about this substitution of an infinite-length test is needed. The notion of sampling examinees from a large finite population idealized as an infinite population of examinees is an almost universally accepted modeling device in the theory of educational measurement (for example, assuming that the latent ability distribution is standard normal). What is much less accepted is another asymptotic aspect to the probability modeling of examinees taking a test, that of a long test, or more abstractly and realistically, that of a long-test manufacturing process (as introduced in Stout, 1990, and nicely articulated in Douglas, 1997). Indeed, individual examinee ability estimation asymptotics is impossible without it. In summary, one derives properties of a virtual test U_∞ of length ∞ whose properties are studied in order to understand properties of a real length n test U_n (n large enough to be thought as a "long" test and hence with its properties well approximated by U_∞).

This shift to studying U_∞ permits the discussion of the asymptotic consistency of ability estimation procedures as test length grows long. It also allows us to easily define the notion of a test being essentially unidimensional even though the actual length n test is multidimensional (and from the carefully reasoning practitioner's perspective cannot be strictly unidimensional). The point is that this infinite-length test abstraction, which in fact is just as reasonable from the modeling perspective as the infinite-population abstraction, provides a practically useful conception of an actually administered finite length test having one dominant or essential dimension.

Consider the case of long test and large examinee sample *joint asymptotics.* A number of important parametric logistic IRF (the dominant parametric modeling choice of practitioners doing IRT modeling) and nonparametric IRF results establishing joint consistency of test structure and examinee latent ability joint estimation require both examinee sample size and test length to cooperatively go to ∞ (see Haberman, 1977; Douglas, 1997; Trachtenberg and He, 2002 for various important milestones in this line of research). In conclusion, joint examinee and item asymptotics is a powerful foundational approach, as discussed further below.

If one takes advantage of an infinite length test formulation, an easy to formulate appropriate and rigorous definition of when a test U_∞ (or more naturally from the applications perspective, a test manufacturing process) is essentially unidimensional with respect to a latent variable θ can be given.

Definition 4: A test U_∞ is *essentially unidimensional* with respect to the unidimensional latent random variable Θ if for all θ, noting that n items produce $n(n-1)/2$ item pairs

$$\frac{\sum_{1\leq i<i'\leq n} |\mathrm{Cov}(U_i, U_{i'}|\Theta = \theta)|}{\binom{n}{2}} \to 0 \tag{5}$$

as $n \to \infty$. □

Here θ is viewed as the dominant dimension intended to be measured. In general, other existing (but asymptotically vanishing) dimensions force conditional covariances to be nonzero with respect to θ. An excellent example is provided by the minor-dimension-producing content dimensions of paragraphs in an essentially unidimensional reading comprehension test.

Out of Definition 4 comes the satisfying theoretical result that total test score, appropriately monotonically rescaled, consistently estimates the *unique* (no ordinally distinct θ' works) essential latent dimension θ as test length $n \to \infty$. Equivalently, the number correct score consistently estimates ability on the latent expected test score (called "true score") scale (Stout, 1990). This uniqueness theorem (modulo the equivalence class of monotone increasing transformations of θ of course) of the latent ability scale is philosophically and practically important because it justifies the intuitive notion that a long essentially unidimensional test really measures just one specific latent ability. This result, of course, does not imply any kind of asymptotic estimation efficiency for the number correct (rescaled) statistic for the most common parametric models like two parameter or three parameter logistic (such a claim being false). But, it is certainly important from the foundational perspective and useful from the NIRT modeling perspective, where the absence of a parametric model precludes using parametrically asymptotically efficient estimators such as a maximum likelihood estimator (MLE).

A delicate foundational issue needs addressing. It is the identifiability question for a unidimensional latent model (MLI1), presuming that the latent distribution of Θ is specified in order to rule out trivial and practically irrelevant causes of nonidentifiability. One first realizes that mathematically some nonidentifiability for the IRFs of a MLI1 model of a finite-length test must exist when the family of permissible IRFs is allowed to be fully nonparametric. That is, two distinct sets of IRFs for a MLI1 model with a specified latent ability distribution can in general produce the given manifest distribution for U_n. By contrast, for the infinite-test-length MLI1 U_∞ formulation, it is reassuring from the foundational NIRT perspective that identifiability holds asymptotically for the infinite set of model-fitting IRFs. For a careful formulation and proof of this result, see Douglas (2001).

The Douglas result provides foundational justification, for a moderately long to long test, for searching for a statistical procedure to jointly estimate IRFs nonparametrically together with the estimation of examinee abilities. In Douglas (1997), joint ability and IRF asymptotic consistency (uniform) is proved for a unidimensional class of kernel-smoothing-based NIRT IRF estimation procedures as test length and examinee sample size jointly approach infinity in an appropriately and reasonable manner and under suitable regularity conditions. Interestingly and predictably, when one thinks carefully about it, this result holds in spite of the finite test length nonidentifiability of IRFs, a barrier removed by letting test length approach infinity.

A nonparametric IRF estimation procedure growing out of this is the Douglas and Cohen (2001) kernel-smoothing approach inspired by Jim Ramsay's TESTGRAF (see Ramsay, 2000) used together with a nonparametric hypothesis testing approach to assess lack of fit for a specified parametric IRT models, such as one of the usually used logistic family models. A NIRT model is fit using kernel smoothing, and its lack of fit to the closest-fitting model among the user-specified parametric IRT models is assessed. Implicit in this work, I would propose, is a hybrid unidimensional statistical model-fitting approach: fit IRFs parametrically for those IRFs for which the fit is sufficiently good and fit the remaining (uncooperative) IRFs nonparametrically using the Douglas and Cohen approach or using Jim Ramsay's functional data analysis approach (see Rossi, Wang, and Ramsay, 2002).

Another foundational asymptotic result, an answer to a question posed by Holland (1990b), is the proof of the asymptotic posterior normality of Θ given examinee response pattern $\boldsymbol{U}_n = \boldsymbol{u}_n$ under any of a very large class of parameterized (that is, the IRFs are specified known parametric functions of θ) MLI1 IRT models, as defined by a broad and unrestrictive class of regularity assumptions (Chang and Stout, 1993).

Brian Junker had been interested, starting with his thesis (Junker, 1993), in finding an empirical characterization of when a MLI1 IRT model is possible (that is, the characterizing conditions are a function of the statistically estimable "manifest" examinee response distribution, as test length goes to infinity). Brian Junker and Jules Ellis (Junker and Ellis, 1997; Ellis and Junker, 1997) applied the concept of conditional association, due to Holland and Rosenbaum (1986), to produce an empirical characterization (that is, one discernable from the data, modulo statistical variability) of when a MLI1 model is possible.

2.3. *Assessing latent multidimensional examinee distribution structure via a geometric interpretation of conditional covariances*

The weakening of strict unidimensionality (the existence of a unidimensional WLI M IRT model, recall) to essential unidimensionality (the existence of an M IRT model with (5) holding) requires item pair conditional covariances for its formulation. In addition to helping statistically assess unidimensionality, it is natural to ask how useful such conditional covariances might be for assessing the important aspects of the multidimensional latent ability structure when many dominant latent dimensions may be present. That is, can we use these conditional covariances to infer aspects about the multidimensional latent structure that is generating the test data?

Zhang (see Zhang and Stout, 1999a) gives a strongly affirmative answer, powerfully showing that such conditional covariances do indeed provide important information about the latent multidimensional structure. Zhang adopts a semiparametric model formation that encompasses a broad class of parametric IRF models in forming the class of *generalized compensatory models.* The major assumptions that define this class of models are latent trait multinormality (one natural way to specify the multidimensional latent distributions), compensatory modeling in the sense that each item's monotone IRF is a linking function of a linear combination of the latent traits of the latent space (thus parameterizing the relative contribution of each latent dimension and in particular assuming that weaknesses in some of the latent dimensions can be compensated for by strength in others) to the probability of correct item responding, and monotone IRFs. Indeed, foundationally, the assessment of the multidimensional latent structure is a mathematically well-defined problem only when certain assumptions, such as requiring generalized compensatory IRFs, are made.

Definition 5: An MLI1 IRT model is *generalized compensatory* provided

$$P_i(\theta) = H_i(\sum_{j=1}^{d} a_{ij}\theta_j - b_i), \qquad H_i(-\infty) \geq 0,\ H_i(\infty) = 1 \qquad (6)$$

where each linking function $H_i(.)$ is required only to be monotone increasing, $\boldsymbol{a}_i' = (a_{i1}, a_{i2}, \mathrm{K}, a_{id})$ is the item's *discrimination* vector (modeling how informative that particular component is in influencing the examinee response) with respect to the d-dimensional latent space indexed by $\boldsymbol{\theta}_d' = (\theta_1, \theta_2, \mathrm{K}, \theta_d)$, and b_i is a centering parameter influencing item difficulty. □

Zhang develops his geometric viewpoint for generalized compensatory models, each item being represented geometrically using its item discrimination vector $\boldsymbol{a}$ as having a direction in the latent ability space with coordinates θ as shown in Figure 1.

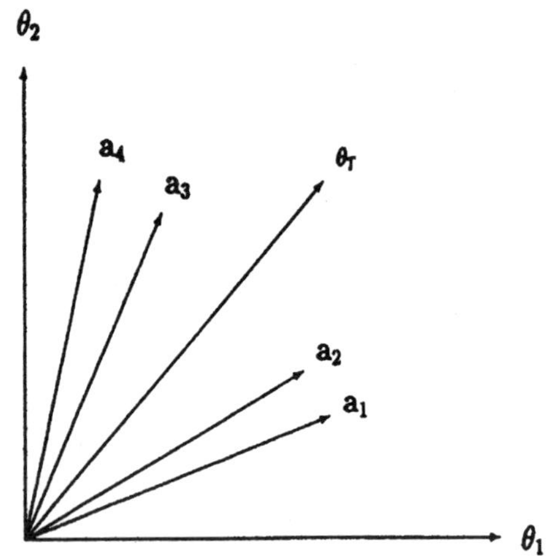

Fig. 1. Geometric representation of a four item two-dimensional test.

The defining of the *direction of best measurement*, $\boldsymbol{\theta}_T$, of a test score $X = \sum_{i=i}^{n} U_i$, or more generally of a subtest score Y, in the IRT model's multidimensional latent space indexed by $\boldsymbol{\theta}$ is key to Zhang's development. In practice, the statistical procedures based on the conditional covariance perspective discussed above must estimate item pair conditional covariances given a direction of best measurement $\boldsymbol{\theta}_Y$ for a specified subscore Y. The usual way to estimate such conditional covariances is to condition on the subtest score Y (often the test score itself) of an appropriately selected subtest, and partition examinees based on their Y value. Then one can estimate the expected conditional covariance $E[\mathrm{Cov}(U_i, U_{i'}|\Theta_Y)]$ by first estimating the covariance for examinees within each partitioning interval (intervals determined by a lattice of y values of Y), namely $\mathrm{Cov}(U_i, U_{i'}|Y \approx y)$, in the usual way of estimating covariances and then taking the weighted average of these estimated conditional covariances using the empirical distribution of the partitioning score Y over the Y-determined partitioning intervals to determine the weights and hence estimate the distribution of Y.

The Zhang and Stout paper then explains how such expected conditional covariances, given a direction of best measurement $\boldsymbol{\theta}_T$ of the test score, reveal, once the conditional covariances are estimated, aspects of the multidimensional latent structure of the test. In a set of theoretical results, striking both for their intrinsic simplicity and their practical usefulness, it is shown that the angles between the various item pair discrimination vectors (see Figure 1) for an item pair as projected on the hyperplane V_{θ_T}, defined as the hyperplane perpendicular to the direction of best measurement $\boldsymbol{\theta}_T$, actually reveal much information about the multidimensional latent structure.

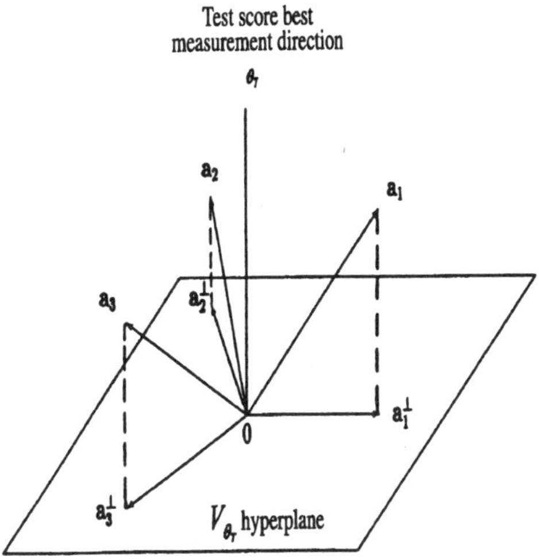

Fig. 2. A three dimensional test with projections of item discrimination vectors onto V_{θ_T} hyperplane.

In particular, when an item pair has a projected angle of less than 90 degrees, the items tend to measure the same dimension, and when the projected angle is greater than 90 degrees they tend to measure distinct dimensions. This flows nicely into a practical, geometrically-based definition of *approximate simple structure* (see Definition 2 of Zhang and Stout, 1999b) such that when the definition holds for the IRT model, items can be statistically partitioned into dimensionally distinct but approximately unidimensional clusters, that is, into clusters such that each cluster consists of items approximately measuring the same intellectual construct and such that different clusters are measuring distinct constructs. For example, the hypothetical items of Figure 3, as shown by their projection onto V_{θ_T}, form a three-dimensional approximate simple structure.

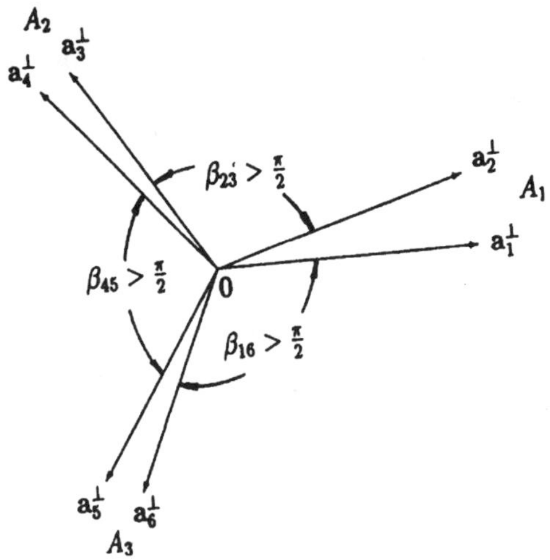

Fig. 3. Projection of item discrimination vectors onto V_{θ_T} hyperplane for a six item three-dimensional approximation sample structure.

Interesting research questions remains to further develop this powerful geometric generalized compensatory approach of Jinming Zhang.

2.4. *Conditional covariance based statistical procedures*

Using Section 2.3's conditional-covariance-based semiparametric compensatory IRT modeling foundation, three useful statistical procedures can be proposed and justified. Together, they constitute a coordinated set of procedures that assess when essential unidimensionality holds and, when it is inferred to fail, can assess important characteristics of the multidimensional latent trait structure.

DIMTEST, a conditional covariance based nonparametric statistical test of unidimensionality, was formulated by Stout (1987) and first refined by Nandakumar and Stout (1993). Its statistical power depends on effective user selection of a set of items called the assessment subtest (AT1) that has been chosen either on substantive or explorational statistical grounds because it seems dimensionally approximately homogeneous and dimensionally distinct from the larger set (usually) of remaining items, referred to as the partitioning subtest (PT). This PT subtest's score is used as the conditioning subscore in the DIMTEST-required conditional covariances. When unidimensionality holds, the approximate unbiasedness of the DIMTEST hypothesis testing statistic depends on the effective selection of a second bias-correcting set of assessment subtest items, denoted AT2. The modern era of DIMTEST, fueled by work of Furong Gao and Amy Froelich (Stout, Froelich, and Gao, 2001) replaces having to choose actual AT2 items, which sharply limited the range of applicability of DIMTEST, by estimating the needed AT2 based statistical bias correction by means of a resampling scheme using nonparametric kernel smoothed estimation of the AT1 chosen IRFs to create a virtual AT2. The latest and by far most statistically effective methods for the choice of AT items (there now being only one actual assessment subtest) use variations of our various conditional covariance based procedures to replace the crude linear factor analysis exploratory procedure for choosing AT1 originally used in DIMTEST (Froelich and Habing, 2002).

Zhang (see Zhang and Stout, 1999a) rigorously defends the capability of DIMTEST to have appropriate statistical power to detect multidimensionality even when there may be three or more dimensions (Stout's, 1987, theoretical defense of DIMTEST's statistical power was more informal and its consideration of statistical power was limited to there being two dimensions).

Louis Roussos developed a conditional-covariance-based hierarchical cluster analysis approach (its proximity measure uses conditional covariances), HCA/CCPROX, to sort items into relatively dimensionally homogeneous clusters (Roussos, Stout, and Marden, 1998) in his thesis. This approach can play a useful exploratory role in discovering the tendency of a set of items to split into dimensionally distinct clusters that are internally dimensionally homogeneous, i.e., tending to form approximate simple structure. Even when approximate simple structure fails, HCA/CCPROX is useful. In fact, the Zhang conditional covariance geometry provides a theoretical justification for HCA/CCPROX's capability to identify relatively dimensionally homogeneous item clusters, whether they are sufficiently dimensionally distinct to produce an approximate simple structure or not.

Hae Rim Kim (see Stout, Habing, Douglas, Kim, Roussos and Zhang, 1996) developed in her thesis the conditional-covariance-based DETECT procedure. In his thesis, Jinming Zhang both refined DETECT and developed a theoretical defense of it (see Zhang and Stout, 1999b). DETECT not only counts the number of latent dimensions, but also provides an index measuring the strength of the test's departure from unidimensionality, and finally, when appropriate, partitions items into approximate simple structure item clusters. We note that the concept of the strength of the departure from unidimensionally is distinct from the number of dimensions: For instance, a two-dimensional structure could strongly depart from the closest fitting unidimensional model, while an 8-dimensional MLI structure could weakly depart from the closest fitting MLI1 model.

The three conditional covariance-based procedures are applied to real data in an integrated manner in Stout, Habing, Douglas, Kim, Roussos, and Zhang (1996), with one high point being the multidimensionality analyses of the analytical reasoning (AR) and reading comprehension (RC) sections of the LSAT. In analyses of these sections of the LSAT, paragraphs strongly displayed themselves as contributing distinct paragraph-based latent dimensions. However, in one of my favorite findings, DETECT combines two paragraphs as producing a single dimension, an apparent error until one discovers that both paragraphs are science-based while each of the other two paragraphs are not science based and are very distinct from each other (this may or may not be the true explanation, but, it is certainly quite compelling). The Stout et al. paper is perhaps the best place to read about the three conditional covariance procedures (DIMTEST, HCA/CCPROX, and DETECT).

Section 2 has summarized two decades of research efforts of the Laboratory to develop effective IRT nonparametric assessment of the possibly multidimensional latent structure that underlies any educational test. Philosophically, the substantive specifications (usually content component based) of what a test measures and the statistical description of the resulting latent IRT structure (especially as viewed through dimensionality focused glasses) of the test should be jointly consistent and each should enhance and inform the other for use in developing future related test construction and test scoring. It has been shown above how foundational work, which draws heavily on an the infinite-item test formulation, and a constellation of conditional-covariance-based nonparametric latent dimensionality assessment procedures, which are theoretically defended, especially the geometrically-based theory for conditional covariances, combine to provide a flexible, theoretically and empirically well supported, informative, and easy to apply set of nonparametric multidimensional latent structure assessment tools. As reported above, a mixture of large-scale simulations and real data analyses that evaluate the procedures have been presented in the psychometric literature. These demonstrate and justify the effectiveness of these conditional-covariance-based dimensionality assessment procedures. In addition to the work reported upon above, much interesting, challenging, and, from the applied measurement perspective, important, work remains concerning conditional-covariance-based approaches to IRT multidimensional latent structure modeling and statistical analyses. One long-term research goal is to develop an effective and relatively complete NIRT item-level factor analysis methodology.

3. Test Fairness

High stakes standardized testing is used worldwide to inform educational admissions, placement, and honors/awards decision-making processes. Moreover, the level of ethnic and cultural diversity in many countries that rely heavily on high stakes testing continues to increase. As a result, issues of test fairness are of vital and increasing importance. The statistical analysis of item-level test fairness is usually called Differential Item Functioning (DIF). The role of psychometrics in providing our understanding of test fairness and thus improving the fairness of tests is all too often inappropriately compartmentalized, minimized, or even bypassed entirely. The challenge to psychometrics has been, and still is, to change this unfortunate state of affairs

By the claim that DIF has been compartmentalized is meant the almost total disconnect that has evolved between substantive (content-based) and DIF (statistical) approaches to the understanding and practice of test fairness. For example, Paul Ramsey (1993), in a review of what was then the ETS item sensitivity (a substantive portion of their test fairness analysis) review process, states, "...there is no consistent effort to inform sensitivity reviewers what we are [statistically] learning from DIF." Reacting to this unfortunate state of affairs, Robert Linn (1993) recommended "taking into account not only what has been learned from DIF analyses but [also] what has been learned from sensitivity reviews" when standardized tests are being designed.

The applied DIF literature is full of examples of the failure to explain substantively why certain seemingly innocuous items display statistical DIF and why certain potentially biased items (an ancient but well known example is a vocabulary item concerning the word "regatta", whose familiarity is certainly linked to social class: this item was expected to and likely did display DIF) do not display DIF. Addressing this very point, William Angoff (1993), in his introductory article to the *Differential Item Functioning* monograph, states, "It has been reported by test developers that they are often confronted by DIF results that they cannot understand; and no amount of deliberation seems to help explain why some perfectly reasonable items [from the substantive perspective] have large DIF values."

In reacting to these enervating problems, our first research goal was to develop a probability model for test bias that leads to an effective integration of statistical and substantive approaches to monitoring and improving test fairness. Thinking about causes of DIF led to developing a theoretical multidimensional IRT model that rigorously captures and explicates the intuitive notion that the cause of DIF in singly-scored tests is the presence of secondary dimensions, denoted by η's, in addition to the primary or essential dimension intended to be measured, denoted throughout by θ as estimated by the score.

A second important goal was to use this model to shift the psychometric DIF paradigm from a totally reactive (removing unfair items only after they have been constructed and pretested and statistically or substantively found wanting) and one-item-at-a-time approach to a potentially proactive (that is, applicable at the test design stage to improve the fairness of the test) and item-bundle-based (that is, looking at sets of items for their combined DIF) approach to DIF. The approach we developed stresses *substantively interpreted* latent-dimensionality-based explanations of causes of DIF that then can contribute feedback information for improving future test design specifications. (Standardized tests are built from very carefully developed blueprints providing considerable detail about the kinds of items allowed and not allowed. These "blueprints" are referred to as the test specifications.) In this manner, substantive and psychometric approaches to test fairness can be unified. To illustrate, if reading-comprehension test items of paragraphs about the physical sciences are discovered to display DIF against women, then the test specifications for future versions of such a reading comprehension test could exclude physical-science-based paragraphs.

Following in the tradition of Holland and Thayer's (1988) theoretically-defended Mantel-Haenszel IRT DIF approach (a foundational and practically important milestone, marking the beginning of the modern era of DIF from the psychometric perspective), a third major DIF goal was to develop a constellation of nonparametric IRT DIF procedures to address specific important issues in conducting DIF/equity analyses that remained unsolved.

These included the detection of crossing DIF where a group experiencing DIF (African-Americans, say), holds for only low ability examinees, say, and switches to favoring the other group for high ability examinees, detection of DIF for polytomously scored items, developing statistically robust DIF procedures, detecting DIF for tests intentionally designed to measure more than one dominant dimension (a basic math test usually is designed to basically measure both algebra and geometry, for example), detection of DIF collectively for a bundle of items– called item-bundle-based DIF, detecting DIF in computer adaptive tests (tests where an experimental design for the sequential administration of items for each individual examinee takes into account what has been learned already about the examinee's ability in order to choose an optimally informative item for the next item to be administered, appropriately defining and estimating model-based theoretical DIF parameters for scaling various kinds of DIF in various settings, and local (in the intended to be measured latent ability θ) DIF.

3.1. *A substantively motivated multidimensional model for DIF (MMD)*

As usual, we model DIF as a phenomenon experienced by a targeted *focal group (F)*, such as African Americans, when compared with a nontargeted *reference group (R)*, such as Caucasians. The commonly targeted focal groups include those defined by race, gender, ethnicity, disability, and first language. Specific settings may call for special choices of focal groups. More formally, DIF for an item is defined to occur at θ when the probability of a correct item response, for examinees all having the same intended-to-be-measured ability level θ, differs because of group membership. It is important to note that, *as is appropriate*, this definition has nothing to do with whether the focal and reference group distributions of the desired to be measured ability Θ are identical or not. So better performance by the reference group does not *by itself* constitute test unfairness against the focal group, although it may help suggest that a DIF analysis be carried out.

In two companion papers (Shealy and Stout, 1993a, 1993b; noting that our MMD model for DIF was actually first presented in Shealy, 1989), Robin Shealy and I laid out our multidimensional model for DIF, abbreviated MMD. The 1993b reference provides a detailed in-depth theoretical description of the model. It carefully derives various interesting consequences of a foundational nature that mathematically follow from the model. The 1993a reference provides a more brief and informal description of the aspects of the model needed to understand the SIBTEST DIF statistical procedure we developed. The most complete description of the MMD model, which breaks important new ground , especially from the applications perspective, occurs in Roussos and Stout (1996a). It should be noted that Kok (1988) independently and concurrently to Robin Shealy and me developed a similar multidimensional modeling approach to DIF. The Shealy/Stout MMD model calls for a new focus on DIF based on three related principles.

First, consistent with the test validity perspective that a test must measure well what it is intended to measure, and noting that educational and training decisions involving examinees are made on the basis of total test scores rather than item scores, the assessment of test fairness should also occur based on the test score level rather than only for individual items. This insight led to the closely related concepts of *differential bundle functioning (DBF)* and *differential test functioning (DTF)*. DBF (analogously, DTF) is defined to occur at θ for examinees having the same intended-to-be-measured ability level θ when the expected item bundle subscore (analogously for DTF, test score) given θ for a bundle of items carefully selected on the basis of apparent substantive and/or dimensional homogeneity, differs across group. DBF measures the combined amount of DIF at the item bundle score level experienced by examinees from different focal and reference groups. Since one-item -at-a-time DIF must usually also be a vital component of assessing test fairness; our approach is to augment and embed DBF and DTF analyses in commonly used one-item-at-a-time DIF analyses.

Second, the explicitly multidimensional nature of the model allows, and exhorts, us to thoughtfully study and understand the necessary role of secondary dimensions in causing DIF, DBF, or DTF. In particular, when several items (perhaps forming a dimensionally homogeneous and substantively interpretable bundle, such as a set of items on a geography test that each require map-reading skills) each depend on the same secondary dimension, the possibility of a large amount of DBF experienced by the focal group at the bundle subtest score level caused by individual item *DIF amplification* (see Nandakumar, 1993) becomes an issue. By contrast, when the influences of multiple secondary dimensions interact, the possibility of *DBF cancellation* (see Nandakumar, 1993, again) of the influence of DBF-producing bundles at the test score level (such as a reading comprehension test where the paragraphs are carefully balanced by content, based on explicit consideration of gender) or cancellation of the influence of DIF-producing items at the bundle score level, becomes important. Because people are cognitively heterogeneous and because items cannot, and should not, be context-free, and hence all items are potential DIF producing objects, the notion of DIF cancellation at the item bundle level and DBF cancellation at the test score level is perhaps more important than casual thought first suggests.

Third, DIF is *correctly* modeled by the MMD model to occur locally in θ. Thus, the DIF experienced by the focal group may vary in amount, even occasionally reversing itself, as a function of the focal group examinee's level of ability on the intended to be measured ability. This seemingly innocuous truth has important and sometimes paradoxical implications. It leads us to "crossing DIF" possibly occurring, where, an item displays DIF against(say) low-scoring focal-group examinees while it also displays DIF against high-scoring reference group examinees. Such varying local DIF at θ is caused by varying conditional distributions of a secondary dimension (for simplicity, we only consider the case of one secondary dimension influencing examinees) H given θ across the focal group contrasted to this distribution for the reference group (here H is the upper case version of η). This is counter to the fairly widely and often uncritically accepted informal model for what causes DIF, which presumes, usually implicitly, that differences in the *marginal* H distributions across the two groups are what causes DIF. Naïve trust in this seductively intuitive, but sometimes incorrect, informal model creates some striking paradoxes as discussed below.

In explaining the MMD of Shealy and Stout, we only present the two-dimensional case, with θ denoting the dimension intended to be measured and the random variable H (taking on values η) denoting an additionally measured secondary dimension. H is often but not always thought of as a "nuisance dimension" that is outside the intended-to-be-measured construct. Indeed, an important issue arises when H seems, at least to some cognitive specialists, to be part of the construct intended to be measured. IRF invariance across the focal and reference groups is assumed to hold for all items with respect to the complete latent (θ, η) space. That is, for all items, the IRFs $P_i(\theta, \eta)$ are the same for the focal and reference groups, this a foundational principle that must hold for all valid IRT models. Thus, when the (likely multidimensional) latent space is completely specified, for every fixed point of the complete latent ability space all group differences in item performances, by definition, disappear.

The focal and reference group Θ_F and Θ_R distributions will in general be different and often stochastically ordered. These possibly differing Θ_F and Θ_R distributions do *not* cause DIF, DBF, or DTF, although such a difference in distributions clearly can contribute to non-DIF group differences in test score distributions.

We now briefly review the essential features of the MMD model. The potential for DIF occurring at θ for an item with (group invariant) IRF $P(\theta, \eta)$ is caused by the conditional distribution of H_g *given* $\Theta = \theta$ differing for $g = R$ versus $g = F$. To see this, we compute the group dependent marginal IRF with respect to θ;

$$P_g(\theta) = \int_{-\infty}^{\infty} P(\theta, \eta) f_g(\eta|\theta) d\eta, \tag{7}$$

where $g = R$ or F and $f_g(\eta|\theta)$ is the density of H given $\Theta_g = \theta$. Then the amount of DIF against F locally at θ is

$$B(\theta) = P_R(\theta) - P_F(\theta). \tag{8}$$

Further, the average (over θ) amount of DIF against F is given by

$$\beta_{\mathrm{UNI}} = \int_{-\infty}^{\infty} B(\theta) f_F(\theta) d\theta, \tag{9}$$

where $f_F(\theta)$ denotes the density of Θ_F. Here UNI indicates that our viewpoint is unidirectional in the sense that we are measuring the average amount of DIF against F. β_{UNI} is the fundamental DIF index of MMD. It is what the SIBTEST DIF statistical procedure estimates and about which it tests hypotheses. β_{UNI} for an item bundle is defined analogously with $B(\theta)$ denoting the difference of reference and focal expected item bundle scores at θ.

Because DIF is due to a difference in the conditional distributions of H_g given θ across group rather than to differences in the marginal H_g distributions across group, on occasion paradoxical situations arise where differing marginal H_g distributions across group fail to translate into DIF and also where nondiffering marginal distributions of H_g nonetheless end up being accompanied by DIF (these and other paradoxical explanations for DIF discovered and explained by Louis Roussos: see Roussos and Stout, 1996a). This just-referenced paper gives a plausible example of this latter and most paradoxical case, using some findings of O'Neill and McPeek (1993) and Douglas, Roussos, and Stout (1996): O'Neill and McPeek found that some types of items, intended to measure verbal reasoning (θ), tended to exhibit DIF in favor of males (R) versus females (F) when the items concerned the construct "practical affairs," (H), for which money is mentioned as a typical component. The finding of DIF in favor of males on these items would indicate that $\mu_{\mathrm{H}_R} > \mu_{\mathrm{H}_F}$ must hold if one accepts the informal DIF model perspective that differing across-group marginal distributions for $P\mathrm{H}_g$ cause DIF, where H denotes familiarity or knowledge of money and other practical affairs. By contrast, and paradoxically, Douglas et al. reported that, contrary to what the informal model for DIF predicts, items intended to measure logical reasoning (θ) where the context is money or finances (H), showed little DIF against either males or females even though $\mu_{H_R} > \mu_{H_F}$ is believed to hold.

We resort to the MMD model to explain this apparent paradox. Supposing that (Θ_g, H_g) are bivariate normal for each group g (a reasonably innocuous assumption), we have for both tests that

$$E(\mathsf{H}_R|\Theta_R = \theta) - E(\mathsf{H}_F|\Theta_F = \theta) = (\mu_{\mathrm{H}_R} - \mu_{\mathrm{H}_F}) - \rho(\mu_{\Theta_R} - \mu_{\Theta_F}) \quad (10)$$

for all θ. A close examination of (10) reveals a possible explanation for the apparent contradiction between observed DIF behavior for the verbal reasoning items and the logical reasoning items. As already stated, for both the logical reasoning items and the verbal reasoning items the first term on the right-hand side of (10) resulting from the nuisance dimension H seems to be positive. It is in fact the second term that explains the paradox. In the case of the verbal reasoning items, because females often test as more proficient in verbal reasoning (θ) on average, the second term on the right hand side of (10) (including its minus sign) is likely positive as well, almost ensuring DIF against females. However, in the case of the logical reasoning (θ) items, in which males seem to have a higher mean proficiency, the now *negative* second term on the right hand side of (10) tends to cancel the influence of the positive first term. Thus, the genuinely paradoxical, but logically correct, conclusion is that DIF against F can be reduced, at least to some extent, if the ability distribution on the primary (intended-to-be-measured) dimension θ also favors R. Note, however, that this is a delicate matter depending among other things on the size of ρ.

3.2. *A parameterization of the amount of DIF*

One important advantage of having a believable and realistic probability model for a complex real world phenomenon is that important and sometimes subtle aspects are quantified within the model and hence can be better understood and inferred from data. In the case of the MMD model, model-based scaling of various kinds of DIF, as the character of the DIF varies, the number of items involved varies, the number of dimensions intended to be measure by the test varies, and so forth, becomes possible.

The fundamental parameter β_{UNI} of the MMD model measures the amount of unidirectional DIF against F averaged over all levels of θ. Similarly, within the framework of MMD, as already remarked, an analogous parameter measuring unidirectional DBF is easily defined (see Shealy and

Stout, 1993a). Parameters measuring the amount of crossing DIF for an item and for a bundle of items are defined by Li and Stout (1996). The polytomous case is handled by using the expected item score given θ as the parameter providing the DIF metric (Chang, Mazzeo, and Roussos, 1996). In Stout, Li, Nandakumar, and Bolt (1997), the parameter measuring the amount of DIF when the test is designed to measure two dimensions is defined carefully by modifying the basic formulation by replacing the unidimensional θ by a two-dimensional (θ_1, θ_2). As already mentioned, the test might be a mathematics test designed to measure geometry, θ_1, and algebra, θ_2.

Last, in an important foundational paper with important implications for certain test fairness applications, Roussos, Schnipke, and Pashley (1999) develop (from first principles and a plausible heuristic asymptotic argument) a formula for the parameter measuring the amount of DIF that is in effect estimated by the widely used Mantel-Haenszel (MH) odds ratio DIF estimator. One practical consequence based on the findings of the Roussos et al. paper is that significant caution should accompany the use of the MH odds ratio as an index of DIF when the 3PL model is used.

3.3. *DIF statistical procedures developed using the MMD model*

Developing a nonparametric estimation and hypothesis-testing procedure to assess item DIF and bundle DBF was the initial methodological challenge. Shealy and I also had the task of producing a robust and theoretically defensible metric for DIF The resulting SIBTEST procedure is described in Shealy and Stout (l993a).

Recall that, by definition, DIF occurs when examinees, matched on the latent variable θ *that the test is intended to measure*, perform differentially depending on their group membership. Hence, the SIBTEST strategy was to first match as best as one could examinees on a *valid subtest* score that measures θ without serious contamination from secondary η_1, η_2, K, "nuisance" dimensions. Here, the MMD model helps us see that the contamination issue is at the subtest or even test score level rather than the individual item level. Although the ideal would be to have only items measuring θ alone in the valid subtest, having items present that are influenced by secondary dimensions is also an option (and likely)as long as their influence roughly cancels out at the valid subtest score level.

The terminology of "valid subtest" was chosen carefully, intended to stress that the matching criterion used by SIBTEST, MH, and other procedures should be selected with validity-based care, rather than automatically using total test score (perhaps with the studied item removed) to match examinees. Potential SIBTEST users sometimes misunderstand this terminology and think that SIBTEST, in contrast to other DIF procedures, cannot be used without its having a truly valid subtest. Rather, the point is that SIBTEST, in common with all other DIF procedures requiring a matching subtest, is only as effective as the matching subtest is in matching examinees on the construct intended to be measured. In this regard, serious contamination of the "valid subtest" caused by secondary, nuisance dimensions must not occur. Our intent in introducing the "valid subtest" terminology was to stress that practitioners' efforts to achieve validity in matching examinees of equal ability on the desired-to-be-measured construct is a vital prerequisite to the success of any DIF/DBF/DTF analysis. That is, at its heart, an effective test fairness analysis requires a prerequisite validity analysis.

How practical it is to find sufficiently valid subtests is a partly experiential question. First, in the prototypical large standardized test setting where one first assesses pretest items (items seeded in tests but not counting, hence the term "pretest", in order to assess their effectiveness in future operation tests where they do count) for possible DIF, matching examinees on the score on the "operational items", which should have previously undergone careful test design and psychometric scrutiny, seems reasonable. If reasonably done, attempts to purify an initially proposed matching subtest by removing DIF items seem appropriate too, this a fairly standard approach. Regarding the selecting the valid subtest items, one should note that the SIBTEST procedure, including its valid subtest terminology strongly encourages the discerning user to actively specify exactly which items constitute the valid subtest.

The fundamental idea of the SIBTEST statistic used to estimate the amount of DIF is simple and intuitive: Let X_V be the score on the user-chosen valid subtest judged to be measuring θ without serious contamination. (Borrowing from our NIRT work, one would hope that the desired to be measured θ lies close to Zhang's direction of best measurement of X_V.) Consider a preselected bundle (could be a single item) of possible DIF items with bundle score denoted by Y. In particular, let $\bar{Y}_{gk}$ denote the average bundle score of all Group g (either R or F) examinees having valid subtest score $X_V = k$. Then the proposed SIBTEST estimator of β_{UNI} is given by

$$\hat{\beta}_{\text{UNI}} = \sum_k (\bar{Y}_{Rk} - \bar{Y}_{Fk})\hat{p}_{Fk} \tag{11}$$

where $\hat{p}_{Fk}$ denotes the proportion of focal group examinees for which $X_V = k$. The most common application, which we hope is modified over time, is where the bundle consists of a single targeted possibly DIF item and $Y_{gk} = 0$ or 1 for each examinee.

In spite of the obvious intuitive appeal of the estimator in (11), it turns out to be seriously statistically biased when, as is often the case, the Θ_R and Θ_F distributions are stochastically ordered. This is caused by the regression to the mean phenomenon: The conditional distribution of Θ_g given test score X_V depends on the differing means of the Θ_R and Θ_F distributions, as Figure 4 shows. Hence, because the distribution of $\bar{Y}_{gk}$ depends on the conditional distribution of Θ_g given test score X_V, statistical bias tending to cause the event $(\bar{Y}_{Rk} > \bar{Y}_{Fk})$ to occur is expected even when there is no DIF at θ if the marginal distribution of Θ_R is stochastically larger than that of Θ_F.

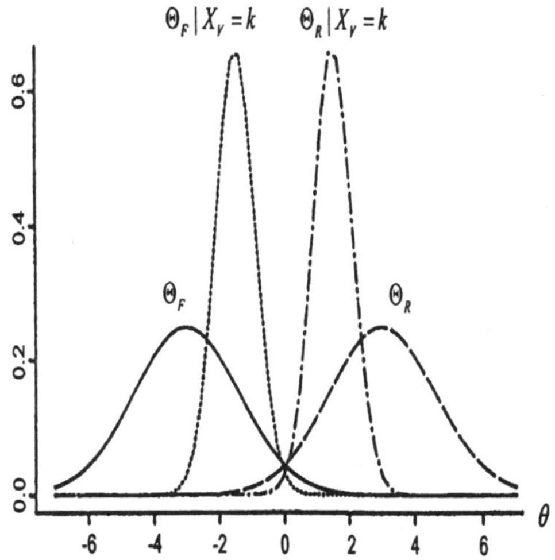

Fig. 4. Comparison of Θ_F and Θ_R distributions with $\Theta_F|X_v = k$ and $\Theta_R|X_v = k$ distributions.

In fact, the proposed SIBTEST statistic $\hat{\beta}_{\text{UNI}}$ is seriously biased unless test length is very large, when in fact the matching (valid) subtest scores become completely reliable as a stand-in for their respective θ's because the regression to the mean influence on $\bar{Y}_{Rk}$ and $\bar{Y}_{Fk}$ becomes negligible. The modification to (10), as detailed in Shealy and Stout (1993a), is to shift each of the $\bar{Y}_{gk}$'s for $g = R$ and F according to a heuristically-justified *regression correction* (undoing the regression to the mean influence). Dividing the regression-corrected $\hat{\beta}_{\text{UNI}}$ by an appropriate estimated standard error produces a useful hypothesis-testing statistic. When there is no DIF, a heuristic argument shows this statistic to be standard normal asymptotically as reference and focal group sample sizes go to infinity. The SIBTEST estimator $\hat{\beta}_{\text{UNI}}$ and its associated hypothesis testing statistic perform well, as predicted by the heuristic asymptotics, as is confirmed in Shealy and Stout (1993a) by a large-scale simulation study showing that the SIBTEST estimator is relatively unbiased, and that the SIBTEST hypothesis testing procedure is powerful and adheres well to the nominal level of significance in its simulated Type I error behavior over a wide range of realistic DIF models, including models allowing up to one standard deviation of difference ($d = (\mu_{\Theta_R} - \mu_{\Theta_F})/\sigma_{\Theta_g}$ in the Θ_R and Θ_F means, where it is assumed for simplicity $\sigma_{\Theta_R} = \sigma_{\Theta_F} \equiv \sigma_{\Theta_g}$).

The large-scale SIBTEST simulation study included both a study of bundle DBF and individual item DIF. This study showed, as predicted, that the hypothesis-testing power is much greater for a homogeneous bundle of DIF items contrasted with DIF items analyzed singly. The SIBTEST DIF/DBF simulation study also demonstrated reasonable valid-subtest robustness against minor contamination of the user-chosen valid subtest by DIF items measuring η in addition to θ. (see Shealy and Stout, 1993a).

A large scale Type I error study was done using the regression corrected SIBTEST in Roussos and Stout (1996b). It demonstrating that SIBTEST regression is more robust than the MH procedure in the presence of sizeable group differences in the Θ_g distribution. However, even for SIBTEST, some of the observed Type I error rates were unacceptably high (when group ability differences are large and an item is highly discriminating, for example), thus motivating the Jiang and Stout nonlinear regression correction version of SIBTEST: In Jiang and Stout (1999), the linear regression correction of Shealy and Stout is replaced by a more effective nonlinear correction, as shown in the paper's simulation study. It is this nonlinear regression correction that is currently used in SIBTEST.

The distribution theory behind SIBTEST is based on a heuristically defended normal distribution large-sample theory. Unfortunately, in order to develop a variation of SIBTEST to detect crossing DIF, asymptotic normality fails and a randomization-test-based hypothesis testing approach was used instead. In Li and Stout (1996), through a series of simulation studies, such a procedure, called "CrossingSIB", is shown to have decent Type I error behavior and good power for detecting crossing DIF or DBF.

A polytomous version of SIBTEST called "polySIB" is developed in Chang, Mazzeo, and Roussos (1996) and its performance studied. In Douglas, Stout, and DiBello(1996), nonparametric estimation of the amount of DIF locally in θ is considered. Being able to assess DIF/DBF locally is important in situations where the test is targeted to measure ability accurately over a limited ability range such as the PSAT when used in America to award National Merit Scholarships to high ability examinees. Using nonparametric kernel smoothing estimation, the amount of DIF/DBF, as given by the function $B(\theta)$ defined above, is estimated in Douglas et al.

A version of SIBTEST is developed in Nandakumar and Roussos (2002), where matching is on $\hat{\theta}$, thus allowing SIBTEST to be applied in computer adapted testing settings. Last , in Stout, Li, Nandakumar, and Bolt (1997), a version of SIBTEST called "MultiSIB" is developed where the ability intended to be measured is two-dimensional and hence two-dimensional matching is required so that matching validity is maintained. Again, a thorough simulation study is done, which showed that DIF/DBF estimator bias is fairly low and that reasonable Type I error behavior and hypothesis-testing power are observed in realistically simulated settings. This procedure is of special foundational interest because it is a particularly compelling instance of the fundamental validity mandate that the matching criterion has to be appropriately developed if the real purpose (achieving test fairness) of the DIF/DBF analysis is to be achieved.

To illustrate, if a test is designed to measure both algebra and geometry, then clearly large DIF might occur if we match on total test score (influenced by a blending of algebra and geometry abilities) , even though the test is in fact measuring exactly what it is supposed to - namely both algebra and geometry. Interestingly, for a test designed to measure both algebra and geometry say, a study where one matches on total score can be interesting from the cognitive perspective; we are, however, not assessing test fairness in the process! To illustrate, in the simulation-based Type I error studies of DIF hypothesis testing with nominal level of significance set at $\alpha = 0.05$, the very acceptable observed Type I error of 0.059 for two-dimensional matching was increased to the totally unacceptable 0.39 when incorrect total score matching was substituted for the valid two-dimensional matching on two subscores.

3.4. *The implementation of our newly developed DIF/DBF procedures*

The development of a DIF model, appropriate DIF parameters to be estimated, and DIF/DBF statistics are all important DIF analysis components. However, it is the full integration of these components that is required to produce a fully developed implementation procedure for effective DIF/DBF analyses. In this regard, the most important are the user embracing of the MMD paradigmatic implications and imperatives for conducting test fairness research and practice. In summary, these include (i) integrating psychometric DIF/DBF SIBTEST analyses and substantive analyses of test bias-in particular addressing underlying substantive causes of DIF/DBF; (ii) using substantive feedback from DIF/DBF SIBTEST analyses to influence future test design and assembly; (iii) matching examinees in DIF/DBF studies in ways that make the matching procedure consistent with modern test validity considerations as espoused by Lee Cronbach, William Angoff, Sam Messick (see Chapters 1-3 of the Wainer and Braun edited *Test Validity*, 1988), and others; (iv) including the carrying out of bundle DBF analyses where the bundles are selected to be homogeneous based on statistical, substantive, or blended grounds; (v) recognizing that the practical test validity implications of DIF/DBF are expressed at the test score level even though the "atoms" of DBF are of course the items causing DIF; (vi) being sensitive to the potential for amplification and/or cancellation of item DIF at the bundle score level and of bundle DBF at the test score level; and (vii) taking advantage of the potential for increased statistical detection power when one works at the item bundle level using a DBF approach versus taking an individual item level DIF approach.

From this test equity paradigm-changing perspective, five papers are quite informative. The first paper, by Terry Ackerman (1992), expresses the Shealy and Stout MMD model for DIF geometrically using the logistic multidimensional IRT model (MIRT) and studies various cases of DIF from the perspective of studying the behavior of

$$E(H_R|\Theta_R = \theta) - E(\mathsf{H}_F|\Theta_F = \theta) \tag{12}$$

as a function of θ under the assumption of bivariate normality of (Θ_g, H_g) for both groups.

Ackerman introduced the potentially useful concept of the *validity sector*. It is meshes coherently with the notion of essential unidimensionality presented in Section 2. The validity sector can be thought of as a hypercone in the (θ, η) space having hopefully a reasonably small vertex angle.

It assumes that the valid subset consists of validity sector items that are hence reasonably dimensionally homogeneous and reasonably close to the θ axis. Figure 5 shows a hypothetical validity sector for a two-dimensional test.

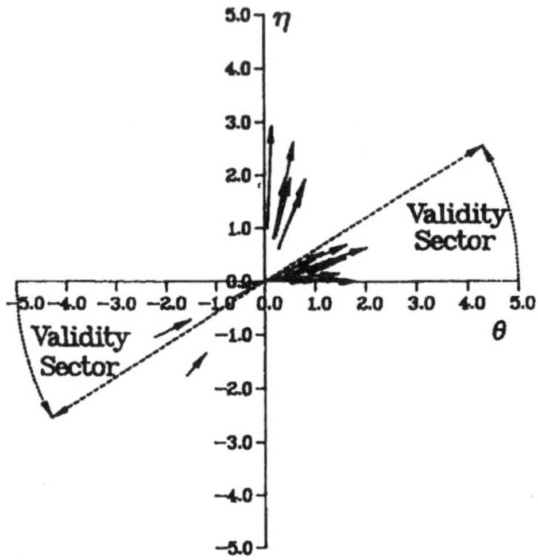

Fig. 5. Item discrimination vectors of a 22 item validity sector.

The detailed Douglas, Roussos, and Stout (1996) methodological paper on doing bundle DBF analyses, the Nandakumar (1993) paper on amplification and cancellation, and the Bolt, Froelich, Habing, Hartz, Roussos, and Stout (2002) paper presenting an in-depth application and further development of the bundle DBF SIBTEST methodology, applied to the GRE-Q (quantitative) exam are each important from the MMD paradigm-changing perspective.

The fifth and perhaps most central paper, Roussos and Stout (1996), provides the clearest and most complete statement of the MMD paradigm. It stresses that a standardized test given periodically can have its level of fairness improved over time through a principled application of substantively and statistically suggested DBF hypotheses when affirmed (for example through a SIBTEST DBF analysis) can then be incorporated into the test specifications of future versions of the test.

There are at least two distinct approaches to forming item bundles hypothesized to display DBF. First, as laid out in Douglas, Roussos, and Stout (1996), one can take a confirmatory approach making use of the opinion of experts in looking for substantively homogeneous bundles. In Douglas et al., this approach is applied to male/female DBF for eight (expert chosen) bundle-defining categories (items involving social issues, the military, technological sciences, health sciences, and so forth). Figure 6 shows the high correlation between the SIBTEST DBF $\hat{\beta}$ index and the panel of experts' own combined substantive index of the bundle's DBF.

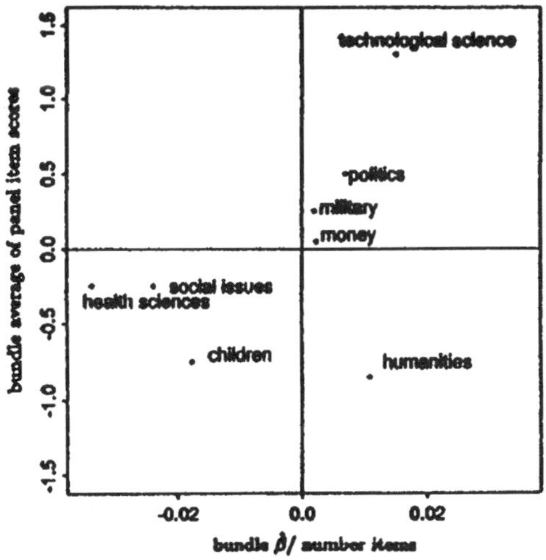

Fig. 6. Panel index versus bundle DBF $\hat{\beta}$/item.

The second approach for candidate bundle DBF items uses DIMTEST and HCA/CCPROX to identify "suspect" bundles together with an expert-driven substantive refinement of these statistically identified bundles suspected of DBF. In an example from the Douglas et al. paper, a bundle of six items on an NAEP history exam, which all pertained to early American documents, is identified by this blended statistical/substantive approach. Then, a SIBTEST bundle analysis of this six-item bundle produced an observed level of significance of 0.002, strongly indicating DBF against women.

Real-data examples of both cancellation and strong amplification are presented by Ratna Nandakumar (1993). Amplification can be of real importance in that many items having slight DIF against the same group can amplify to have a highly deleterious effect at the test score level: One question on baseball may not matter much on an American sixth grade math test taken by boys and girls, but ten questions (not that any teacher would be so clumsy) likely will.

The SIBTEST-based GRE bundle DBF analyses (Bolt, Froelich, Habing, Hartz, Roussos, and Stout, 2002) are of particular interest because of the very careful and detailed use of the bundle DBF approach as a possible aid to GRE test design from the test fairness perspective. The GRE study is the most complete example of applying the MMD paradigm in real data settings. For further interesting analyses of SIBTEST-based bundle DBF studies, the reader is referred to Gierl, Bisanz, Bisanz, Boughton, and Khatiq (2001); Gierl and Khaliq (2001); and Gierl, Bisanz, Bisanz, and Boughton (2002).

Much remains to be done before proactive psychometric bundle-based approaches to DIF/DBF will have been fully utilized in achieving test fairness by practitioners. In fact, most of the components of what we suggest above as a needed psychometrically driven paradigm shift in how to view and practice test equity has not yet been implemented.

4. Probability Model Based Skills Diagnosis: a New Standardized Test Paradigm

The long dominant summative-assessment-focused and single-score-based testing paradigm of psychometrics/educational measurement that unidimensional IRT probability modeling so effectively has addressed has begun to be challenged in the last decade or so. Summative assessments focus on broad constructs such as mathematics, reading, writing, physics, etc. and typically assess individual educational achievement based on a single test score as examinees exit one system and make the transition to another, such as from high school to college or from college to job. Currently, the educational standards and instructional accountability driven drumbeat intensifies in America for formative assessment testing, whose primary goal is assisting the teaching and learning process *while it is occurring.* This creates the psychometric challenge of converting through a probability-model-based statistical analysis each examinee's test response pattern (showing which questions are answered correctly and which incorrectly) into a multidimensional student profile score report detailing the examinee's skills learned and skills needing further study. The new testing paradigm that I claim is overdue calls for *a blending* of summative and formative assessments, with sometimes the same test being used for both purposes and sometimes new tests being developed for formative assessment (skills diagnostic) purposes alone.

Skills-based formative assessments of examinees, often aggregated over classroom, school, school district, state, or other collective, as appropriate, can be used by students, teachers, curriculum planners, government departments of education, and others involved in the educational process, to facilitate teaching and learning. In particular, using feedback from a test-based formative assessment, each student learns what skills he or she has mastered and what skills need his or her further attention. To illustrate, a student's number right algebra test score is supplemented or replaced by a skills profile indicating mastery of linear equations, nonmastery of factoring, and so forth, such information hopefully leading to informed remediation. At the classroom level, the summative-assessment-based classroom test score average is supplemented or replaced by formative-assessment-based classroom proportions of masters and nonmasters for each targeted skill. Used effectively, this can lead to changes in instruction for the immediate future for the just-tested classroom and more distant future changes in instruction for future classrooms on the same subject. Skills-level formative assessment can also help students meet the State government educational standards that are receiving so much current emphasis in America. The concept of a "skill" as used here is generic from the cognitive psychology perspective in the sense that it refers to any postulated mental quality whose possession improves cognitive functioning, especially in a test setting and should not be associated with any particular cognitive psychological model.

In American education there is currently a legislation-mandated push to periodically conduct standards-based skills diagnostic assessments of students from kindergarten to the end of high school (K-12). The biggest push comes from the U.S. government's recently passed "Leave No Child Behind" legislation, which mandates periodic standards-based testing for all American K-12 students. The emphasis on formative assessment skills diagnosis is emphasized in the U.S. Department of Education's widely referred-to regulations for this legislation: "A State's academic assessment system must produce individual student interpretive, descriptive, and diagnostic reports that...allow parents, teachers, and principals to understand and address the specific academic needs of students." In this section, after a brief historically focused literature survey, we survey the Reparameterized Unified Model (RUM) psychometric modeling approach we have developed to the skills diagnosis of standardized tests.

4.1. *A brief survey of psychometric skills diagnostic models*

The psychometric cognitive modeling of standardized testing of examinees at the individual skills level had its early and important psychometric modeling pioneers, including Gerhardt Fischer, Edward Haertel, Robert Mislevy, Susan Embretson, and Kikumi Tatsuoka. I'd like to indicate a few milestones in the psychometric history of cognitive modeling. The first skills-level (or "cognitive") psychometric model was Gerhardt Fischer's (1973) linear logistic trait model (LLTM). The LLTM model is a unidimensional latent ability one parameter logistic IRT model. As such, it is not designed to model, and hence support standardized test data-driven measurement (assessment) of multiple examinee skills. However, it does factor each item's difficulty parameter into skills-based difficulty subcomponents, thereby providing a skills-based structural IRT model of the test.

To provide a skills diagnosis modeling approach that postulates that examinees possess a multidimensional latent skills structure, Susan Embretson developed a series of multidimensional continuous latent variable (each such variable viewed as a skill having a continuum of levels) noncompensatory (in fact, conjunctive, which presumes that an examinee must have a high level of mastery on every required skill for an item in order for the examinee to have a high probability of getting the item right) logistic IRT models (Whitely, 1980; Embretson, 1984, 1985). Susan Embretson's diagnostic models are probably the first models capable of statistically undergirding multidimensionality-based skills profiling using examinee test data.

The key insight that the item/skill test structure (that is, which items require which skills) is important and useful information for effective skills diagnostic focused statistical inference that should be built into one's skills diagnostic model was fully exploited by Kikumi Tatsuoka (1990, 1995). The Tatsuoka Rule Space pattern recognition approach dichotomizes skills as mastered (assigned a latent skill level of $\alpha = 1$) versus nonmastered (assigned a latent skill level $\alpha = 0$), in the process forming a latent skill vector α with binary components. The Rule Space approach requires a moderately sized user-supplied list of skills judged by experts as needed for performing well on the test being analyzed. Then, the item/skill structure (usefully thought of as the skills-level test design) is represented by a user-supplied item by skill $\mathbf{Q}$ *matrix* of 0s and 1s, where the 1s in the ith row identify which of the user-provided list of skills are needed for solving the ith item. The underlying cognitive processing assumption of the Rule Space and certain other diagnostic models using the $\mathbf{Q}$ matrix approach (such as the Unified Model discussed below) is that correct examinee responding is *conjunctive* but random: The examinee noisily tends to get an item right if he or she has mastered *all* the skills identified by the $\mathbf{Q}$ matrix as required for the item and otherwise noisily tends to get the item wrong. It is a point of research contention among cognitive psychologists and educational measurement specialists whether conjunctive modeling is universally appropriate for cognitive processing of test questions modeled to depend on a specified set of skills. Nonetheless, it seems appropriate for many and perhaps most test settings where psychometric modeling of test performance based on skills mastery/nonmastery is needed.

The most basic conjunctive modeling approach using such item/skill structural information is Edward Haertel's (1989) restricted latent class model, which assumes local independence, conditional upon an examinee's discrete latent class, and dichotomizes each examinee relative to each item by assigning two probabilities of correct item response, namely $g_i < 1 - s_i$, where g_i is the probability of correct item response for an examinee whose latent class is such that the examinee is a nonmaster of at least one of the required skills (as specified by Q) for item i and $1 - s_i$ is the probability of correct item response for an examinee who has mastered all the required skills for the item. Here, for linguistic convenience, g is often called a guessing probability and s a slip probability.

The Rule Space model uses a continuous two-dimensional latent examinee ability space (θ, ζ) representation to facilitate inference about the examinee's latent skill vector α Rule-Space-based data fitting is accomplished by augmenting the standard unidimensional IRT logistic modeling and statistical inference approach. Inference about examinee skill mastery/nonmastery profiles are reduced to inferences based on an examinee's standardly estimated latent logistic model-based $\hat{\theta}$ and a continuous "caution index" $\hat{\zeta}$, which measures how Guttman-response atypical (overly inconsistent or overly consistent) the examinee response pattern is for an examinee of estimated ability $\hat{\theta}$. To clarify, the deterministic Guttman approach hypothesizes that each examinee gets all of the easier items right and the harder items wrong, with the cut varying from examinee to examinee. Of course, for any reasonable probability model of examinee responding, one expects a somewhat Guttman-like pattern. The caution index quantifies this in such a way that the $\hat{\zeta}$ value, in addition to $\hat{\theta}$, is helpful in classifying an examinee's skills profile. Thus, the basic assumption is that these two continuous indices, once estimated, can be used to estimate an examinee's latent skill vector α effectively. This approach has a sort of principal components flavor, with two components needed

A version of the Rule Space approach developed by ETS scientists, especially Lou DiBello and Kikumi Tatsuoka, is in current operational use on the widely used College Board's PSAT and as such seems to be the first psychometrically sophisticated large-scale standardized-test-based skills diagnostic application, a major pioneering milestone, both from the psychometric and the formative assessment perspectives.

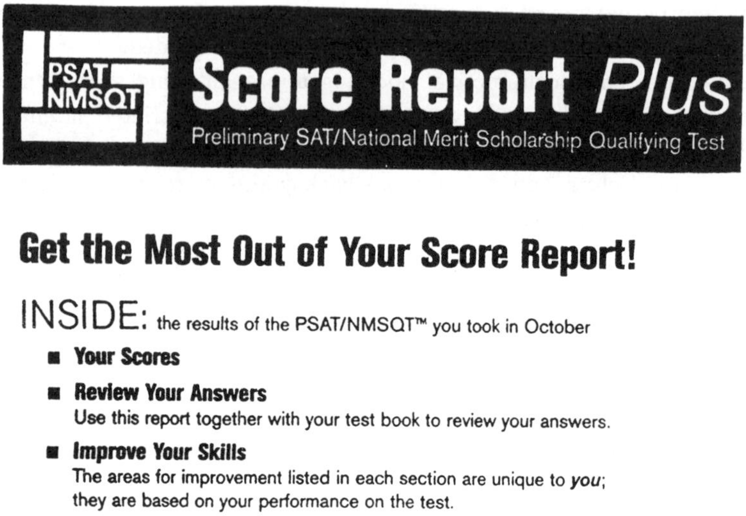
PSAT NMSQT

Score Report *Plus*

Preliminary SAT/National Merit Scholarship Qualifying Test

Get the Most Out of Your Score Report!

INSIDE: the results of the PSAT/NMSQT™ you took in October

- **Your Scores**
- **Review Your Answers**
  Use this report together with your test book to review your answers.
- **Improve Your Skills**
  The areas for improvement listed in each section are unique to *you*; they are based on your performance on the test.

Fig. 7. PSAT Score Report *Plus* Skills Mastery Reporting.

For example (see Figure 7), an examinee taking the Math PSAT might be given the following description of skills to improve on: "organizing and managing information to solve multistep problems" and "applying rules in algebra and geometry." In fact, every PSAT test-taker is given advice on improving up to three skills identified by the ETS (Educational Testing Service) Rule Space algorithm provided to the College Board. This major advance by ETS scientists, undergirded by the Tatsuoka Rule Space research, is an example of practically important and intellectually challenging psychometric probability modeling research stimulated by an important test applications problem and in the process making an important societal contribution to educational testing.

The Bayes net approach to skills diagnosis is the product of a team-oriented research effort that has been vigorously led by Robert Mislevy and colleagues (especially Russell Almond and Linda Steinberg). An excellent source is Bob Mislevy's (1994) Psychometrika Society presidential address paper. This Bayes net approach to cognitive diagnosis has been applied in a variety of applied diagnostic settings, including the assessment of problem-solving skills of dental hygienists (Mislevy, Almond, Yan, and Steinberg, 1999) and airplane repair training. A recent exposition, which includes a description of Bayes nets, appears in Mislevy, Steinberg, and Almond (in press).

The Bayes net approach represents a joint probability latent cognitive model of examinee test responding as a carefully sequenced recursive product of conditional probabilities, where the order of multiplied factors is carefully chosen to simplify the model representation as much as possible through local independence assumptions:

$$p(X_1, X_2, \mathrm{K}, X_n) = p(X_n|X_{n-1}, X_{n-2}, \mathrm{K}, X_1)p(X_{n-1}|X_{n-2}, X_{n-3}, \mathrm{K}, X_1) \times \Lambda p(X_2|X_1)p(X_1). \quad (13)$$

Graph theory (in particular directed acyclic graphical networks) is used as a tool for discovering and representing the conditional dependencies and the simplifying conditional independences used to simplify (13) to the point of being tractable.

The University of California Berkeley Evaluation and Assessment Research (BEAR) embedded-assessment system is another important psychometric approach worth recognizing. It uses a graded-response multidimensional one parameter per dimension logistic approach and has been extensively developed under the leadership of Mark Wilson. This system has been successfully applied in precollege science learning settings and draws upon the unidimensional one parameter logistic model (see http://bear.berkeley.edu/pub.html for a list of relevant papers).

Two excellent surveys on the contributions of statistics and psychometric modeling to skills-level assessment, both underscoring the importance of probabilistic modeling of skills diagnosis, are (the first survey) Chapter 4 of the National Research Council's *Knowing What Students Know* (ed. by Pellegrino et al., 2001) , which in fact is based in part on a (the second survey) more technical survey by Brian Junker (1999). A nice foundational paper about skills-level latent probability modeling by Eric Maris (1995) is also highly recommended.

4.2. *The unified model and useful enhancements*

Lou DiBello, Louis Roussos, and I set out to develop a **Q**-matrix-based parametric IRT skills diagnosis probability model with easily interpretable parameters for both users and psychometricians. Moreover, its parameters were required to describe the major sources of stochastic departures from the deterministic conjunctive pattern-recognition-responding model predicted by the **Q** matrix. Here, a simplified version of the Unified Model (UM) we developed (for details see DiBello Stout, and Roussos, 1995) is given.

The UM characterizes each examinee by his or her multidimensional latent ability $(\boldsymbol{\alpha}, \theta)$ in the traditional latent modeling sense that LI given latent ability $(\boldsymbol{\alpha}, \theta)$ is assumed. $\boldsymbol{\alpha}$ is a binary component latent vector characterizing an examinee's mastery profile (with 0,1 nonmastery, mastery binary components) on a set of skills specified as important by the user. As such, estimating $\boldsymbol{\alpha}$ is the main goal of most skills diagnostic assessments.

From the statistician's perspective, the amount of statistically recoverable skills-level information about examinees provided by a standardized test of dichotomously scored items is limited by its intrinsic nature. In particular, if a skills-based psychometric probability model such as the UM is to be used to undergird a statistically effective and useful skills diagnosis, the model must obey the principle of statistical parsimony, that is, introduction of only a statistically reasonable number of parameters, and hence skills to be assessed. It is thus obvious that, in contrast to traditional cognitive psychology modeling of human intellectual performance (where there is no penalty for model complexity: as an illustration, see Koedinger and Maclaren, 2002, for an example of a complex cognitive model developed in pursuit of a deeper understanding of human problem-solving behavior in the "early algebra" stage), many of the skills influencing item performance must be intentionally left out of the psychometric model. This has somewhat the flavor of multiple regression modeling.

It is this profoundly simple insight, viewed from the psychometric perspective, that mandates a certain level of parametric simplicity of our skills diagnostic models. Thus, the skills diagnosis probability modeling challenge is to achieve enough parametric model complexity to provide a useful skills diagnosis while using interpretable, informative, and useful (for the practitioner) parameters and in particular avoiding so much complexity that statistical and computational tractability are sacrificed.

The influence of all the intentionally omitted (and usually numerous) skills upon item responding is made explicit in the UM by the introduction of a "residual" continuous ability θ that unidimensionally summarizes the examinee ability level on the large and unspecified collection of skills intentionally and unavoidably left out of the UM. That is, θ compensates for the cognitive-processing *incompleteness* of the specified skills vector $\boldsymbol{\alpha}$.

Even though the UM dichotomizes each skill as being mastered or nonmastered by an examinee, the UM was developed from the philosophical position that an examinee correctly labeled as a nonmaster of a skill can still (noisily) apply it correctly in attempting to solve an item. Similarly an examinee correctly labeled as a master of a skill can (noisily) fail to apply it correctly. This phenomenon of deviating from the mastery/nonmastery patterns as predicted by $\mathbf{Q}$ for an examinee with skill vector $\boldsymbol{\alpha}$ is called *positivity.*

Although some might view positivity as literally accounting for guesses and slips, in most cases something much more cognitively subtle and important is being modeled. That is, the presumed dichotomization of examinees into masters and nonmasters of a skill is an oversimplified idealization that partially breaks down in reality. To illustrate, a particular algebra item may require such challenging uses of the "rules of exponentials" skill that many "masters" of the skill will fail to apply it correctly and thus answer the item incorrectly. Likewise, for a particular item requiring use of exponents, the correct application of the skill may be so routine that many "nonmasters" of the skill will tend to apply the skill correctly for the item. It is this *heterogeneity* of each skill across items of the role of examinee mastery and nonmastery of each skill that leads to the positivity parameters of the probabilistic UM.

Modeling positivity and incompleteness led to the UM's evidence model for an item IRF:

$$P(U_i = 1|\boldsymbol{\alpha}, \theta) = \prod_k \pi_{ik}^{\alpha_k} r_{ik}^{(1-\alpha_k)} P(\theta + c_i), \tag{14}$$

where the product is over the attributes k required for item i as specified by $\mathbf{Q}$, $0 \leq c_i \leq 3$ is the completeness parameter, $\alpha_k = 0$ or 1 denotes skill nonmastery or mastery of skill k respectively, and, by definition,

$$\pi_{ik} = P(\text{Attribute } k \text{ applied correctly to item } i|\alpha_k = 1), \tag{15}$$

$$r_{ik} = P(\text{Attribute } k \text{ applied correctly to item } i|\alpha_k = 0), \tag{16}$$

and the logistic function is given by

$$P(x) = \frac{\exp(1.7x)}{1 + \exp(1.7x)}. \tag{17}$$

To remove irrelevant issues of scale indeterminacies, centering and variation are controlled by assuming θ is a standard normal distribution. Note as with all IRT models that the correct applications of different skills are assumed independent given latent ability $(\boldsymbol{\alpha}, \theta)$ in (14).

In the case of an item i for which the set of required α_k's are relatively complete in the sense that the role of other residual attributes as captured via θ is minor, c_i will tend to be large (close to 3: the choice of 3 an arbitrary convenience). Clearly an item with small r's, large π's, and a large c will be a highly discriminating item in its large capacity to separate masters from nonmasters of the skills required by the item, and hence the item will be highly desirable for skills diagnostic purposes.

The positivity and completeness item parameters of the UM promise to be useful in providing a model for skills diagnosis test performance evaluation and just as importantly in providing a model for skills-level test design purposes. This dual usage is analogous to the difficulty and discrimination parameters of commonly used logistic IRT models (conventionally used for unidimensionally scaled tests) being helpful for both assessing the effectiveness of existing tests and new test construction intended to measure a single construct, like algebra. This second capability of providing useful test design information is very important because skills-based testing is a "whole new ball game" for which practical principles of item construction, test design, and test performance evaluation are totally undeveloped. In particular, the impressive body of knowledge about item construction, test design, and test performance evaluation for unidimensionally scaled summative tests appears to not be very transferable to the multidimensional discrete-skills formative assessment test setting. In summary, having estimable parameters that measure diagnostic effectiveness at the fine-grained item/skill level lays a sound foundation for evaluating skills-level test performance and for designing skills level tests.

The 1995 version of the UM with its considerable parametric complexity foundered on the shoals of the credit/blame problem (that is, discerning which skill masteries deserve credit for a correct item response and which skills nonmasteries deserve the blame for an incorrect item response) in that examinee item level correct/incorrect response data is intrinsically not rich enough in information to render all the π's and r's of (14) identifiable. In her thesis, Sarah Hartz (2002) reparameterizes the UM in a manner that produces identifiable and still nicely interpretable and highly useful parameters, these parameters continuing to quantify the key concepts of incompleteness and positivity.

In her thesis, Hartz also recast the UM as a hierarchical Bayes model and wrote an MCMC algorithm to successfully calibrate the resulting reparameterized UM. This was a major required step forward, making the model statistically tractable with highly interpretable parameters, useful to the practitioner designing or psychometrically evaluating the effectiveness of skill diagnostic tests. Henceforth, the reparameterized (R) Bayes UM of

Hartz will be referred to as the "Bayes RUM."

Large scale simulation studies by Hartz (2002) showed effective estimation of the model parameters by her MCMC procedure. In a 1500 examinee, 7 skills, 40 item, 2 skills/item on average setting with highly skills-discriminating items assumed ("*strong cognitive structure*"), the correct classification rate achieved by the MCMC algorithm averaged 95% across the seven skills with only about 4% of the examinee/skill pairs left unclassified on average across the seven skills because of lack of statistical information. Further, the evidence was strong that MCMC convergence to the stationary distribution for the non burn-in portion of the generated Markov chain had occurred. Interestingly and as expected, when a weak cognitive structure setting (low item discrimination of skills) was simulated, then the classification accuracy dropped to 84% and the proportion of examinees not classified rose to 14%. Item parameters still tended to be well estimated.

A big concern was that the prerequisite specification of the **Q** matrix might produce a high degree of nonrobustness. That is, if the **Q** matrix was inaccurate by a minor amount (as it surely must be if the test is moderate to long in length and the number of skills is at all sizable!), then the diagnostic accuracy could be seriously compromised. Indeed, we tend to view a user-supplied **Q** matrix as a carefully formulated expert hypothesis, setting up a confirmatory situation where good statistical practice would allow data-driven minor modifications of the **Q** matrix.

With this in mind, Hartz carried out a number of simulation studies where the **Q** matrix was incorrectly specified. In all cases, the deterioration in skills classification accuracy was acceptably minor. In an interesting finding, the presence of θ in the UM compensated nicely for 0s erroneously placed in the MCMC's presumed **Q** matrix - see Hartz (2002) for details. Further, the data-driven parameter-reduction statistical approaches built into the MCMC Bayes UM analysis, designed to remove unneeded parameters, nicely removed the parameters corresponding to 1s erroneously placed in the MCMC's presumed **Q** matrix. The terms "erroneously placed" and "incorrectly specified" mean that the (correct) **Q** matrix entry of the simulation model generating the data differed from the corresponding entry in the **Q** matrix (incorrectly) presumed by the MCMC statistical analysis of the simulated data.

4.3. *The Bayes unified model applied to PSAT data*

It is a sort of psychometricians' joke that one can prove almost anything using simulation studies. We were thus pleased when ETS asked us to try out the Bayes UM on experimentally obtained PSAT test/retest data. The PSAT math test had 40 items and a **Q** matrix based on 16 skills (these skills carefully developed for ETS by teachers, educational psychologists, psychometricians, and cognitive scientists), having approximately three skills per item.

Of particular interest is the Bayes UM methodology's capacity to assess skill, and item, level performance aspects of a test being used for skills diagnostic purposes. Indeed, the Bayes UM parameters assess the discrimination of every item/skill pair. Thus, an item's effectiveness across the set of skills it is purported to measure, and a skill's effectiveness across the set of items purported to require the skill, can both be assessed for all the test items and user-specified set of skills. To illustrate, according to the Bayes UM analysis of the PSAT Math test conducted using the Hartz MCMC approach, three items on one math form failed to display a useful amount of skill discrimination on any of the skills the **Q** matrix claimed they were measuring.

This result is not surprising in that the PSAT is designed to scale examinees on mathematics achievement unidimensionally (it is a summative assessment), rather than being designed to diagnose the 16 post-test created mathematics skills. Thus, these three items were contributing to the desired unidimensional scaling by measuring mathematics skills other than those specified by the **Q** matrix. From the skills test-design perspective, where one might want to modify a test to improve its diagnostic power on a set of specified skills, such skills-level information about particular items would be highly useful. The capability to assess test performance for each skill and for each item will be very useful for future skills-level test design, where one wants assurances that each targeted skill is being effectively measured.

The ETS PSAT setting was an experimental setting that went beyond operational data by providing test/retest data. This allowed the use for research purposes of the performance criterion of skill assessment agreement across the two tests each student took. That is, skill/examinee assessment agreement occurs when an examinee is independently assessed by both tests to have mastered the skill or to have not mastered the skill. Although it is not the same as observing how reliably examinee/skill pairs are *correctly diagnosed*, as can be done in a simulation study, it is an excellent stand-in. Naïvely, one might decide that anything above 50% agreement (the naively presumed chance agreement rate) across tests would provide evidence that the procedure is effectively classifying examinee skill masteries based on PSAT examinee performances. But, this is wrong of course: the correct baseline rate depends on the baseline rates of assigning skill mastery on each of the two tests (for example, 100% agreement across tests can be obtained by artificially and foolishly labeling every examinee a master of all skills on both tests): Let p_i denote the assigned proportion of masters of a skill on Test i. Then the baseline chance agreement rate (assigning (0,0) or (1,1)) is

$$b \stackrel{\text{def}}{=} p_1 p_2 + (1 - p_1)(1 - p_2). \tag{18}$$

Let a be the observed across-test agreement rate proportion for a skill. Then the percentile rank of a relative to the chance rate b should be a good adjusted agreement rate index:

$$100\,\frac{a - b}{1 - b}. \tag{19}$$

Using this index, one version of the Bayes UM analyses achieved a percentile index of 70% fairly uniformly across the 12 skills (out of a possible16) judged to be effectively measured-very respectable for a test not designed for skills diagnosis.

From the goodness of fit perspective, one could wonder whether the labeling of the examinees as master versus nonmaster on each of the entire set of skills required on an item is consistent with overall examinee performance on the item, in the sense that "masters" of an item should perform well and nonmasters (those judged to lack at least one required skill) poorly. For, if either the Bayes UM fails to fit the data well or, even when it does, the ensuing statistical MCMC analysis fails to calibrate the model well or to use the well-calibrated model effectively to predict examinee skills well, then one would expect the statistically inferred masteries and nonmasteries of examinee skills to likely fail to be strongly consistent with examinee correct/incorrect performance on the item.

With the above concern in mind, for each item/examinee combination, we classified the combination as an *item mastery* or an *item nonmastery combination depending* on whether or not the MCMC analysis assessed the examinee as having mastered all the skills the **Q** matrix shows as required for the item. Further, the nonmastery category was split into two subcategories depending on whether an examinee nonmaster had mastered less than half or at least half of the required skills for the item, producing high and low nonmastery item/examinee combinations. In this regard, analysis of both forms of the PSAT math and both forms of the PSAT writing test produced excellent results. For example, on the V2 math form, item mastery resulted in an average 85% item correct rate, and item nonmastery in an average 27% item correct rate, which split into 42% and 18% for high nonmaster and low nonmaster examinee nonmasters respectively.

Hence, on average, examinees estimated to be item masters performed very well on the items and those estimated to be item nonmasters did relatively poorly, with low nonmasters doing extremely poorly. This result is very encouraging. It then seemed relevant to ask whether such behavior holds up item by item. In this regard it is interesting to contrast two simulation studies performed by Hartz (2002), one assuming strong cognitive and one with weak cognitive structure. Even the weak cognitive structure case effectively produced high-performing masters and low-performing nonmasters for each of the items in the simulated data, with reasonable variation from item to item. The results for the strong cognitive structure were even better.

A real-data-based (for model calibration purposes) simulated reenactment of the actual College Board PSAT student reporting process described above was of particular interest, where each student taking the PSAT has up to three nonmastered skills reported. This was carried out on the ETS PSAT data set using the Bayes UM model as calibrated from PSAT data via the MCMC statistical inference approach. The results were especially satisfying. About 90% of the test takers had at least one nonmastered skill "reported". Among these examinees identified as lacking at least one needed skill, correct performance proportions (using the actual test data) on items involving skills examinees were judged to have not mastered was around 0.21 on average over the four tests (two math forms and two writing forms), a pleasingly low number. The most authentic aspect was that cross-validation was carried out, in the sense that we recorded the proportion of times a skill that was "reported" as nonmastered by the examinee on one test was inconsistently "reported" as mastered on its paired test. These two error-rate proportions (viewing each test of a test/retest pair as the report-generating test produces two numbers) for the PSAT Writing test were both around 0.06. On the PSAT Math test, for whatever reason, the resulting pair of test/retest error rates were still very good but less so, averaging 0.13.

4.4. *Skills diagnosis: The new paradigm?*

Is formative assessment skills diagnostics the new test paradigm, as I suggested in the opening paragraph of Section 4? I think so, and I hope the brief and somewhat informal description in this section convinces my probabilistic and statistical colleagues that this claim may be valid. Further, I am optimistic that the Bayes UM MCMC approach will play a significant role , according to the stated criterion at the beginning of this paper that psychometric research is valuable when it is effectively used in actual educational practice. Of course, many vital research challenges remain in the skills-level formative assessment arena, this hopefully of interests to probabilistic and statistical readers casting around for interesting new research problems.

5. Dimensionality, Equity, and Diagnostic Software

The purpose of this very brief section is to give a partial listing of who to contact concerning methodology software for procedures mentioned in the article. The conditional covariance based dimensionality assessment software, namely DIMTEST, HCA/CCPROX, DETECT, is available from Assessment Systems Corporation (http://www.assess.com/) Many of the multiple variations of SIBTEST discussed are available from Assessment Systems Corporation (http://www.assess.com/). The MCMC Bayes UM skills diagnostic software can be formally requested from the Educational Testing Service for research purposes.

6. Concluding Remarks

I wrote this invited paper with several purposes in mind. First, I wanted somewhat informally to survey progress made in three major psychometric research areas heavily dependent on sophisticated probability modeling that I and colleagues worked on over many years. These areas have not only been intellectually interesting areas to work on from the psychometric research perspective, but I see them as very important from the applied test measurement perspective. In summary, the three areas are (i) nonparametric latent ability structure probability modeling and dimensionality assessment from the nonparametric perspective, with the emphasis being on the use of item-pair conditional covariances, the conditioning being on an appropriately chosen subtest score, (ii) the nonparametric probability modeling and statistical assessment of test fairness with the emphasis being on uniting substantive and statistical approaches to test equity via the MMD multidimensional probability model for DIF/DBF/DTF and on the SIBTEST family of DIF/DBF/DTF procedures growing out of the MMD model, and (iii) the parametric skills diagnostic IRT probability modeling and associated Bayes Unified Model (RUM) based skills diagnostic methodology for carrying out skills-level formative assessment and skills-level test design.

Second, I wanted to suggest one appealing, and I think, effective, approach for selecting and carrying out psychometric probability modeling research problems. This consists of selecting one's research problems as motivated by important applied educational testing and assessment research challenges, then bringing sophisticated probabilistic modeling and modern statistical thought to bear upon the research problems selected, and last, making the effectiveness of the research in improving educational measurement practice the ultimate criterion for judging its worth.

Third, I wanted to stress the great power and enjoyment that can result when a team of dedicated and talented researchers, as I have had the great privilege to be associated with regarding the work described in this presidential paper, cooperatively and collegially attacks carefully selected research problems in a determined manner, this collegial approach being natural for me after its imprinting on me by Y.S. Chow.

Fourth, I wanted to stress that there is a major paradigmatic shift broadening the nature of large scale educational testing and assessment that I believe has major implications for future psychometric IRT research, especially in its encouragement of the development of parametrically complex probability models. Put simply, the total score summative assessment paradigm for standardized testing is being supplanted by a new blended summative assessment and formative assessment paradigm. For those interested in perhaps catching a ride on this research train, it promises to be an exciting and eventful trip, both because it is intellectually fascinating and challenging to psychometricians and statisticians and because it is of great importance to the future of education and training.

Fifth, although the acorns described above are indeed rather far from the Y.S. Chow probability limit theorems oak tree, I appreciate and am honored by the editors' encouragement of its inclusion and in particular want to acknowledge that the work above and the careers of mine and my colleagues owes its existence to Y. S. Chow's patience, inspiration, encouragement, and wisdom (for which I am very thankful) regarding one rather naïve Purdue Ph. D. student whose undergraduate degree was not even in mathematics or statistics.

References

1. Ackerman, T.A. (1992). A didactic explanation of item bias, item impact, and item validity from a multidimensional perspective. *Journal of Educational Measurement*, 29, 67-91.
2. Angoff, W.H. (1993). Perspectives on differential item functioning methodology. In P. W. Holland & H. Wainer (Eds.), *Differential Item Functioning* (pp. 3-24). Hillsdale, NJ: Lawrence Erlbaum Associates.

3. Bolt, D., Froelich, A.G., Habing, B., Hartz, S., Roussos, L., & Stout, W. (2002) An applied and foundational research project addressing DIF, impact, and equity: with applications to ETS test development. ETS Technical Report.
4. Chang, H., Mazzeo, J., & Roussos, L. (1996). Detecting DIF for polytomously scored items: an adaptation of the SIBTEST procedure. *Journal of Educational Measurement*, 33, 333-353
5. Chang, H., & Stout, W. (1993). The asymptotic posterior normality of the latent trait in an IRT model. *Psychometrika*, 58, 37-52.
6. DiBello, L., Stout, W., & Roussos, L. (1995). Unified cognitive/psychometric diagnostic assessment likelihood-based classification techniques. In P. Nichols, S. Chipman, and R. Brennen (Eds.), *Cognitively Diagnostic Assessment.* Hillsdale, NJ: Earlbaum. 361-389.
7. DiBello, L., Roussos, L., & Stout W. (invited book in preparation; Ed of series: Wim van der Linden) *Skills Diagnosis: Issues and Practice.* New York, NY, Springer Verlag
8. Douglas, J. (1997). Joint consistency of nonparametric item characteristic curve and ability estimation. *Psychometrika*, 62, 7-28.
9. Douglas, J.A. (2001). Asymptotic identifiability of nonparametric item response models. *Psychometrika*, 66, 531-540.
10. Douglas J.A. & Cohen A. (2001). Nonparametric ICC estimation to assess fit of parametric models. *Applied Psychological Measurement*, 25, 234-243.
11. Douglas, J., Kim, H.R., Habing, B., & Gao, F. (1998). Investigating local dependence with conditional covariance functions. *Journal of Educational and Behavioral Statistics*, 23, 129-151.
12. Douglas, J., Roussos, L., & Stout, W., (1996). Item bundle DIF hypothesis testing: identifying suspect bundles and assessing their DIF. *Journal of Educational Measurement*, 33, 465-484.
13. Douglas, J., Stout, W., & DiBello, L. (1996). A kernel smoothed version of SIBTEST with applications to local DIF inference and unction estimation. *Journal of Educational and Behavioral Statistics*, 21, 333-363.
14. Ellis, J.L., & Junker, B.W. (1997). Tail-measurability in monotone latent variable models. *Psychometrika*, 62, 495-524.
15. Embretson, S.E. (1984). A general latent trait model for response processes. *Psychometrika*, 49, 175-186.
16. Embretson, S.E. (Ed., 1985), *Test design: developments in psychology and psychometrics.* Chapter 7, (pp. 195-218). Orlando: Academic Press.
17. Fischer, G.H. (1973). The linear logistic test model as an instrument in educational research. *Acta Psychologica*, 37, 359-374.
18. Froelich, A.G. & Habing, B. (2002). A study of methods for selecting the AT subtest in the DIMTEST procedure. Paper presented at the 2002 Annual Meeting of the Psychometrika Society. University of North Carolina at Chapel Hill.
19. Gierl, M. J., Bisanz, J., Bisanz, G., Boughton, K., & Khaliq, S. (2001). Illustrating the utility of differential bundle functioning analyses to identify and interpret group differences on achievement tests. *Educational Measurement: Issues and Practice*, 20, 26-36.

20. Gierl, M.J., & Khaliq, S.N. (2001). Identifying sources of differential item and bundle functioning on translated achievement tests. *Journal of Educational Measurement*, 38, 164-187.
21. Gierl, M.J., Bisanz, J., Bisanz, G.L., & Boughton, K.A. (2002). Identifying content and cognitive skills that produce gender differences in mathematics: a demonstration of the DIF analysis framework. Paper presented at the annual meeting of the National Council on Measurement in Education, New Orleans, LA.
22. Haberman, S.J (1977). Maximum likelihood estimates in exponential response models. *The Annals of Statistics*, 5, 815-841.
23. Habing, B. (2001). Nonparametric regression and the parametric bootstrap for local dependence assessment. *Applied Psychological Measurement*, 25, 221-233.
24. Haertel, E. (1989). Using restricted latent class models to map the skill structure of achievement items. *Journal of Educational Measurement*, 26, 301-321.
25. Hartz, S.M. (2002). *A Bayesian framework for the Unified Model for assessing cognitive abilities: blending theory with practicality.* Unpublished doctoral dissertation: Department of Statistics, University of Illinois, Urbana-Champaign.
26. Holland, P.W. (1990a). On the sampling theory foundations of item response theory models. *Psychometrika*, 55, 577-601.
27. Holland, P.W. (1990b). The Dutch identity: a new tool for the study of item response models. *Psychometrika*, 55, 5-18.
28. Holland, P.W., & Rosenbaum, P.R. (1986). Conditional association and unidimensionality in monotone latent variable models. *The Annals of Statistics*, 14, 1523-1543.
29. Holland, W.P. & Thayer, D. T. (1988). Differential item performance and the Mantel-Haenszel procedure. In H. Wainer & H. I. Braun (Eds.), *Test Validity* (pp. 129-145). Hillsdale, NJ: Lawrence Earlbaum Associates.
30. Jiang, H., & Stout, W. (1998). Improved Type I error control and reduced estimation bias for DIF detection using SIBTEST. *Journal of Educational and Behavioral Statistics*, 23, 291-322.
31. Junker, B.W. (1993). Conditional association, essential independence, and monotone unidimensional latent variable models. *Annals of Statistics*, 21, 1359-1378.
32. Junker, B.W. (1999). *Some statistical models and computational methods that may be useful for cognitively-relevant assessment.* Prepared for the National Research Council Committee on the Foundations of Assessment. http://www.stat.cmu.edu/ brian/nrc/cfa/. [April 2, 2001].
33. Junker, D.W., & Ellis, J.L. (1998). A characterization of monotone unidimensional latent variable models. *Annals of Statistics*, 25(3), 1327-1343.
34. Junker, B. W. & Sijtsma, K. (2001). Nonparametric item response theory in action: an overview of the special issue. *Applied Psychological Measurement*, 25, 211-220.
35. Koedinger, K. R. & MacLaren, B. A. (2002). Developing a pedagogical domain theory of early algebra problem solving. CMU-HCII Tech Report 02-100.

36. Li, H. & Stout, W. (1996). A new procedure for detecting crossing DIF. *Psychometrika*, 61, 647-677.
37. Kok, F. (1988). Item bias and test multidimensionality. In R. Langeheine & J. Rost (Eds.), *Latent Trait and Latent Models*, 263-275, New York: Plenum Press.
38. Linn, R.L. (1993). The use of differential item functioning statistics: A discussion of current practice and future implications. In P.W. Holland & H. Wainer (Eds.), *Differential Item Functioning* (pp. 349-364). Hillsdale, NJ: Lawrence Erlbaum Associates.
39. Lord, F. M. (1980) *Applications of Item Response Theory to Practical Testing Problems.* Hillsdale, NJ: Lawrence Erlbaum Associates,
40. McDonald, R.P. (1994). Testing for approximate dimensionality. In Laveault, D., Zumbo, B. D., Gessaroli, M. E., & Boss, M. W. (Eds.), *Modern Theories of Measurement: Problems and Issues.* Ottawa, Canada: University of Ottawa, pp. 63-86.
41. Maris, E. (1995). Psychometric latent response models. *Psychometrika*, 60, 523-547.
42. Mislevy, R.J. (1994). Evidence and inference in educational assessment. *Psychometrika*, 59, 439-483.
43. Mislevy, R.J. Almond, R.G., Yan, D., & Steinberg, L.S. (1999). Bayes nets in educational assessment: where do the numbers come from? In K.B. Laskey & H.Prade (Eds.),*Proceedings of the Fifteenth Conference on Uncertainty in Artificial Intelligence* (437-446). San Francisco: Morgan Kaufmann.
44. Mislevy, R. Steinberg, L. & Almond, R. (in press). On the structure of educational assessments. *Measurement: Interdisciplinary Research and Perspective.* Hillsdale, NJ: Lawrence Erlbaum Associates.
45. Mokken, R.J. (1971). *A Theory and Procedure of Scale Analysis.* The Hague: Mouton.
46. Molenaar , I.W., & Sijtsma, K. (2000). *User's manual MSP5 for Windows: A program for Mokken Scale Analysis for Polytomous Items. Version 5.0* [Software manual]. Groningen: ProGAMMA.
47. Nandakumar, R. (1993). Simultaneous DIF amplification and cancellation: Shealy-Stout's test for DIF. *Journal of Educational Measurement*, 30, 293-311.
48. Nandakumar, R. & Roussos, L. (2002). Evaluation of CATSIB procedure in pretest setting. *Journal of Educational and Behavioral Statistics.*
49. Nandakumar, R. & Stout, W. (1993). Refinements of Stout's procedure for assessing latent trait unidimensionality. *Journal of Educational Statistics*, 18, 41-68.
50. O'Neill, K. A., & McPeek, W. M. (1993). Item and test characteristics that are associated with differential item functioning. In P.W. Holland & H. Wainer (Eds.), *Differential Item Functioning* (pp. 255-276). Hillsdale, NJ: Lawrence Erlbaum Associates.
51. Pellegrino, J. W., Chudowski, N., & Glaser, R. (Eds.). (2001). *Knowing what students know: The science and design of educational assessment.* Chapter 4, (pp. 111-172) Washington, DC: National Academy Press.

52. Philipp, W. & Stout, W. (1975). *Almost sure convergence principles for sums of dependent random variables.* American Mathematical Society Memoir No. 161, American Mathematical Society, Providence, RI.
53. Ramsey, P.A. (1993). Sensitivity review: the ETS experience as a case study. In P. W. Holland & H. Wainer (Eds.), *Differential Item Functioning* (pp. 367-388). Hillsdale, NJ: Lawrence Erlbaum Associates.
54. Rossi, N., Wang, W. & Ramsay, J.O. (2002). Nonparametric item response function estimates with the EM algorithm. *Journal of Educational and Behavioral Statistics*, in press.
55. Roussos, L., & Stout, W. (1996a). DIF from the multidimensional perspective. *Applied Psychological Measurement*, 20, 335-371.
56. Roussos, L., & Stout, W. (1996b). Simulation studies of the effects of small sample size and studied item parameters on SIBTEST and Mantel-Haenszel Type 1 error performance. *Journal of Education Measurement*, 33, 215-230.
57. Roussos, L.A., Stout, W.F. & Marden, J. (1998). Using new proximity measures with hierarchical cluster analysis to detect multidimensionality. *Journal of Educational Measurement*, 35, 1-30.
58. Roussos, L.A., Schnipke, D.A., & Pashley, P.J. (1999). A generalized formula for the Mantel-Haenszel differential item functioning parameter. *Journal of Educational and Behavioral Statistics*, 24, 293-322.
59. Shealy, R.T. (1989). *An item response theory-based statistical procedure for detecting concurrent internal bias in ability tests.* Unpublished doctoral dissertation, Department of Statistics, University of Illinois, Urbana-Champaign.
60. Shealy, R., & Stout, W. (1993a). A model-based standardization approach that separates true bias/DIF from group ability differences and detects test bias/DTF as well as item bias/DIF. *Psychometrika*, 58, 159-194.
61. Shealy, R., & Stout, W. (1993b). An item response theory model for test bias and differential test functioning. In P. Holland and H. Wainer (Eds.), *Differential Item Functioning* (pp. 197-240). Hillsdale, NJ: Lawrence Erlbaum Associates.
62. Sijtsma, K. (1998). Methodology review: nonparametric IRT approaches to the analysis of dichotomous item scores. *Applied Psychological Measurement*, 22, 3-32.
63. Stout, W. (1987). A nonparametric approach for assessing latent trait unidimensionality. *Psychometrika*, 52, 589-617.
64. Stout, W. (1990). A new item response theory modeling approach with applications to unidimensionality assessment and ability estimation. Psychometrika, 55, 293-325.
65. Stout, W., Froelich, A.G., & Gao, F. (2001). Using resampling to produce an improved DIMTEST procedure. In Boomsma, A., van Duijn, M.A.J., Snijders, T.A.B. (Eds.) Essays on item response theory. (pp. 357-376). New York: Springer-Verlag.
66. Stout, W., Habing, B., Douglas, J., Kim, H.R., Roussos, L., & Zhang, J. (1996). Conditional covariance based nonparametric multidimensionality assessment. Applied Psychological Measurement, 20, 331-354.

67. Stout, W., Li, H., Nandakumar, R., & Bolt, D. (1997). MULTISIB - A procedure to investigate DIF when a test is intentionally multidimensional. *Applied Psychological Measurement*, 21, 195-213.
68. Suppes, P., & Zanotti, M. (1981). When are probabilistic explanations possible? *Synthese*, 48, 191-199.
69. Tatsuoka, K.K. (1990). Toward an integration of item-response theory and cognitive error diagnosis. In N. Frederiksen , Glazer, R., Lesgold, A., & Shafto, M.G. (Eds.), *Diagnostic Monitoring of Skill and Knowledge Acquisition* (pp. 453-488). Hillsdale, NJ: Lawrence Erlbaum Associates.
70. Tatsuoka, K.K. (1995). Architecture of knowledge structures and cognitive diagnosis: a statistical pattern recognition and classification approach. In P. Nichols, S. Chipman, and R. Brennen (Eds.), *Cognitively Diagnostic Assessment.* Hillsdale , NJ: Earlbaum. 327-359.
71. Trachtenberg, F., & He, X. (2002). One-step joint maximum likelihood estimation for item response theory models. Submitted for publication.
72. Tucker, L.R., Koopman, R.F., & Linn, R.L. (1969). Evaluation of factor analytic research procedures by means of simulated correlation matrices. *Psychometrika*, 34, 421-459.
73. Wainer, H., & Braun, H.I. (1988). *Test Validity.* Hillsdale, NJ: Lawrence Erlbaum Associates.
74. Whitely, S.E. (1980a). Multicomponent latent trait models for ability tests. *Psychometrika*, 45, 479-494.
75. Zhang, J., & Stout, W. (1999a). Conditional covariance structure of generalized compensatory multidimensional items. *Psychometrika*, 64, 129-152.
76. Zhang, J., & Stout, W. (1999b). The theoretical DETECT index of dimensionality and its application to approximate simple structure. *Psychometrika*, 64, 213-249.